A Hist

Frank Dumont presents personality psychology with a fresh description of its current status as well as its prospects. Play, sex, cuisine, creativity, altruism, pets, grieving rituals, and other often neglected topics broaden the scope of this fascinating study. This tract is imbued with historical perspectives that reveal the continuity in the evolving science and research of this discipline over the past century. The author places classic schemas and constructs, as well as current principles, in the context of their socio-political catalysts. He further relates this study of the person to life-span developmental issues and to cultural, gender-specific, trait-based, genetic/epigenetic, and evolutionary research findings.

Personality psychology has recently reconciled itself to more modest paradigms for describing, explaining, and predicting human behavior than it generated in the nineteenth and twentieth centuries. This book documents that transformation, providing valuable information for health service professionals as well as to teachers, researchers, and scientists.

FRANK DUMONT, now Professor Emeritus, was Full Professor in the Department of Educational and Counselling Psychology and Director of the Ph.D. program in his field at McGill University. He was the co-editor of *Six Therapists and One Client* (2000) and a co-editor of Corsini's *The Dictionary of Psychology* (1999). He gratefully acknowledges that much of this book on personality psychology was researched and drafted while he was Visiting Fellow at Wolfson College, Cambridge University (2005–2006).

A History of Personality Psychology

Theory, Science, and Research from Hellenism to the Twenty-first Century

Frank Dumont

CAMBRIDGE
UNIVERSITY PRESS

CAMBRIDGE UNIVERSITY PRESS
Cambridge, New York, Melbourne, Madrid, Cape Town,
Singapore, São Paulo, Delhi, Mexico City

Cambridge University Press
The Edinburgh Building, Cambridge CB2 8RU, UK

Published in the United States of America by
Cambridge University Press, New York

www.cambridge.org
Information on this title: www.cambridge.org/9780521133265

First published 2010
First paperback edition 2012

A catalogue record for this publication is available from the British Library

ISBN 978-0-521-11632-9 Hardback
ISBN 978-0-521-13326-5 Paperback

To my spouse, Rita,
and our three children,
Rachel, Caroline, and Marc

Contents

Tables and figures

Preface

This text presents a contemporary description of personality psychology in an historical perspective and makes projections as to the future of its developmental course. As this broad field of inquiry has a rich, venerable, and storied past, I thought it necessary not only to devote the introductory chapter to it, but also to place the material in the individual chapters in their relevant cultural contexts. Academics and professionals typically acknowledge the importance of understanding the history of their disciplines and their scientific evolution (see Cunningham and Napier, 2008, in their touching obituary to Anne Elizabeth Kelley [1954–2007]). Paradoxically, however, ahistoricism is still rampant, not only in textbooks of the social sciences and psychotherapy, but also in journal articles and reference works. In consequence, I have considered not only the products of current researchers in personality psychology[1] but also those of their distinguished predecessors in times past. Over the past two centuries alone there have been sea changes in scientists' conceptualizations of major aspects of human personality. Such changes continue and need to be situated in their historical contexts.

The thriving sectors of inquiry into personality psychology that we witness today, and which require a multi-volume encyclopedia to delineate adequately, have their origin in the science revolution that was rekindled in the late Renaissance and effloresced in the West in the eighteenth-century Enlightenment. That spirit of untrammeled inquiry, which gave birth to modern rationalist and scientific enterprises, spread like a prairie fire over post-Renaissance Europe and, through trade and mercantile structures, to remote points of the globe.

[1] Referring to this social science as *personality psychology* rather than *personology* reflects the usage of many experts in this field. Some, indeed, prefer to call it simply the science of *personality*. Whether to prefer one over the others has proved a vexing issue throughout the book, not only because personology is stylistically simpler and more elegant, but also because nuances of meaning for these terms, which will be addressed, seem to call for one term rather than another.

On a macro-level, attention has been paid in this book to the social, economic, and political forces that have shaped our twenty-first-century world and the way in which we have conceptualized the individual's place in it. On a more narrowly focused level, I have looked at the development of the earlier laboratory-based procedures and brass-instrument technologies, largely developed in German research institutes (where large numbers of young American psychologists went in the nineteenth century for their *praktikantship*) as well as in other regions of the Continent and the United Kingdom. Space is devoted to the historical development of statistical methods that mediated the birth of a science of individual differences, and its idiographic, as opposed to nomothetic, approach to the study of personality. The influence of mental health professions on current and previous models of human personality is sketched. In addition, I define the politico-cultural matrices that have fostered our science and provided the context for our personological paradigms.

Attention is paid to such factors as the mass production of mirrors and looking glasses and the presumed contribution that this has made to the individualism, if not narcissism, of the West. Political movements have thrust us into current socio-cultural, feminist, and intensely humanistic conceptualizations of human nature. Thus, I have a chapter on the shift from illness to wellness models of human nature, another on culture and personality, and a third on gender and personality. Attention is given in these and other chapters to the intersection of personality psychology with other psychological disciplines. This field, frequently referred to as "personology" since Henry Murray coined the term, would be the poorer without the underpinning principles of those psychologies termed developmental, social, neuropsychological, and evolutionary, among others. This text presents in broad strokes the contributions of each of these disciplines to personality psychology.

This book, then, is less a snapshot of personality psychology, time-stamped 2010, than a thematically organized panorama of its evolution to date. Attention will be given to the contributions of nineteenth- and early-twentieth-century psychiatry to later formulations of personality, especially those associated with the names of Janet, Freud, and Jung. Though their "grand systems" have been in large part superannuated, what survives and has been transformed will be integrated into those chapters in which their contributions are particularly relevant and significant. For example, developmental psychology, regarded as one of the foundational disciplines of personality psychology, has rendered obsolete many of the principles of nineteenth- and early-twentieth-century stage-based psychiatry. The divergence of contemporary from earlier developmental psychologies needs to be clearly delineated where that is possible. To explain personality while apparently subscribing to discredited developmental notions entails two negative sequelae. First, it deforms the cognitive structures that readers, especially students, use for assimilating current nomothetic notions of personality development. Second, it distorts their idiographic perspectives as well – an outcome particularly damaging for those who will continue their studies in the science of individual differences and possibly in one of the mental health professions.

To help readers grasp more fully the dimensions of the wellness models of the *human* and the "positive psychology" that has gained ascendancy in the twenty-first century, I have sketched, by contrast, the illness models that reigned supreme in the systems of Schopenhauer and Freud. As well as studying those developmental processes that are commonly regarded as healthy and self-actualizing, I found that it was necessary to detail the heuristic usefulness of studying disease (that is, pathognomonic) processes to gain an understanding of the well-adjusted and healthy human. Attention to the work of nineteenth-century psychiatrists, on the one hand, and to such recent psychoneurologists as Antonio Damasio, Joseph LeDoux, Michael Gazzaniga, Erik Kandel, Bradley S. Peterson, and Oliver Sacks, on the other hand, provides support for this latter purpose. Attention, for example, to such precursors of *positive psychology* as William Stern, Alfred Adler, Kurt Goldstein, and Carl Rogers is useful for the wellness perspectives. Advantages, and possibly pitfalls, in either approach are noted, analyzed, and illustrated. Another example: I have described current research and theorizing in trait psychology, but have taken the liberty of tracing the origins of one of its more important tools, factor analysis, to the humble product–moment correlation associated with the names of Galton, Pearson, and others in the British empirical tradition. The analysis of personality psychology, viewed as a continually evolving discipline rather than as a static one, can be seen at a glance by scanning the table of contents of this book.

The history of the philosophy and science of personality is given some initial chapters of its own to counter the ahistorical penchant that modern students, and even some noteworthy scholars, evidence in addressing this discipline. This description represents its dynamic evolution from principles that have been tested and retested by generations of committed, gifted researchers who dared to go beyond the fashionable templates of their day to those principles that are now widely accepted. Twelve topical chapters, despite their historical lens, have relevance to the current practice of mental health professionals, human resources managers, vocational counselors, public relations officers, advertising agents, and others who work in the public domain. Scholars consider science to be morally, if not politically, neutral and extend this view to the science of personality psychology. This, of course, is a contentious issue. Readers will be guided by their own principles in making use of the ideas proposed in this volume.

My biases need to be made apparent from the start. As I have noted (2000):

On entering the 21st century, the *Zeitgeist* will favor, it appears, theories of personality that are life-span developmental ... more persons-relational, constructivist, process-oriented and dynamic, that is, Heraclitean, holistic, teleonomic, evolutionary, genetics-based, gender-equal, emic, sociological, and idiographic. Clearly, personality theories that predate the Second World War are not, by and large, consistent with these descriptors. It appears that the era of the "grand systems" is past. Personology will reconcile itself to more modest paradigms to describe, explain, and predict human behavior. (p. 2093)

Two points, one substantive, one stylistic, must be made evident here. First, I have tried to integrate frequently neglected topics in the evolution of personality psychology, such as the developmental character of play activities in middle and later adulthood, the individual differences that exist in the importance given to food preparation in the lives of individual families, as well as bereavement and grieving, and the personality penchants that nudge individuals toward and away from religious practice. Second, I have tried to walk a narrow path *between* (a) sounding like a pedant and (b) talking down to readers. I trust that seasoned professionals and scholars will not be offended by explanatory notes and definitions that I have offered in footnotes for the benefit of students. For the same reason, as well as in the interests of completeness, I have included the commonplace notions of, say, Freud's psychosexual stages of development and details bearing on the Big Five Model of temperament.

Finally, I ask for clemency from the reader who, I trust, will reasonably cut me some slack given the breadth of this project. I did not *fully* appreciate before I began this odyssey how difficult it was going to be to trace in a balanced way the impact on personality psychology of its (ancient) historical antecedents, developmental psychology, self-psychology, and such topics as gender, culture, emotion, biology, trait psychology, psychometry, and certain domains in which *I do not* consider myself an expert. Fine scholars spend their lives investigating slivers of any one of them. (I have spent much of my professional life in the domains of educational, developmental, and counselling psychology [at McGill University].)[2] In that light I welcome any corrections and the righting of any imbalances that readers (to be given due credit) will chance to find in this volume.

[2] I was tenured, full professor at McGill University, Department of Educational and Counselling Psychology, until retirement in 2006. This Preface and large segments of this book were written in 2005–2006 while a Visiting Fellow in Wolfson College of Cambridge University.

I wish to thank the community of scholars, both known and unknown to me – seasoned researchers and unseasoned students too numerous to name – who share in the "distributed cognition" that this book represents. They have functioned as the intellectual prostheses that we all need as we progress in our careers. I have welcomed their feedback, both corrective and positive, to the contentious positions here articulated in many areas of this discipline. This book is the product of ideas that have been generated by legions of scientist-thinkers and which, like open-source software, can be stitched together in varying patterns. Few of the ideas *per se* are original with me. I have simply examined them through different lenses.

In particular I wish to thank those who have reviewed this text in various parts, or in whole, and made important contributions to it, most notably, Danny Wedding, Michael Wertheimer, Edward Shorter, Daniel J. Driscoll, Bradley Peterson, Marilyn Fitzpatrick, Andrew Carson, and Millard Susman. I would be remiss if, further, I did not thank the Editor, Hetty Reid (the Commissioning Editor, Psychology) and her staff, for their prompt, helpful, and ready response to my every query and concern. I express a special thanks to Lyn Flight, who diligently read the manuscript from start to finish and presented a great deal of helpful professional advice on documentational and stylistic matters. I am also grateful to the learned reviewers of the manuscript who offered extremely helpful ideas and suggestions for the book's improvement. Their contributions enhanced the overall integrity of the manuscript and facilitated its production in book form. The author has had the last say on its contents and is therefore responsible (a) for flaws in the way the ideas have been configured, (b) for significant lacunae in his arguments, and (c) for the fragile sustainability of his hypotheses and conclusions. In the final analysis the book's virtues belong almost entirely to the scholars who have preceded him in the varied fields broached in this book

FRANK DUMONT
Wolfson College, V. F. 2005–2006
Cambridge

Notes on tables

1 Historical precursors of personality theory

> [A person is] a flow of powerful subjective life, conscious and unconscious; a whispering gallery in which voices echo from the distant past; a gulf stream of fantasies with floating memories of past events, currents of contending complexes, plots and counterplots, hopeful intimations and ideals ... A personality is a full Congress of orators and pressure groups, of children, demagogues, communists, isolationists, war-mongers, mugwumps, grafters, log rollers, lobbyists, Caesars and Christs, Machiavellis and Judases, Tories and Promethean revolutionists.
>
> Henry Murray, 1940, pp. 160–61

A characteristic of contemporary times is that many students – and more than a few scholars – think that the core principles of psychology were developed and validated in the twentieth century, indeed, in their lifetime. Students are often reluctant to read journal articles that are more than a few years old. But this is not just characteristic of the young. In his masterful history of dynamic psychiatry, Henri Ellenberger (1970) wrote that great system builders and theoreticians like Jean-Martin Charcot and Sigmund Freud also suffered from a surprising ahistoricism. He stated, "these men shared an illusion, which has by no means disappeared today, that everything they brought forth was new" (p. 750). This bias can obscure students' views of the wellsprings of modern psychology and cause them to spurn the work of those scholars who laid the foundations of their disciplines. The fact of the matter is that much that Charcot, Freud, and others in the psychiatric tradition proposed as their own was the legacy of scholars, philosophers, and scientists who preceded them, in some cases by thousands of years.

An intellectual history of personality theory does exist, and to ignore it is to truncate our understanding of the very principles on which we wish to base our studies and research. Those who ignore history lose a precious perspective on what they are actually thinking.[1] Reading history, however, presents its own problems. The works of Pierre Janet, William James,

[1] To focus exclusively on the "culture of the present," without regard to the "cultures of the past" that have contributed to it is to be temporally encapsulated. Monocultural modernism impoverishes our vision of a field that prides itself on its multicultural perspectives on reality.

Edward B. Titchener, or Sigmund Freud need to be understood in terms of the *Zeitgeist* that permeated the societies in which they were bred and educated. We tend to interpret their words, and give meanings to the events that they describe, in terms of our own social schemas, values, belief systems, and cognitive structures. This is all the more problematic when the texts we study were written in a language with which we are not familiar and the translation of which does not capture the precise meaning that they communicated in the original. The more the worldview of thinkers whom we are studying is alien to our own, the more we are apt to misunderstand them, and the more cautiously should we apply the principles they fashioned to explain their own psycho-social ambience and conception of the human, let alone our own.

In Martin Gardner's (2000) introduction to his book, *The Annotated Alice*, we are warned of the problems in fully understanding the text that describes Alice's adventures in Wonderland. Gardner writes:

> In the case of Alice we are dealing with a very curious, complicated kind of nonsense, written for British readers of another century, and we need to know a great many things that are not part of the text if we wish to capture its full wit and flavor. It is even worse than that, for some of Carroll's jokes could be understood only by residents of Oxford, and other jokes, still more private, could be understood only by the lovely daughters of Dean Liddell.[2]

We characterize the psychological works of the ancients as philosophy, a term that literally signifies "love of wisdom" as readers know. But this was an umbrella term for a multitude of disciplines. As the scholarship of the nineteenth and twentieth centuries became increasingly specialized, broad fields of study splintered into numerous sub-disciplines. The terms we used for a global domain, such as philosophy, were replaced by new designations. Logic, political science, rhetoric, physics, ethics, cosmology, epistemology, sociology, and psychology, among others, all formerly fell under one rubric, philosophy. Much of the psychology we treasure as our heritage from past epochs derives from the studies of scholars who regarded themselves as philosophers (Hilgard, 1987; Mahoney, 1991). Edna Heidbreder (1933, p. 20), in an influential and widely cited book, reminded us that prior

[2] Hermeneutics, the scientific discipline that creates a methodological framework for understanding text that was created by individuals whose perspectives and personal experiences were very different from our own, was first developed to interpret sacred Scripture. It is no less important in the historical texts bearing on psychology. Each of the chapters that follow this Introduction needs to grapple with the "hermeneutical problem" insofar as the subject matter relates to principles that proceed from a historical and cultural context significantly different from our own. Of course, the readers of the chapters that follow differ among themselves with regard to the mediating schemas they employ to interpret historical texts, as well as the actual text they are reading. They will make their own adjustments.

to the twentieth century psychology was likely to be encountered as *epistemology*. A cursory review of the past will facilitate understanding the process by which psychology has arrived at its relatively mature state.[3] More specifically, it will provide an understanding of human nature and its diverse phenotypic expressions, which constitute, indeed, the subject matter of personality psychology and the science of individual differences.[4] This chapter will reach only to the threshold of a scientific personality psychology that effloresced in the early twentieth century.

Matrix of eastern Mediterranean science

Ancient intellectual systems

Reaching back in time to briefly examine the sources of our knowledge base in personality psychology, which is, of course, rooted in our common nature, casts light on our present understanding of human behavior. The way we think today, our categories of thought, and the unconscious assumptions we make about the world in which we live derive in great measure from the work of towering intellects that addressed the natural mysteries of the world and gave us systems of thought that have not been surpassed – only challenged.

The earliest literate civilization that contributed to a Western understanding of human nature (of which we have a record) is that of the eastern Mediterranean, which flourished between 2,000 and 3,000 years ago. Prior to the cultural and military hegemony of Rome in the Mediterranean basin there was a burst of philosophical and scientific inquiry that centered in Athens, the jewel of the Mediterranean world, a city set on the Attic peninsula. That city, along with the other coastal and island cities of the Aegean Sea, created the major pillars of Western civilization.[5] Among the most remarkable psychologists who have ever lived was a cluster of thinkers

[3] Strangely, the multitude of degree designations that were created to correspond to these free-standing disciplines are increasingly being collapsed into the formerly unitary Doctor of Philosophy (PhD) degree and Master of Arts (MA) degree; the most notable exception is the Doctor of Medicine degree (MD), which is actually a second-cycle degree like the MA rather than a third-cycle degree like the PhD or the DPhil. In the United Kingdom the Bachelor of Medicine (MB) and Bachelor of Surgery (MB BS) are equivalent to the MD in North America.

[4] Individual differences are differences with respect to *something*. Discrepant ideas of what human nature is can confuse a discussion of individual differences and those stable aspects of *Homo sapiens sapiens* on which the differences bear.

[5] Other great civilizations (for example, among many others, the Babylonian, Chinese, Indian, Minoan, Persian, and Semitic) existed in this period, but their influence on the scientific, artistic, and especially the psychological disciplines of the West were not as important.

who dwelt, studied, and taught in a city that is still regarded as the birthplace of Western democracy. The city, Athens, in a dark "fierce world," in the words of Edith Hamilton ([1958] 1993), became "a little centre of white-hot spiritual energy" (p. 13).

Although much of the classical Greek legacy has been lost to us, enough has endured so that *we who are inheritors of that tradition think like Greeks to this day*. Heidbreder (1933) states, "even in the early Greek cosmologies, before the distinction had been made between mind and body, many of the conceptions found in modern psychology, many of its characteristic ways of dealing with its material, were present and in general outline mature" (p. 21). Nevertheless, the number of texts that have been lost is of tragic proportions. For example, of the 120 plays of the great tragedian, Sophocles, who gave inspiration in his Theban trilogy to Freud in his conceptualization of the *Oedipal Conflict*, as well as to feminists in his portrayal of Antigone, only seven have survived. Epicurus, whose renown pervaded the ancient world, was highly esteemed in our era by Nietzsche, especially for the value he placed on friendship as the source of the greatest human pleasure. Of his numerous volumes we have only a few fragments of text. Of the works of the poet, Pindar, who so inspired Rudyard Kipling and other modern poets, relatively few verses have survived. Nevertheless, what has survived from this era has had a profound influence. We think and feel differently because of the systems of thought that flourished in that period.[6]

The influence of Hellenism was not restricted to philosophy or literature. The thinkers of Ancient Greece made towering contributions in architecture, physics, geometry and mathematics, formal logic, rhetoric, statecraft, political science, ethics, astronomy, medicine, music and drama, epic narrative, sculpture and pottery, and, of course, competitive, "Olympic" sports, and psychology. This brilliant efflorescence of the human spirit and its influence on the thought of contemporary humankind is hard to comprehend fully.

Egyptians and numerous ethnically diverse Semitic societies of the eastern Mediterranean, as well as the Minoans and the Babylonians, and the

[6] Scraps and shards of that epoch continue to come to light. Most recently fifty-three fragments of papyrus, dating from the first century, bearing some text of Empedocles, were found in a library in Strasbourg, France. Aristotle cites Empedocles as the inventor of rhetoric and Galen calls him the father of Italian medicine. Although until 1999 we had no extant text of this major thinker at our disposal, these fragments have given us insight into the mind of one of the seminal thinkers of that era. The *National Post* (Canada), January 29, 1999, reported that Peter Parsons, Britain's leading papyrologist, stated that the find was astounding. He said that Empedocles is a major figure in thought whose work has been lost since the fall of Rome. And now we have a substantial stretch of this philosophical text. It is as if there had been a figure called Shakespeare, of whose work we knew nothing bar a few quotes in other works, and then we discovered ten full pages of one of his plays.

Persians further to the east, all made significant contributions to Western thought (Dumont and Carson, 1995; *cf.* Bernal, 1987). Unlike the Greeks, however, these societies were principally focused on issues of the spirit, the afterlife, and the placation of unseen powers. Their daily activities were largely ordered to please divinities jealous of their status among mutually hostile ethnic communities (see Hamilton, [1958] 1993). On the other hand, the Pythagoreans, Heracliteans, Democriteans, Platonists, Aristotelians, Epicureans, and most post-Socratics, including that towering genius, Lucretius, who wrote in Latin, focused their gaze on a world of *natural laws* rather than one that was controlled by angelic spirits and divinities. They trusted in the *reason* of humans to create order in a "rationally" structured society, rather than in the precepts of a priestly bureaucracy and a theocratic state. Curiously, although they extolled the virtues of a democratic society for themselves, they did not hesitate to bring back prisoners from their foreign wars to serve them. But because the Greek ethos was uncongenial to profit-making enterprises and disdainful of retailers and tradesmen, as well as of manual labor, a broad window of opportunity was created that allowed these subservient classes to advance in that society, and the lower socio-economic classes of the Hellenic and succeeding periods were upwardly mobile. Eventually, the servants of the free citizens of Athens were assimilated into the body politic and assumed the rights of citizens to participate in the governance of the state.

The pre-Socratics

Among the pre-Socratics, physiological theories of the development of personality can be traced at least to Empedocles. Later, Hippocrates (fourth century BCE[7]) and Galen (second century CE) postulated that the features of temperament were related to the physical humors of the body: blood, black bile, yellow bile, and phlegm were linked, as these nouns suggest, to sanguine, melancholic, choleric, and phlegmatic temperaments, respectively. Although the notion that these bodily humors determine the expression of personality traits is now relegated, but not entirely accurately, to pre-scientific and discredited folklore, the principle that personality has certain humoral and physiological determinants is very much alive.

7 The Euro-American convention of designating points in time with the birth of Jesus of Nazareth as an anchoring reference point has been a global convention of convenience for geopolitical and commercial purposes. However, out of respect for the great majority of the human race that does not resonate to the expression "Before Christ" or "Year of the Lord" (the Latin of which is abbreviated to AD), we have adopted the convention among some English-speakers of using the expressions, "Before the Common Era" (BCE) and "Common Era" (CE) to replace BC and AD.

As important as these physicians and biologists were for the intellectual bases of Greek culture, the greatest influence was that of one man, Pythagoras (sixth century BCE), whose school provided the impulse for understanding the world in terms of numbers. The school was fascinated by the correspondence of whole numbers and ratios to the realities of their world. As lovers of music, the Pythagoreans were fascinated by the fact that halving the length of a lyre string provided a tone one octave higher than the original. Pleasurable musical chords required lengths that were in simple whole number ratios of 3 to 2, or 4 to 3. Aside from the seminal mathematical discoveries that Pythagoras made, his school promoted an intellectual focus on harmony, order, rhythm, and system, to which our world of the twentieth century still resonates. Modern psychologists are bathed in a Pythagorean ethos that rivals the influence of Plato, Epicurus, Aristotle, and their progeny, the great scientists and thinkers of the eighteenth century (see Murphy, 1968, pp. 3–22).

Character and career psychology

Career psychology, which is rooted in personality theory, is firmly grounded in the Pythagorean philosophy that found its finest expression in the work of Plato. The Greek approach to career emphasized the endowments, inclinations, talents, and worldly aspirations of the individual. If Western culture is individualist more than it is collectivist, we can thank (or blame, as one's value system may dictate) the thinkers of this ancient culture. Dumont and Carson (1995, p. 373) wrote, "In the measure that a society prizes the individuality of its members and recognizes the unique constellation of aptitudes and interests that each person represents, in that measure is there respect for the individual's right to progress in his or her … career." Twenty-seven centuries ago, the great "law-giver," Solon, stated that the lowliest and poorest citizen of Athens had the right and the capacity for self-determination (Hamilton, 1957, p. 24). Herodotus, whose *History* found recent acclaim in the novel (and film), *The English Patient* (Michael Ondaatje, 1993), stated that the members of this commonwealth "obey only the law." And Pericles proclaimed to the Athenians of the fifth century BCE, "You live in the only free city in the world. In Athens alone the state does not interfere with a man's daily life. The individual can be trusted" (Hamilton, 1957, p. 25). Individuals *qua* individuals were allowed, indeed encouraged, to do the extraordinary, to strive to reach the heights in their chosen profession. Hamilton ([1958] 1993, p. 24) cited the classic Greek definition of happiness: "The exercise of vital powers along lines of excellence in a life affording them scope." The ethic of "*excelsior*," allied to a fierce competitiveness, propelled this society to stunning heights of achievement in several brief centuries. Hannah Arendt (1958), marveling at the eruption of this cornucopia of scientific, literary, and artistic products, proposed that the notion of a free *polis* (city state), combined with the principle of permitting all individuals, as individuals, to achieve what otherwise would have been impossible, "multiplied the chances for everybody to distinguish himself, to show in deed and word who he was in his unique distinctness" (p. 197). This *meritocratic society* was programmed for success. Arendt stated that, "one, if not the chief, reason for the incredible development of gift and genius in Athens, as well as for the hardly less surprising swift decline of the city-state, was that precisely from beginning to end its foremost aim was to make the extraordinary an ordinary occurrence of daily life" (p. 197).

The scaffolding of personality psychology is a sound developmental psychology (see Chapter 3, below). The principles of classical developmentalism, elaborated by (among others) Plato and Aristotle, the Stoics, and the austere Epicureans, have subtly and pervasively permeated the ethos and the thinking of contemporary Westerners, and the tacit belief systems that underpin their intellectual discourse. When the classical Greek and Roman writings, such as survived, were resuscitated in the Renaissance, they cascaded like rapid streams into the thought patterns of the major philosophical and political writings of Europe and infused the thought of what we now know as the Enlightenment.

Educational principles

Plato articulated a large number of principles that are fundamental to twentieth-century developmental psychology. Their implications for personality theory seem self-evident. For example, he stressed in more than one work that the development of character begins in childhood: "You know also that the beginning is the most important part of any work, especially in the case of the young and tender thing; for that is the time at which the character is being formed and the desired impression is more readily taken" (Plato, [*c.* 370 BCE] 1991, [377], p. 320). The child's success as an adult depends on the education he or she has been given. This education includes gymnastics, music, literature (both fiction and nonfiction), and other cognitive disciplines. Interestingly, Plato's description of the dangers of allowing children to hear or read stories that will shape a lazy or dangerous character are reminiscent of current concerns about violence in the media and its pernicious influence on youth.

That the character of individuals is shaped by the experiences of their infancy was also well established in Hellenic thought, although that principle (as it bears on temperament) is being challenged today. In *Laws*, for example, one of Plato's later works, he proposed an environmental perspective on the deformation of character that is occasioned by parents' over-reactions to children's spontaneous and immature behaviors.

> The privacy of home life screens from general observation many little incidents, too readily occasioned by a child's pains, pleasures, and passions which are not in keeping with a legislator's recommendations, and tend to bring a medley of incongruities into the characters of our citizens. (Book 7, section 788)

Plato asserted that severity in disciplining children could break their spirit and compromise their ability to function in a civil society. He stated:

> While spoiling of children makes their tempers fretful, peevish and easily upset by mere trifles, severe and unconditional tyranny makes its victims spiritless, servile, and sullen, rendering them unfit for the intercourse of domestic and civic life. (Book 7, section 791)

Plato does not exclude consideration of a genetic component, as the following passage from the same work attests, but he cautions that discipline and control must be folded into the childrearing mix:

> Now of all wild young things a boy is the most difficult to handle. Just because he more than any other has a fount of intelligence in him that has not yet 'run clear,' he is the craftiest, most mischievous, and most unruly of brutes. So the creature must be held in check, as we may say, by more than one bridle – in the first place, when once he is out of the mother's and the nurse's hand, by attendants to care for his childish helplessness, and then further, by all the masters who teach him anything. (Book 7, section 808)

It should not be surprising that a nation that was surrounded by enemies and that had frequently been threatened with domination from the north as well as from the east would place great value on the skills and prowess of the warrior. The Greeks memorialized the victories of their warriors at Marathon and Salamis, as they did those recounted in the Homeric epics of ancient times. In addition to their respect for military ability, the Greeks admired the heroes of the Olympic games, and the achievements of these young men were celebrated in processions and civil ceremonies in which they preceded all others.

The dark side of this respect for military and athletic prowess is the generally subservient role held by women in Greece in the millennium that preceded Athens' golden age. The mythical Pandora was a woman, and she is a symbol of the ambivalence with which women were regarded in Greek society. Plato reacted to these misogynistic values by affirming the potential of women to contribute to the wealth and the civility of a free society.[8] His ideal society was conceived as a meritocracy. He promoted the doctrine that an individual's status in the Republic would be determined by his or her competence, and he was an enthusiastic advocate of universal education:

> Children must attend school not only if their parents please but if they do not please; there will be compulsory education of all and sundry, as far as this is possible ... My law would apply to females as well as males; they shall both go through the same exercises ... Nothing can be more absurd than the practice which prevails in our own country of men and women not following the same pursuits with all their strength and with one mind, for thus the state, instead of being a whole, is reduced to a half, but has the same imposts to pay and the same toils to undergo. (Plato, [*c.* 350 BCE], 1991, [804], p. 722)

Plato also repeatedly reaffirmed his conviction that the child's adult character was shaped in the earliest years of its life by the care and instruction given by parents and teachers, a principle that would later come to be associated with Rousseau, Freud, and many twentieth-century psychologists. In *Laws*, Plato wrote:

> According to my view, anyone who would be good at anything must practice that thing from his youth upwards, both in sport and earnest, in its several branches ... The teacher should endeavor to direct the children's inclinations and pleasures by the help of amusements to their final aim in life. The most important part of education is right training in the nursery. The soul of the child in his play should be guided to the love of that sort of excellence in which when he grows up to manhood he will have to be perfected. (Plato, [*c.* 350 BCE] 1991, [643], p. 649)

The Greeks appeared to possess something quite similar to a modern conception of *aptitude*: "We are not all alike; there are diversities of natures among us which are adapted to different occupations" (Plato, [*c.* 370 BCE] 1991, [370], p. 317). Thus, the principle of *aptitude–occupation fit*, so prominent in recent career psychology, was proposed by Plato 2,400 years ago.

Moral character

Plato, in *The Republic*, assessed the needs of the *polis* for a police force that could discharge its responsibility to protect the citizenry from foreign invaders and enforce respect for the law within the walls. He stated that: "the higher the duties of the guardian, the more time, and skill, and art, and application will be needed by him..." (p. 319). He continued: "will he not require natural aptitude for his calling? ... Then, it will be our duty to

[8] It should be noted that women of Athens in the fourth century BCE practiced careers that far transcended the usual skills of "husbandry" and home economics (Lefkowitz and Fant, 1982, pp. 27–31).

select if we can, natures which are fitted for the task of guarding the city. And the selection will be no easy matter" ([374], p. 319).

Probity of character must also enter the process of selection:

> And a man will be most likely to care about that which he loves? ... and he will be most likely to love that which he regards as having the same interests with himself? ... Then there must be a selection. Let us note among the guardians those who in their whole life show the greatest eagerness to do what is for the good of their country, and the greatest repugnance to do what is against her interests. (Plato, [*c.* 370 BCE], 1991, [412], p. 339)

Plato deliberated about what was most highly esteemed by Athenian society and required, for example, in the character of its police force. A first determination was whether the candidates for the force would make the interests of the society the rule of their lives. A second determination of the suitability of candidates was their personal history, from youth onward. These young men would be asked to perform tasks in which they could be deceived by others, and it was imperative to determine if they had the astuteness of judgment to suspect and uncover these deceptions: "And there would also be toils, and pains and conflicts prescribed for them, in which they will be made to give further proof of their qualities" (p. 340). Lastly, "they will be tried with seductions – that is the third sort of test – and see what will be their behavior" (p. 340).

The higher the office and responsibilities, the more careful should one be in the selection of those individuals who would fill the office:

> Candidates should be proved more thoroughly than gold is proved in the furnace, that we may discover whether they are armed against all seduction, and of a noble bearing always, guardians of themselves and of the music which they have learned, and retaining under all circumstances a rhythmical and harmonious nature, such as will be most serviceable [to others]. ([414], p. 340)

Writing about physicians, Plato asserted that "the most skillful among them are those who from their youth upwards have combined with the knowledge of their art the greatest experience of disease" (p. 337). It is especially useful for the physician to have personally experienced illness; he would then be far more able to understand his patients. (Plato says that the physician differs in this respect from the judge. It would not be appropriate that a judge should have had the experience of evildoing in order to better judge those whom he must condemn.) Moreover, a physician does not have to be healthy to be a good physician, for he cures the body of his patients with his mind; but "the mind which has become and is sick can cure nothing" (Plato, [*c.* 370 BCE] 1991, [408], p. 337). It seems likely that Plato would side with those directors of clinical training programs who argue that personal counseling or analysis (and, with it, personal self-understanding) is a necessary prerequisite for anyone who presumes to enter one of the mental health professions.

In an earlier passage in *The Republic*, Plato stated that people require a natural aptitude for the various projects they intend to undertake: "It is for those who engage them, to select the ones best fitted for the position" ([374], p. 319). He repeatedly lamented that making a selection is no easy task. In the twenty-first century we still find that it is difficult to make decisions in the context of uncertainty. Judging human character, determining the qualities of mind and spirit that make people suitable for certain careers, understanding the temperament of an individual one proposes to marry, selecting collaborators or business partners, and gauging the stamina and probity of a candidate for an advanced degree are all examples of the vexing decisions we are obliged to make in everyday life.

Plato was a Pythagorean, rationalist to the core, as his Socratic discourses and the above passages demonstrate. Over the portal of his academy in Athens he famously wrote, "Only mathematicians enter here," which is emblematic of his appreciation of mathematics – and key to his understanding of ideal and eternal forms. Above all, Plato believed that the republic (that is, society) must provide its citizens all the resources necessary to promote the noblest motives and the strength and nobility of character that would best serve the citizen and his society. Harmony, order, reason, balance, rhythm – all reflecting the eternal order of nature – were the essential properties of the fulfilled human being. Many of the principles for a contemporary personality psychology are embedded in Plato's oeuvre.

Rationalism

Plato's student, Aristotle, was a confirmed environmentalist. Although the principle of *tabula rasa* (blank tablet) is often associated with the writings of the seventeenth-century empiricist, John Locke, priority must be granted to Aristotle, for he clearly affirmed this principle in his treatise, *On the Human Mind* (often given its Latin title, *De Anima*).

> Mind is in a sense potentially whatever is thinkable, although actually it is nothing until it has thought. What it thinks must be in it, just as characters may be said to be on a writing-tablet on which, as yet, nothing stands written. This is exactly what happens with the mind. (Book 3, chapter 4, section 430)

Like his mentor, Plato, Aristotle and other Hellenists prized rationality above other qualities of the human being, and Aristotle defined the human as "*animal rationalis.*" This notion, to the chagrin of many contemporary psychologists, has been dominant in Western thought ever since. We must also remember that Aristotle characterized humans as *political* animals, and in no domain is reason more valuable than in the political. Though the "primacy of the cerebral" has fallen into disfavor among many post-modernists, it can be argued that it is the exercise of human reason that made possible much of the glory of Hellenism – as well as the achievements of modern science.

There is, however, a twofold problem at issue here, and it is in part a value-laden one. First, if humans are to regard themselves as superior to all other species on the phylogenetic scale, is it by virtue of their *intellect* or by virtue of their *emotions*, *conscience*, *consciousness*, *feelings*, *sense of self*, or, less plausibly, their *virtues*? The ancients regarded humans as superior to all other species, as none appeared to be as intellective and rational as humans, though it should be noted here that Plato is widely reputed to have said, "a dog has the soul of a philosopher." This idea clearly meshed with the ethos of a Christian religion that established itself in the declining centuries of the Roman Empire and in the regions adjacent to, and associated with, that empire. The emotions that were seated in the gut or the heart, emotions that also were believed to exist in lower species, were perceived as inferior to clear-headed human rationality. Ultimately, rationality was the way by which one could come to understand and believe in the Divinity. This bias, reflecting the prevailing *Zeitgeist*, pervaded Western Europe and fixed its philosophical values for the next 2,000 years.

Second, the notion of rationality rested on the presumption that the manner in which an organism negotiated the world external to it was predicated on a generally accurate perception of that world. This was not an uncontested notion among Greek philosophers. Democritus, the atomist philosopher, was concerned that obstructions between our human senses and the events of an external world could distort our perceptions. Diaphanous atomic veils could "mistify" and distort sounds and sights before they were captured by one's senses. Democritus wrestled with the problem of a simple realism. How do we explain illusions, hallucinations, and the perceptions of a sleepwalker? Likewise, Heraclitus, Parmenides, and Empedocles all voiced concerns about the unreliability of our senses (*cf.* Watson, 1963, pp. 7–8). An extreme and more philosophical position on the unreliability of sense data is Plato's *Allegory of the Cave*. The message of that tale is that what we observe is only a pale imitation of the "real" universe – that of ideal forms. Analogously, there is ample evidence that they were not oblivious to the distortions that can compromise our perceptions of what others are thinking or doing. Greek tragedies offer numerous examples of the flawed emotional reactions and erroneous judgments made by intelligent men and women.

It is interesting to speculate about the extent to which these thinkers would have embraced the constructivist philosophy of post-modern psychologists, who posit that every organism generates an idiosyncratic "reality." (All forms of constructivism begin with the assumption that conceptions of reality begin with an observer, and that knowledge is a function of one's "constructions" as much as one's experiences.) What is certain is that the science of the Greeks focused on this world rather than a preternatural one. Aristotle in particular was fundamentally empirical. For example, at the age of 40, Aristotle went to the island of Lesbos where he spent two years engaged largely in the study of natural history and marine biology. Before writing his treatises on political science and statecraft he studied the constitutions of 156 nation-states. Aristotle's fundamental assumption was that what you perceive exists and is *there* – at least in the measure that you can perceive it. One clearly needed to study the world before one could make assertions about its properties. This empirical perspective is confirmed in one's day-to-day secular existence.

Aristotle ([*c*. 320 BC] 1991) gave some attention to personality types in both his *Rhetoric* and *Ethics*. But it was his successor in the Academy, Theophrastus, who developed the best-known tract on this subject. He compiled two sets of essays, one on good characters and the other on bad. The former has been lost, but the latter exists, providing numerous and detailed descriptions of various character types.[9]

With the decline of the Greek city-states and the emerging hegemony of Rome, the character of the psychology that was taught in the academies of the time changed dramatically. The self-discipline and the martial virtues that were extolled and rewarded in Rome and its domains provided a different bias to the view of the good and the virtuous than that provided

[9] Theophrastus ([4th century BCE] 1929, pp. 43–44) describes the flatterer as follows:

> Flattery might be understood to be a sort of converse that is dishonorable, but at the same time profitable, to him that flatters; and the Flatterer will say as he walks beside you "Are you aware how people are looking at you? No man in Athens gets such attention"; or this, "You were the man of the hour yesterday in the Porch; why, although there were more than thirty present, when the talk turned to who was the finest man there, the name that came to every lip both first and last was yours." And while he says such things as these, he picks a speck from your coat; or if so be a morsel of chaff be blown into your beard, plucks it out and then says with a smile "D'ye see? because you and I be not met a whole day, your beard's full of gray hairs – though I own your hair is singularly dark for your age." He will desire silence when his friend speaks, or praise the company for listening to him; when he comes to a stop, he will cry in approbation "Quite right"; and if he make a stale jest will laugh, and stuff the corner of his cloak in his mouth as if he could not hold his merriment. Moreover, any man that comes their way is bidden stand awhile till the great one be gone past. He will buy apples and pears and bring them in for the children, and giving them before their father will kiss them and cry "Chicks of a good strain." When he buys shoes with him at the cordwainer's, he will tell him that the foot is shapelier than the shoe. And if he go visiting a friend of his, he will run ahead and tell him he is coming, and then face round and say "I have announced you."

by the Athenians.[10] The Romans ruled (themselves and others) with an iron fist. They were a highly ordered society, and their codes of civil law have shaped the governance of Western Europe for 2,000 years. They were also supreme builders, constructing bridges, aqueducts, arenas and theaters, temples, and public monuments. But the thinkers of Rome, mostly Stoics and Epicureans, were acolytes of the seminal thinkers of Ancient Greece. The Athenians, on the other hand, rewarded excellence in all fields of human endeavor. They strove for extraordinary achievement – Plato's counsel was, "Do one thing and do it well" – and they reached their goal. But the psychology of the ancient Greeks was not preserved in Rome so much as in North Africa and the Near East.

The post-Hellenistic world

As Greek societies declined, to a great extent as a result of their internecine wars, the centers of Greek learning shifted to Cyprus, Egypt, and Syria, as well as to Sicily and the southern Italian peninsula to the west, and Baku (in contemporary Azerbaijan) to the east. The center of this diaspora was Alexandria in Egypt, over which Ptolemy I, a Macedonian general who served under Alexander the Great, established his rule. The intellectual treasures of Greece eventually fell into the custody of Arabic societies, which made their own contributions to this legacy. This was fortunate as many of the manuscripts, not to mention the works of art of "pagan" Greece, were considered to be inimical to the doctrines of the nascent Christian religion. Indeed, historians believe that the tragic destruction by fire of the great Library of Alexandria was the work of disaffected Christians.

Following the fall of the Roman Empire, Arabs established control over a region larger than that of the Roman Empire at its zenith. It stretched from Persia in the east to the western edge of Africa, and it surged into Sicily, Spain, and eventually France. Persians (most notably Avicenna), Arabs, and later Jewish commentators brought with them a deep appreciation of Aristotle and the Platonists. They had not only thoroughly assimilated this intellectual tradition, but had also made substantial contributions to it. The most notable of the Arab commentators were al-Kindi, Alfarabi, and Averroës. The contributions of Jewish scholars who worked in thriving Jewish enclaves in Egypt and Spain are often neglected

[10] For example, if one does a comparative analysis of the Greek Olympic games in which there was wide public participation and the spectacles that were staged in, say, the Roman Colosseum, one will realize that the former extolled grace and beauty, harmony and balance, coordination and poise, as well as strength, speed, and endurance, whereas the latter had the character of blood sports which the populace witnessed, but in which they did not participate.

in historical accounts. Pre-eminent among the significant number of Jews in the Islamic empire who made philosophical contributions that ultimately flowed into medieval Europe was the Spaniard, Moses Maimonides, who is largely credited with hellenizing Judaic thought. Maimonides contributed to the stream of thought that flowed into the work of the notable Catholic Schoolman, Thomas Aquinas. Richard E. Rubenstein (2003) provides a panoramic and well-documented overview of this 1,000-year interlude between the Golden Age of Hellenist science and the stirrings of the rebirth of an Aristotelian worldview.

The mood that developed in the chaos following the collapse of the Greek city-states, as well as the crumbling Roman civic structures (the atrophying sinews maintaining Roman military hegemony in this Greco-Roman world), was one of despair and helplessness. As the conditions of daily life became insufferable the populations bordering the northern shores of the Mediterranean basin began to turn increasingly to the unseen world that offered consolation in this life and the hope of a better existence in an afterlife. There were legions of the meek and the poor, slaves, fishermen, soldiers, and the lowest level of tradesmen, who were instructed to see themselves as the inheritors of the "Good News"[11] that zealous Christian missionaries proclaimed to them. In that perspective, Murphy and Kovach (1972) wrote:

> The one kind of psychology which did take shape after the demise of the Greek creative spirit was that of the Church Fathers, based upon despair regarding this world, including despair regarding the goodness and the value of the living body, and appropriately saturated with two dominant motifs: first Plato's mind–body dualism, and second, a moralistic prejudice against everything in human nature which seemed to bespeak a relation with the animal kingdom. (p. 14)

Essentialism in the Dark Ages

The ethos and moral structures of Western society were largely shaped by the Christian Church, which, following the collapse of the Roman Empire, burgeoned into a continent-wide institution. The Merovingian and Carolingian dynasties that dominated the Frankish Empire were Roman Catholic, and propagated the doctrines of the Church in pre-medieval times. The religious view of the human race held by the societies of that period was that it was intrinsically flawed – and had been since the original transgressions of Adam and Eve. The biblical dictum, "I was shapen in iniquity and in sin did my mother conceive me" (Psalm 51), shaped this

[11] The word gospel means "good news," a translation of *eu-angelikon*, or "glad tidings." The phrase "Good News" was often used synonymously with the four Christian gospels.

traditional Christian view, a view found in the epistles of the Apostle Paul and the Church Fathers and one that is promoted, in a less strident tone, to this day. This notion was based less on objective analysis of human behavior than on the religious notion of "Original Sin." This is a principle that knows no cultural or temporal bounds from the point of view of those who propound it. It is a nomothetic principle, par excellence: that is, it has the force of a law of nature that applies to all humans in all times and places.

The exaltation of the individual that we witness in the writings of the Ancient Greeks had faded into obscure and faint pastels by the fall of the Roman Empire. Thomas H. Leahey (2000) writes that in this period "there was no conception of the individual as an important object of concern or study. This lack is part of the Neoplatonic *Zeitgeist* that dictated that the human intellect knows only universals, not individuals. The rational mind thus knows another person only as an essence – humanness – not as an individual defined by the characteristics that make each person unique" (p. 117). Consequently, there was little interest in *individual differences*, a bias that endured among theorists until the late nineteenth century. It would be an error, however, to attribute this failure to Plato, for as we have seen, both in his *Republic* and in *Laws*, he spoke of the necessity of discerning individual aptitudes and qualities of character when selecting people to fill the multitude of roles required for a well-ordered society.

Original Sin

The primordial corruption of human nature was integral to the model of human personality that emerged in early medievalism. To pick up a different thread, Original Sin was clearly fashioned by the Fathers of the Church and the prevailing supernaturalism of the time. By the power and authority of his intellect, Saint Augustine (given his sexual adventures and the trailing vestiges of Manicheism to which he had been an adherent) cast a pall over human sexuality. The message of the Church at this time was that one needed to turn away from the things of this world and fix one's gaze on the unseen world to which all were called and ordained. This was a dramatic transition from Hellenic humanism to a Christian religionism. The fulfillment of "man" was scheduled for the afterlife. This was not an unwelcome vision to the populations of that time, foundering in the debris of a crumbling empire and prey to the waves of often hostile migrants from the north looking for a better life (see Leahey, 2000, pp. 101–20).

This model of personality was profoundly opposed to sex for pleasure, and Christian clerics compared the convulsive joys of intercourse with epileptic fits ("seizure disorders") in which the individual was seized by Satan himself. Saint Jerome professed that "in view of the purity of Christ's body, all sexual intercourse is unclean" (Pagels, 1989, p. 94, cited in Leahy, 2000, p. 120). Likewise, the following description of Saint Augustine's youthful sexual adventures leaves little doubt about the threat that human sexuality posed to any individual seeking peace in this world and salvation in the next:

> So I muddied the stream of friendship with the filth of lewdness and clouded its clear waters with hell's black river of lust. And yet in spite of this rank depravity, I was vain enough to have ambitions of cutting a fine figure in the world. I also fell in love, which was a snare of my own choosing. My God, my God of mercy, how good you were to me, for you mixed much bitterness in that cup of pleasure. My love was returned and finally shackled me in the bonds of its consummation. In the midst of my joy I was caught up in the coils of trouble, for I was lashed with the cruel fiery rods of jealousy and suspicion, fear, anger, and quarrels. (Augustine, [400 CE] 1991a, p. 16)

The Augustinian tradition, one that blended the *neo-Platonism* of that epoch with the nascent theology of the Roman Catholic Church, dominated Europe for the next eight centuries and, indeed, ramified in various circles, Catholic and Protestant, well into the modern era. The major Christian authorities of the time believed that flesh and its pleasures were at best unclean, and, at worst (and quite often), evil. In addition the primordial seducer, Eve, posed the primary threat to the virtue of men.[12] Peter Abelard (1079–1142) later posed a countervailing position to this notion as well as to the doctrines of Original Sin and the need for a divine redeemer, but his views did not gain enduring traction as we shall see (Rubenstein, 2003, pp. 88–126).

Augustine is of interest to psychologists and theorists for the deeply introspective accounts he gave of the conflicted soul and the image of the Human – images Augustine projected onto the socio-cultural canvas of a Europe straining toward a renaissance. There are many poignant passages in his work in which he wrestled methodically with doubt centuries before Descartes made it a centerpiece of his philosophy. He described indecision, tearing conflicts, and introspective issues as sensitively as did William James, 1,500 years later:

> And one has attempted to establish this, and another to establish that. Yet who ever doubts that he himself lives, and remembers, and understands, and wills, and thinks, and knows, and judges? Seeing that even if he doubts, he lives; if he doubts, he remembers why he doubts; if he doubts, he understands that he doubts; if he

[12] That most lapidary of legal codes, the Ten Commandments, enjoins us from coveting our neighbor's wife, but not from coveting our neighbor's husband. (Tangentially it may be noted that this code also enjoins us to honor our father and mother, but there is no reciprocal commandment that one treat one's children with respect and refrain from abusing them.)

doubts, he wishes to be certain; if he doubts he thinks; if he doubts, he knows that he does not know ... (Augustine, [*c.* 400] 1912, p. 133)

Disdain of the flesh and this material world (an attitude one can trace back by various threads to the early Gnostic sects of North Africa) found expression in the nascent Church – and radically so in the Middle Ages in the religious life of the Cathars and the Albigensians. It assumed eremitical and monastic forms in which adherents committed to lifelong celibacy and a most austere life style. The more mainstream *cult of the Virgin*, which inspired millions to embrace a life of celibacy, also inspired the most brilliant artistic creations of the high Middle Ages (eleventh to fifteenth centuries): the magnificent cathedrals with their radiant windows and soaring, vaulted arches, as well as the works of pen, brush, and chisel. Its contrast to the secular, material, natural order of science and industry has been eloquently analyzed by Henry Adams (1838–1918) in "The dynamo and the Virgin," Chapter 25 of his autobiography, *The Education of Henry Adams* ([1900] 2000).

The Middle Ages and the rise of individualism

The individual in love

Romantic love is by its very nature selective, personal, and idiosyncratic; it is born of the eyes and the "heart" of *these* lovers and no others. It is the meshing of two distinct personalities in a bond that purports to be indissoluble and eternal. An iconic drama that prefigured the acceptance of this principle was the historic love affair of Héloise Garlande and Peter Abelard, two brilliant scholars working in twelfth-century Paris. As Abelard was a renowned scholar, in minor orders and committed to celibacy in the service of the Church, news of their liaison erupted like a Roman candle in the night. They became ensnared in the political and clerical intrigues of the time, and though they maintained a lifelong devotion to each other, family and Church extinguished their carnal relationship. Abelard's intimate account in his autobiography of their love "was not a generic morality play but a story emphasizing the individual personalities of the lovers" (Rubenstein, 2003, p. 100). As cited in Rubenstein, the historian Norman F. Cantor (1993) referred to this historic episode as "the critical turning point in the twelfth-century rediscovery of personality" (p. 332). An appreciation of the individual in all of his or her particularities was becoming salient in the culture of the time. Abelard's book, no less than his remarkable philosophical and theological treatises, portended vast social changes in the ensuing centuries.

The most poignant personological epiphanies of the Middle Ages and the dawning Renaissance were revealed in the works of troubadours, the wandering minstrels of Provence and, later, the northern reaches of Europe. They wrote and sang the poems and heart-wrenching songs of courtly love that frequently ended in loss and bereavement, of which Augustine could have intoned: "I told you so." The love–death of *Tristan and Isolt* is an exemplary tale of this genre (see Campbell's spell-binding *The Masks of God*, [1968] 1991, pp. 175–256). Romantic unions, themselves the result of an appreciation of the distinctiveness of the other, were rarely the basis of marriage in previous eras (Leahey, 2000, p. 121). Up to that time, nubile males and females[13] were widely regarded as simply fungible entities that could be paired in marriage on the basis of political and economic considerations.

Scholars of the thirteenth and fourteenth centuries launched a movement that we can justly call the seedbed of humanism. No one better represents this than the classicist Petrarch. It was his passion for Laura, and the beautiful poems and the *canzoniere* that he composed in her honor, which laid a foundation for the individualism which would later find its finest expression in the plays attributed to William Shakespeare. Petrarch's passion was doomed as Laura was already married when he first became smitten by her beauty. Lore has it that Petrarch retreated to a hermitage near the *Fontaines de la Vaucluse*, deep in the forests of northern Provence, to mourn his loss and write some of his classic lyric poems.

What Petrarch did for the Italian language, *Chaucer* did for Middle English. No less a classicist than Petrarch, he wrote *Troilus and Criseyde*, a tale of the Trojan Wars. The prologue of *Troilus and Criseyde* alerts us to the pain, grief, and bitterness that necessarily accompany the ecstasy of consummated love. The sharp, brilliant delineations of character that sparked an interest in the individual and all his or her distinctness are eloquently expressed in the love songs of the poets, troubadours, and German *Minnesingers* of Chaucer's era. We will return to this theme repeatedly, for just as sexuality pervades all of human nature (and that of innumerable other species), so, too, it shapes and defines the contours of our personality. It is strange that normative sexuality has generally received so little attention in the analysis of personality, except, of course, by those who are not professional psychologists: the poets, novelists, epicists, and playwrights who have seen deepest into the strengths and fault lines of the individuals they described. The most notable exceptions to this in modern European psychology can be found in the works of Schopenhauer and Freud.

[13] Our modern "social clock" has redefined the outermost limits of the nubile condition. Students have challenged me to define the threshold leading beyond nubility.

A reading of Chaucer's *Canterbury Tales* reveals the richness of personality and subtlety of motivation with which the author invested his characters. Campbell ([1968] 1991) wrote: "The Renaissance of delight in this world had begun to refute, in its own immediate way, the Gothic system of disparagement. Petrarch (1304–1374), directly following Dante ... is followed in England by Geoffrey Chaucer (1343–1400), in whose *Canterbury Tales* the wakening interest in portraiture, the features, character, motives, and delights of living individuals, comes to the foreground" (p. 599). Chaucer's work ([*c.* 1386] 1990) is filled with vivid, brilliant, but often subtle personality profiles of characters who are mostly career-designated. The miller, the knight, the prioress, the physician, the merchant, the wife, and the clerk – the characters of all of these individuals were described with humor and insight. The cook, for example, describes:

> A prentice living in our town
> Worked in the victualling trade, and he was brown,
> Brown as a berry: spruce and short he stood,
> As gallant as a goldfinch in the wood.
> Black were his locks and combed with fetching skill;
> He danced so merrily, with such a will,
> That he was known as Revelling Peterkin.
> He was as full of love, as full of sin
> As hives are full of honey, and as sweet.
> Lucky the wench that Peter chanced to meet.
> At every wedding he would sing and hop,
> And he preferred the tavern to the shop. (p. 324)

A budding individualism

This is *not* the creation of a neo-Platonist. Chaucer was the harbinger of a tradition of finely executed characterizations that found its epitome in the works of Shakespeare. Indeed, a science of personality can be built on the characters, more than 120, that the Bard fashioned from the lives that he studied and recreated. Shakespeare held a mirror up to the human throng and etched the most memorable personalities that have graced the theater – and the psyche – of the West. The personality theorist could do no better than to study the characters of his plays. Shakespeare has even given us the terminology necessary to discuss the problems we confront in understanding ourselves. For example, the ancient dichotomy for conceptualizing the relative contributions of heredity and personal history is presented in Shakespeare's characterization of Caliban in *The Tempest*. Prospero refers to him as "a born devil, on whose nature Nurture can never stick."

The two greatest personalities to flow from the mind of Shakespeare, Falstaff and Hamlet, present us with two very different and very distinct views of self. Bloom (1998) tells us "for Hamlet the self is an abyss, the chaos of virtual nothingness. For Falstaff, the self is everything" (p. 5). Facets of their personalities emerge in the innumerable personages who follow them in the pageant of history. We see them about us – and we find their values, their doubts, their features, and their weaknesses in ourselves. They set the stage for a riot of characters that populate the landscape of Western Europe's creative arts. The corpus of Shakespeare's plays virtually launched the individualistic ethos that came to permeate, for better or for worse, the culture of the West.

The rising tide of humanism, one of the dominant characteristics of the Renaissance, found early expression in the tension between the power of institutions, on the one hand, and the strivings for freedom of expression and the valuing of the individual, on the other hand. Scholars like Desiderius Erasmus (1466–1536), Thomas More (1477–1535), and Michel de Montaigne (1533–92) epitomized the surge to freedom of expression and response to the dictates of the individual's conscience. Each in his own way asserted this freedom of conscience vis-à-vis the state, and to a lesser extent the institutional Church.

The movement away from *essentialism* and the universal forms of the Platonists toward the *existentialism* of the humanists can be found in the writings of all these scholars and legions of their fellows. The introspectionism of an Augustine and the honesty and integrity of a Thomas More or a Martin Luther can be found in the essays of Montaigne, and it repays all psychologists to read these essays attentively. We find brilliant sketches of Montaigne's ([1580] 1990) personality in descriptions of his reactions to the feelings and experiences of others, as is evident in the following passage:

> A strong imagination creates the event, say the scholars. I am one of those who are very much influenced by the imagination. Everyone feels its impact, but some are overthrown by it. Its impression on me is piercing. And my art is to escape it, not to resist it. I would live solely in the presence of gay, healthy people. The sight of other people's anguish causes very real anguish to me, and my feelings have often usurped the feelings of others. A continual cougher irritates my lungs and throat … I catch the disease that I study, and lodge it in me. I do not find it strange that imagination brings fevers and death to those who give it a free hand and encourage it …
>
> We drip with sweat, we tremble, we turn pale and turn red at the blows of our imagination; reclining in our feather beds we feel our bodies agitated by their impact, sometimes to the point of expiring. And boiling youth fast asleep, grows so hot in the harness that in dreams it satisfies its amorous desires, "So that as though it were an actual affair, They pour out mighty streams, and stain the clothes they wear [states Lucretius]." (pp. 91–92)

Clearly secularization of scholarship and, to a lesser extent, the visual and dramatic arts, was emerging in the 1500s. Although God-centered activities were still a privileged sector of human art and scholarship, human-centered perspectives, like an eccentric wheel, spun with increasing speed. "Man" became an object of study in itself. The Ancient Greek aphorism, "Man is the measure of all things" (attributed to Anaxagoras), came into fashion in more secularized circles and burst into full blossom in the Enlightenment of the 1700s. Humans were admired for their own qualities and not simply as reflecting glory on the Creator. For example, Shakespeare has Hamlet (Act II, scene ii) exclaim:

> What a piece of work is a man! How noble in reason!
> How infinite in faculty!
> In form and moving how express and admirable!
> In action how like an angel!
> In apprehension how like a god!
> The beauty of the world, the paragon of animals!

Though these are the words of a depressive who finds delight in neither woman nor man and who has "lost all his mirth," they are illustrative of the intrinsic dignity with which humans in the Elizabethan age invested themselves.

A precursor of this exaltation of the human, a century earlier, was the youthful polymath, Pico della Mirandola (1463–94), who wrote, "Who is there that does not wonder at man?" And one need only look at the statuary and paintings of the Italian Renaissance to appreciate that, although the paintings often had a biblical theme, it was the human body itself that was admired, studied, exalted, and, indeed, venerated. Botticelli's (1446–1510) renderings of women are exquisite paeans to the beauty of the human form (as are Michelangelo's renderings of men). These Florentine masters, at the behest of their patrons, the Medici family, found inspiration in the ancient ideals of the human form that is epitomized in Hellenic civilization. And in a life that partially overlapped with Shakespeare's and Locke's, Bernini (1598–1680), the finest sculptor–portraitist of that era, sculpted busts of patrons in exquisite detail and individuality.[14] A critical mass of thinkers, artists, and humanists had emerged by the sixteenth century that repudiated the "contempt for the

[14] Holland Potter (2008, August 7) wrote in the *New York Times* of Bernini's work on display at the J. Paul Getty Museum, "In his best marble portraits, every inch of the surface has been … chiseled and smoothed, tapped, scraped and brushed. Every facial feature sings, every fall of cloth is a luscious little aria. Each detail – the freshly shaved cheek, rolls of flesh under eyes, moisture gathered at the corners of lips – adds to the vivacious ensemble."

flesh" that characterized the Gnostic/Manichean ethos that suffused the early Christian era.

This is the era that witnessed a serious effort to understand nature in terms of natural principles. Although a Divine Creator was (and still is) largely recognized as being the prime mover of the *celestial mechanics* that were increasingly an object of study, God and other unseen powers were no longer seen by the intelligentsia as intervening in all the events of life and in its cosmic framework. As scientific inquirers were increasingly able to explain natural phenomena by recourse to natural causes, a *nature philosophy* began to emerge. God was relegated to the status of "god of the gaps" (that is, he was called on to explain the gaps in human knowledge). Thus, angelic spirits did not move the planets in their orbits, thunder was no longer thought to be his voice, seas raged or subsided, tides surged and receded, eclipses of the sun or the moon occurred, disease and famine ravaged populations, not as a function of God's emotional response to the vagaries of human behavior, but through the deterministic operation of natural laws and principles. The notion that supernatural and unseen powers personally intervened in the daily lives of individuals and governments was increasingly relegated to the domain of theology (see, for example, the classic work of mystic, de Caussade ([1751] 2001), *Abandon à la Providence Divine*).

A gathering storm in psychology: the great cleavage

Psychology was on the cusp of its own revolution in the seventeenth century – although it was still regarded as a philosophical discipline. The rational psychology of Thomas Aquinas continued to be dominant in much of Western, especially Catholic, Europe and, as a result, the science of human thought and behavior continued in this era as a dualistic discipline. René Descartes is the foremost spokesman for this stance. Its "universe of discourse" is the immaterial mind; the body is only the physiological substrate that underpins it. Organismic explanations for mental behaviors began to appear and were substituted for psychological ones. New fields of inquiry were increasingly able to explain "psychic" events through a better understanding of their neurological and endocrinological causes. The construct of "intellectus agens" proposed by Aquinas to enable the transit of sensory experience into an immaterial psyche was increasingly abandoned. Psychology as a discipline of the psyche became, in its own way, "the discipline of the gaps" – and scientists in the twenty-first century are still wrestling with this problem as later chapters will explain.

The rationalist expression of our flawed and godless nature was most pungently stated by Thomas Hobbes (1588–1679). In his seminal work, *Leviathan* ([1651] 1990), he characterized human life as "lonely, poor, short, nasty, and brutish." The religious expression of this secular dictum can be found in various vehicles of social thought and intellectual discussion of that time, not least in the works of the American fundamentalist divines of the eighteenth century. Hobbes moreover was a signal precursor of the Enlightenment, skewering as effectively as Voltaire, but not as fiercely or acerbically, religious apologists for established religions (Lilla, 2008). He reasoned that political and religious conflicts – two sides of the same coin – surge from the fountain of our shared human nature. Humans need religion and good governance to daunt the demons of fear and pride, but vigilance must be exercised to keep the secular art of government separate from the "canting" and power-seeking of theologians and preachers (see chapters VIII and XII in *Leviathan*).

The philosophers of the Enlightenment in eighteenth-century Europe (most notably, Jean-Jacques Rousseau, Stoicist in spirit as he was) were more positive about human nature, yet inveighed against comparable Hobbesian evils. They were not sufficiently self-aware, however, to view themselves as by-products of the linear, teleological worldview that divines and philosophers continued to propagate and which adumbrated in the following centuries in political and other secular writings. Rousseau and his contemporary, Montesquieu, reasoned that societies, and a political way of life, could not only corrupt their members but also ennoble them.[15] In Charles Taylor's (1989) words, political structure "provides the matrix within which [individuals] can be the kinds of human beings they are, within which the noble ends of a life devoted to the public good are first conceivable" (p. 196). But the notion that human nature was essentially corrupt persisted, as we shall see, in many of the systems of philosophy and psychotherapy that flourished in the nineteenth century.

John Locke (1632–1704), an English philosopher, had a profound influence in setting empirical standards for inquiries into human affairs. His celebrated work, *Concerning Human Understanding* ([1690] 1990), is a

[15] During the Renaissance, and the Enlightenment of the eighteenth century that followed, the doctrine of Original Sin was widely repudiated by men of letters – but not by a multitude of divines, most notably Jonathan Edwards – who adopted a rationalist rather than a religious approach to understanding human nature. Where it was accepted, it was often seen to be less the result of the primordial sinfulness of a couple who figure prominently in the Judeo-Christian scriptures (Adam and Eve) than of the corrupting nature of the societies that humans have created and in which they find themselves. Rousseau asserted that humans are, indeed, born noble and uncorrupted. Humans' flawed character must be attributed to a pathogenic environment.

delight to read: straightforward, lucid, and simple. Though he is considered to be an apostle of common sense, by no means did he endorse all the ideas that were commonly held to be valid in his time. Indeed, he stated that "universal consent proves nothing ..." (p. 96). Locke took a fresh look at the traditional wisdom that had been handed down in the rational psychology of the medieval Schoolmen, and then went his own way. He is reputed to have differed from Descartes on the issue of whether humans are born with innate ideas. However, if we read carefully their statements on this matter, it is difficult to find a difference. As Leahey (2000) has stated, the difference between Descartes and Locke on the issue of "innate ideas" is virtually nil (p. 158).

That individuals are born with different dispositions (nature) *is* a Lockean principle. Nevertheless, the development of our abilities depends on the opportunities that life affords us (nurture). Locke stated, "Men then come to be furnished with fewer or more simple ideas from without, according as the objects that they converse with afford greater or less variety" (p. 122). Thus, our environment determines to a great degree the ideas we will entertain and the expansion of the personality that was seeded in us at birth. This focus on the individual in Locke's philosophy is paralleled in the plastic arts by his contemporaries, most notably the Florentine, Bernini. Locke reminds us of the importance of a cross-cultural perspective by pointing out that the values and customs of human society, qualities that certainly contribute to the stereotypical personalities of its citizens, differ from country to country. We also find in his essays the embryonic principles of a utilitarianism associated largely with the late eighteenth-century philosopher, Jeremy Bentham. Bentham argued that humans are motivated both by pain and by pleasure (which no one would dispute – only that they are our exclusive motivators), and they work hard to secure the means to prevent the former and attain the latter.

It is interesting to note that the entire oeuvre of Locke, which has, with that of Francis Bacon (1561–1626), laid much of the philosophical foundation for an empiricist approach to studying and understanding the world in which we live, is almost entirely the work of an armchair philosopher. We find none of the empirical research that characterized, say, Aristotle's studies of marine life on the island of Lesbos nor the meticulous experiments of Galileo (who died shortly after Locke was born). Locke's passages on the difference between "idiots and madmen," for example, are a model of simple speculation, and he makes only a passing reference to Bedlam, an institution housing both groups, which he could have studied (p. 146). Nevertheless, Locke had an enormous influence on the world of Western science as we know it, and his work is a major tributary to the studies that we will address in this book.

The modern era of psychology

People's personalities, as distinguished from their basic temperaments, are largely shaped by the culture in which they have been born, educated, and raised. Although there is a great deal of within-culture variance in the personality of members of any culture, the typical behavior patterns of individuals within a particular culture can be strikingly different from those of another. Italians, for example, are different from the Japanese in the way in which they express their emotions, both publicly and privately. Scandinavians, in general, present a public *persona* that is remarkably different from that presented by Egyptians, say, or Spaniards. The realization of the power of a society to shape the personalities of its members and the modalities by which they arrive at an understanding of their individual "truths" was expressed in the writings of philosophers who antedated the work of contemporary constructivists by several centuries.

Giambattista Vico (1668–1744), widely regarded as the father of historiography, may also be regarded as the precursor of the constructivist philosophy that has laid claim to large swaths of the social sciences in the twentieth century. This obscure professor of rhetoric at the University of Naples claimed that truth is "fabricated," and to know something is to have created it. Vico believed that history is merely the embellished record of a nation's self-creation. Although each society is the product of many causal strands stretching back into the mists of pre-history, its distinctiveness, and the distinctiveness of each of its members, is the product of its self-creativeness. We can arrive at knowledge of anything, including ourselves, only by *creating* this knowledge. In short, to know is to make (that is, *verum ipsum factum est*). Vico anticipated many of the ideas that constructivists of the modern era have integrated into our concepts of the self and of society. Vico was not alone. As so often happens, there is a simultaneity of inventions and new philosophies that well up from a common source, as individuals independent of each other reflect on the same historical body of knowledge and create similar responses to events of which they are all aware.

Johann Gottfried Herder (1744–1803), born the year that Vico died, seemed to recapitulate many of his ideas, although Vico's works were apparently unknown to him (Leahey, 2000, p. 192). Herder, as well as many other German thinkers of the time, rejected the individualism of the Enlightenment, a product in great measure of eighteenth-century French rationalism and the *Encyclopedists*. Shorter (2008) has noted that "'individualism' is not the key to unlocking Romantic thought, as in many ways [Romantic philosophers] were highly collectivistic, admiring the *Volksseele* and the organic nature of village life, for example." Although the German philosophers of this period created the infrastructure of the Romantic tradition whose ideas flowed into the works of nineteenth-century dynamic psychiatry, individualism ultimately gained ascendancy over more collective expressions of human well being. According to this philosophy, to be individualist and socially independent in Western (especially North American) cultures was to be psychologically healthy. To be collectivist, dependent, and socially "enmeshed" was to be psychologically compromised (for example, Dwairy and Van Sickle, 1966; Garrison, 1987) and governed by "group-think."

Humanism and the Enlightenment

The Enlightenment, that burst of collective genius that catapulted continental Europe and Great Britain into a radically rationalist and empiricist mode of inquiry, was inspired by the scientific and technological achievements of the seventeenth century. This had a number of profound effects on the way philosopher–psychologists conceptualized human beings. One dominant stream of thought – *human beings as complex machines* – arose from the work of the seventeenth-century physiologist, William Harvey, as Pinker (1997) reminds us. Harvey revealed the complex nature of the circulatory system that distributes blood throughout the body.[16] The heart as pump became a cognitive heuristic for understanding the nature of the human body as well as a model for industrial advances. The most celebrated work in this tradition was the counter-dualist book by the physician Julien de la Mettrie (1709–51) titled, *Man a Machine*. The notoriety of this materialist work overshadowed the significant advances he made in medicine. Yet it is important in retrospect as it serves as an historical marker for one Western departure from the Cartesian dualism that continues to coexist with it. Pinker (1997) notes that since that era "we have understood the body as a wonderfully complex machine, an assembly of struts, ties, springs, pulleys, levers, joints, hinges, sockets, tanks, pipes, valves, sheaths, pumps, exchangers, and filters ... a contraption of tiny jigs, springs, hinges, rods, sheets, magnets, zippers, and trapdoors, assembled by a data tape whose information is copied, downloaded and scanned" (p. 22).

[16] A Spanish physician, Michael Servetus (1511–53), preceded Harvey in this discovery. Interestingly, Servetus was burned at the stake in Geneva for his heretical ideas on the Trinity. He had been duly warned by Calvin to stay out of Geneva, lest they do precisely this to him.

The conscious and the unconscious

The eighteenth century witnessed the attention that a gifted scholar, Gottfried Wilhelm Leibniz (1646–1716), gave to a construct that was to plow a very deep furrow for our understanding of human personality. Empiricist that he was, Leibniz proposed that perceptions could be measured, and he speculated on the role that subliminal perceptions could play in our mental life. Leibniz labeled these perceptions *petites perceptions*. The sensations that gave rise to these perceptions were insufficiently intense and lengthy to cross the threshold, that is, the *limen* into consciousness – hence they were *subliminal*. The scientific interest in the unconscious and the role it plays in the expression of human personality can justly be stated to originate in the work of Leibniz.

Johann Friedrich Herbart (1776–1841) continued this stream of thought. Although he was not the intellectual equal of Leibniz or Locke (Wolman, 1968, p. 29), Herbart gave an impetus to ideas that were to percolate through psychology for the next two centuries. Most relevant to personality theory are Herbart's notions of the unconscious and his efforts to develop exact mathematical formulae to account for the processes by which conscious thoughts could be displaced into the unconscious, and conversely emerge into consciousness (Boring, 1950, pp. 254–61). It is self-evident that the psychic domain of the unconscious would play a major role in the theories of personality that were to evolve in the following centuries.

Conflict of ideologies: nineteenth-century personality theory

In the nineteenth century psychology had not yet distinguished itself as a distinct and separate discipline. To be called a psychologist was tantamount to being called a philosopher or a psychical researcher. With far greater reason there were no *personologists*, an appellation coined by Henry Murray toward the middle of the twentieth century. Indeed, Harvard University did not establish a department of psychology until 1938, this discipline having theretofore always been nested in its original home, the Department of Philosophy.[17] It is useful to recall in this light that some of the greatest psychologists of the nineteenth century regarded themselves, and were

[17] The first psychology laboratory was, arguably, established at Harvard in 1875 by William James. The more celebrated laboratory established by Wilhelm Wundt in 1879 at the University of Leipzig today generally marks the birth of the *science* of psychology. Other anniversaries compete for this distinction, not least the publication in 1860 of Gustav Theodor Fechner's *The Elements of Psychophysics* and the even earlier publication of Hermann von Helmholtz's two great works, *A Handbook of Physiological Optics* and *Sensations of Tone*.

regarded by their readers, as philosophers. Nietzsche is a good example. He is widely regarded today as the greatest theoretical psychologist of the nineteenth century, although his contemporaries considered him a philologist (his first field of expertise), and then simply a "philosopher."

The American community of psychologists seems to have chosen the establishment of a psychology laboratory at the University of Leipzig by Wilhelm Wundt in 1879 as the birth year of their science, rather than the dates of the remarkable publications of Fechner or von Helmholtz – publications that surpassed in genius and originality those of Wundt (*cf.* for example, Heidbreder, 1933, pp. 92–100). One can only speculate that his creation of an institutional context for doing bench work, using the most sophisticated "brass-instrument" technology that he could find or devise, and his unrelenting and uncompromising demand for the experiment over any other modality of research, weighed more in the balance than any of his, or anyone else's, numerous publications. Humans have a need to commemorate their origins and the benchmarks in their cultural and personal pasts. What event could be more emblematic of its values to psychologists than the establishment of a laboratory by an order-preoccupied scholar of the stature of Wundt? The irony of this for students of the science of personality is that although the German laboratory tradition of the nineteenth century set the standard for serious research (in a positivist mode), it ultimately had less impact on the work of personality psychologists than the work of several Britons and the psychiatrists of Vienna and Paris.

The human sciences that began to flourish in the late eighteenth and early nineteenth centuries reflected the complex social and political desiderata of the time. Philosophers and sociologists of science have long propounded the notion that science, and the research that generates a canonical knowledge base in one field of inquiry or another, is highly ideologized (for example, Mannheim, 1936; Knorr-Cetina, 1981). Society determines what is useful and what is irrelevant to its interests. As a consequence, society sanctions, supports, and funds the research that responds to its needs as well as its values.[18] The social imperative that gave birth to psychiatry was the need to create custodial asylums that were not simply institutions for incarcerating large numbers of troublesome, self-

[18] In the grammar of twenty-first-century scientific discourse, science is considered by many thinkers to be socially constructed, although many laboratory-based researchers (for example, Levitt, 1999) passionately oppose this notion. However, the history of ideas suggests that the science that conforms to social expectations and economic needs is believed and fostered, while the science that is inconsistent with such needs and expectations is neglected and withers. Successful scientists in particular are prone to reject this position. However, John Dupré (2000) reminds us that "Imre Lakatos once remarked that scientists typically understand science about as well as fish understand hydrodynamics" (p. 41).

destructive, socially embarrassing, and clearly deranged individuals. What were needed were therapeutic institutions and a medical profession structured for healing rather than simple confinement (Shorter, 1997, pp. 8–26). But in order to heal, it was necessary to understand the disease. To understand the disease it was necessary to understand the human psyche and its neurological substrate. This endeavor gave birth to dual visions of human personality, which like two molten streams have never fused and which to this day cause tension in all the human sciences. The *organicists* stressed the neurological and other physiological causes of human behavior. In the German tradition these scientists were called *Somatiker*, and they exercised considerable influence. Those who stressed the psychosocial factors that shaped human personality were called the *Psychiker*, or, alternatively, the dynamicists.

The upshot of this dual path to understanding human personality is that both visions gave rise to an understanding of humans that derived from considerations of their failures rather than their strengths. The distortions were more pronounced on the part of the dynamicists, especially those like Freud and Jung, who were immersed in the German Romantic tradition. To draft a model of the human personality and the developmental psychology that gives rise to it by studying only such defective exemplars as walked into one's clinic almost ensures a biased theory. On the other hand, there was widespread belief in the (now well established) heritability of "madness" in the eighteenth century. For example, Pinel (as cited in Shorter, 1997, p. 29) wrote, "It would be difficult not to concede a hereditary transmission of mania, when one recalls that everywhere some members of certain families are struck in several successive generations." This conviction was prevalent in England as well as on the Continent.

The issue of heritability of personality traits almost immediately evokes the name of Sir Francis Galton (1822–1911), cousin of Charles Darwin. Galton was a formidable intellect in his own right and, conducting his inquiries in the Darwinian culture that bathed England at the time, he investigated the heritability of human traits and, most saliently, human intelligence. He was interested in the matter of individual differences in human personality. It was not surprising that his work, touching as it did on what has become the increasingly politically sensitive matter of eugenics and the comparison of races, has found more criticism than acclaim. Nevertheless, his importance for the science of personality must not be underestimated, for he not only gave impetus to the study of personality differences, but also developed a methodology for measuring these differences. The methodology was twofold: questionnaires, tests, and other instruments were developed for distinguishing the level of task performance of several or more individuals; and a statistical device was developed for assessing the degree of association between two sets of data, data considered to be theoretically relevant to one another. This latter instrument was the computational technique that generates an "*index of co-relation.*" The procedure, refined as it was by his student, Karl Pearson, is known today by psychology students around the world as the *Pearson product–moment correlation*.

Everything that Galton accomplished, and that the school and tradition that he founded promoted, was anathema to the great German experimentalists of the nineteenth century who were founding a basic, not an applied, science of psychology. Helmholtz, Fechner, and Wundt were fervid experimentalists, who in principle had little truck with speculative psychological constructs, yet the latter two often found themselves conflicted by the materialism of Helmholz. Fechner[19] indulged frequently in non-experimental studies of a philosophical character, and Wundt in historical, folk psychology (*Völkerpsychologie*) studies. Wundt insisted on the replicability (and replication) of all research, which severely restricted his introspectionist studies to the most simple of mental acts (Leahy, 2000, p. 255). These Mandarin élite of German intellectuals had given birth to an organicist and experimental stream of research that was strongly anti-introspectionist in character (Danziger, 1997, p. 81), yet often found themselves in tension with the pull of the humanist *Geisteswissenschaft* tradition. Researchers in the physiological and experimental psychology tradition opposed the philosophical, armchair psychology of William James that was person-centered, functionalist, and introspectionist.

The American poster boy for Wundtian structuralism and a hard-nosed, laboratory-based empiricism was initially Edward B. Titchener, an expatriate Briton who spent most of his professional life in the United States. He reveled, as did others in the Wundtian tradition, in the "brass-instrument" approach to psychology. Titchener spent most of his professional life at Cornell University, but seemed to have absorbed less of the ethos of Cornell University in those 35 years than he did of the ethos of the University of Leipzig in his two years as a student (Heidbreder, 1933, p. 114). Titchener had only contempt for the pragmatic psychology of an entrepreneurial culture that was generating applied psychologies as fast as an explosive

[19] Fechner had a spiritual side that led him at various periods of his life to explore philosophical issues. E. G. Boring (1950) wrote: "This then was Fechner. He was for seven years a physiologist (1817–1824); for fifteen a physicist (1824–1839); for a dozen years an invalid (1839 to about 1851); for fourteen years a psychophysicist (1851–1865); for eleven years an experimental estheticist (1865–1876); for at least two score years throughout this period recurrently and persistently, a philosopher (1836–1879); and finally during his last eleven years, an old man whose attention had been brought back by public acclaim and criticism to psychophysics (1876–1887)" (p. 283).

industrial economy and a massive immigrant population required it. Mental testing, concern with individual differences, vocational and career psychology, psychotherapy and behaviorism, industrial and organizational psychology – in fact, all such disciplines in the service of commerce and industry and human suffering rather than the service of science – served as his whipping boys. According to Titchener, psychology is the rigorous, laboratory-based, value-free, uncontaminated, factual, nonspeculative, and parsimoniously defined study of "experience dependent on an experiencing person." The introspective study of "an experiencing person" is a significant departure from the purity of Wundtian objectivism. Titchenerism eventually moved from a constricted and rigorously Wundtian introspectionism to a pronounced mentalistic and elementarist approach. This approach, purged of course of the common sense that leads us to think that the mind is, in his words, "an unsubstantial mannikin living inside the head," found a formidable antagonist in Wertheimer (for example, [1925] 1938) and other founders of the Gestalt school. He also distanced himself from the younger Wundt's physiological psychology as much as from the anti-mentalist, John Broadus Watson.

The models of personality development that proved most durable and influential in the twentieth century were, nevertheless, those that were generated by the dynamicist psychiatrists working in *fin de siècle* Vienna, Zurich, and Paris. From the early 1880s to the early 1920s, "personality, abnormal, social, and clinical psychology had deep roots … in the French–Swiss–English-American psychotherapeutic alliance before the emergence of psychoanalysis as an international movement" (Taylor, 1998, p. 119). This is evident if one reads the works of the great nineteenth-century psychiatrists who developed most of the seminal ideas that were to permeate twentieth-century psychotherapy. As Ellenberger (1970) has demonstrated in his monumental work, *The Discovery of the Unconscious*, many of the principles that were presumed to be original with Charcot, Freud, Jung, and Adler had long been current in European philosophy. Whether consciously or not, they integrated them into their formulations, giving scant credit to those who had originated them.

The important role of psychiatry in the development of notions of the self and of personality is intuitively plausible. After all, psychiatrists treated and studied disorders of personality, and they needed a coherent theoretical base for conceptualizing the abnormalities that they faced on an ongoing basis in their clinics. Unfortunately, their theories of personality and human development grew step-by-step with their flawed understanding of their patients' disordered psyches and behavior. And psychiatrists' notions of personality development were subtly, but profoundly, influenced by the vagaries of their practice and the assortment of their clientele. The theories of personality that underpinned their clinical work were not always explicit – nor are they today. One could arguably make a case that all clinicians anchor their practice to principles of personality that are partly tacit, implicit (and consequently operative) in their diagnostic and treatment methods.

The dominant personality theorists of the nineteenth and early twentieth century, with few exceptions, were clinicians with medical training. (It must be noted that it was difficult at this time to establish oneself as a psychotherapist unless one had a medical degree.) Ellenberger (1970, p. 340) notes that Pierre Janet, one of the great system builders of this period, along with Sigmund Freud, Carl Gustav Jung, and Alfred Adler, felt he could not continue with his research in psychopathology unless he acquired an MD degree. This was in spite of the fact that Janet was already celebrated for his work in psychodynamic "psychiatry."

The training of these great theoreticians in medicine and psychopathology largely *preceded* the construction of their architectonic systems – systems which were predicated on a conception of human nature, personality development, and developmental psychology that was derived from the study of emotionally and mentally troubled patients. Their professional background led them to stress what typically goes wrong, rather than what one can expect to go right, in the development of the human personality. As a consequence, their theories of personality and personality development gave prominence to the causes of deviant behavior rather than to the conditions for normal growth. Models of human personality were imbued with speculation about aggressive and narcissistic behavior. For example, Freud in his more mature writings designated the aggressive instinct as one of the two pillars of his drive theory; the other, of course, was the erotic, a notion conceived in variously narrow or broad terms.

Janet, Freud, Jung, and Adler were children of the nineteenth century. They were influenced by its industrial and scientific technology, shaped in large part (with the exception of Janet, who was a child of the Enlightenment) by the Romantic German *Nature Philosophy* that colored the thinking of many influential psycho-philosophers in the German-speaking nations. Indeed, one finds very little that is original in the work of these system builders except some novelties that have since been discredited. The antecedents of most of their principles can be traced to the work of numerous pioneers in clinical hypnotism, dream analysis, and psycho-philosophy, as we shall see. However, what *is* notable in their work is that they created self-consistent, integrated systems that resonated not only among many fellow professionals and academicians, but especially among the middle and upper classes, and those who ardently sought to be indoctrinated by the vulgarizers of these systems – for example, the novelists, playwrights, poets, journalists, bistro-philosophers, and cabaret entertainers of this era. The psychologies of Freud and (especially) Jung not only made a good fit with the *Zeitgeist* of Western Europe, they helped create that *Zeitgeist*.

As occasionally happens when the science of world-class scholars gains wide recognition, such persons are tempted to expatiate with authority in domains that transcend the bounds of their expertise. Freud, for example, ventured into the uncharted realm of paleo-anthropology and speculated about the influence of fire, as artefact, in human culture.[20] Freud and the other great system builders became celebrities in a way that laboratory-based scientists rarely did. The scientist as celebrity is a social phenomenon that generates its own hazards, not only for celebrities themselves but also for their uncritical adulators, the rank and file that take much that they read as true.

Conclusion

This brief historical overview of some of the foundational principles of the science of personality brings us up to the twentieth century. It is in the second third of this century that personality theory, as a field of inquiry, took flight on the wings of some giant intellects – launched by the work of thousands of conscientious doctoral students and their skilled mentors. It will be interesting to observe how the emergence of a dynamic feminism and the swirling tides of global immigration will alter international politics and consequently a science that has been shaped and influenced by these political movements.

The following chapters will present personality theory topically, integrating, as appropriate, the work of the pioneers as well as the most notable systems builders and theoreticians of our time. However, the science of personality is a protean field that embraces every dimension of human experience. I will focus on the most important of these dimensions, some of which have been neglected, confident that readers will share the fascination of many psychologists with the intellectual struggles that shape the development of contemporary personality theory.

[20] In *Civilization and Its Discontents* ([1930] 1961), Freud noted that males were able to extinguish fire by urinating on it. Because of the phallic shape of the flames, they experienced an erotic and homosexual exhilaration in their contest with these flaming "phalli." This, of course, was an anatomical capability that women did not enjoy. The renunciation of this pleasure purchased numerous benefits: the use of fire for practical purposes such as cooking, heating, glazing pottery, and smelting ore. Just as daring was his view that females invented clothing because they wanted to hide their lack of a penis. Freud went further and suggested that women's pubic hair inspired the invention of weaving (1933, as cited in Ellenberger, 1970).

2 From illness to wellness models of human nature

> Science ... is a creative human activity, its geniuses acting more as artists than as information processors. Changes in theory are not simply the derivative results of new discoveries but the work of creative imagination influenced by contemporary social and political forces.
>
> Stephen Jay Gould (1977a)

Changing conceptions of the human

In the practical world of human affairs, we often ignore those things that work to our satisfaction and are quick to give attention to those that break down. Unfortunately, the human psyche[1] is just as prone to break down as the body of which it is a product (or in the view of some, in which it is merely embedded). To appreciate the theory and research that are the foundation of contemporary personality psychology and that continue to influence our thinking about it, we need to study nineteenth-century psychiatry, which was preoccupied, not surprisingly, with the ways in which the psyche could malfunction. The work of the psychotherapists of that period shaped current notions of human mentation and mental health. In addition, those ideas had a profound influence on the direction and the focus of their research and, consequently, defined theoretical schemas used by many modern personality psychologists and psychotherapists.

One must not discount the work of nineteenth-century laboratory-based psychologists as noted in Chapter 1. Using a model of research that had been developed and found to be highly successful by investigators in the natural sciences, they laid a solid foundation for the psychological

[1] *Psyche* is a convenient mental construct (that is, a useful theoretical figment) for contrasting human mentation to the somatic components of the person that generate it. Readers may not wish, then, to regard psyche as having real existence ontologically distinct from their body. Other readers may not share this view.

research that flourished in the twentieth century. But, as is so often the case, the tools available for research limit and thus determine the problems that are addressed. By and large, these men (and, indeed, they were all men), like their illustrious successor, B. F. Skinner, grappled with those natural phenomena that they could measure and rigorously manipulate, and shunned the introspective and anecdotal approach that characterized the field of psychiatry.[2] Such psychologists have been derisively caricatured as *brass-instrument methodologists* by those who are more humanistically and philosophically oriented. This is not entirely accurate.[3] Even one of the most ardent proponents of this approach, Edward Bradford Titchener, changed, as we are prone to do as we age, into an introspectionist. Nevertheless, their work was less influential in personality psychology than that of the psychotherapists whose inquiries and research were shaped by their classical education.

The transition from "exorcism" to dynamic psychotherapy, which Franz Anton Mesmer (1734–1815) epitomized, was influenced by the spirit of the times and especially by the writings of the great thinkers of the eighteenth century. Indeed, the history of modern psychotherapy has been directly traced by Ellenberger (1970) to the work of Mesmer and the psychological system that this physician created (pp. 57–69). Mental illness had been largely "decriminalized" by the mid-eighteenth century and was no longer regarded as punishment for sin. Mesmer treated mental illness as an abnormal but natural event that should be treated with natural rather than supernatural remedies. Although many of the practices and accoutrements of his therapy seem bizarre to us today – think of his patients interconnected to a "*baquet*," a tub of water which he had "*magnetized*" – he introduced seminal ideas that shaped the thinking of psychotherapists for the next 200 years. These ideas influenced the practice of psychotherapy as well as speculation about the nature of the human psyche.

Although science is dedicated to the sober acquisition of fact, facts have a way of becoming invalidated by time and subsequent experience.

[2] Eugene Taylor (1998) wrote that "exactly 100 years following the publication of [William] James's *Principles of Psychology*, the late Gregory Kimble read the book and claimed that he could not find a single scientific principle in it" (p. 102). Kimble is highly regarded as an investigator in the positivist tradition. On the other hand, the demise of the positivism that Kimble represented has often been announced. A review of the experimental journals published under the auspices of the American Psychological Association (as well as those of British and Continental provenance) attests to the fact that this tradition is still alive and flourishing.

[3] One has only to read the biography of Gustav T. Fechner (Boring, 1950, p. 253) to understand the hazards of historical generalizations. Fechner blended an interest in philosophy and mysticism into a career which achieved remarkable results in psychophysics. Edwin Garrigues Boring (1950) wrote that Fechner had a spiritual side that led him episodically to explore philosophical issues.

Psychology in particular is filled with examples of "riotous thinking [that] has produced systems of thought for which the factual proof is admittedly lacking and will be obtainable only in the remote future" (Heidbreder, 1933, p. 12). Systems of thought are not so much expressions of canonical knowledge as they are heuristic tools that support the generation of useful and *temporarily* meaningful facts. Although the systems may themselves eventually fall into disrepair and be discarded and forgotten, some of the findings that were generated within their scaffoldings continue to serve a scientific purpose and become integrated into newer and more robust paradigms. Mesmer's system is a case in point. Although the baquet and animal magnetism were relegated to the rubbish heap of quaint and antiquated ideas, the seminal ideas of rapport, therapeutic suggestion, hypnotism, the influence of the personality of the therapist, and the appurtenances of power – that is, the socio-cultural credentials of the therapist – that this system incarnated have remained with us. In the clinical as well as the political domain, it is an ill wind that blows no good.

Medical influences. Much of the theorizing that was done in the nineteenth and early twentieth centuries about the nature of humans, and specifically the structure of personality, was the work of medical practitioners. These were physicians who had left their laboratories and were practicing their professions in the clinics and asylums of western and central Europe. Like all the intelligentsia of that period, they had received a classical education and were steeped in the great literature of Greece and Rome. Most of them had studied Latin and Greek in the *Gymnasien*, colleges, and *lycées* of their respective societies. They were familiar with the poets, philosophers, playwrights, and encyclopedists who had created the intellectual foundations of this larger "Caucasian" culture. Though they differed among themselves in vivid and important ways, they nevertheless shared a common intellectual heritage – one that gave a peculiarly freewheeling and philosophical cast to their theory building.

On the other hand, these theorists also shared a medical background that provided the common (but limited) framework for much of their theorizing. Rigorous training in laboratory techniques, meticulous measures, isolation of contaminants, control of irrelevant variables, and small-scale mechanistic and reductionist research projects, as well as a conviction that all human behavior could ultimately be reduced to biophysical laws, both shaped and limited the theories they constructed. Their training gave a deterministic, biology-based cast to their studies of personality and the agentic mind that powered it. As Havens (1993) observed of those who have followed in that tradition, "by utilizing the individual or biological elements as the point of departure [for studying human nature] human behavior is attributed to fixed, internal qualities such as drives, instincts, neurochemicals and genes" (p. 510). It is likely that these nineteenth- and twentieth-century system builders – specifically Pierre Janet, Sigmund Freud, Carl Jung, and Alfred Adler – experienced a rending yet creative tension given their bifurcated intellectual formation: on the one hand, classically humanist (even religious in some cases) and physio-psychological, on the other hand.

The philosophical tradition that comprises the major works of the Western canon was heavily imbued with theological notions of the nature of "Man." Although it had been purged of much of its negativism by the French secularist encyclopedists and polemicists, as noted in Chapter 1, pessimistic notions of human nature persisted. The main proponent of this position was Arthur Schopenhauer, who furnished Freud with the philosophical template with which he formulated psychoanalysis. The perspectives taken by both men were determinist, irrationalist, pansexual, and pessimistic. Their views continued well into the twentieth century and, of course, form in part the scaffolding of the therapies still evolving from them. Winter and Barenbaum (1999) state that as late as the 1920s the term "personality" was still used predominantly in discussions of abnormal psychology. This stemmed from the fact that the psychiatrists of the nineteenth century developed theories of personality to enable them as clinicians to explain what had gone awry in their patients' lives. Their clinical case studies constituted the empirical basis of their theories. Most experimental and more positivistically oriented psychologists have regarded this as an unsatisfactory method for arriving at an understanding of the normal person. On the other hand, this approach was defended by Freud and the majority of psychodynamic practitioners.

The pessimism that pervaded the world of Schopenhauer was not restricted to the psychiatric community. William James ([1892]1963a), who was not a psychotherapist, observed, "We are all potentially sick men," which suggests that the human race can be classified into those who are disabled and those who are not yet disabled. Rieber (1998, p. 364) calls our attention to this Jamesian viewpoint:

> The sanest and best of us are of one clay with lunatics and prison-inmates. And whenever we feel this, such a sense of the vanity of our voluntary career comes over us, that all our morality appears as a plaster hiding a sore it can never cure, and our well-doing as the hollowest substitute for that well-being that our lives ought to be grounded in, but alas! are not.

Guilt and the pathogenic secret. Although the notion of sin was largely expunged from the psychiatric literature of the past two centuries, the construct, guilt, which is its corollary, was not. Considered to be remorse that results from having committed real or imagined injustice to others and possibly oneself, it had long been regarded as a source of illness and unhappiness. The advantage of the notion of guilt to the secularists of the nineteenth century was that it did not have the theological connotations of "sin." But not all those in the psychiatric tradition were secularists. We have only to think of Carl Gustav Jung, the most overtly religious of the great theoreticians of that era. Jung, like O. Hobart Mowrer more recently (for example, 1976), appreciated the value of "confession" as a way of purging the psyche of festering regrets, guilt, and self-reproach.[4] The notion that confession is "good for the soul" has historical roots that can be traced back to the monastic communities of the (Semitic) Essenes. In a communal confession rite, they acknowledged the violations of the mores and rules of their community. This practice was seamlessly woven into the fabric of early Christian congregations and has continued to this day in various organized religions.

The notion that openness and social transparency are healthful personality traits captured the attention of Jean-Jacques Rousseau and now contemporary psychologists, but its corollary – that secrets can be depressing and sickening – has been widely accepted for centuries in the religious pastoral literature. This view gradually filtered into the secular culture through the work of the early hypnotherapists (the so-called "magnetists" who worked in a Mesmerian tradition) as well as through the writings of Nathaniel Hawthorne, Henrik Ibsen, and other popular novelists and playwrights. Moritz Benedikt (1835–1920), one of the great pioneers of dynamic psychiatry, systematized in the last third of the nineteenth century

[4] Ellenberger (1970) recounts an experience that Jung described in his autobiography. It bears on the clinical inspiration that fostered his routine search for a pathogenic secret in the early sessions of his psychotherapy. When Jung was a young resident in the Burghölzli at the University Psychiatric Hospital in Zurich, he was charged with the care of a profoundly depressed woman whose prognosis was dismal. But Jung doubted the accuracy of the diagnosis, and based on her responses to the Word Association Test and an analysis of her dreams, he became convinced that she was concealing a terrible secret. She subsequently shared her secret with Jung:

> She had been shocked to learn that the man she had wished to marry who had not seemed to be interested in her had actually been in love with her. But there was no possible remedy because she had married someone else and had two children. Then she let her small daughter suck a sponge [filled] with polluted water, and even gave her small boy a glass of the same water to drink. When the little girl died of typhoid fever, her mother became so disturbed that she had to be committed. Jung explained to her how it was her secret that was making her ill, and a fortnight later she [was able to] leave the hospital, cured. But Jung decided that he had to keep the matter secret from his colleagues. He had the opportunity to reiterate such cures and concluded that the possibility of a pathogenic secret was to be systematically considered in every case. (p. 714)

the then current knowledge of the pathogenic secret by demonstrating (to the satisfaction of many) that the cause of neuroses in many cases was a painful secret, often of a sexual character. Its treatment consisted in divulging the secret to the therapist, and working through the psychological issues that resulted as a consequence of sharing this personal information. The most notable contributor to this literature has been Carl Gustav Jung, who made it central to his psychotherapy. When one considers the constraints that bind the clinician to respect the confidentiality of the darkest secrets from the client's past, one can recognize the similarity of the confessional to the therapist's office – and the dynamics operative in both.

Recent research evidence suggests that simply expressing troubling thoughts and recounting traumatic episodes to others (and, indeed, to oneself by keeping a *journal intime* or a diary) has been found to correlate with measures of mental health (for example, Pennebaker, Kiecolt-Glaser, and Glaser, 1988; *cf.* Marlo and Wagner, 1999). Although the notion of the intrinsic wickedness of humans that was predicated on the doctrine of "Original Sin" was not introduced into the dynamic psychiatry of the nineteenth century, it was suggested by the notion that morally fallible humans needed to admit their malfeasances to other humans in order to achieve – even maintain – health. A modern expression of this belief developed in the second half of the twentieth century in the therapeutic community (TC) movement. Synanon, Daytop Village, Phoenix House, and the Portage Program are examples of rehabilitation centers where each day begins with a morning meeting in which the members of the "family" publicly confess their delinquencies of the previous day. Mowrer (1976) related the troubled residents' need to cleanse themselves of their guilt (a "need" they often didn't recognize) to the ancient rite of *exomologesis* (p. 12). This was a practice in which primitive Christians, gathered in the homes of fellow believers, integrated communal confession into their religious rites. Readers can make a personal judgment as to whether humans require this level of transparency for the maintenance of health (*cf.* Jourard, 1971).

Nineteenth-century psychiatry and visions of "Man"

The influence of psychiatry

The theories of personality that were developed by the practitioners of dynamic psychiatry in the nineteenth and early twentieth centuries were largely derived from the Romantic tradition and the Nature Philosophy that paralleled the stream of thought flowing from the scientific tradition of the Enlightenment (Ellenberger, 1970, pp. 202–23). Freud and Jung could both quote long passages from Goethe's *Faust* (Ellenberger, 1970, p. 727), which suggests that both men were influenced by the dilemmas, tensions, and despair embedded in the tragedy of Faust,[5] who sold his soul to the devil to achieve the Mephistophelian power that had eluded him. Goethe was also the author of the novel, *The Sorrows of Young Werther* (a suicidal figure), that had seized the imagination of generations of European intellectuals. Faust, Werther, and a host of other desperate, melancholic figures who struggled with the destructive forces of nature and their own character weakness set the stage for much of the theorizing of nineteenth- and early twentieth-century psychiatry. In fact, it appears Freud identified much more with Goethe than he did with his mentors, Josef Breuer (1842–1925), Jean-Martin Charcot (1825–93), and Wilhelm Fliess (1858–1928).

Romantic philosophers[6] included the unconscious as a central concept in their system of thought. In their view the realm of the unconscious connects humans in a spiritual way with the rest of the universe. Further, it was for them a wellspring of creativity, of poetic and artistic inspiration. In retrospect, it relates more to the creative unconscious of Jung connecting humans in a spiritual way with the rest of the universe (or later of Milton H. Erickson) than the libidinal unconscious of Freud. It was the human link to all that was most noble and virtuous in nature. Hypnotism (especially "magnetic somnambulism" in the Mesmerian tradition), mystical ecstasy, mediumistic activity, dreams, and other psychic avenues to this psychic domain presumably made it possible for humans to access understandings about themselves that they otherwise could not.

Flournoy and the unconscious. The numerous investigators of the unconscious in the nineteenth century included Théodore Flournoy (1854–1920), a Genevan who became a close friend of William James. Flournoy's investigations, particularly of one Hélène Smith, a well-known medium, led him to believe that the unconscious had several positive and uplifting functions. Parts of ourselves allow us to play tricks on one

[5] In Goethe's poetic epic, Faust sold his soul to Mephistopheles (a devil in medieval legend) in exchange for the magical powers that he craved. In its theatric prelude a dominant theme is evoked when the director addresses the Poet and the Comedian as follows: "You two, Sirs, who have been my stay, In many a time of storm and stress." Goethe ([1832] 1952) has been credited with inaugurating the personological as well as socio-political concept of *Sturm und Drang*. This has become a shorthand phrase used by psychologists to characterize periods in personal development that are filled with turmoil and struggle.

[6] Romantic Philosophy, a particular historical philosophical and literary movement of the eighteenth and nineteenth centuries that thrived especially in the German culture, is capitalized to distinguish it from more generic expressions of "romantic." Of historical note is that the poetic icon for this movement is Schiller's *Ode to Joy*, a poem that inspired the last movement of Beethoven's ninth symphony.

another – in fact, mediums who pretend to be in touch with pure spirits and the dead are essentially engaged in *ludic behavior*. Their primary intent is not to deceive, but to have fun. This is like little children who fabricate falsehoods. Their intention is not to lie, but to express what they wish were happening – that which they yearn to be true, but is not. The impulse to lie becomes an imperative when a punishment is expected. The boundaries between fantasy, wishes, and reality can be fuzzy and easily crossed without one's awareness. False memories are one example of this kind of wishful thinking. Rarely has one social scientist, Flournoy, learned so much from such a small-sample study.

The unconscious in Flournoy's view is *protective*. It results in spontaneous behaviors that in retrospect we realize prevented a blunder, avoided a misstep, stifled the recovery of a destructive memory, or provided us with some modicum of comfort and peace of mind that comes from we-know-not-where. The unconscious is also *compensatory*, resulting in adventures and romances that occur at a subliminal level as wish fulfillment. Though we are not ordinarily conscious of these fantasies, they are occasionally revealed in dreams and hypnotic states. Finally, and perhaps most importantly, the unconscious is *creative*. This aspect of subconscious life (proposed by Galton and C. S. Myers as well as the earlier Romantic philosophers [Murphy, 1949, p. 204]) was revisited in the late twentieth century, especially in the work of Milton H. Erickson (Lankton and Lankton, 1983; O'Hanlon and Weiner-Davis, 1989; de Shazer, 1988; Zeig, 1980). Flournoy illustrated the creative aspects of the unconscious by describing the philosophical discourses of a young mother who in a wakened state would have had neither the interest nor the sophistication necessary to make these learned observations (Ellenberger, 1970, p. 317).

This psychiatric view of the unconscious reveals a personology that seems both modern and positive. Everyone's life unfolds on various levels of awareness, and our natural playfulness is much more closely linked to our unconscious than to our conscious deliberations. The unconscious appears to have a mythopoietic function also, as we are all more or less spontaneously creative. We are instinctively and fiercely self-protective, of ourselves as well as of our children and others with whom we identify. In the nineteenth century, "the assumption that disturbing tendencies were forced into the unconscious was a matter of course" (Murphy, 1949, p. 204). This notion flowed most significantly from the writings of Nietzsche, but also from the studies of Jean-Martin Charcot, Alfred Binet, Adolf Myer, and many other pioneers in this field.

The historical record shows that a pessimistic view of the Human was prevalent if not dominant in nineteenth-century Europe. What is less clear is why this vision persisted well into the twentieth century, for many theoreticians (with the exception of John Broadus Watson – and a nascent behaviorism) continued to focus on morbid views of human nature. Certainly the horrendous large-scale carnage of the First World War could only have reinforced this view. It motivated Freud to add a second, aggressive, drive to his model of the Human.[7] Westen and Gabbard (1999) make this same point by citing Otto Kernberg, an able theoretician in the psychoanalytic tradition: "According to Kernberg, personality organization – that is, the enduring ways people perceive themselves and others, behave interpersonally, pursue their goals, and defend against unpleasant feelings – can be understood on a continuum of pathology" (p. 64). Positivist eddies, however, and later strong countercurrents to this Schopenhauerian ethos had begun to emerge as noted above.

Memories and anecdotal evidence. There are numerous problems that result from anchoring one's personality theory in patients' recollections (for example, Kihlstrom,1994; Loftus, 1979, 1993; Loftus *et al.*, 1989; Loftus and Ketcham, 1994). First, there is the issue of clinical bias. Early personality theorists developed paradigms that grew out of data gathered in clinical interviews. There is an abundance of evidence today that memories of the sort that arise in a clinician's office are profoundly altered by the expectations of therapists and the language and theoretical constructs they use to communicate with patients (for example, Bowers and Farvolden, 1996; Dumont and Fitzpatrick, 2001; Laurence, Day, and Gaston, 1998).[8]

[7] Of historical interest is the fact that earlier Alfred Adler, when a member of the Vienna Psychoanalytic Society, had proposed (in Nietzsche's wake) that an aggressive drive was inherent to human nature. This became a major source of disagreement between Freud and Adler and contributed to Adler's alienation from the Society. Following the First World War they reversed their positions; the former endorsed and the latter repudiated it. (See Ellenberger, 1970, pp. 274–77, for a synopsis of the origins of "aggressive instinct.")

[8] A classic passage exists in the collected works of Sigmund Freud that demonstrates the power of the clinician to shape the memories of his or her patient. We cite Freud, not because he was unique in this respect but precisely because he manifested this human tendency to a salient degree in his clinical procedure. He wrote: "The work keeps on coming to a stop and they [the patients] keep on maintaining that this time [no memory] has occurred to them. We must not believe what they say, we must always assume, and tell them too, that they have kept something back … We must insist on this, we must repeat the pressure and represent ourselves as infallible, till at last we are really told something … There are cases, too, in which the patient tries to disown [the memory] even after its return. 'Something has occurred to me now, but you obviously put it into my head' … In all such cases, I remain unshakably firm. I … explain to the patient that [these distinctions] are only forms of his resistance and pretexts raised by it against reproducing this particular memory, which we must recognize in spite of all this" (1955, pp. 279–80). Another example: "Quite often we do not succeed in bringing the patient to recollect what has been repressed. Instead of that, if the analysis is carried out correctly, we produce in him an assured conviction of the truth of the construction which achieves the same therapeutic result as a recaptured memory" (1964b, pp. 265–66).

In fact, therapists assist their clients in confabulating memories that are consistent with their theories. Qualitative researchers also work with small numbers of "unrepresentative" subjects, interviewing them and conducting focus groups. Even the most able researchers, who make a studied effort to isolate their own preconceptions from the testimony of their research subjects, are not immune to this source of error – which they candidly admit. The research setting, whether the clinician's office or laboratory, covertly elicits in subliminal whispers the responses that promote the mindset of both the researcher and clinician. Popular lore is replete with accounts of how the Freudian patient excels in remembering incestuous childhood urges or, for example, the female patient's yearnings for a penis like her brother's. (Freud's ([1905] 1953a; [1931] 1962b) analysis of human female sexuality is, of course, complexly intermeshed with his theory of the Oedipal complex.) The Jungian patient obliges the therapist with dreams containing assorted images of archetypes, religious symbolism, and the rich accoutrements of the collective unconscious. Even when the requisite admissions are not forthcoming, they are often dragged out of the unwilling patient. Freud, in assisting in the treatment of Little Hans, wrote of "'the castration complex,' the presence of which we are so often obliged to infer in analyzing neurotics, though they one and all struggle violently against recognizing it" ([1905] 1953b, vol. VII, p. 7).

Westen and Gabbard (1999), who believe that psychoanalysis has a great deal to offer personality psychology, have argued that skilled analysts have learned how to teach others "a way of listening to the manifest content of a patient's communications and recognizing patterns of thought, feeling, motivation, and behavior that seem to co-occur and become activated under particular circumstances" (p. 71). In some contexts this appears to be a plausible and sound line of argument, which indeed is at the root of much scientific research. They continue, noting that the methods of inquiry used by psychoanalysts "aim less at asking people to describe themselves than to express themselves, and in so doing to reveal themselves" (p. 71). But as Thomas Aquinas noted six centuries ago, "whatever is received is received in the mode of the receiver" (that is, *quidquid recipitur, ad modum recipientis recipitur*). All of our perceptions (including therapists' construal of what their clients have told them) are shaped by the cognitive structures each of us has formed – the products of our personal history and cultural heritage. If this were not true, why, one may ask, do thinkers belonging to varied philosophical and clinical schools make such different things out of the same clinical material? In any event, the assumption that people know what they are thinking and feeling better than anyone else does is the safest default assumption we can make.

Memory: a fickle resource When patients begin to give a therapist an account of their life, they are likely to report events as if they were viewing a film. But our memory is more like a palimpsest, for the events of our personal history have not only been encoded but over the years have been embellished, purged of unflattering features, erased, and overwritten. Each retelling involves the encroachment of distortions and novel details that appear to blend plausibly into the narrative. Some critical events can be forgotten, for there is no compelling evidence that an ineradicable memorial record of our life history exists in our cerebrum. Indeed, there is good evidence that we create our memories. Inevitably, the reconstruction of our life involves a good deal of fabrication (consult the corpus of Elizabeth Loftus's work on memory). As Laurence, Day, and Gaston (1998) have pointed out, "Everything we can recall about our own past is the result of some accurate information about events gone by, and a healthy dose of filling in the details." Forgetting is not *per se* evidence of pathology. It becomes pathological only when it seriously compromises individuals' ability to function authentically in their everyday life.

Second, these psychiatrists tended to extrapolate their findings to the entire human race, not simply to their "spoiled" clients and the wealthy, socially privileged classes of their own social milieu. Viennese psychiatrists of the era seemed blithely oblivious to the impact of the socio-cultural variables that shaped even the most important of the developmental patterns of behavior. This was consistent with the notion that the important patterns of human behavior were biologically determined. Destiny was implicit in biology. To be a member of the species *Homo sapiens* was to be heir to certain inescapable determinants of social as well as biological activities. The cultural encapsulation of most of the influential theoreticians of that era is striking. But as every knowledgeable historian and hermeneuticist knows, we must beware, in this electronic age, of imposing our frame of reference onto the products of a previous age, except, of course, to illustrate the contrasts in the meanings and the perspectives of the two historical periods.

Jung's mentors were Eugen Bleuler (1857–1939), Théodore Flournoy (1854–1920), Pierre Janet (1859–1947), and Alfred Binet (1857–1911). They were notable exceptions to the pathognomonic (that is, disease descriptive) orientation assumed by nineteenth-century psychiatry. Jung in particular rejected the pansexualism, pessimism, and radical determinism that were promoted in the Viennese Psychoanalytic Association to which he belonged for a time. He had a different family history and cultural background from that of Freud, and thus did not find compelling personal reasons to support the notion of Oedipal complex, or the other "sexist" complexes that Freud hypothesized. Further, he repudiated Freud's hostility to religion. On the contrary, Jung (1957) posited that humans are naturally religious. He stated that he never had a patient who had passed the "midpoint" of his or her life whose principal problem was not a religious one. His therapy involved an analysis of his patients' ideations and dreams in terms of the myths, archetypal images, and symbols that he found in the ethnographic literature. It is this search for universal principles animating humans among the cultures of the Afro-Asian world as well as those of Europe and the Americas that distinguishes Jung from the generality of psychiatrists of the past. He attempted to transcend the bounds of his own upbringing through his study of Zen Buddhism, the various yogas, the Gnostic sects of primitive Christianity, the mystic physician, Paracelsus, the spiritual exercises of Saint Ignatius Loyola, the medieval alchemists, and mystics of all eras. He reveled in the *Tibetan Book of the Dead*, the *I Ching*, and the ancient Chinese tract, *The Secret of the Golden Flower*.

Some have argued that Jung's openness to the world simply masked a selective reading of the ethnographic literature whose purpose was to find conceptual artefacts that fit a preconceived paradigm of his own. It is possible that he escaped the cultural enclaves of Basel and Zurich only to explore a larger, global enclave of religious and transpersonal semiotics. In any event, this work was an early foray into the multicultural and nomothetic aspects of personality development. Jung affirmed that all people had to participate in the public activities and governance of their society (societies, nevertheless, that typically had not yet given women the right to vote). Foreshadowing Erik Erikson's description of the sixth and seventh stages of human development, Jung stated the need of people to be educative, generative, and constructive in their social environment – without a heavy emphasis on sexuality (Erikson, 1963, pp. 263–66). The social isolate was not a fulfilled human being. In a similar vein, and at approximately the same time, Alfred Adler made "community interest" (in German, *Gemeinschaftsgefühl*) the centerpiece of psychotherapy and socio-political philosophy. Adler and his wife, Raïssa, were actively committed to socialist causes.

Jung's life span psychology of personal individuation inspired Erik Erikson's creation of the later stages of life that went beyond those in the Freudian schema. Departing still further from a pathognomonic template for understanding the development of the human personality, Jung asserted that humans use the mature stages of life for fully individuating themselves – claiming and reclaiming their undeveloped potential. A major contribution to his theory of individuation was his paradigm of psychological types (Jung, [1921] 1971). The two foundational types are *extraversion* and *introversion*. These are continuous variables (admitting of greater or lesser degrees of intensity) describing characteristic ways in which individuals respond to their ambient world as well as to their internal and intrapsychic milieu. Extraverts are disposed to engage large numbers of their fellows in group activities, communicate fluently, make friends easily, enjoy parties and other convivial events. Introverts are self-reflective, have few but deep friendships, cultivate a rich inner life, and enjoy solitude. These two constructs were borrowed by Hans-Jürgen Eysenck to found his typology of human traits. It is fair to say that Jung's portrait of the normal human being springs less from a pathognomonic than from a wellness template, and this marks him as a modern. He developed a psychological rather than a psychiatric approach to the study of *Homo sapiens*.

What the pathognomonic can help us understand: the pros and cons

The cons of this issue

Raymond B. Cattell (1905–98), a renowned trait psychologist and psychometrician, is representative of a large school of contemporary psychologists who have maintained that the study of disease processes does not provide a sound foundation for personology – an issue treated below in Chapter 5. He energetically advanced the idea that the clinical approach to this problem was not, *as such*, science. It had the additional "drawback that it produces its personality theories from data gathered from abnormal, diseased processes rather than normal ranges" (1965, p. 22).[9] Cattell, and others who shared this position, presumed that psychological diseases are typically entities that involve a unique causal process – suggesting that there is a radical discontinuity between a mentally disordered state and a normal state.

The continuum problem. The crux of this nettlesome question, which will be expanded upon in Chapter 12 on disordered personality, is

[9] That we can learn about the normal from a study of the abnormal has long been current among biologists. This principle is illustrated by a quote from Shorter (1997, p. 264): "The great French physiologist Claude Bernard [1813–78] articulated the notion of using drugs to study the brain: 'Poisons represent a means of analyzing the qualities of the nervous system, a kind of physiological scalpel much more delicate and subtle than ordinary scalpels ... ' [1869]" with which to delineate and treat disorders of the mind and brain. Using such a "scalpel" to suppress normal functions of neurological systems is one means of determining what those functions are.

whether individuals with a well-defined mental illness are simply those in the tails of a normal distribution of behavior, or whether they constitute a category of disordered people – a population apart. If we think that mental illnesses are not disease entities but simply normal processes that have gone more or less awry, then we can conclude that we can learn a good deal about human nature by studying them. After all, behavior whether functional or dysfunctional is *human* behavior. Extrapolating from excessive or deficient human behavior to normal, presumably healthy behavior seems less problematic, it has been argued, than extrapolating from the healthy behavior of other species to normal behavior in humans.

The position one takes on this issue depends in part on one's philosophical bias. Simply put, if one is a *constructivist*, one states that normality and abnormality are socially constructed, and disease may be simply in the eye of the clinician – and one's neighbor. And how observers construe the social world in which they are enmeshed depends on the social standards of behavior in which they and the patient have been indoctrinated. For example, what constitutes paranoia in one setting is considered due diligence and sound business practice in another. On the other hand, if one is an epistemological *realist*, one can affirm that a disease entity exists, independent of our social conventions, permitting us to dichotomize populations into those who suffer from the disease and those who do not. The *pragmatist*, however, will say, "Fie on your philosophical squabbles. There is an abundance of evidence that we can learn a lot about healthy behavior and how to prevent disorders by studying disease processes and conditions. Whether these conditions are continuous or discontinuous matters less than whether or not we can build an internally consistent model of personality psychology – and use it to help people individually and collectively." Still further removed from the fray is the opportunistic taxonomist who may say, "I'll design you anything you want. Tell me the kind of classification system that will suit your research or institutional objectives and I'll custom make it for you." While waiting for the next generation of answers to such questions – answers that may never come as the questions verge on the metaphysical – readers are encouraged to take a position which they are prepared to defend.

There are ancillary methodological problems that can result from this viewpoint. Psychiatric professionals were *historically* focused on socio-behavioral rather than on biologically oriented explanations of disorders (*cf.* Shorter, 1997, p. viii). The upshot was that this specialty, nominally medical, was shunted away from developing physio-psychological treatments of disorders toward mentalistic and behavioral treatments – that is, healing people by talking with them rather than medicating them.[10]

A more insidious threat to the usefulness of some theories is that as one's professional life becomes wholly involved in the diagnosis and treatment of mental disease one's notion of what is natural and healthy becomes slowly distorted. This begets a pessimism that gets built into the conceptual structure of human nature. Humans are viewed as being prone to pervasively sexualized, narcissistic, inauthentic, deceptive, self-seeking, and aggressive behavior patterns. This is all the more problematic if those involved in the theory building are undergoing or have suffered severe psychological disorders of their own.

The pros of this issue

In his book, *An Anthropologist on Mars*, Oliver Sacks (1996) presented evidence that the study of disease *can* give us insights into the nature of the organism that we might otherwise not suspect. In this sense we become aware of a capability only when we, or an individual we are observing, is deprived of it. Moreover, "defects, disorders, diseases ... can play a paradoxical role, by bringing out latent powers, developments, evolutions, forms of life, that might never be seen, or even be imaginable, in their absence. It is the paradox of disease, in this sense, its 'creative' potential, that forms the central theme of this book" (p. xvi). V. S. Ramachandran and Sandra Blakeslee (1998) quoted Laurence Miller relative to the value of studying the diseased or abnormal to understand the normal: "By the deficits we may know the talents, by the exceptions we may discern the rules, by studying pathology we may construct a model of health. And – most important – from this model may evolve the insights and tools we

[10] We must recognize that this was not the only orientation to healing the mentally disturbed, although it had become the dominant one in the early twentieth century. Edward Shorter, a respected historian, has voiced a strong anti-psychodynamic position on this matter. He wrote (1997): "Psychiatry has always been torn between two visions of mental illness. One vision stresses the neurosciences, with their interest in brain chemistry, brain anatomy, and medication, seeing the origin of psychic distress in the biology of the cerebral cortex. The other vision stresses the psychosocial side of patients' lives attributing their symptoms to social problems or past personal stresses to which people may adjust imperfectly" (p. 26). This produced an anomalous result, according to Shorter's acerbic analysis, for the "American Psychoanalytic Association initially insisted that only MDs could be trained as analysts, and later that only psychiatrists could do analysis. In retrospect this insistence was bizarre, for psychoanalysis required no more medical training than astrology, and the attempt to impose a medical monopoly over Freud's technique was a self-interested ploy to exclude psychologists, psychiatric social workers, and other competitors ... " from the field of psychotherapy (p. 146).

need to affect our lives, mold our own destinies, change ourselves and our society in ways that, as yet, we can only imagine" (p. xv).

Life history and personal schemas. Developmentalists, such as Bleuler, Flournoy, Forel, Freud, and Jung, whose medical careers were shunted to the treatment of psychiatric populations, had had pre-professional lives of their own in which they had developed tacit schemas of what normal humans were like. These schemas existed long before they started treating patients, and it is not as if their understanding of personality, and of human nature in general, derived exclusively from the study of disease processes. There was always a "compare and contrast" analysis operating as they studied their clients and made inferences about how their disorders might shed new light on their childhood understanding of the Human. But herein lay the problem. It was the interaction of their idiosyncratic view of the human deriving from their personal life history with their also highly personal, schema-driven view of the life history of their patients that produced their divergent developmental theories. Whatever distortions existed in the mutually contradictory schemas of those generations of psychiatrists congealed and hardened in the privacy of their clinical office. These views, with supporting data, were not usually subjected to the scrutiny of a community of scientists. They were proposed as *doctrine* to their acolytes and to select groups that read each other's papers and evaluated the merit of each paper by the extent to which it conformed to the core principles formulated by the founder of the system.[11]

Political movements still influence the kinds of research that psychologists do and even the nature of the results that they are inclined to find (Knorr-Cetina, 1981; Redding, 2001). The contemporary corrective for this is to subject research findings to scrutiny by the hawk-eyed editorial boards of the journals in which they wish to publish, and to make their raw data available for reanalysis and reinterpretation for at least five years following publication. Additionally, psychologists understand that disciplines are interlinked, and that they need to forage for core principles

[11] An historical example is the psychoanalytic movement, which was at least as much of a political movement as a psychological or medical one. This is demonstrated (Rieber, 1998) by the expulsions from the society Freud founded of members who either diverged from him in matters of theory or even leaned too far to the left in their politics. Rieber states that "Freud was basically opposed to Marxism and socialism ... [and] had Wilhelm Reich ostracized from the psychoanalytic establishment for this reason" (p. 360). This is a contentious historical issue, and students are advised to suspend judgment until they have had a chance to read relevant primary sources. In any event, Freud considered psychoanalysis a science apart from psychology and medicine, a judgment shared by most experimental and clinical psychologists if not the medical faculties of North America.

across various psychological specializations. The personality psychologist of today reads widely in social psychology, developmental psychology, neuropsychology, perceptual psychology, evolutionary psychology, and behavior genetics, among other specialties.

What disease processes can teach us. Experimentation in psychology has classically, and in its simplest form, taken a random sample of a population of interest, randomly divided it into two equal groups, a control group and an experimental group, and then altered the entire experimental group by introducing in a consistent way a single intervention, known as an *independent variable.* The classic paradigm is quite simple, although research design has become multivariate and highly complex. A disease process, on the other hand, is an experiment in nature – and one that we cannot ethically conduct on humans no matter how great the prospective benefits. Calamities strike humans, often in a random way. We can study the psychological results of these calamities, learn a good deal about the deficits in functioning they entail, the compensatory adjustments that the victims can make, the individual differences in resilience and adaptiveness to traumatic sequelae, and procedures for their remediation. In brief, we can learn a lot about human nature. The literature on post-traumatic stress disorder (PTSD), for example, yields much useful information about the long-term personological effects of traumas of varying severity.

Neurological disease as well as head injuries can also be considered unfortunate experiments in nature. Cognitive, motor, and sensory functions are compromised by injuries and lesions to the central nervous system (CNS), and we can learn a great deal from studying individuals who have sustained such injuries. For example, we have learned that consciousness and emotion are not separable (Damasio, 1999, p. 16; see also Chapters 2, 3, and 4), for "the fine scalpel of neurological disease" reveals the consequences to emotion of impairment of conscious processes. The results of these "experiments" in neuroscience are of a single-case variety and are generalizable to the rest of the species for the simple reason that the architecture (and most of the functions) of the human brain do not differ significantly from person to person.[12] Vygotsky ([1932] 1997) quoted Engels in this respect: "One steam engine demonstrates the law of transformation of energy no less convincingly than 100,000 engines" (p. 309).

[12] This is a simplification, of course, as brains are as idiosyncratic as people's faces, though species-wide both are similarly structured.

Damasio (1999) illustrated the point we wish to make with brilliant clarity (pp. 62–67). He discussed the role of the amygdalae in mediating human emotion (see also LeDoux, 1996, on this subject). A young woman who had been suffering from minor seizures arrived at his clinic for help. Her brain CT scan revealed that her amygdalae were completely calcified and nonfunctional, a congenital condition in her case that typically begins in infancy. Now it had long been known that the amydalae, which are found in both hemispheres of the primate brain, were related to the expression of fear. It was also suspected that it enabled primates to learn new facts. What was more curious, however, is that her emotional demeanor was unfailingly positive, upbeat, optimistic, and cheerful. In social situations she was outgoing, unreserved, exceptionally receptive to personal (but not scandalous) encounters. She was exceptionally devoid of social inhibition and had no concerns about personal rejection. A colleague in Damasio's laboratory, Ralph Adolphs, demonstrated that the lopsidedness in the range of her emotional expressiveness was due to the impairment of her ability to experience *the single emotion, fear*. She was unable to develop an appropriate aversion to dangerous, or simply unpleasant, situations (Adolphs, Tranel, and Damasio, 1994). Clearly, this experiment in nature illuminated our understanding of human personality and the biological mediators of socialization. What is important to a personality psychologist is that failure to abstain from inappropriate behaviors could be directly linked to a part of the brain that was not functioning properly. Results like these must make us wonder to what extent high risk-taking behavior or extreme shyness and fearfulness are a function of hyperactive or hypoactive amygdalae. The efficacy of talking therapies when such organicity exists is a matter to be empirically investigated.[13]

There were other cognitive and affective aspects of this team's work that the reader may wish to explore. Their studies are but a few recent examples of the insight that can emerge from a study of disease processes. However, an increasingly favored approach to achieving insight into personality development is through the study of healthily functioning human beings – although standards of healthy functioning vary from culture to culture. Several generations of scientists have now taken that approach.

[13] The amygdalae are complexly interconnected with the prefrontal cortices of both hemispheres. Altering or overriding the well-entrenched neuronal circuitry that favors certain behavioral patterns is not easy as these circuits may have been years in use (see, for example, Grawe, 2007). For those with some background in neurobiology interested in the involvement of the amygdalae and hippocampus in trait anxiety, bipolar disorder, attention-deficit/hyperactive disorder, and other behavioral problems, consult, for example, the work of Bradley S. Peterson's laboratory at Columbia University (for example, Plessen *et al.*, 2006).

The wellness models

Later twentieth-century theoreticians, principally university-based experimentalists, developed personological models focusing on wellness rather than pathology. The shift toward wellness models was gradual, but the seeds of this movement had begun to germinate centuries earlier. In fact, the model of "Man" that Jean-Jacques Rousseau proposed in the eighteenth century emphasized what was noble and virtuous in humans. Humans are born good and malleable; the wickedness we witness in the conduct of human affairs results from the influence of noxious and corrupting social environments. In their pristine state, humans are presumably *noble savages.*[14] The sorry condition in which they find themselves is due to the societies that they have created. This was a secularized version of the Garden of Eden myth and the Fall of Man. Rousseau's argument was that our redemption can be achieved by a new social covenant. His vision included the reorganization of civil society and the institution of a humane and respectful system of education for the young. Rousseau valued countryside over city, nature over technology, affect and feeling over pure reason; these values provided (only in part) the intellectual framework of the Romantic tradition that was to flourish in the German states and to radiate throughout Europe.

Humanistic psychology

The term *humanistic psychology* may strike the reader as an unusual label for the study of *certain* aspects of the human person only. But such is the case. The reasons for this are largely historical and go back to the Renaissance. Those scholars who earned the sobriquet, "Humanist" valued individualism, the rule of conscience over that of the institution, the dignity of each person, and the principle of self-determination. This social philosophy, largely espoused by secularists, evolved over the centuries into humanistic psychology, a broad field of inquiry that emphasized the right of individuals to make their own choices, define themselves, develop their own style of life, and actualize their own potential.

Strangely, an antipathy developed between those who were committed to a laboratory-based, frequently experimental, approach to understanding the structures of the mind and human patterns of behavior versus those who preferred to focus on issues of human freedom, intuition, value construction, meaning-making, social influence, personal growth, and

[14] Why congenitally noble savages would corrupt themselves by repeatedly creating noxious social and political structures within which to dwell is the mooted question.

self-actualization. The former tended to be "tough-minded," reductionist, measurement-oriented positivists; the latter were discursive, hypothetico-deductive, often laboratory-aversive, eclectic inquirers who allowed their creative muses to soar – often just to meander. These two camps spent a good deal of time in rancorous debates, each denigrating the intellectual path they had not chosen to follow, their champions facing off in public arenas like larger-than-life gladiators (for example, Rogers and Skinner, 1956).[15] Scientists have rarely been dispassionate seekers after truth (Mahoney, 1976); in fact, they have generally been ardent proselytizers. They have needed not only to be right, but also to be acknowledged as right (even when they were unknowingly wrong).

Existentialism. Humanistic psychology has its roots in the nineteenth-century existentialist philosophy of Søren Kierkegaard and other thinkers who had become disenchanted with the abstract, Platonic theorizing of rationalist philosophers who preceded them. The term *existential* is contrasted with the term *essential*, both of which derive from an Aristotelian paradigm for conceptualizing the world. An *essentialist* view of the world focuses primarily on ideal forms: that is, the nature or essence of things, which are analogous to blueprints. When the blueprint or "form" of an object has become actualized, it receives a concrete existence. Those who wish to attend more to the concrete existence than the abstract essence of things fall into the *existentialist* camp. This, of course, takes on great importance when one is less concerned with defining human nature than with examining the concrete realities of flesh-and-blood human beings, with all the joys, travails, anguish, responsibilities, dilemmas, and ultimate concerns that they must face: in their bio-physical world, their interpersonal environment, and their interior life of consciousness, emotion, volition, and feeling[16] (which is different from emotion, as we shall see). From this foundational concept, many different construals of human mentation and the human condition arose, principally from the work of philosophers such as Friedrich Nietzsche, Edmund Husserl, Karl Jaspers, Martin Heidegger, and Jean-Paul Sartre, novelists such as Albert Camus and Fyodor

[15] Roger W. Sperry (for example, 1993, 1995), the neurologist, joined this debate in the last decade of his life. To the great disappointment of many of his fellow neurologists, he renounced "the former stark, strictly physical, value-empty, and mindless cosmos previously upheld by science" (1995, p. 506). See also the May 2001 issue of *American Psychologist* (for example, Leary, 2001) that assesses the legacy of Sigmund Koch, who was central to this debate over the past half-century.

[16] These three constructs correspond to the *Umwelt*, *Mitwelt*, and *Eigenwelt* in Martin Heidegger's existential philosophy.

Dostoevsky, and others who worked principally in the mental health professions. Phenomenology is one such school.

Phenomenology.[17] Immediate experience is the raw data with which phenomenologists work. They focus on people's experience of themselves and the world in which they live their daily lives. They are less concerned with *a priori* models of our psychic lives than with what individuals tell them they are experiencing. They listen carefully, trying not to impose their own schemas on what they hear others telling them. This can be only partially successful, as we have no other way of understanding what others tell us than using the very cognitive structures that allow us to make contact with them in the first place. In other words in this perspective *there are no raw data*. In the very act of capturing the data, we transform them. In a deconstructionist and post-structuralist view, one recognizes that everything one sees is shaped by individualized and cultured "lenses." This, of course, is not just a problem for psychologists. Mahoney (1991) stated that "from astrophysics to quantum mechanics, scientists have increasingly encountered themselves as 'variables' or 'factors' in their own inquiry" (p. 6).

Edmund Husserl (1859–1938), who is regarded as the founder of contemporary phenomenology, brought these problems to the fore, although he purportedly wished to design a system for accessing absolute verities. The various principles of Husserl's phenomenology were formulated in part by others who preceded him, as is often the case.[18] Hume, Kant, Hegel, and Brentano are among the more notable of the modern thinkers who wrestled with the problem of accessing the thing-in-itself (Kant's *Ding an sich*) – the reality that lies outside, and unadulterated by, the observer.

What are the implications of this for the personologist? There are several, the most important of which is that to understand the personality of an individual one needs to gain access to his or her private world. It is not enough to place an external and "objective" template on those being observed and then give them a "score." It requires a painstaking and empathic tuning-in to the psychic processes of the one being observed, and then teasing out in as self-conscious and unbiased a way as possible the ideas and feelings that well up from the deepest springs of their interior

[17] See footnote 12, Chapter 6.

[18] Creative thinkers usually try to establish themselves as the originators of their ideas. Tycho Brahe, for example, was greatly exercised by his (in retrospect plausible) conviction that Kepler was pilfering his data. In this struggle for priority, the spoils of public acclaim often go to those who can gain recognition for the work of others. If interested, see Robert K. Merton's principle, *The Matthew Effect*, bearing on misdirected credit.

world, their *Eigenwelt*.[19] The only person who knows what he or she is experiencing is the person who is experiencing.

Self-actualization. The code word for a wellness approach to the study of the Human is self-actualization. Often, and rightly, associated with the theorizing of Carl Rogers (1947; 1951; 1959; 1961) and Abraham Maslow (1962), it formed a major element in the broader field of humanistic psychology that flourished in the generation that followed the Second World War. This cluster of approaches to the study of the Human, which Maslow called the "Third Force" in contemporary psychology,[20] shared a common set of principles. Among these principles are: (a) holistic approaches to understanding humans improve prediction of their behavior more than reductionistic and atomistic ones; (b) people's life goals and purposive behavior are more important for understanding them than is exhaustive exhumation of their past; (c) anthropomorphic, person-centered models of the Human are superior to mechanomorphic models for representing the experiences of people; (d) humans enjoy the ability to creatively and spontaneously create and express themselves; (e) humans cannot be understood without examining their subjectivity – the interior psychic dimensions of human experiencing, including the tacit (unconscious) realms; and (f) social aspects of human existence are meaningful factors in personality development and recovery from misfortune (*cf.* Raskin and Rogers, 2000, pp. 135–36).

There were modern precursors to this orientation to the study of humans including: the social psychologists, J. C. Smuts, William L. Stern, Kurt Goldstein, Alfred Adler, Erich Fromm, Abraham Maslow, Andras Angyal, Karen Horney, Harry Stack Sullivan, Carl Rogers, and Fritz Perls (several of whom were primarily psychotherapists); the Gestalt psychologists, Max Wertheimer,[21] Wolfgang Köhler, and Kurt Koffka; the phenomenological

[19] This German word comes from the Heideggerian lexicon that literally means "self-world," a universe of feelings, ideations, and perceptions that can never be fully accessed by anyone else. Our awareness that we are isolated in a vast universe engenders the angst that we all must suffer.

[20] The term "Third Force" was widely used in the 1960s and 1970s to designate an alternative to the "other two forces," considered by some at that time to be psychoanalysis and behaviorism. The term was more rhetorical than scientific even in that period of psychology's evolution. Psychology is a protean field, with specializations and disciplinary alliances changing from day to day. This schema is anachronistic in twenty-first-century usage.

[21] Rudolf Arnheim (1989), in a tribute to Max Wertheimer, showed that he prefigured the positive psychology that is in vogue in the twenty-first century: "To be good meant for Max Wertheimer: to live in conformity with natural law. That is why he objected so passionately whenever what is wrong, evil, or deformed was presented as the rule, as the main object of the scientist's attention" (cited in King and Michael Wertheimer, 2005, p. 8).

and existential psychologists, such as Gordon Allport, Rollo May, and Victor Frankl; and a host of seminal European thinkers who laid the foundations for contemporary existential theorizing, among whom are Husserl himself, Heidegger, Boss, Marcel, Merleau-Ponty, Binswanger, and Sartre (consult also Raskin and Rogers, 2000, p. 136). These pioneers and paradigm busters were generally opposed to rubricizing human beings and placing diagnostic labels on them. They were disposed to treat human beings as unique individuals, not only in their superficial characteristics but in their most profound subjectivity. Pathology was conceptualized as a departure from normality rather than a component of it. This movement was clearly in reaction to the more mechanistic and drive-based systems represented by the psychodynamic and behavioral approaches that had dominated the mental health fields of the pre-Second World War Western world. Unlike many of the conflict-based conceptualizations of nineteenth- and early-twentieth-century psychodynamic models, existential theories gave full recognition not only to the sombre and most troubling issues of the human condition, but also to its free, creative, generous, and joyful life expressions.

Alfred Adler and the shift toward the socio-political. Among the great system builders, Janet, Freud, Adler, and Jung, it is Adler who is saliently emblematic of the modern trend to a "positive psychology." A closer look at his system in this frame of reference will be rewarding.

Before Adler and Freud met, they had fashioned approaches to the world that foredoomed a working relationship. Their rivalry can be characterized as a clash of titans (see Ellenberger, 1970, pp. 571–656 for an incisive analysis of their relationship and the divergent paths that they took). Although the fiction persists in popular psychology that Adler was a disciple of Freud who became disgruntled and spun off a deviant psychoanalytic system, it is difficult to imagine two more divergent conceptions of human development.[22] Before Adler met Freud he had already begun to promote the core principle of his psychology: to wit, mental health depends on personal involvement in the society in which one is embedded. Concern for one's fellows and practical interest in promoting the peaceful, harmonious, and equitable socio-political management of the social environment is not only a sign of good health, it is a means of

[22] "Contrary to common assumption," states Ellenberger (1970), "neither Adler nor Jung is a 'psychoanalytic deviant,' and their systems are not merely distortions of psychoanalysis. Both had their own ideas before meeting Freud, collaborated with him while keeping their independence, and, after leaving him, developed systems that were basically different from psychoanalysis, and also basically different from each other" (p. 571).

acquiring it. Community interest – *Gemeinschaftsgefühl* – is in the foreground of the Adlerian system.

Occupational safety and social responsibility. In 1898, Adler published a monograph that illustrated his interest in social medicine. It is also emblematic of his interest in the sociological determinants of psychological development. Entitled *Health Book for the Tailor Trade*,[23] it deplored medicine's neglect of the social conditions that were the seedbed for social diseases. Adler transferred this interest in social conditions of health into his theory of human development and psychotherapy. He believed that there are many familial and constitutional[24] determinants of personality. After all, infants' physical constitution and family constellation are the first realities with which they must work. Nevertheless, the society in which we as social beings live, work, and love is what requires the greatest professional attention. Working to build a sane society will help to form sane individuals whose values and "needs" are shaped by it – and who in turn make their own contributions. Preventive, community-based educational and psychological programs generate more benefits than recuperative ones. Preventing train wrecks makes more sense than trying to repair the consequences of them.

Schooling and character. Adler's belief that not only the nuclear family but also larger social and environmental factors shaped personality led him to give more attention to schooling and its structures. Whereas Freud gave all of his time to working with "neurotic" adults (it was primarily his daughter, Anna, who adapted psychoanalysis to the treatment of children), Adler organized child guidance clinics and teacher consultation programs in which he worked with parents, children, and their teachers for resolving learning and behavior problems. By the late 1920s, there was a network of twenty-six schools just in Vienna in which he conducted his consultations. In these sessions he taught the principles of his "Individual Psychology," the most important of which were, on the one hand, to communicate to children that they were in the care of a group that genuinely sought their best interests and, on the other hand, that they were being given the freedom to pursue their own interests and exploit their creative impulses. When they broke the few rules that existed, or

[23] The original edition was cited by Henri Ellenberger (1970, p. 651): "*Gesundheitsbuch für das Schneidergewerbe ...*" Berlin: Carl Heymanns, 1898.

[24] When psychologists use the term "constitutional," they are referring to the physiological basis of personality and behavior. This includes congenital temperament, and other musculoskeletal, endocrinological body-based influences in human development.

violated the rights of others, they were gently but firmly corrected. Barring seriously destructive acts, which if anticipated were forcibly prevented, they were allowed to suffer the logical consequences of their behavior.

Following ten years of negotiation with school authorities, several of Adler's followers established, in 1931, a remarkable experimental high school in an impoverished quarter of Vienna. Large classes of thirty to forty students were divided into smaller work groups of five to seven children, who were encouraged to help each other. The entire class engaged in a group discussion once a week to promote community spirit. Monthly teacher–parent meetings were scheduled to discuss how school and home could avoid working at cross-purposes. Children were allowed to choose from an array of school-designed activities, but once engaged in an activity they were obliged to pursue the task to the end. Although this has a modern ring to it, it was a radical practice for the time.[25] We note even in these few details some of the key tenets of Adlerian psychology: first, the centrality of mutual aid and community interest in shaping the character of the young; second, respect for freedom of choice within conventional bounds; third, goals drive humans' most enduring endeavors, that is, the future is more important than the past; and fourth, the collaboration of home, school, and community can provide an integrated, nurturant, and nonconflictual environment in which children can grow to adulthood. In Adler's educational philosophy, the "aggressive instinct" of his earlier theorizing had been cast aside.[26] Further, we do not find the pansexualism that he had never found acceptable (even in his collaborative years with Freud). This educational model was a mirror image of his system of psychotherapy, which we will look briefly at below.

A political philosophy. Adler endorsed a leftist political philosophy, a bias that did not endear him to Freud (see footnote 7, above). His theory of personality and human development was another obstacle to a smooth working relationship with the other more "orthodox" members who participated in Freud's Wednesday evening discussion circle (precursor to the Viennese Psychoanalytic Society) that met for years in Freud's apartment at Berggasse 19. It was radically different from what was proposed by Freud. As Ellenberger (1970) stated, "When studying

[25] A contemporary model that is reminiscent of this Adlerian program and incorporates many of its features is *Individual Education* (IE) developed by Raymond J. Corsini. Corsini's IE schools have been established in various parts of the United States, Canada, the Netherlands, and Israel.

[26] As noted in footnote 7, this was a major source of contention between Freud and Adler. During their early collaboration Freud rejected the notion of an aggressive instinct (additional to the sexual).

Adler, the reader must temporarily put aside all that he learned about psychoanalysis and adjust to a quite different way of thinking" (p. 571). Adler was an existentialist, concerned with the practical, concrete challenges that faced humans, and the knowledge of how they could best cope with them (*cf.* Frankl, 1970, p. 38). His was an intuitive, common-sense approach in which he did not attempt to reify notions of the unconscious, mind, will, emotion, or complex. Indeed, he rejected the notions of Oedipal complex and derivative constructs, while minimizing the importance of libidinal drives.

Inferiority complex and the male protest. The overarching motif in Adlerian psychology is power deriving from competence, the ability to achieve parity with others in terms of social influence and mutual respect. When individuals sense a deep inadequacy in themselves and an inferiority relative to their reference groups, they attempt to compensate for it. The more inferior they feel, the more they are impelled to compensate in ways that bring on emotional disturbance and self-destructive behavior. This will to power is not Nietzschean in the sense of seeking power as an end in itself. It is an expression of a need for completeness in a social sense, an intense desire to be regarded as *significant* by one's fellows. This Adlerian principle has much in common with the views of Ralph Waldo Emerson, the brilliant American essayist, whose essays "On Self Reliance" and "Self Confidence" inspired generations of young people with the courage to transcend their self-imposed limitations. The maladaptive form that the compensation takes is the *masculine protest*. It manifests not only among men but also among women. The blatant assumptions of male superiority, and the personality traits associated with dominant males, form the template for not only the prejudicial treatment of women by men, but also of men by men. Clothing oneself in the persona of the inordinately assertive, aggressive, self-aggrandizing, confrontational, challenging, and self-vindicating male becomes the formula for precipitating conflicts that ultimately benefit no one. Social rules exist in the home for resolving the rivalries that spring up among siblings. "Rules" also exist in the ethological realm of the kennel, the bullpen, and the henhouse. Humans have developed conventions that allow them to engage in rivalries without destroying one another.

Childhood and personality. Although a group of children may all grow up under a common roof, exposed to a single (but evolving) familial culture, they nevertheless develop very different personalities. There are genic determinants of these differences, of course. Even fraternal (dizygotic) twins differ significantly in personality traits. But everyone's position in the family is different; each is born in a different historical moment and context; each elicits different responses from the parents as well as from each other by the same gesture or behavior. On any given day each is more or less competent in a variety of tasks, depending on developmental age as well as ordinal (age-graded) position. Adler believed that, in addition to biological factors, these systemically mixed and shifting variables accounted for the idiosyncratic character and lifestyle that each adopted over time. The influence of birth order is still under investigation (for example, Hoffman, 1991; Lasko, 1954; Sulloway, 1996; Zajonc and Mullaly, 1997). If one grants that in any social system it is difficult to isolate the influence that any event, constellation of historical circumstances, or set of ideas has on all the other variables of the system – after all, that's how fads, fashions, recreational trends, and political parties get their start, not to mention religious movements – then it is difficult to reject the notion that birth order plays *some* part in the formation of personality. In a systems view of the cosmos, given enough time, everything affects everything else, although in often unpredictable ways. A stone tossed in a lake sets in motion chains of causality that affect all the molecules of water and other constituents of that ecological system.

Constructivism, Vaihinger, and Adler. As noted in Chapter 1, a constructivist epistemology can be conveniently dated beginning with the work of the noted founder of historiography, Giambattista Vico (1668–1744). This is something of a fiction as there have been notable figures throughout history who have postulated that the human intellect is an intrinsically fallible instrument for representing the reality that it perceived. (For example, Democritus, over two millennia ago, doubted the exact correspondence of our mental representations of reality with what they purportedly imaged.) Vico, however, gave reasoned and compelling arguments that the human mind shapes, embellishes, and fabricates its own images of the world as well as idiosyncratically construes what other people have communicated to it about past events. As Mahoney (1991) stated, "In his 1725 book, *Scienza Nuova,* he emphasized that humans create order in their experience by projecting familiar categories onto unfamiliar particulars. Vico said that 'to know' is to 'make' (*facere*), anticipating by more than two centuries the assertion of the constructivist Jean Piaget that to know an object is to act upon it" (p. 98). Christian Herder and Immanuel Kant may have been unfamiliar with the writings of this Neapolitan scholar, but they espoused the substance of his position and established a tradition that flourished in the twentieth century.

One disciple of Kant who exerted a powerful influence on Adler was the constructivist, Hans Vaihinger (1925), whose book, *The Philosophy of "As*

If", was published in 1911, the year that Adler took leave of the Viennese Psychoanalytic Society. "The mind is not only merely appropriative, it is also assimilative and constructive," wrote Vaihinger (p. 2). This philosophy nicely served the principal point of divergence (among many other differences) between Adler and Freud: to wit, that people are not so much driven by their past as drawn by their hopes and expectations of the future. It is this intentionality and purposiveness in everyone's behavior that explains that behavior. This teleological[27] notion characterizes the psychology of Adler, which emphasizes that one's ambitions, goals, specific aims, and "memories of the future" are more determinative of what one will do than the residue of childhood experiences that have receded into the deepest storage bins of the unconscious. This principle places Adler, like Jung, firmly in the camp of the self-actualizationists. For both of them, life is a work of self-individuation and the actualization of the potential that resides in each one of us.

Holism and individual psychology. Adler named his system "Individual Psychology." He used the term "Individual" in its etymological sense – *not divided* (the "in" is a privative prefix as in "intrepid," "inadequate," or, indeed, "indivisible"). Adler always affirmed the basic unity of the person, rejecting the topographical schemas of nineteenth-century psychiatry. For example, he found it unnecessary to posit the existence of an entity called "the unconscious." There are only unconscious dynamics; no superego, just moral and ethical judgment and behavior. Thus, he espoused a principle of dynamic "use" rather than static "possession." We do not *have* an unconscious, and we are not "a composite of part functions" (Bugental, 1963). The individual is a psychologically unpartitioned *person* who functions in ways that maximize the likelihood of achieving valued ends. Adler repeatedly emphasized that it was critical to remember that when we speak of functions as if they emanated from a "part" of the psyche, we are simply using a figure of speech rather than designating an entity.

It would be an error to suggest that Adler initiated the movement toward a holistic view of the Human. The assault on reductionism and an exclusive bio-physicalism had already begun. For example, Jan Christian Smuts had already made a significant contribution by the publication of his book, *Holism and Evolution* (1926). The Gestalt psychologists had already stressed the importance of this construct in their research

[27] "Teleology" derives from a Greek root, *telos*, that means "end," and "far" and *teleios*, that means perfected. This led to the concept in psychology of a process that is linked to goals of self-realization.

and publications. No doubt Adler was influenced by all of them. Holism had entered and formed the Zeitgeist, and Individual Psychology became the mirror of that spirit.

Psychotherapy and personality. Adler liked to say, rather optimistically, that the neurotic person is a discouraged person who seeks by hook or by crook to be pampered (Ansbacher and Ansbacher, 1956). Such a person needs to learn a new way of behaving (see commentary by Dreikurs, 1950; also consult Adler, *Understanding Human Nature*, [1927] 1959). His was a therapy of encouragement, an educational program in which he led patients to take concrete steps in tasks in which they could succeed. As one would expect, Adler emphasized the social realm, an area in which patients could utilize all their available personological resources. But these resources were not the fixed personality traits that some theorists postulated had been formed in infancy and toddlerhood. He abhorred the determinism that was implicit in this view of the Human. On the contrary, Adler taught his clients that they are not possessed by their traits – they use them to escape the responsibilities that a sound mind and body would otherwise oblige them to accept. He recognized what has been called defense mechanisms in the psychodynamic literature, but he conceptualized them as rather more or less maladaptive problem-solving devices (*cf.* Mosak, 1995).

Adler's approach sprang from a conviction that personality is not rigidly fixed in childhood, and immutable in more mature stages. Nor do people have traits with which they have been passively imprinted and which can be altered only by lengthy and intensive psychotherapy. What they have done is develop lifestyles that afford few options, and dysfunctional ones at that. It is the job of the mental health professional to teach these individuals a more adaptive style. Although Adler fashioned a set of questions intended to probe the early childhood recollections of his patients, it was not to learn the details of their personal histories. After all, these recollections were, in his view, largely fictions. What he learned from these memories was the self-image and the current lifestyle that masked their ambitions and despair. The most important question was not "Where have you been?" but "Where are you going?" Accordingly, Adler analyzed clients' style of life to find the embedded hopes and limitations that portended either success or failure.

The creative self. What enables individuals to actualize themselves is their innate creativity. "Every child is born with potentialities different from every other child," states Adler. But as the same circumstances are apprehended, analyzed, and used in individualized ways by each person, he drew the conclusion that creativity was at the core of adaptive human coping. "We have been impelled to attribute to the child a creative power, which casts into movement all the influences upon him and all his potentialities, the movement to the overcoming of an obstacle," he concluded (Ansbacher and Ansbacher, 1956, pp. 176–77). Hall, Lindzey, and Campbell (1998) wrote that Adler searched for an explanatory principle for human strivings and achievement and "found the creative self ... This concept is Adler's crowning achievement as a personality theorist. When he discovered the creative power of the self, all his other concepts were subordinated to it" (p. 135).

Theorists of self-actualization

The construct, self-actualization, refers to a process, some would say a drive, whereby one strives to exploit, to a greater or lesser degree, the inherent talents and potential of the species. Ansbacher and Ansbacher (1956, p. 105) cite a passage from Kurt Goldstein who concluded from his work with traumatized German war veterans that *the* basic drive in humans is striving to actualize their individual capacities. Hall, Lindzey, and Campbell (1998) state (more strongly), "This is Goldstein's master motive; in fact it is the only motive that the organism possesses. What appear to be different drives such as hunger, sex, power, achievement, and curiosity are merely manifestations of the sovereign purpose of life, to actualize oneself" (p. 441). Adler clashed with Freud on this point as on so many others. He stated that "the striving for perfection is innate in the sense that it is a part of life, a striving, an urge, a something without which life would be unthinkable" (p. 104).

On an even more basic level, William Stern stated that the total organism is governed by a tendency to "self-unfold," to blossom, as it were (Kreppner, 1998, p. 319). Karl Bühler (cited, for example, by Maslow, 1962, p. 22) wrote that humans engage in many cognitive and psychomotor activities for the sheer pleasure of it. Humans delight in acquiring new skills and revel in their repeated exercise. There is no ulterior motive, be it monetary or simply sensual, though such may be by-products of these activities. Bühler termed this drive performance pleasure (*Funktionslust*). We see evidence of this all about us, even in the earliest years of life. Watch children as they try to master the skill of tying their shoelaces, manipulate a Rubik's Cube, dribble a basketball, or negotiate the struts, bars, and poles of a jungle gym. Older people derive pleasure from solving puzzles; some will attack difficult crossword puzzles each day; others have hobbies that bring no other reward than the intrinsic pleasure of building, playing, or problem solving. On the other hand, many salaried workers avidly look forward to their work, especially if it's work of a skilled and creative nature. Their pleasure is immediate, internal, expansive, self-referred. The more talent one has in a particular activity, the more satisfaction one gets from honing, exploiting, and repeating the activity. This seems evident in the work of professional golfers, chess players, architects, performing artists, and scientists but is also true for conscientious window-washers, cab drivers, landscapers, and house painters (Ryan and Deci, 2000).

Adler ([1927] 1959), Allport (1955), Angyal (1951), Fromm (1947), Goldstein (1939), Horney (1950), and Jung (1933) all made contributions to the construct of self-actualization (though each used somewhat different language and lines of argument to promote its acceptance). They all advanced the shift away from pathognomonic models and toward wellness models of human personality. However, the latter-day theorists and researchers who caught that wave and whose names are most closely linked to this seminal approach are Carl Rogers and Abraham Maslow.

Maslow's psychology of health. Maslow (1962) began his influential book, *Toward a Psychology of Being*, with the following words: "We have, each of us, an essential biologically based inner nature, which is to some degree 'natural,' intrinsic, given, and in a certain limited sense, unchangeable ... Since this inner nature is good or neutral rather than bad, it is best to ... encourage it rather than suppress it. If it is permitted to guide our life, we grow healthy, fruitful, and happy" (p. 3). Consistent with the viewpoint of Raymond B. Cattell, among others, Maslow thought that if you wish to understand human nature it is best to begin by studying individuals who are healthy, productive, creative, and to all appearances, happy and optimistic. As Hall, Lindzey, and Campbell (1998) stated, Maslow "believed that if psychologists study crippled, stunted, neurotic people exclusively, they are bound to produce a crippled psychology" (p. 453). "Freud supplied us with the sick half of psychology," he stated "we must now fill it out with the healthy half" (p. 5). Maslow addressed this task and spent much of his professional life studying individuals who were healthy and exceptionally self-actualized. (See chapters 6 and 7 of his book, *Toward a Psychology of Being*, relative to "peak experiences," personal events of the most exalting and exhilarating kind.)

Although Maslow's research focus, and that of others who placed themselves in the "Third Force" of psychology, was principally on highly self-actualized humans, it would be a misinterpretation of the "wellness" movement to label it as elitist. No less than Adler and Goldstein before him, Maslow was convinced that all humans are born with an intrinsic drive to constructively become everything that they can become. We are competency and mastery motivated (*cf.* White, 1959). He characterized the needs that this drive attempts to satisfy as instinctoid, that is, species typical (1970, pp. 27–28, 103) and ineluctable. That people often failed to develop skills that they were fully capable of, that they rejected responsibility and opportunities for growth, and that in extreme cases they wallowed in apathy and sullen defiance of all convention is self-evident. *But the latter is not normal. We are born to excel.* A contemporary statement of a "normative" standard of human nature can be found in Ryan and Deci's (2000) article outlining their Self-Determination Theory:

> The fullest representations of humanity show people to be curious, vital, and self motivated. At their best, they are agentic and inspired, striving to learn; extend themselves; master new skills; and apply their talents responsibly. That most people show considerable effort, agency, and commitment in their lives appears, in fact, to be more normative than exceptional, suggesting some very positive and persistent features of human nature. (p. 68)

The fact that none of the seminal thinkers on whose work this short article is partially founded is cited by Ryan and Deci is an indication of how seamlessly this tradition has been knitted into contemporary personality theory[28] – and of its robust health. References for the obvious are neither needed nor valued.

Human needs and potential. Maslow founded his system on two orders of needs, which he structured into a *hierarchy.*[29] The first order is physiological. Every organism has basic needs that enable it to survive and thrive. At the most basic and physiological level there is the need for a balanced diet, an abundance of clean water to drink, fresh, unpolluted air to breathe, and clothing and shelter to protect one from the elements. The second order of needs is largely psychological. There are four distinct levels within the psychological needs. The lowest level of needs in this spatial metaphor (the second level) is the need for safety, security, a stable and structured domestic universe, and, for the very young and dependent, familiar surroundings (especially the constancy of primary caregivers). At the next higher level is the need for relationships with people who respect and love us, the need to be fully integrated into the family, and a need for

[28] Ryan and Deci have a specific task in mind, which is to demonstrate that variance in personal efforts to become more autonomous and competent is a function of extrinsic motivational forces. These can often depress the natural agentic impulses of humans.

[29] The term *hierarchy* derives from the Greek word, *hierarchia*, meaning rule by priests. Rising in the hierarchy signified rising in an order of governance and dominance. This term needs to be used with caution, as in "hierarchy of needs." Higher needs do not necessarily override lower needs in this spatial metaphor. In fact, the opposite often occurs. "Hierarchy" is only a template, a creation of the human imagination that helps us understand the world about us and within us.

belongingness. We are gregarious animals from the moment of birth, wanting to relate to others with intimacy and acceptance. The fourth level (the penultimate one) is the need for respect from others and for their esteem. All humans wish to *be* and to *feel* significant among their fellows. If Charles Horton Cooley, George Herbert Mead, and the other Symbolic Interactionists are right (as noted in Chapter 6, below) in asserting that our sense of self derives from the appreciation we receive from our fellows, the respect of others is a precondition for self-esteem. Respect for ourselves is as necessary as is respect from our peers. When the subordinate goals of life have been largely met, a nameless yearning and discontent will haunt those individuals who have not undertaken to achieve their noblest ambitions: "A musician must make music, an artist must paint, a poet must write, if he is ultimately to be at peace with himself" (Maslow, 1970, p. 46). At this ultimate level is the need to be fulfilled and to realize one's potential, whether that potential is modest, average, or extraordinary. To be the greatest, as Muhammad Ali proclaimed himself, is not an achievable or reasonable goal for many humans. But to be *the best that one can be* is not only reasonable, but an imperative, in the view of Maslow: "What a man can be, he must be" (p. 46). This aphorism applies with equal force, of course, to women.

The drive for self-actualization is (at least embryonically) manifest at all levels of the needs hierarchy. After all, the route to self-actualization begins with a single act – when a newborn first reaches for the mother's breast. What Maslow insisted on was that one could not progress to the higher levels of need satisfaction without satisfying basic physiological and psychological needs. However, it is possible to falter at any level. The drive that propels us to the highest level "is not strong and overpowering and unmistakable like the instincts of animals. It is weak and delicate and subtle and easily overcome by habit, cultural pressure and wrong attitudes toward it" (1962, pp. 3–4). The intrinsic motivation that propels humans at lower levels of the hierarchy is more immediately compelling and urgent, for it touches on the very sources of life, survival, and health. Physical health is imperiled by inadequate feeding and drinking. Likewise, mental health is imperiled by deficiencies in love, nurturance, and a caring environment. Deprivation in this domain can compromise our bodily well-being as well. An extensive literature on *anaclitic depression*, *marasmus*, and other problems arising from sensory deprivation and emotionally abusive neglect of children provides empirical support for this principle.

Deficiency motivation. Maslow's position began to take shape as he reflected on the origin of neurosis and other psychopathology. His studies convinced him that mental illness is a deficiency disease. The more humans are deprived of the constituents of sound health, the more mentally sick they become. (There are also, of course, problematic physiological sequelae.) He stated, "I concluded ... that pathology was born out of being deprived of certain satisfactions which I called needs in the same sense that water and amino acids and calcium are needs, namely that their absence produces illness ... [Most neuroses involved], along with other complex determinants, ungratified wishes for safety, for belongingness and identification, for close love relationships and for respect and prestige" (1962, p. 19). Those who have been damaged by severe deprivation of familial love, nurturance, and acceptance are deficiency motivated and impelled to recover their health by a nameless, haunting need to recover lost opportunities. Effective psychotherapy is largely concerned with enabling this process and compensating for the most searing personal losses. When such "patients" succeed in recouping their losses, their illness and personality deficits tend to disappear.

Rarely does anyone manage to satisfy all of his or her basic needs to an optimal level, for satisfaction is not an all or nothing proposition. Nor must one be fully actualized at one level before proceeding to a higher level. Success at higher levels is facilitated by higher achievement at lower levels. But those who are no longer primarily deficit motivated can be meta-motivated by higher internal values and loftier goals. No longer obsessing about the needs that were denied satisfaction and the external goals that eluded them, they turn their attention to becoming a fully competent person.

In the decades in which Rogers and Maslow were advancing their principles of self-actualization, seminal work was being done in the teratogenic effects of severe emotional stress during pregnancy. The Vedas and the Hippocratic doctrines of the ancients warned of these effects (Ferreira, 1965) as have, more recently (in the 1950s and 1960s), the research findings of Marcelle Geber and Mary Ainsworth among the children of Uganda. Greater public exposure has been given to the effects of postnatal maternal deprivation and the baneful immediate as well as long-term influence this has both on physical as well as personological development. The various deprivational effects of warehousing infants in institutions (for example, marasmus, anaclitic depression, hospitalism, kwashiorkor, and failure to thrive) were demonstrated in the immediate post-Second World War years (explore, for example, the developmentalist research of René Spitz, William Goldfarb, John Bowlby, Margaret Ribble, Harry Harlow, and the literature on the Infants of the Crèche). This important subject goes far beyond the scope of this chapter and can be pursued by the interested reader. Its relevance to personality development is evident.

Carl R. Rogers: on becoming a person

Carl Rogers' life work can be summed up as a profound, assiduous, wide-ranging pursuit of the conditions for personal growth and health, and an inquiry into the nature of the organism that governs those conditions. Much of his theorizing and subsequent research emerged from his experiences as a therapist, which seems anomalous in the light of the emphasis that fellow humanistic psychologists of his era placed on the need to study the healthy rather than the ill.[30] Rogers' view of the Human was that we are genically ordained to mature to adulthood in such a way that all of our potential is realized. This was a position that he articulated in the 1940s (1942, 1947), before Maslow but in the wake of the great pioneers, Jung, Adler, and Goldstein. Indeed, this belief can be traced back to the biology-based theories of nineteenth-century psychiatrists.

Organismic valuing process. The human zygote encodes a "blueprint" for human growth just as an acorn encodes one for an oak tree. Monomorphic genes program us to look like and, in important ways, behave like a human. They govern the structure and the growth of our central nervous system, perceptual organs, and even our five-fingered hands and our eyebrows. There are a relatively small number of polymorphic genes that determine many of our individual characteristics such as height, color of eyes, fluid intelligence, and temperamental characteristics. Given no serious anomalies in our personal genome, an optimal environment will permit us to develop a well-rounded personality, a minimum of dysfunctional quirks, and no serious personality disorders. If our social environment, primarily and initially our family, does not oblige us to adopt values, behavior patterns, attitudes, and self-concepts that are at variance with our personally experienced needs, we will mature into healthful and authentic human beings. There is an *organismic valuing process* springing from our very nature that impels us to an adaptive rather than a maladaptive way of behaving and thinking. When we are forced to act in violation of that valuing process, we become inauthentic, and the seeds of personality disorder begin to germinate.

Congruence and authenticity. The unfortunate reality is that parents and the larger family and neighborhood promote values and cognitive schemas that conflict with the needs of the authentic human

[30] As we noted above, the big questions in science are rarely either–or questions. There are always elements of truth on both sides of important issues. Neils Bohr is reputed to have observed that the opposite of a small truth is an error, but the opposite of a big truth is another big truth. In any event, there are few absolute dichotomies in nature.

being. These values and schemas become internalized. They are transmuted, as it were, into our second nature. They can take any number of forms. The child may be taught, for example, "You must always give in to your younger brother; after all he's not well," or "when your sister wants that toy, you must not be selfish," or "you must not allow your curiosity about your body to take over – stop touching yourself *there*, that's not nice," or "you must beat the other guy at all costs." Some of these schemas are pro-social, some anti-social; all have some effect on the manner in which we interact with our social and physical environment. We internalize innumerable taboos and moral directives (many incongruent with the needs and yearnings of the organism) that shape the person each of us eventually becomes. High levels of incongruence between what the organism naturally values and what the maturing human is conditioned to think about herself or himself leads to disabling incongruence. It is this inauthentic existence that breeds anxiety and maladaptive ways of interacting with one's environment – and that motivates people to seek therapy or counselling.

The experiencing organism. Rogers always treated individuals as the single best authority on what they were experiencing, because at the deepest levels their experience was an expression of their nature. If given their freedom, humans would always make choices that would enhance their healthful development. This Rousseau-like notion implied that if the environment was not a coercive and corrupting influence people would make the best choices afforded by the environment into which they were "thrown" (to use Heidegger's expression) and had chosen to live. For this reason it is necessary that we listen attentively to what individuals are saying for, as noted above, they are their own best authority on themselves. Given freedom, they will ultimately choose to move in the direction of wholeness and fulfillment. Phenomenologist that he was, Rogers insisted that all of us have no other reliable contact with reality than the experience provided by our bodies. Our sense of *self* has no sounder basis than the organism that is the ground of our *existence*. His counsel to all humans was this: trust your experience.

The actualizing tendency. Like Adler before him, Rogers believed that there was only one drive, the drive to self-fulfillment. He shunned the notion of specific drives. The psychoanalytic libido theory and other complex psychodynamic drive theories were not useful in his view. An American pragmatist, he advocated that we pay attention to what works. And what works is not a grand system, a complex assemblage of nebulous constructs, but an examination of what is going on. He attended to "the facts," for the facts are always friendly in his estimation. He listened carefully to what was going on – by attending to what others told him from the depths of their experiencing. Therapy was the transaction of "deep calling unto deep," and being intensely attentive to that call.

We conclude this brief segment on Rogers (1980) by recording his "voice" as we hear it in the following passage:

> The actualizing tendency can, of course, be thwarted or warped, but it cannot be destroyed without destroying the organism. I remember that in my boyhood, the bin in which we stored our winter's supply of potatoes was in the basement, several feet below a small window. The conditions were unfavorable, but the potatoes would begin to sprout – pale white sprouts, so unlike the healthy green shoots they sent up when planted in the soil in the spring. But these sad spindly sprouts would grow 2 or 3 feet in length as they reached toward the distant light of the window. The sprouts were, in their bizarre, futile growth, a sort of desperate expression of the directional tendency I have been describing. They would never become plants, never mature, never fulfill their real potential ... I often think of those potato sprouts. So unfavorable have been the conditions [my clients] have developed that their lives often seem abnormal, twisted, scarcely human. *Yet the directional tendency in them can be trusted.* (added emphasis) (pp. 118–19)

Catching the wave

The drift toward wellness models of human nature and a positive psychology of personality has accelerated in the last generation. A number of factors have propelled this phenomenon. First, the influence of nineteenth-century psychiatry, which was largely centered on the study of illnesses that were prevalent in Europe at the time, has declined. Examples of these are *vapors*, *neurasthenia*, *fugues*, *hysteria*, and *obsessive-compulsive disorders* (OCD). Before the First World War, Freud energetically repudiated Adler's thesis that humans were driven by an innate aggressiveness that "civilized" nations had difficulty in controlling. Conflict over this issue was one of the reasons underlying their alienation from one another. Freud, however, having witnessed the carnage that engulfed millions of armed men in institutionalized aggression and the ritualized trench warfare of the First World War, later accepted the Adlerian position and it became the second pillar of his drive theory. Ironically, by the end of the war Adler had already discarded that position and gladly ceded ownership of it to Freud. Within the second third of the century, as the pessimism of classical Freudism and its underpinning drive theory began to recede in popularity, acceptance of the holistic, self-actualizing, social–psychological orientation of Adler and others of that era began to swell (Allen, 1971).

Second, the social–psychological foundations of behavior were increasingly given prominence. Apart from the importance that Jung and Adler gave to a healthy involvement in civil society, on the one hand, and post-war disenchantment with a narrow bio-physical conception of human behavior on the other, the *social sciences* were coming into their own in the first half of the twentieth century. *Cultural anthropology* and *sociology* were emerging as important disciplines in their own right. Edward Sapir, Ruth Benedict, Franz Boas, Margaret Mead, Clyde Kluckhohn, Alfred Kroeber, and Ernst Cassirer among others had become icons for the inter-war and post-Second World War generation of undergraduate social science students open to a less medicalized, more relativized approach to understanding human behavior. The same can be stated about the sociological work of James Mark Baldwin, Gustave LeBon, Max Weber, Georg Simmel, and George Herbert Mead. Their influence permeated the work of social psychology and other psychological disciplines in such a thorough, often subtle, manner that few student psychology majors are even aware of their presence. Wellness and illness are conditions that society both promotes and defines.

Third, psychoanalysts of the pre- and post-Second World War eras, many of whom could be more aptly labeled post-Freudian rather than neo-Freudian, began to depart in ever more radical ways from the classical views of psychoanalysis. The *object–relations theorists*, such as Melanie Klein, D. W. Winnicott, W. R. D. Fairbairn, and Harry Guntrip, gave importance to the relational aspect of human functioning rather than to the libidinal. If one grants that humans are driven, they are driven more by their need for meaningful relationships than by the search for pleasure. When the reverse is true of anyone, it is a sign of poor mental health. This sociologizing of Freudism resulted in freeing the psychoanalytic community from a conception of the Human straitjacketed by a remorseless, unrelenting libido and a prisoner of a personal infantile history. The constructive and healing properties of human relationships were acknowledged and utilized in therapy. In another, but overlapping, sector the so-called ego psychologists, such as Heinz Hartmann, Melanie Klein (again), Heinz Kohut, and M. M. Gill, rethought human agency and the "conflict-free" area of the self. Humans, in the view especially of Kohut, have one basic, essential need: the presence of caring, loving, nurturing, empathic fellows – whether parents, spouses, children, neighbors, or simply friends. Clearly this departure from the pessimistic, irrationalist, pansexual, and determinist system of Schopenhauer and his disciples was a movement toward a normative wellness model of the Human.

There have been many other developments in the broad field of psychology that have exercised an indirect influence on this secular drift toward a positive psychology. Although behaviorism is still a powerful, admittedly neutral, ideological force in contemporary psychology it has assimilated many cognitive features into its governing paradigms. Arguably the most influential psychotherapy of a new century is the cognitive–behavioral (note the work of Aaron Beck, Albert Ellis, Stanley Rachman, and Donald Meichenbaum, among a host of others). Applied cognitive psychologists have given an added impetus to a more optimistic, mindful conception of the human. Although there have been other, cautionary voices that have made psychology pause in its enthusiastic endorsement of a humanist psychology (for example, Koch, 1971; note as well the work of evolutionary psychologists), the trend toward wellness models of human personality appears to be continuing.

An entire issue of *American Psychologist* (2000, vol. 55(1): 2–183) has been devoted to this theme. In a lead article Seligman and Csikszentmihalyi (2000) deplored psychology's traditional focus on a disease model of the human. They stated:

> This almost exclusive attention to pathology neglects the fulfilled individual and thriving community ... Positive psychology at the subjective level is about valued subjective experiences: well-being, contentment, and satisfaction (in the past); hope and optimism (for the future); and flow and happiness (in the present). At the individual level, it is about positive individual traits: the capacity for love and vocation, courage, interpersonal skill, aesthetic sensibility, perseverance, forgiveness, originality, future mindedness, spirituality, high talent, and wisdom. At the group level, it is about the civic virtues and the institutions that move individuals toward better citizenship: responsibility, nurturance, altruism, civility, moderation, tolerance, and work ethic. (p. 5)

Conclusion

It can be argued that over the past two centuries psychology of personality has shifted from being a science of pathology, victimology, weakness, and trauma toward being a science that gives precedence to wellness conceptions of the Human and to developmental paradigms for positive and adaptive private and civic fulfillment: "It is about work, education, love, growth, and play" (Seligman and Csikszentmihalyi, 2000, p. 7). The proponents of this rather Adlerian vision affirm the need to conduct the pursuit of their scientific goals with the same methodological rigor that underpins other human sciences. But transcending this is the need for a reworking of the philosophy that has traditionally framed our understanding of what it means to be human. Demonstrating that this tectonic shift began generations ago, and continues in a new century, has been the principal burden of this chapter.

As Ellenberger (1970) pointed out in numerous passages, the theories of the system builders, both major and minor, are in large measure autobiographical. But he also added that like the artist and the novelist, each of us draws on specific talents and aptitudes, as well as the accumulation of our critical life experiences, in fashioning our conception of the world (p. 888). Those readers who will decide to become mental health service providers or personality psychologists will choose a system of procedures and tools as well as a specific audience or clientele they wish to serve that fits their personality.

3 Developmental perspectives on personality: from youth-based to life-span models

Creatures of a day, what is any one? What is he not?
Man is but a dream of a shadow.
But when there comes as a gift of heaven a gleam of sunshine,
There rests upon men a radiant light and, aye, a gentle life.

From an ode of Pindar (sixth century BCE), celebrating the victory of an athlete at the Pythian games

Developmental psychology and personality theory

To understand any theory of personality we need to understand the developmental psychology in which it is grounded.[1] Further, we need to appreciate the enormous advances that the science of human development has made over the past century. The theories of personality that were predicated on obsolescent developmental models have become outdated themselves. Stating the obvious, personality theories that reflect the most recent research in developmental psychology inspire the most confidence, everything else being equal. This chapter explores the relationship between the postulates and science of these two disciplines and what they contribute to each other.

Current science informs us that human development is a life-long process, a conclusion we can draw from the recent research of a host of developmental psychologists (for example, Baltes, Reuter-Lorenz, and Rösler, 2006a; see also the balanced and well-documented article by Paul B. Baltes, 1997). Freud (1856–1939), on the other hand, affirmed

[1] Developmental psychology is a branch of general psychology that focuses on the processes of growth and change in the Human over the entire life span. In its holistic sense human development does not exclude any particular dimension of human ontogeny – physical, behavioral, personological, cognitive, social, perceptual, or psychomotoric. It is concerned with individual variation in all aspects of development as well as with its nomothetic principles (see Scarr, 1992).

that personality was largely "set in plaster"[2] by the time the Oedipal conflict of the pre-schooler was resolved (or not resolved as the case might be). We have here in stark contrast two opposing views on the developmental psychology that underpins some Euro-American conceptions of personality. It is not surprising, then, that as our understanding of how individuals develop has advanced over the past century, it has necessitated corresponding adjustments in our understanding of the dynamics of human personality and the possibilities for life-long change and growth.

Static (as opposed to developmental) views of personality are analogous to a snapshot of a person in process of change. Such views are obsolescent at birth. What mitigates the distortions that are apt to flow from this perspective is that changes in personality typically take place very slowly. The hazard in this is that those who are committed to this line of research are *inclined* to find significant stability in the variables that they are studying. If one grants that humans are not born with an evolved personality but develop such over their passage through life, then it follows that the more soundly based the developmental psychology on which scientists ground their personality theory, the more sound that theory will be. The fact of the matter is that there are both genomic and epigenetic determinants of human development, including the temperamental substrate of many personality dimensions.

Humans, as well as other vertebrates, may, however, be said to be born with inchoate temperamental traits, such as emotional lability or stability, placidity or irritability, rhythmicity of biological functions as in sleep patterns, and quality of mood (Thomas and Chess, 1980, pp. 68–76). It is a commonplace of personality science that personality, as distinguished from temperament (discussed in Chapter 5 below), is complex and generally in flux – although there are lengthy periods of relative stability. The pattern of responses that individuals typically make to the world about them (and to their *internal milieu*) evolves continually over time. This evolution takes place in a cyclical fashion and at varying speeds, depending on the transitions, biological and social, through which individuals move.

[2] The metaphor, "set in plaster," entered the developmental literature with the publication of William James' *The Principles of Psychology* ([1890] 1990), which is discussed in Chapter 6. He wrote: "Already at the age of twenty-five you see the professional mannerism settling down on the young commercial traveler, on the young doctor, on the young minister, on the young counsellor-at-law. You see the little lines of cleavage running through the character, the tricks of thought, the prejudices, the ways of the 'shop,' in a word, from which the man can by-and-by no more escape than his coat sleeve can suddenly fall into a new set of folds ... In most of us, by the age of thirty, the character has set like plaster, and will never soften again" (p. 79).

Every parent understands that an infant exhibits some relatively stable temperamental traits as it matures. These traits appear in the first days of life. Parents with more than one child often comment on the remarkable differences among their children, as well as the variety of behaviors that distinguish them from each other and from children in other families. Thomas and Chess (1977) demonstrated in their widely cited New York Longitudinal Study that temperamental traits, the core constituents of most models of personality, appear early in life, and they provide evidence that, in large measure, these qualities persist into mature adulthood. Caspi and Silva (1995) have demonstrated the same principle in their longitudinal study of 3- to 18-year-old individuals.

On the other hand, the repertoire of behaviors that children develop as they approach adulthood becomes increasingly diversified, individuated, and, normally, more effective. They learn different roles for different social situations. *They actively create themselves*, learning to cope with the demand characteristics of the many situations into which they enter. And to this end they develop varied and subtly differentiated personas, personas that are played out differentially as they assess the scope and possibilities inherent in each new situation they enter.

This is one (contemporary) view of the development of personality. Before we proceed to look briefly at other views it will be instructive to look for a moment at the stage-based structure utilized in many models of human development.

Stages of development in human consciousness

We can look at the development of organisms as a smoothly progressing, incremental process unpunctuated by any sudden changes. Alternatively we can conceptualize it as a process that moves from phase to phase with qualitatively distinct spurts of growth, and sensitivity to specialized stimulatory fields. The reader may think, "One of these two perspectives has to be wrong." Unfortunately, for those who dislike such ambiguity, both are legitimate perspectives.

Defining a stage. A stage is a period in the life of each human when a pattern of behavior emerges that is qualitatively different from that of the periods that precede and follow it. That pattern needs to satisfy a number of criteria. First, to be useful, the pattern cannot be trivial; it must be salient, of cardinal importance, and relevant to a large range of behaviors. Behaviors characteristic of a particular stage include activities of the organism that are both interior to it (such as covert emotional and cognitive functioning) and interactive with its environment (such as social behavior). Puberty, following the onset of menarche in girls and the development of secondary sex characteristics, would fit this definition. Second, the transitions from one stage to another should be *relatively* distinctive if not abrupt. A degree of discontinuity distinguishing, say, pre-pubertal and post-pubertal behavior, must be present. We do not have a stage if the pace of behavioral change within a stage is the same as the pace of change ushering in the stage (see Corsini, *Dictionary of Psychology* [1999], p. 938). A threshold phenomenon followed by a plateau of sorts needs to be present. Just how salient the threshold is or how constant are the defining behaviors of the stage must be subjectively determined.

The reader will observe that some corollaries of this concept make it a target of criticism for various schools of personality theory. The concept of stage is usually founded *in large part* on the assumption of endogenous (that is, genetic, physiological) variables. This means that the stages have a universal (that is, nomothetic) character. They have cross-cultural validity and historical continuity. Although the expression of stages will be culturally molded to fit the prevailing values of the society in which individuals live, and though powerful historical currents can diminish or enhance their vividness, stages possess core similarities and can be recognized across cultures. The Piagetian paradigm, for example, would aver that the *formal operations* stage of cognitive development is recognizable as a *stage*, even though there is a good deal of variability in the content and timing of its appearance from culture to culture, from (secular) age to age, not to mention from individual to individual within any culture. The countervailing argument is that stages are popular mental constructs that exist only "in the eye" of observers or, alternatively, that if they do exist it is only because they are pure artefacts of fashion and our social institutions, which promote their appearance.

Theorists, sages, and scientists, nevertheless, working independently in different languages, cultures, and historical periods have formulated multi-stage models of human development. This universality suggests that there is a foundation in reality for such models. Freud's and Erikson's stage theories of human development are only the most currently notable of many such theories that we can find throughout history. Stage theories are found both in scholarly works and in classical literature. Note, for example, the humorous template that Shakespeare fashioned in Jaques' soliloquy in *As You Like It*,[3] or, respectively, Jean Piaget's and

[3] All the world's a stage,
And all the men and women merely players:
They have their exits and their entrances;
And one man in his time plays many parts,
His acts being seven ages. At first the infant,

Lawrence Kohlberg's widely endorsed stage theories of cognition and moral reasoning.

Such formulations are not just artefacts of the Western world. *The Great Chain of Being*, of Eastern inspiration, implicitly describes the development of human consciousness in stage terms. Other life-span models of human growth and development patterned on the contemplative traditions of the East are largely stage-based. Wilber (for example, 1986) described some of these developmental systems along with the dominant ones of the West. Among the Eastern traditions that provide developmental schemas for understanding human consciousness and personality is the paradigm developed by the celebrated Indian psycho-philosopher and sage, Aurobindo, which parallels in part the Piagetian cognitive stages. Wilber has discussed other stage-based systems such as Sufi, the Yoga Sutra (Hindu) tradition, the Mahayana (Buddhist) model, and Kabalah, among others (for example, pp. 57–65). The reader who wishes to explore these stage cartographies of the East will find that they have many features in common with Western structural models of personality development.[4]

That stages exist in human development appears to be self-evident. But is this concept grounded in human nature or is it simply a function of socially constructed patterns for organizing our familial and civic life? Is it first and foremost a convenient way for humans to conceptualize and

Mewling and puking in the nurse's arms.
And then the whining school-boy, with his satchel
And shining morning face, creeping like snail
Unwillingly to school. And then the lover,
Sighing like furnace, with a woeful ballad
Made to his mistress' eyebrow. Then a soldier,
Full of strange oaths and bearded like the pard,
Jealous in honour, sudden and quick in quarrel,
Seeking the bubble reputation
Even in the cannon's mouth. And then the justice,
In fair round belly with good capon lined,
With eyes severe and beard of formal cut,
Full of wise saws and modern instances;
And so he plays his part. The sixth age shifts
Into the lean and slipper'd pantaloon,
With spectacles on nose and pouch on side,
His youthful hose, well saved, a world too wide
For his shrunk shank; and his big manly voice,
Turning again toward childish treble, pipes
And whistles in his sound. Last scene of all,
That ends this strange eventful history,
In second childishness and mere oblivion,
Sans teeth, sans eyes, sans taste, sans everything. (Act II, Scene vii)

[4] See Roger Walsh (2000) for the psychotherapeutic properties of certain of these traditions.

make sense of their progression through life? These are legitimate questions and are arguable in one direction or another. Behaviorists insist that patterns of behavior that "constitute" a stage are really under environmental stimulus control and are no more durable than the social context that has shaped them. Stages are artefacts of history, they point out, and they argue that "adolescence" was only discovered by nineteenth-century sociologists. A reading, for example, of Ariès' (1962) classic work, *Centuries of Childhood*, makes it evident that the notion of adolescence was not widely recognized in the pre-industrial era of the West.

Some who are more biologically grounded would argue that stages are not theoretical confections but result from the complex dynamics of the human genome and epigenome. This is why one finds comparable cognitive-stage structures in various linguistic and cultural groups around the globe. When Piaget addressed this issue he replied that whether one "observes" stages or not boils down to a matter of scale. If one takes a broad, molar view of the human life course, stages come into view. The boundaries may be fuzzy but they delineate phases that we can recognize as distinct from one another – in short, phases of development. On the other hand, if one takes a molecular view of very limited and arbitrary segments of the life course, the transitions will "disappear." It would be like getting too close to an extensive landscape painting. If one stands with one's nose 6 inches (15 cm) from the canvas, one no longer sees the hills, but one may gain a fine appreciation of the texture of a swatch of oil or the filigree in the lace-like frost on a leaf. Like so much else in life, what we perceive as reality depends on where we are standing – and, to a certain extent, on what we have been taught to see, or even simply want to see.

Finally, a word needs to be said about stage models themselves. In each of these theoretical formulations, it is assumed that the order in which the individual transits the various stages never varies. Freud posited the oral, anal, and phallic stages as the first three. This order never changes. As each stage builds on the acquired capabilities of the preceding one, it is virtually impossible, granted the validity of the schema, for a person to by-pass any stage, or to invert the purportedly natural order of progression. Moreover, if one gets stuck in any stage, it is impossible to make progress in the more advanced stages. The child who has not learned to count cannot learn to tell the time (say, by "reading" a clock or sun dial). Of course, as these stages are not hermetically sealed categories, there can be some overlapping of the capabilities inherent to one or the other. A person who has developed, for example, some ability to think abstractly may occasionally revert to more primitive, concrete methods of problem conceptualization, even when theoretical planning would be more effective.

Stage-based versus nonstage-based paradigms. There are few true dichotomies in nature, although they abound in primitive thinking and in the homespun algorithms of the rank-and-file. Because dichotomies are simple and intuitively appealing, formulating models and paradigms in such terms hides serious pitfalls. In nature multiple categories and continua are the rule, perhaps even in such a simple matter as the division of the human species into two genders (Fausto-Sterling, 2000, insists there are five). It is comforting to know, in any event, that when one runs into a wall presented by a mental construct the experience is more forgiving than a wall of biological constraints.

This is true of models of human development. Although the early models of human development were rigorously stage-based, many that followed were not. Among those models that were, wide differences appear in (a) their structural content, (b) age of onset of similarly defined stages, (c) the degree of plasticity accorded the organization of behaviors and cognitive capabilities, (d) their relative importance vis-à-vis each other, (e) the flexibility of their boundaries, and (f) the constructs deemed critical or even admissible in their conceptualization. Relative to (f), the concept of *self*, for example, received little attention in the more behaviorally oriented models of human development, such as those of Clark Hull and Kenneth Spence, but it is critical in the mentalistic approaches of J. Mark Baldwin and William James. The notion of the *unconscious* plays a salient role in the grand systems of the late nineteenth and early twentieth centuries, but tends to be less important in contemporary systems. Modern interest in the unconscious has been revived and is vigorously investigated by such psychoneurologists as Antonio Damasio. Likewise, stages of *emotional development* were an essential part of the nineteenth- and twentieth-century systems of Darwin, Janet, Binet, and Freud. After having waned in the mid-twentieth century, the concept *self* has come back in full force in contemporary research, figuring prominently not only in the developmental literature, but also in psychoneurology and psychotherapy.

Arrested-development models and the unconscious

Psychoanalytic theory is emblematic of those systems that stipulate that human personality is largely developed (and sometimes fixed) in the earliest years of life. Psychoanalysis was not the first system to make this assumption, but Freud developed a comprehensive theory of human development which succeeding theories could treat as a point of reference as well as a foil. As we have noted in Chapter 1, Plato affirmed 2,400 years ago that the manner in which children were raised resulted in citizens who were variously suited for domestic and civic life. The privacy of the home, he asserted in *The Republic*, screens from general observation many dramas that can introduce a "medley of incongruities" into an individual's character. Freud would endorse this view in spades.

As Ellenberger (1970)[5] has documented, the work of Freud springs more from the German Romantic tradition than from the spirit of the eighteenth-century Enlightenment. Though he was trained as a physician and did much of his youthful research in the laboratories of noted physiologists of the time, few of the principles of those disciplines found their way into the structure of his personality theory. His first laboratory experiences were in comparative anatomy, but he soon gravitated to the laboratory of the celebrated physiologist, Ernst Brücke, where he worked for six years. What is notable about Brücke, whom Freud venerated above all his other teachers of that period of his life, was that he reduced all psychological processes to physiological processes; and these in turn he reduced to physical and chemical laws. In the world according to Brücke, there was no room for vitalism, for spiritual variables, for finalism, or even for free will. There is no reason to doubt that Freud ever disavowed Brücke's approach to science, but there is a strange disconnect, as we shall see, between Brücke's assumptions and many of the principles Freud developed in psychoanalysis. At most Brücke's framework served as a metacognitive backdrop for Freud's theorizing, whose grand theory of personality development is distinctively psycho-social. There are certain latent neuropsychological assumptions, which give a deterministic tone to his work, but they do not enter operatively into the dynamics of his schemas. We can assume, however, that the deterministic character of Freud's theory of personality development sprang, by contrast, from his earlier laboratory-based training (as well as from the influence of the celebrated philosopher, Schopenhauer).

Romanticism, especially as it developed in Germany, is regarded as a reaction to the French rationalism of the Enlightenment. But the Romantic philosophical tradition took root in France and England as well as Germany in the first half of the eighteenth century and influenced many of the intelligentsia of these three countries (see Shorter, 1997, pp. 29–32). It rejected the celebration of pure reason and the primacy of

[5] Henri F. Ellenberger's *The Discovery of the Unconscious* (1970) is a towering achievement in the history of nineteenth- and early-twentieth-century psychiatry. Ellenberger (1905–93) was the pre-eminent historian of that period. Students are encouraged to read the chapters of this volume dealing with the origins of the grand systems, including that of Sigmund Freud. This volume was published in 1970; as archived documentation bearing on Freud's work comes to light in the twenty-second century, revisions may have to be made to some of Ellenberger's conclusions.

society and, instead, exalted individuals and the emotionally ennobling and non-rational in their lives. The Romantics had a veneration of nature and delighted in studying the intellectual and emotional development of individuals. It was probably this tradition that fostered the detailed clinical case histories in which psychiatrists reviewed the entire life course of their patients (Ellenberger, 1970, p. 201). Freud was a child of this tradition.

Allied to the Romantic philosophy was the Philosophy of Nature, which exalted human empathy. The Romantics resonated to the notion of "rapport," the bond between the patient and the healer, which found a salient role in the work of the hypnotherapists and others in the Mesmerian tradition. Polarities were very important in this tradition, and poets, novelists, philosophers, scientists, and psychiatrists sought them throughout their disciplines. In this connection, Ernest Jones (1995), the most visible and widely cited biographer of Freud, observed that Freud's concepts of mental life were dominated by polarities. He evidenced throughout his life "a proclivity to dualistic ideas" (for example, dualism of instincts, polarities of subject–object, pleasure–unpleasure, activity–passivity, eros–thanatos, and so forth). The Romantics, as well as scientists of the eighteenth and nineteenth centuries, gave particular significance to the unconscious, and the study of the unconscious (or the "*sub*conscious" as Pierre Janet labeled the notion) became one of the dominant motifs of this intellectual tradition. If one wished in that era to understand human development, one needed to gain insight into this obscure and elusive foundation of emotion, feeling, human mentation, consciousness, self, and behavior. Moreover, studies of the "unconscious" and the neurophysiological substrate of this psychological realm have taken on a great importance at the turn of the twenty-first century. It will pay us to examine the concept more closely.

The psychology of the unconscious

The realm of the unconscious was studied intensively throughout the nineteenth century, and it figured prominently in the genesis of theories of personality that flourished in the first half of the twentieth century (see Whyte, 1960, for a comprehensive overview of the many "discoverers" of the unconscious that preceded our era). Their approach to this construct was rather different from the intuitive approach of Saint Augustine ([400 BCE] 1991a). Augustine saw the unconscious as a "spacious palace, a storehouse for countless images … [and] thoughts by which we enlarge upon, or diminish or modify" our perceptions. It was a repository that "also contains anything else that has been entrusted to it for safekeeping, until such time as these things are swallowed up and buried in forgetfulness" (p. 93). By the nineteenth century the concept of the unconscious had become a natural "faculty" to which the scientists of that era could apply the methods and tools of science to achieve an understanding of human mentation. The concept would eventually lead to the "bicameral mind" (posited by Julian Jaynes, 1990), which purportedly allowed the gods to speak to us and direct our lives.

Leibniz (1646–1716) is commonly regarded as the first scientist to study the "unconscious" scientifically (he developed the notion of subliminal perceptions), but greater credit must be given to Herbart (1776–1841), who advanced our understanding of unconscious perceptions and developed a theory to support his investigations. He mathematicized in exact formulas the processes by which conscious ideas were repressed into the realm of the unconscious. Although his efforts were disputed and ended in failure, the thrust he gave to a natural sciences approach to studying psychic phenomena was enduring and useful. Herbart suggested that "ideas struggle with one another for access to consciousness, as dissonant ideas repel one another and associated ideas help pull each other into consciousness or drag each other down into unconsciousness" (Leahey, 1987, p. 168; see Lowry, 1971, p. 72; *cf.* Klein, 1970, pp. 762–69). This idea found its way into the work of all the major theoreticians of the nineteenth century including Esquirol, Griesinger, Benedikt, Richet, Fechner, Nietzsche, Bleuler, Janet, Freud, and Flournoy, to mention only a few.

The laboratories and the bedrooms

There were three principal streams of scholarship in the pre-Freudian era that influenced later theories of the unconscious and of human psychic development. The contributors to these streams were (a) experimentalists and other systematic empiricists,[6] (b) philosophers of nature, and (c) clinician-researchers. Notable among the first was Fechner (1801–87). Reminiscent of the objectives of Herbart, Fechner addressed the problem of the intensity of stimulation necessary for a perception to cross the boundary from the unconscious into the conscious realm. The qualities of the dreaming state, as contrasted with the waking, were one of the questions he tried to answer. He recognized that these two cognitive theaters could reveal a good deal about each other's limitations and resources. Another experimentalist, von Helmholtz, "discovered the phenomenon of 'unconscious inference,'" which he perceived "as a kind of

[6] See Edward Shorter's (1997) chapters, "The First Biological Psychiatry," an informative analysis, and "Nerves" for the evolution of nineteenth-century European psychotherapeutics.

instantaneous and unconscious reconstruction of what our past taught us about the object" (Ellenberger, 1970, p. 313). Earlier in the nineteenth century, studies were done on the unconscious manipulation of "ouija boards," divining rods, pendulums, and other such instruments by individuals who were unaware of the personal role they were playing in moving these inert objects.

These studies of the unconscious were largely conducted in research institutes. Independent, but equally remarkable, studies of dreams laid the groundwork for the later theorizing of Freud and Jung. These investigators, whose data were gathered in bedrooms, their own and those of their subjects, included Scherner, Maury, and Hervey de Saint-Denis (see Ellenberger, 1970 for a wealth of detail on these and other dream investigators). After long and tedious studies of dream life, in some cases their own, they deduced principles bearing not only on the meaning of dreams but on the nature of the unconscious. Their work is important not because of the reliability of their findings (most of which have been discredited), but because of the influence that their work was to exert on later theorists. Scherner's "The Work of the Dream" (1861) illustrates the fascination that dream interpretation held for those who lived in that period, and the penchant for florid, anecdotal, and speculative interpretations that were to be integrated into the theory-building of later dream analysts. A few examples taken from Ellenberger (1970) follow:

> According to Scherner, there is one basic dream symbol: the image of a house, which is the expression of the human body, parts of the house representing parts of the body. Scherner relates the story of a lady who went to bed with a violent headache and who dreamed she was in a room in which the ceiling was covered with cobwebs swarming with large and disgusting spiders. A dozen pages in Scherner's book are devoted to symbols related to the sex organs. As male symbols he mentions high towers, pipes, clarinets, knives and pointed weapons, running horses ... ; among female sex symbols Scherner mentions a narrow courtyard and a staircase that one must climb up. (p. 305)

The philosopher-psychologists

The work of the brass-instrument methodologists and dream scholars pales in significance by comparison with that of the psycho-philosophical writers of the first half of the nineteenth century, whose ideas had the greatest impact in the theory-building of Freud and Jung. They constitute the second stream of scholarship that shaped twentieth-century thinking on the unconscious.

It is difficult to appreciate the extent to which the ideas of Arthur Schopenhauer (1788–1860) infiltrated the nooks and crannies of popular as well as intellectual thought by the second half of the nineteenth century. But that century witnessed a host of psycho-philosophical writers, of whom he was but one of the foremost. Like Carl Gustav Carus (1789–1869), Eduard von Hartmann (1842–1906), and Friedrich Nietzsche (1844–1900), Schopenhauer wrote copiously and cogently about the unconscious. His major work, known to generations of American students through their universities' "Great Books" programs, is *The World as Will and Idea*, published in 1819. It was an expression of a pessimistic, pansexual, irrationalist, and deterministic view of humankind. His representation of the unconscious (which he termed "will") was that of a blind driving force, essentially irrational and thoroughly imbued with sexual energy. The irrational and driving character of the unconscious determined one's behavior and left no room for the exercise of a free will. Despite its pessimism, Schopenhauer's work gained an escalating popularity in the second half of the century and patently informed the work of legions of imitators including, most notably, Freud. Ellenberger (1970) wrote:

> The similarities between certain essential teachings of Schopenhauer and Freud have been shown by Cassirer, Scheler, and particularly by Thomas Mann. Mann, who during his youth had been deeply immersed in the metaphysics of Schopenhauer, declares that, while becoming acquainted with Freud's psychoanalysis, he "was filled with a sense of recognition and familiarity." He felt that Freud's description of the id and ego was "to a hair" Schopenhauer's description of the will and the intellect, translated from metaphysics into psychology. (p. 209)

Carl Gustav Carus (1789–1869), a physician, presented one of the more sophisticated theoretical models for the unconscious. Carus distinguished three layers of the unconscious: first, an absolute unconscious to which one could never gain access; second, a relatively obscure unconscious that controlled the autonomic functions of the body, regulating respiration, heart rate, and other organic functions as well as the emotional expressions of these systems. This layer of the unconscious interacted fluidly with one's conscious mentation. Third, there was a layer of the unconscious that comprised the totality of our thoughts, feelings, and emotions that had previously been conscious. In our interactions with other human beings, there is a complex interaction of our various levels of unconscious with the consciousness of others. Thus, the conscious of one person interacts with the unconscious of another, and so forth as in the unconscious with the unconscious, the conscious with the conscious, and the unconscious with the conscious. Anticipating Jung's theory of the collective unconscious, Carus believed that the individual unconscious of each person is related to the unconscious of all people.

The clinician-researchers

The tracts of Schopenhauer and Carus set the stage for von Hartmann's and Nietzsche's (see Torrey, 1992, p. 6; Whyte, 1960, pp. 153–76) influential writings on the unconscious. These in turn greatly influenced the theorizing of a new generation of professionals, an extensive array of brilliant psychiatrists who flourished in Vienna, Zurich, Paris, and elsewhere on the Continent. Among the most notable is Moritz Benedikt, a Viennese clinician, whose name is usually associated with the "doctrine" of "the pathogenic secret," a secret usually larded, in his view, with sexual content. Ambroise Liébault (1823–1904), with Hippolyte Bernheim (1840–1919), is regarded as the founder of the Nancy School of hypnotherapy. Auguste Forel (1848–1931), another notable, was professor of psychiatry at the University of Zurich, a job that enabled him to conduct far-reaching reforms of the mental hospital known as the Burghölzli. Forel mentored the renowned and innovative clinicians, Adolf Meyer (1866–1950)[7] and Eugen Bleuler (1857–1939), and Bleuler in turn mentored the young Carl Gustav Jung. Bleuler studied "dementia praecox" extensively and coined the term *schizophrenia* to designate it. Contemporaries of these men included Sigmund Freud and Josef Breuer, luminaries among many others on the lively Viennese scene.

At the Salpêtrière of Paris, one of the eminent nineteenth-century psychoneurologists, J-M. Charcot[8] (after whom Freud named his son, Jean-Martin), demonstrated the influence of the unconscious in the waking life of his patients. In his weekly magisterial presentations given in a spacious hall or amphitheater, he induced trances in his "well-schooled" patients and implanted post-hypnotic suggestions. These subjects-on-exhibit, after having been aroused from their trance and executing these suggestions, would then fabricate *post facto*, in all apparent sincerity, irrelevant but plausible reasons for their actions. The implication of this for Freud and the other students, scientists, and practitioners in the amphitheater who witnessed it was that "the heart [read the unconscious] has reasons of which reason [read the conscious] has no knowledge." Charcot also argued on the basis of his clinical experience that the hysteria

[7] Adolf Meyer was an extremely influential, if not distinguished, American psychiatrist of whom Shorter (1997) stated that "he indiscriminately advocated all forms of psychiatric thinking from the biological to the psychoanalytic ... later plunging "the United States into its psychoanalytic adventure" (p. 110).

[8] A curious historical fact is that the syndrome for so-called Lou Gehrig's disease, technically Amyotrophic Lateral Sclerosis (ALS), was first described by Charcot and was long known by the eponym, Charcot's Syndrome.

manifested by his female patients often had a sexual etiology.[9] This was another principle that Freud and many other psychiatrists of the period uncritically took to heart.

Sulloway (1979) has observed that "These dramatic demonstrations ... first revealed to Freud the remarkable circumstance that multiple states of consciousness could simultaneously coexist in one and the same individual without either state apparently having knowledge of the other" (p. 32). In his later years Freud ([1925] 1959) recalled: "I received [from Charcot] the profoundest impression of the possibility that there could be powerful mental processes which nevertheless remained hidden from the consciousness of men" (p. 13). To credit Charcot exclusively for these "insights," as these two quotations would suggest, would be a mistake, as Nietzsche, the philologist-errant, had already written extensively on the unconscious as the realm of libidinous impulses that sought outlet in dreams, illness, and overt, often symbolic, behaviors. These views were discussed in the bistros and salons of Europe as well as its newspapers and magazines. No educated person, including Freud, could have been ignorant of them. An even greater departure from the historical record is Kohut and Seitz's (1963) affirmation that "Freud [made] the discovery that man's consciousness illuminates only a narrow and limited part of his own mental activities" (p. 118).

Nineteenth-century psychiatry and psycho-philosophy were preoccupied with the notion that human cerebration was like an iceberg, whose tip corresponded to conscious mental processes. The point that psychotherapy required an active exploration of the subconscious was made by legions of therapists beginning in the era of Johann Christian Reil, who authored the clinical volume, *Rhapsodien*, in 1803. It stretched to Pierre Janet (1859–1947), the most influential of the theorists of the 1890s, a contemporary of Freud. He developed a more rationalist, less pessimistic view of the human than Freud, but gave equal importance to the analysis of the subconscious. He dubbed this process *psychological analysis*, which, in his words, consisted among other things in "making the subconscious conscious" (a turn of phrase later adopted by psychoanalysts).

Sigmund Freud was deeply embedded in this psychiatric tradition as he began his effort to develop a psychotherapy that was distinguished from those of his predecessors. We can note at this point that Ellenberger speaks of the current legend that "attributes to Freud much of what belongs, notably, to Herbart, Fechner, Nietzsche, Meynert, Benedikt,

[9] La Salpêtrière, where Charcot did much of his research, was exclusively for female patients, as distinguished from the other major asylum of Paris, Le Bicêtre, that was reserved exclusively for males.

and Janet, and overlooks the work of previous explorers of the unconscious, dreams, and sexual pathology" (1970, p. 548). It is interesting to speculate about how this popular lore took root. A fully satisfactory explanation can be given only by historians, but part of the answer lies in Freud's appreciation of the importance of "branding" the system that he was to develop. For that reason, he shortened Pierre Janet's *psychological analysis* to *psychoanalysis*, a "brand" that has become as recognizable as any in the twentieth century. Freud wanted initially not only to create a scientific system; he wished to create a movement. He and his followers succeeded beyond their dreams.

Freud's theories of personality[10]

The evolving nature of psychoanalysis. Three points need to be made clear before we undertake an overview of Freud's theory of human development. First, although it is customary to treat Freud's grand system as if it were a monolithic and static conception that took shape in a brief historical "instant," the facts about its genesis are rather different. It began in earnest in the 1890s when Freud was in his late thirties and suffering from serious medical and emotional problems. His theory continued to evolve well into his older years and did not take a definitive shape until after the First World War. A major metamorphosis of his first formulations occurred in 1915 and thereafter with the publication of *Introduction to Narcissism* in which he reconceptualized his theory of drives. This was followed with new formulations of the pleasure–unpleasure principle, the dynamics of repression, a new, dual classification of the instincts (sexualized instincts, on the one hand, and the death instinct, on the other hand), and other innovations. And his turn-of-the-century topographical model of the mind, which was predicated on the traditional nineteenth-century model of the conscious, pre-conscious, and unconscious, was supplemented in the 1920s by his structural model comprising the id, ego, and superego.

An embryonic developmental psychology. Second, the nascent developmental psychology of that time, in which his personality theory and psychotherapy were grounded, was in its earliest formative stages. The limitations and flaws in that psychology are well known to us today,

[10] Because Freud's developmental constructs, more than those of any other theoretician, pervade the academic, artistic, journalistic, artistic, and "pop psych" ethos of the West, a disproportionate part of this chapter is dedicated to a review of his paradigms.

and we will allude to them below. Although Freud himself made an effort to contribute to that developmental psychology, the effort was largely of a speculative nature and based on his self-analysis during a personally troubled period. It has not withstood the empirical tests posed by the scientists who paralleled and followed him. It has been largely discarded by revisionists within the psychoanalytic school itself, and is taught in academia, if at all, for its historical interest. This will be dealt with below in somewhat more detail.

The myth of Victorianism. Third, the myth has been widely propagated that the turn of the century was one of Victorian prudery and that Europe was shocked by the exclusively sexual causal–explanatory paradigm that Freud first presented as the basis for his therapeutic approach. The Victorian spirit was principally of British inspiration, and took its name from Queen Victoria at a period of history when Britain was at the zenith of its power and prestige. One aspect of this spirit, which held a fascination for some segments of the various Continental cultures, was a certain modesty, not necessarily in speech, dress, or home furnishings, but notably in sexual matters. It had faded away on the Continent by the mid-decades of the nineteenth century, and the newspapers, magazines, and books of the time displayed a flamboyant eroticism. It was everywhere. Ellenberger notes that a critic of the day, Jules Claretie, wrote the epitaph, "Here rests the pornographic year 1880" (1970, p. 282). Eroticism dominated the literature of the last decades of the century. It found expression not only in the popular press and the paintings, say, of Gustav Klimt and Toulouse-Lautrec, but in the brilliantly crafted novels of Guy de Maupassant, Gustave Flaubert, and Arthur Schnitzler. (One of the latter's novels was the basis for a Stanley Kubrick film, *Eyes Wide Shut* – neither a novel nor a film for the prudish.) Krafft-Ebing's *Psychopathia Sexualis*, published in 1886, enjoyed a wide and enthusiastic renown and went through numerous editions. So it is an error to think that psychoanalysis was beset in its infancy by assaults of Victorian prudishness. And to suggest that the medical profession as a whole was shocked by the sexual content of Freud's theory when he published his *Three Essays on Sexuality* in 1905 is patently absurd. On the contrary, the acceptance of his theory was abetted by the libertinism that flourished at the *fin de siècle*.

Freudian developmentalism

The psycho-sexual stages of development. The first thing to note about Freud's developmental theory is that it is a *stage* theory. More specifically it is a *psycho-sexual* stage theory. It is difficult to exaggerate the importance of the erotic libidinal drive in the psychoanalytic model. It is a current flowing through all aspects of Freud's oeuvre. One finds it in his anthropological writings[11] as well as in his abnormal psychology, in his developmental psychology as well as in his psychotherapeutics. For example, in 1896, he submitted a paper for publication sketching the etiology of neuroses. He stated categorically that all the major neuroses "*have as their common source the subject's sexual life, whether they be in the disorder of his contemporary sexual life, or in important events in his past life*" (p. 149, original emphasis). He divided these into actual neuroses (those arising from current pathogens) and psychoneuroses (those that have their origins in historical pathogens). The pathogens are all sexual in nature: actual neuroses, more specifically, neurasthenia (dysthymia) and anxiety neurosis, were caused, respectively, either by masturbation or by *coitus interruptus* (or some other form of sexual frustration). Psychoneuroses, more specifically, hysteria and obsessive disorders, result from sexual molestation by an adult of the patient during infancy. If the infant acquiesced to these venereal pleasures (mostly evident in males) the result was obsessiveness; if the child did not take pleasure and was more passive (mostly evident in females), the result was hysteria.

Development of the child

The basic motivating forces in the child are also sexual. Indeed, the personality of the child that will eventually take shape in later toddlerhood is exclusively a product of the manner in which the child learns to manage his sexual drives.[12] For this reason, scholars routinely refer to Freud's theory of human development as one of psychosexual stages. The theory stands or falls on the validity of this conceptual schema. But what are the origins of these drives? The answer is that they are blind forces that emerge from the organic nature of the human. The organism seeks gratification of needs that are related to its survival. It gravitates to those energy sources that reduce the tensions arising from all of its organic needs. Needless to say, the primary source is the mother.

[11] In *Civilization and Its Discontents* ([1930] 1961) Freud speculated that primeval man first learned to control fire by urinating on it. As flames have a phallic shape, he experienced the erotic feelings implicit in a "homosexual" contest: his phallus against the phalli of the fire. The first man to renounce this pleasure was able to bend the fire to his own practical ends. Women on the other hand, lacking a phallus, could neither extinguish fire this way nor sublimate its pleasures. That's why she was consigned to manage fire where it was held captive: in the hearth and home (p. 90, fn. 1).

[12] We say *his* advisedly, as the Freudian system is quintessentially a template for understanding male sexual development.

One has to keep in mind that the nineteenth century was one that was enthralled with the Industrial Revolution and the principal tools that had been invented for the production of the wealth that societies as a whole were beginning to enjoy, albeit, in that epoch, very unequally. The steam boiler, the locomotive, the other thermodynamic systems that were powering the engines and vehicles of production served as the metaphor for conceptualizing the human (see Gigerenzer, 1991). The human was regarded in part as a sophisticated and complex system that needed to maintain energy levels in a state of homeostasis. An excessive build up of energy levels created tension that was uncomfortable (as in dammed-up sexual energies) and in the extreme was experienced as painful (as in thirst and hunger). The cause of the tension was, in the last analysis, organic and tissue-based.

The psychosexual stages.[13] In the Freudian schema, the psychic dimension of the visceral, humoral, and other organic subsystems is the *id*, which contains the reservoir of energy that drives the individual to reduce painful tension levels. This drive component of the id is commonly referred to as the *libido*. And all these libidinal drives have a sexualized character. The child enters the world dedicated to satisfying those drives. He is a thoroughly narcissistic being, and because the primary object of his desire is the breast, he enters the first of the psychosexual stages – the *oral stage*. One can readily appreciate that the principal sources of sexual satisfaction in this theory are localized in the erogenous zones of the body. The first to receive attention is the mouth.

All activities surrounding the mouth, the throat, tongue, teeth, and vocal apparatus are implicated in traversing the oral stage of development. Mastering the requirements of this stage, which involve not only sucking and swallowing but appropriate biting (note, however, certainly *not* the mother's nipples), and angry or joyful vocalizing, among other oral activities, allows the child to move to a higher level of psychic development. The infant's level of success in negotiating the challenges of this stage will have lasting consequences for his personality. If it is not done well or if there are major frustrations in resolving the hurdles of the stage, the child is left with permanent characterological quirks.

These quirks (fixations), by virtue of a certain semantic representativeness, resemble the biological activities of the oral orifice. For example,

[13] This section deals with material that will be familiar to most readers. These readers can skim over it or scan it for possible errors and distortions. An interesting exercise would be to explore how this schema and its literature can be matched with the explanatory notions, say, of Nisbett and Ross (1980), Tversky (1977), and other attribution theorists.

there is a colloquial expression, "he'll swallow anything," that characterizes the person who is highly suggestible. Such a person is considered to be an *oral-sadistic* personality, and it suggests something that has gone awry in that oral phase of development. A person who engages in *biting* sarcasm is also thought to have an oral personality. The semantic linking of a figure of speech to an anatomical part is construed as evidence of a developmental glitch involving that part. The same heuristic is used in the other stages. For example, in the following stage, the "anal," an unsatisfactory resolution of the challenges of that stage involving the appropriate retention and expulsion of feces can result in the quirk of "hoarding" or socially "dumping" on others. The extensive use of such metaphors by Freud is a hallmark of his system.[14] The psychoanalytic literature is a veritable thesaurus of such metaphorical allusions (as is the literature on dream analysis).

The stage that follows the oral stage is the *anal stage*. Freud hypothesized that the major developmental task of the child in this stage is the control of his bowels. He viewed the anal region as an erogenous zone from which one derives enormous satisfaction in the retention and release of organic wastes from the body. This developmental task and its attendant pleasures overshadows all others for the child. The socialization of this psycho-organic process purportedly has a profound effect on personality formation. If toilet training, for example, was rigorous and premature, the child could react by either spitefully retaining its feces – being stingy with them, so to speak, or by expelling them in a rage. Anal stinginess, if not outgrown, resulted in adult obstinacy and stinginess with possessions. In a similar way, anger expressed in the second option was transmuted as an adult into expulsive traits in which this person would vent frustrations in anti-social behaviors, disorderliness, cruelty, and wanton destructiveness.

Freud labeled the third stage the *phallic stage*. The term phallus refers to the male organ, the penis, the third major erogenous zone in the human. The choice of the term phallic stage suggests that the basic template for defining the development of the human is the male. It was an infelicitous choice in the view of many, but it was one that Freud never regretted. Be that as it may, the manner in which the male negotiates this stage of his development is portentous as it ushers in, between the ages of 3 and 5, the monumental drama of the Oedipal conflict. The level of success with which the male resolves the conflict is heavy with consequences for his future happiness and the soundness of his adult personality.

[14] Nisbett and Ross (1980) and Tversky (1977), among others who have worked in the area of attribution theory, have extensively researched this type of argument, known as the "representativeness heuristic."

At this point in the child's development Freud introduced the keystone to his theory of personality development, namely, the *Oedipal complex*. What is it exactly? We must note that the child's nascent sexuality has been developing since birth. He initially derives sexual pleasure, according to this theory, from sucking at the mother's breast. But as the years elapse the focus of this libidinous drive shifts from the mouth, to the anus, and later to the phallus. The autoerotic character of this drive shifts from the male himself as object to his mother as object. It is this dreadful appetition for forbidden fruit, a lustful union with the boy's very own mother, that Freud captured with the truly fabulous character of the *Theban Trilogy*, Oedipus Rex.[15] The toddler regards his father as a rival for the mother's sexual attentions but fears reprisal from the father. The particular form of the feared and wholly imagined reprisal is castration. This is the genesis of "castration anxiety," which is accompanied by feelings of hostility for the father. There is only one natural way for this conflict to be resolved and that is for the toddler to repress his lustful feelings for his mother as well as his hostility for his father. The upshot of this drama is that the child not only identifies with the father but accepts the code of honor that his father represents. In short this marks the full-term birth of the child's conscience.[16]

This drama, with its momentous consequences for personality development, is very different for the female. In what is referred to as the pre-Oedipal phase, the infant girl *cathects* to (that is, emotionally invests herself in) the mother, no less than does the infant boy. The pleasure that she originally found in her clitoral masturbation is sooner or later "spoilt by the influence of penis-envy. She is wounded in her self-love by the unfavorable comparison with the boy who is so much better equipped ... and [at length] repudiates her love towards her mother" (Freud, [1932] 1952, p. 860). This develops because she becomes aware not only that she has been castrated but also that her mother has been castrated as well (p. 860). (From an abnormal psychology perspective, Freud affirms in this same passage that "masturbation is the executive agent of infantile sexuality,

[15] Those who have read Sophocles' *Oedipus Rex* will realize that there is little parallel between the *King* and the typical male child. Oedipus killed his father, whom he took to be a perfect stranger, in a lonely conflict that took place at the intersection of two roads, one of which led to their ancestral village, Thebes. The conflict had nothing to do with rivalry for a woman's affections, still less, that of the wife and the mother of these two protagonists. Further, Oedipus did not even know his mother at the time of this homicide and his future marriage to her was one that occurred by chance. Finally, it was a liaison he found so abhorrent that when the truth was made known to him he plucked out his eyes.

[16] Millions of boys raised in fatherless homes by single mothers need to resolve an Oedipal conflict, so essential to the development of a mature moral sense. How this hiatus in their development is repaired still needs satisfactory explication.

from the faulty development of which [adult neurotics] are suffering" (p. 860). Freud always considered masturbation a pathogenic activity.) It is at this juncture that she turns to her father, impelled by the wish for a penis which her mother has denied her, a wish, Freud adds, that is soon replaced by a wish for a child, especially a male child.

There are a number of anomalous outcomes, from a contemporary point of view, stemming from these conflicting psychodynamics. First, and most importantly, the female's moral development is compromised. The male develops a rigorous, severe conscience, which Freud later called the *superego*. As the *ego*, which manages the child's relations with "reality," emerges from the id, the superego arises from this embryonic personality structure as he resolves the Oedipal conflict and related moral issues by capitulating to his father and adopting the father's internalized set of rules. "What happens in the case of the girl is almost the opposite ... She enters the Oedipus situation, as though it were a haven of refuge" and she remains in it for an indefinite period and only "abandons it late in life, and then incompletely" (Freud, [1932] 1952, p. 861). As a consequence, the development of a woman's moral sense is stunted, her sense of justice does not fully evolve, and she has less capacity for sublimating her instincts. That such an outcome defining "normal femininity" has not found acceptance either among later generations of women or even among many psychotherapists and developmentalists, who still regard themselves as working within the Freudian paradigm, is not surprising.[17]

Genital stage. In the interests of completeness we need to refer to the Freudian period of *latency* before treating the *genital stage*, although few give the former any credence in the twenty-first century. It follows the phallic stage of psycho-sexual development and is regarded as a period of relative sexual quiescence that precedes the turbulence introduced by the physiological changes of puberty. Freud consequently paid it sparse attention. The genital stage has its onset with puberty and is extremely important. In this stage the individual, building on the foundation of the prior psychosexual stages, evolves into a mature, integrated person. Adolescence is not without its pitfalls, nor is early adulthood, as we all

[17] A footnote to this issue concerns the Electra complex, which Freud always repudiated. In ancient Greek mythology, Electra was a strong and principled woman who plotted to avenge the murder of her father, Agamemnon. The provenance of this complex, which gives a more balanced and egalitarian cast to the development of the female personality, can be traced to Carl Gustav Jung. Freud ([1931] 1962), on the other hand, resisted this notion, although the "pop psych" literature often mistakenly attributes it to him. He stated, "We are right in rejecting the term 'Electra complex' ... which seeks to emphasize the analogy between the attitude of the two sexes" (p. 229).

know. The immature, self-absorbed, conflicted, and searching child struggles over these difficult years like a butterfly, to emerge from its chrysalis into the bright daylight of adult independence and self-directedness. Sexuality in this final phase, as in all previous phases of human development, remains the florescent thread that helps us to trace the individual's progress to full maturity. For this reason Freud labeled this stage the genital stage, the final integrative psychodynamic of human personality. Internal consistency, as we note, is the salient virtue of this system.

Freud as scientist. The question of the scientific validity of the system that Freud developed has often been debated. It is, of course, an important question. Writers in the psychoanalytic tradition have traditionally supported and justified the largely anecdotal approach that Freud used to buttress the principles of his developmental psychology and the many derivative principles that constitute his ambitious psychotherapeutic system. Despite Freud's youthful years of training in the physiology laboratories of Ernst Brücke, Theodore Meynert, and other esteemed masters of the German research institutes, he later showed a remarkable lack of interest in their rigorous, experimental approach to establishing the validity of his own findings. He shifted from a natural sciences stream of tightly controlled, rigorously empirical, cautious, reductionist, and carefully documented research into the more speculative, philosophical, free-wheeling German Romantic tradition. In a letter to his friend, Fleiss, in 1900, he wrote, "I am not really a man of science, not an observer, not an experimenter, not a thinker. I am nothing but by temperament a conquistador – an adventurer if you want to translate the word – with the curiosity, the boldness, and the tenacity that belongs to that type of being" (cited in Stone, 1997, p. 37).

Alan A. Stone (1997), a psychoanalyst and former president of the American Psychiatric Association, has stated that in one of his own research projects he spent most of a year reading much of Freud's published work, along with all of his letters and biographical materials. What he discovered "is that Freud had a new hypothesis every day. It is astonishing to see how little evidence he needed; a single patient hour was enough to launch a whole new theory of mental illness" (p. 37). In his work, *Three Essays on the Theory of Sexuality*, "Freud (1905) provides almost no evidence and no direct observational data for his sweeping conclusions" (p. 37). Two renowned, late twentieth-century philosophers of science, Karl Popper (1963) and Thomas S. Kuhn (1989), were in agreement that psychoanalysis is not science. Notwithstanding its many virtues, it does not satisfy the criteria for valid science, any more than does

Marxist historiography. Writing from a British perspective, Kline (1986) avers that "in academic psychology, certainly in Great Britain ... Freud is relegated to the historical portions of courses. His influence is admitted *mais ce n'est pas [la] science*" (p. 295). A generous view of his oeuvre is that it is the fruit of his qualitative rather than quantitative research – but without the rigor of later (say, Strauss and Corbin, 1990) qualitative methodologies. A fuller perspective on this matter can be gleaned from Hall, Lindzey, and Campbell (1998, pp. 57–73).

Psychoanalysis is a dramatic example of the power of theory to evoke supporting evidence. Freud built a powerful, coherent, evocative theory. As with many of us who work within a single impermeable paradigm, Freud's mind was better prepared to perceive evidence that supported his theory than evidence that disconfirmed it.[18] Good examples of this can be found in the case history of Little Hans. Freud's theory asserts that all male toddlers develop a castration complex, the pervasive fear that their father will amputate their sexual parts in retaliation for coveting the mother. Like the "analyst" Freud was supervising in this case, he attributed this fear to Little Hans. When Hans's mother spied her 3½-year-old child "*touching* [original emphasis] his [sexual] member" (p. 7), she warned him that if he did this again she would send for the family doctor to cut it off. Freud observed that "this was the occasion of his acquiring 'the castration complex,' the presence of which we are so often obliged to infer in analyzing neurotics, though *they one and all struggle violently against recognizing it*" (added emphasis). That Freud never found a single client who recognized this fear, yet attributed it universally to human males, reveals the coercive power of a theory. Freud was not immune to this heuristic pitfall.

Derivative psychoanalytic theories

The richness of the Freudian schema was such that it allowed a large number of theories to emanate from it. This has been its great strength as well as its weakness. The strength of psychoanalysis was that it was a great synthesis of much of nineteenth-century thought. (Ironically, it is precisely what is original in Freud's work that has been largely discarded. What he inherited from Herbart, Fechner, Carus, Schopenhauer, and Nietzsche, among others, has a better survival rate. Examples are Carus' topographical model of the mind with its varying levels of the

[18] It is for this reason that double-blind experiments, in which neither the subjects nor the data gatherers are aware of the hypotheses being tested nor, optimally, of the experimental condition to which the participants have been assigned, are most secure from bias.

unconscious, as well as notions of sublimation and repression, and other unconscious defenses, popularized by Nietzsche.) The weakness consists in the inability of his disciples to use it as a scaffolding for developing a cumulative discipline. Freudism has fragmented into a multitude of conflicting models with different emphases and principles. In Kohut's words, the literature had become a "thicket of similar overlapping, or identical, terms and concepts which, however, did not carry the same meaning ... " (see Stone, 1997, p. 36). Indeed, some contemporary psychoanalytic theorists have departed so radically from the original that one hardly recognizes them as psychoanalytic. One example is the work of Karen Horney. Though she is often characterized as psychoanalytic, the resemblance of later Horneyan thought to the root system lies principally in its vocabulary rather than in its content. We will have occasion to refer to some of the more radical offshoots of psychoanalytical thought in Chapter 8, on feminist approaches to human development.

The life-span perspective

Towards the end of the twentieth century, psychologists and other social scientists gravitated to the view that human development, and the correlative personality features that were the observable psychosocial expressions of that development, continued throughout life. Personality is not set in plaster by the age of 5. Although developmentalists recognized a stream of continuity in all the strands that personality comprises ("Man does not become evil or good overnight" has been lore over the ages), one characteristic of the human condition is that nothing is static including personality structure. Baltes (1987) noted a number of reasons why the focus of research has shifted to a consideration of the *entire* life course. Most notably developmentalists have "discovered" a large and clamorous constituency – an aging cohort in the West that has not accepted the traditional perspective.[19]

Vygotsky averred that there are few commonalities that link the mentation of the toddler and that of the adult. The introspection of an adult is not a good basis for inferring the cognitive processes of an infant. The principal reason for this is the prepotency of changing life tasks and the social imperatives that shape our thinking and self-image at various seasons of our lives, and the way in which we respond to them. This is why cohort effects are so important to consider when comparing population characteristics. Cohort effects are those a specific generation of people,

[19] This is further evidence, if it were needed, that science responds to socio-political pressures, especially when those pressures are abetted by funding agencies.

baby-boomers for example, experience from growing up in a society in which *singular* socio-political and economic realities have gripped them. Consider the effect that the increasing life expectancy of populations in the developing nations of the world is having on life span development. In the early decades of the twentieth century, Carl Gustav Jung considered the age of 35 to be middle-aged and theorized that it marked a watershed for the integrative, especially spiritual, reorientation of the human. In his view, the critical issues of how individuals "fit" in the universe and address their mortality surface vividly at this point in their life. Thirty-five is no longer considered middle-aged in the developed world; in fact, large numbers of professionals have hardly completed their career preparation or begun their families by that age. This changing social clock has obliged developmentalists to reconsider the possibilities for growth rather than decline at this arbitrary marker of the aging process. Similar secular shifts in determinants of human behavior and personality came to be noted in other social science domains. This analysis is further complicated by the fact that scientists have varying "takes" on development as a function of whether their focus is on the emotional, attitudinal, moral, behavioral, or strictly cognitive aspects of development.

The notion, however, that life-span approaches to human development only appeared at the end of the twentieth century is an error. They have always coexisted with the notion that the broad structural properties of the human personality were largely determined by the age at which the child presumably achieved "the use of reason," say, 5 to 7. Lore attributes to Jesuit educators the ancient aphorism, "Let *us* instruct the child to the age of seven, and you may have him thereafter." But Baltes (1983) has reminded us of major theoretical contributions made by such thinkers as Tetens in 1777, Carus in 1808, and, especially, Quetelet in 1835, who presented balanced life-span perspectives on human development and shunned the notion that somehow the structuring of personality is completed in early childhood. G. Stanley Hall (1844–1924), a developmentalist (and, incidentally, the first president of the American Psychological Association), published a pioneering work on gerontology shortly before his death (Hall, 1922). His views are eerily reminiscent of the views of C. G. Jung. Of course, each era appears to make one or another period of human development a privileged focus of its investigations, as Ross D. Parke *et al.* illustrated (1994, pp. 29–30). In any event, the absence of an historical perspective in such matters leads to the impression that what we are thinking today is brand new, whereas it rarely is. The implicit lesson is that science is to a large extent a socio-cultural enterprise, and we need multicultural perspectives that are longitudinal as well as cross-sectional.

Life-span developmental models

Among the major life-span approaches to the study of the human, there is none that has enjoyed greater, albeit currently declining, popularity than that of Erik Erikson – among lay people and developmentalists alike. Erikson was not only a psychologist and psychotherapist, he was a cultural anthropologist, and among the pioneers in psychobiography. This breadth of intellectual interest is one of the qualities that has given his work a wide appeal. Although he sprang from a psychoanalytic tradition, having been trained by the daughter of Freud (Anna Freud), he truly altered this paradigm. There were numerous assumptions of the original system that he discarded as he progressed through the life span. The structure of personality was not, in his vision, fixed for better or for worse in the earliest years of life. Personality is a dynamic system that progresses through eight stages from infancy to death (see Table 3.1). He

Table 3.1 *Erikson's eight psychosocial stages of development*

Psychosocial stage	Age	Positive outcomes	Negative outcomes
Basic trust versus mistrust	1	Feelings of inner goodness, trust in oneself and others, optimism	Feelings of inner badness, mistrust of self and others, pessimism
Autonomy versus shame and doubt	2–3	Exercise of will, self-control, able to make choices	Rigid, excessive conscience, doubtful, self-conscious, shame
Initiative versus guilt	4–5	Pleasure in accomplishments, activity, direction, and purpose	Guilt over goals contemplated and achievements initiated
Industry versus inferiority	Latency	Able to be absorbed in productive work, pride in completed product	Sense of inadequacy and inferiority, unable to complete work
Identity versus role diffusion	Adolescence	Confidence of inner sameness and continuity, promise of a career	Ill at ease in roles, no set standards, sense of artificiality
Intimacy versus isolation	Early adulthood	Mutuality, sharing of thoughts, work, feelings	Avoidance of intimacy, superficial relations
Generativity versus stagnation	Adulthood	Ability to lose oneself in work and relationships	Loss of interest in work, impoverished relations
Integrity versus despair	Later years	Sense of order and meaning, content with self and one's accomplishments	Fear of death, bitter about life and what one got from it or what did not happen

Source: Adapted from Lawrence A. Pervin 1996. *The Science of Personality*. New York: Wiley, p. 166. Reprinted with permission of John Wiley.

construed each stage as opening possibilities for growth as well as for stagnation.

In the Eriksonian system, there is no selective search for the sexual wellsprings of human aspirations. He regarded people as richly endowed with aptitudes, interests, spiritual and ludic impulses, vocational attractions, concern for community and social embeddedness, the search for a defining identity, and fascination with the unknown and the existential realities of their common mortality. Further, healthy personalities are actively dedicated to managing their accurately perceived environments in self-consistent ways. This view of the human so far transcends the monochromatic sexualism of the Freudism in which he was nurtured that one can hardly, in the final analysis, call him a member of that tradition. (This is not a trivial matter, as one needs to place oneself in a frame of reference as one embarks on an examination of a paradigm.) It is possible that Erikson, himself, did not realize the extent to which he was an intellectual émigré. His native home became a far country to which he could no longer return.

Erikson (1969) reminded his readers that they must always consider the *epigenetic principle* when they try to understand growth. Epigenesis is the principle that organisms, beginning as a largely undifferentiated mass, become differentiated and increasingly complex as they progress through a series of transforming stages. This means that "anything that grows has a *ground plan*, and that out of this ground plan the *parts* arise, each part having its *time* of special ascendancy, until all parts have arisen to form a *functioning whole*" (original emphases) (p. 31). The potentiality for all the stages is present from the beginning, and their proto-expressions become evident long before they take their turn at center stage. Of course, their more mature forms continue to flourish long after they have stepped out of the limelight. Later, changing forms of trust, autonomy, and initiative, for example, remain in the integrated personality throughout its life.

The *first* stage is "infancy," involving oral–respiratory, sensory–kinesthetic needs and experiences at the forefront. The satisfaction of these needs by a maternal person through feeding, warmth, tactile stimulation, and dry clothes is the solid foundation for trust in the individual's nascent personality. Erratic provision of these "services" sows the seeds of basic mistrust in the infant's personality, which reveals itself as withdrawal from social intercourse in later life. As survival and well being depend on the succorance of others, the infant, like no other, needs dependability in life. This creates a solid foundation for hope and optimism, which become part of the furnishings for the child's psyche. Erikson regarded the corollary in the social order of this developmental principle as "cosmic order."

The *second* stage, "early childhood," occurs in the second and third years. It is characterized by toilet training, but it involves cultivating autonomy over a wide range of human activities beyond bladder and bowel control. This is encouraged by all parental figures in the child's life. Success in this phase results in a sense of autonomy and the associated strength of a certain willfulness. Failures in this domain elicit feelings of shame and self-doubt. The associated core pathology is compulsivity, which is the vestigial reaction to the failures of cleanliness and competence that normally blossom in this period. The child leaves this stage with the sense that social relations require "order" and that they are rule-governed. Any ritualism that emerges from this stage in later life has a judicial and legal character.

The *third* stage, which he labeled, "play age," occurs in the fourth and fifth years and involves the taking of initiatives in a widening radius of activities, both within and outside the home. The significant relations involve the entire family as distinguished from the parental figures alone. The child's activities are principally of a play character in which he or she assumes a variety of roles and purposes. The purposiveness of the initiatives it takes fosters a growing sense of competence that will blossom in the next stage. The child "can gradually develop a sense of moral responsibility, where he can gain some insight into the institutions, functions, and roles which will permit his responsible participation" (Erikson, 1963, p. 256). Failure in this stage allows a sense of guilt to establish a beachhead in the child's psyche. This inference about guilt would seem to flow more from Erikson's residual attachment to an Oedipal explanation of the play schemas than from failures in the child's initiatives *per se*.[20]

The *fourth* stage, which Erikson labeled "school age," bridges the period from the sixth year of life to puberty. Although the home is still important, new cognitive skills and social relationships develop in the child's school and neighborhood. The psychosocial crisis that is presented in these contexts is industry versus inferiority. Studies begin in earnest in a school setting. The fundamentals of technology are mastered. This involves the pleasure of diligent application to challenging age-appropriate tasks. The child will acquire the mental discipline that future studies will require, and the foundational skills for the sophisticated subject matter that will follow in adolescence. The child also addresses in a ludic spirit the mastery of the

[20] Ausubel, Sullivan, and Ives (1980) have stated that "Erikson's developmental stages still emphasize Freudian notions of character and psychosexual development and other products of armchair speculation unsupported by any naturalistic development data whatsoever" (p. 171).

neighborhood games that are homologues of the complex social interactions in which he or she will engage as an adult. When the child succeeds in the tasks and skill acquisitions that this stage presents, the mental set of "industry" is locked into place. When there is broad-based failure in this period, the child begins to feel inferior and then consistently acts that way.

The *fifth* stage, which Erikson labeled "adolescence," is one in which childhood comes to an end. With the onset of advanced schooling and the physiological turmoil and growth that they experience, adolescents address two enormously serious questions. First, who do I want to become careerwise, and do I have the aptitudes and human resources to get there? Second, how do I appear to others vis-à-vis how I experience myself? Closely related is the question, "What kind of a person do I want to become?" The psychosocial crisis that is precipitated here is that of "identity" versus "identity confusion." This is a crisis that is not easily resolved, can drag on for 10 to 20 years, and requires the fighting of many of the same battles over and over again. It can be seriously complicated by doubts about one's sexual identity. "The adolescent mind is essentially a mind of the *moratorium*" (p. 262) in which one is ever questioning, and commitment to an ideology is suspended. The basic strength that emerges from this stage is loyalty to family, friends, group, and, ultimately, a belief system. The characteristic failures of this stage are social pathology and cynicism.

Erikson called the *sixth* stage "young adulthood." Having successfully engaged in the search for identity and sensing clarity on that issue, young adults are normally "eager and willing to fuse [their] identity with that of others" (p. 263). They are prepared to make commitments to others, to engage in a relationship of trust, and to remain loyal, even at considerable self-sacrifice. The basic strength to emerge from this stage is the ability to engage in a mature love, love in its manifold manifestations involving not only genital commerce, but the nurturance and care of others. A failure in this stage is characterized by exclusivity and isolation, not only from an intimate other, but from the rest of humanity. This is often indicated by ritual expressions of *elitism*. Erikson acknowledged that psychoanalysis went too far "in its emphasis on genitality as a universal cure for society and has thus provided a new addiction and a new commodity for many who wished to so interpret its teachings" (p. 266). He added that there are partnerships which amount to "an *isolation à deux*."

The *seventh* stage, labeled "adulthood," is predicated on the model of fully mature humans who are prepared to take responsibility for individuals they have procreated. Adults, however, need to be needed, and children furnish them with a multitude of occasions for satisfying this need. The psychosocial crisis that arises in this stage is "generativity versus stagnation." Those who flee from the responsibilities of proper childrearing (or the care of others for whom they have assumed some responsibility) stagnate. Their stagnation consists in a failure to further the worthy objectives of education and tradition, the consequence of which is the stultification of their own growth. The basic strength emanating from this stage is nurturance. The failure is rejectivity and a folding back on themselves as if they were their own only child.

The *eighth* stage, labeled "old age," involves the capture of order and meaning in some cosmic and spiritual sense. It is the full philosophical acceptance of the one and only life cycle that has been allotted to us. This stage does not involve doing so much as being, and requires coming to terms with the limited successes and defeats of the previous seven stages. Thus, the reality of the stage is intrapsychic rather than interactional. As the first seven stages of one's life are reviewed, there is a sense that the life lived has been a mitigated success more than a qualified failure. It represents a state of integrity that is imbued with quiescence and self-satisfaction, peaceful resignation, and the will to stamp *paid* to the multitudinous claims that life has made. The prospect of dying is not welcome, but not terrible either.[21] Wisdom is the salient virtue in successfully transiting this stage, while true failure in this stage is characterized by bitterness and despair.

Status of this vision. The Eriksonian corpus of scholarship is highly significant for a number of reasons. Although all systems become dated with time, and never so rapidly as in the past generation, there is a certain robustness to the perspectives that Erikson generated, which assure them a continuing popularity. First, it transcended the simple "love and work" formula for life tasks that Freud had bequeathed to his acolytes. A richness of analysis beyond the dualism of eros and thanatos in which he was schooled characterizes all his work. Indeed, an instinctive drive toward death and destruction gets short shrift in his work. Even in

[21] Ivan Illich (1976), in his scholarly treatment of the role of medicine in modern times, recounts an episode in the life of the great French *cuisinier*, Brillat-Savarin. "Brillat-Savarin attended his 93-year-old great-aunt when she was dying. 'She had kept all her faculties and one would not have noticed her state but for her smaller appetite and her feeble voice. "Are you there nephew?" "Yes, aunt, I am at your service and I think it would be a good idea if you had some of this lovely old wine." "Give it to me my friend, liquids always go down." I made her swallow half a glass of my best wine. She perked up immediately and turning her once beautiful eyes towards me, she said, "Thank you for this last favour. If you ever get to my age you will see that death becomes as necessary as sleep." These were her last words and half an hour later she was asleep forever.'" This narrative is suffused with the peace and integrity that Erikson and others would wish for in the last stage of life.

his earlier work (1963), Erikson alludes to the concept of *thanatos* as Freud's "commitment to a mythology of primeval instincts" (p. 68). Second, it is a life-span approach to understanding human development, and although early experiences are important (as they are in all stage theories), change is possible at all ages and stages. His work, therefore, is optimistic rather than pessimistic and presents us with a psychology of health rather than of self-destruction. Third, he gave great significance to the conscious, reasoning determinants of human behavior. This allowed him to escape the hard determinism of blind libidinal forces. Fourth, although his schema reflects a worldview that is more Western than Asian, more individualist and identity-focused than collectivist, less tradition bound than entrepreneurial, his work among the Amerindians reveals a cultural sensitivity that has only recently come front, center, and vociferous in the work of social psychologists and psychotherapists. Fifth, his work in psychohistory and psychobiography reveals the wide-ranging character of an intellect that tried to transcend the limitations of boxed-in modernity. Finally, he accorded importance, like Alfred Adler and other teleologists, to the power of planning and the anticipation of achieving well-thought-out goals. Erikson realized that people are guided not only by their memories of the past but also by what Damasio calls our memories of the future, the cherished schemas that we labor to realize.

Other stage models

Models abound that try to make sense of personality development in terms of stages across the life course. Major models that paralleled the psychoanalytic are those of G. Stanley Hall (see White, 1992), James Mark Baldwin (to begin, see Broughton and Freeman-Moir, 1982; Cairns, 1994), Carl Gustav Jung (see Ellenberger, 1970), and, of more recent vintage, Harry Stack Sullivan (begin with Sullivan, 1953), Jane Loevinger (1976), and Ken Wilber (1986), the last of whom attempts a cross-cultural synthesis of Asian and Western stage theories.

Jean Piaget, the Swiss cognitive psychologist – and another stage theorist – set the standard for research in his field for over two generations. We are passing over his ground-breaking studies at this point for two reasons: first, although his work is related to personality development in the sense defined for treatment in this text, it is narrowly focused on how the child traverses a number of cognitive developmental stages in arriving at adulthood; and second, a keen student can study him extensively in courses in cognitive psychology. We will have occasion, however, to examine Piaget's studies, related as they are to the work of J. M. Baldwin in genetic epistemology and Kohlberg's model of moral judgment. These matters are treated in the context of chapters dealing with the *self* (Chapter 6) and with *motives* and *emotions* (Chapter 9).

Critical periods. We cannot leave the subject of stages of development without referring to the controversial issue of "*critical periods*" (referred to of late as *sensitive periods*) and the contributions that embryologists, child psychologists, and ethologists have made to stage theory. The critical period is a unique phase in the ontogeny of members of a species in which they must experience (or be spared) some exogenous (environmental) pressures in order for normal development to take place. For example, in the embryological and fetal stages of development, human and infra-human, severe emotional distress in the mother can partially arrest neurological development. In short, the adrenocortical (stress) hormones that the mother produces as a reaction to severe threat to her well being cross the placental barrier. These can be toxic to the embryo, as it does not have a sufficiently evolved liver to metabolize such steroids. If the trauma to the mother is severe enough, even the unborn infant's skeleto-muscular morphology can be compromised. Research done on guinea pigs by comparative psychologists indicates that the formation of the dental arch, heart valves, and optical system, for example, can be temporarily arrested or otherwise compromised by the injection of maternal stress hormones. The particular organs or subsystems most at risk depend on which ones are in their most rapid phase of development at the time. When the stressor is turned off, the organ development resumes at the point where it *would have been* had it proceeded normally. It appears that the organism cannot fully repair itself.

Imprinting: a special case. Stressful human pregnancies entail for the infant multiple post-natal socio-psychological sequelae as well as physical liabilities. The "anxious" fetus is born an anxious infant, but how enduring such effects are is still uncertain. In the view of developmentalists sensitive, if not critical, periods for human development exist pre-natally as well as post-natally. *Imprinting* is one developmental feature that demonstrates the importance of critical periods, a phenomenon noted in the sixteenth century by Saint Thomas More ([1516] 1989) in his treatise, *Utopia*. Imprinting has historically been thought to involve the persistent pursuit by the young of many species of the first moving object they perceive (for example, ravens, bitterns, goats, sheep, etc.) during early and quite specific days, weeks, even months of life. This results in psychological bonding to that object – and later attachment *to the species* represented by that object. This temporal window of opportunity (or vulnerability) is different for each species and, of course, is subject to some individual variation. The pursuit behavior targets *any moving object*, not necessarily the mother, during a critical period of development. This phenomenon (described in surprising detail by More) is most closely associated in modern psychology with the work of ethologists (for example, Lorenz, 1952, 1960; Tinbergen, 1939) who have studied the social behavior of animals in their natural habitat. Experimental comparative psychologists, who work with guinea pigs, rhesus monkeys, and other species under controlled conditions rather than in field settings (for example, Harlow 1958; Harlow and Harlow 1969; Hess 1959), as well as researchers in human development (for example, Ainsworth, Blehar, Waters, and Wall, 1978; Bowlby 1946, 1969, 1973; Nash, 1978, pp. 121–53), have studied this phenomenon extensively. If one assumes that there is no radical developmental discontinuity between humans and the species of vertebrates immediately below us on the phylogenetic scale, one can plausibly speculate that there are homologous mechanisms enabling this kind of attachment in humans.

Separating the nanny goat from her kid even for a few hours after birth may compromise the ability of the mother to label and later recognize her kid as her own. Separating a lamb from the ewe and hand nursing it for a period of weeks causes the lamb to become imprinted to a human and presumably to identify itself with the human species rather than its own. There is apparently a maturational, physiological substrate such that, if the proper environmental stimuli, often called critical or salient events, do not impinge on an organism within a specific time frame, the programmed bonding or skill acquisition for which the species is hard-wired will not occur. An example of this is the acquisition of linguistic skills between the ages of two and fourteen (Lenneberg, 1967; *cf.* Snow, 1987) – that is, the ability to pronounce phonemes of one's mother tongue, properly accented, as well as learn its grammar, syntax, and subtle inflections.[22] A declining ability to learn human language after the age of 5 or 6 makes it imperative to diagnose hearing deficits before that age. As the neural basis for the acquisition of specific abilities (for example, vision) develops, appropriate stimulation is necessary or the neurological "space" will be

[22] Bornstein (1989, p. 188) points to one German youth who appears to be an exception to Lenneberg's hypothesis that the sensitive period for language acquisition ends at *approximately* 14 years of age. Individual differences in the exact onset and offset of any sensitive period will be normally distributed. To find one 17-year-old who is an outlier in such a distribution cannot be said to invalidate a strong form of such a hypothesis. Even in the less flexible timing of simpler organisms such as birds, one cannot expect these critical periods to be universally in evidence. Needless to say, finding exceptions to the "rule" does not invalidate the findings on which rules are founded. That some children are relatively resilient should not lessen concern for the many who are not.

co-opted into the service of some other sense modality or skill. Severe visual deficits are an example of this.

As there is more developmental plasticity in humans than in simpler organisms, some developmentalists have preferred to call critical periods *sensitive periods* when speaking of humans (Pervin, 1996, pp. 169–70). This labeling emphasizes the resilience of humans in recovering from deprivation effects, the flexibility of the temporal boundaries of the period, and the gradual character of transitions from sensitive to post-sensitive periods. But whether one calls a particular stage critical or sensitive, or even if one sees a "period" at all, would depend on the skill being acquired and the particular level of central nervous system (CNS) functioning that is involved. Where this issue becomes seriously problematical is in the domain of personality development. The question has been asked: are there any kinds of social deprivation in infancy which, if prolonged enough or traumatic enough, can result in character disorder and long-term susceptibility to emotional problems? It would appear that there is. Although the term imprinting has been abandoned in the literature on the earliest human attachment behavior, an analogous process is called, appropriately enough, "infant attachment."

Harlow demonstrated that infant monkeys that have been deprived of their mother (or the sustained and nurturant attention of another caregiver) during their earliest periods of socialization suffer catastrophic emotional, sexual, cognitive, and social disorders. The nonimprinted monkey becomes dysfunctional and a subsequent social pariah. Bowlby (1973) and others (see Nash, 1978, pp. 153–57) have speculated that "nonimprinted" human infants reveal comparable deficits when there has been an abrupt and unreplaced loss of the caregivers during their primary socialization. One thinks of children who have been orphaned during the first year of life. These psychologists have asked: "is it possible that the psychopathic personality, a disorder so refractory to therapy, is rooted in the failure to have identified in a profound and tacit way with the human species, within the time frame that the neural substrate was primed for this?" They speculated that the sensitive period for this basic attachment to occur was between the appearance of the first social smile in the early weeks of post-natal life and the appearance of stranger anxiety about seven months later. The loss or the deprivation of a caregiver in this period entailed a proportionate impairment of the socialization of the individual. This view links the phenomenon of loss in humans with that of loss in other primates, and roots both in biology.

There are many conceptual and methodological problems with this theory, but a softer-edged version, which has found wide acceptance, lies in attachment theory as proposed by Mary Ainsworth *et al.* (1978).

They studied infant attachment to the primary caregiver (we'll call this person the mother, though it can be anyone who assumes primary responsibility for the care and nurturance of the infant), as evidenced by its behavior when separated from the mother in a strange locale. The child appears secure, comfortable, carefree, exploratory, and tranquil, even in such a situation, as long as it is aware of the presence of the mother. Ainsworth classified the character of the child's attachment on the basis of varied infant responses when the mother is removed from the child's presence and replaced by a strange person. Normally, the well-attached child does not cling to the mother, but interacts freely with its environment. On the other hand, when the mother is replaced in a strange locale by a stranger, the child becomes somewhat distressed. When the mother returns, the child is immediately comforted and resumes its usual composure and behaviors.

In earlier studies sponsored by the World Health Organization, Ainsworth (1962, see pp. 153–54 for her conclusions) found that severe deprivation in the first three years of life, as well as the abrupt (often tragic) rupture of the bonds that link a child to its mother can result in long-term personality and cognitive impairment. Such traumas can result from the death of the mother, institutionalization of an orphan, placement in a succession of foster homes, or negligent or brutal treatment by family members (Bowlby, 1973). The findings are complicated by such variables as nutritional deficits, the maturational level of the infant at the time of a trauma, the length and severity of the deprivation experience, the potentiation and compounding effects of several destructive experiences, the timeliness and effectiveness of the succorance and therapy supplied to the victim, and coincidental compensating experiences. Severely damaging effects on personality development are difficult to reverse by later therapeutic programs. In this matter, as in all others, there are individual differences in vulnerability to pathogenic experiences (for example, Garmezy, 1983). It is in this context that another developmental principle must be called into play: neoteny.

Neoteny. The biological evolution of the human species has afforded us a number of saving graces, few more notable on a macro-level than *neoteny*. Neoteny refers to a number of dynamics bearing on the prolongation of childhood in humans. Developmentalists have argued that the retention into human adulthood of characteristics of earlier (juvenile) developmental stages, not only of humans but also of our evolutionary forebears, is related to the prolonged and flexible development of children. Although the influence of childhood experiences, both positive and negative, on one's adult personality is not in dispute, the partial reversibility of these effects is asserted by Michael Rutter, Jerome Kagan, and others. The variables at issue, for example, bonding in infancy to a primary caregiver or stimulus deprivation in toddlerhood, are, after all, continuous and complex. The strength and duration of the variables, and their interaction with genetic variables of an idiographic, and often unknown, character produce personality effects that are not predictable with any accuracy. There are advantages that this evolutionary advance provides.

First, humans retain the physical features of childhood well into adulthood, remaining in some sense permanent children. "In neoteny," Stephen Jay Gould (1981) wrote, "rates of development slow down and juvenile stages of ancestors become the adult features of descendants" (p. 333). Second, relative to human life expectancy this juvenile period is relatively long. Although neurological losses are always occurring even in childhood, neurological development is never complete, reaching relative maturity only after 20 years or more. Third, humans retain their neural plasticity throughout life, although the central nervous system degrades in significant degrees in the latter decades of life. Fourth, flexibility is a hallmark of neoteny, as is the enjoyment of problem-solving and the capacity for play. "In other mammals," Gould continues, "exploration, play, and flexibility of behavior are qualities of juveniles, only rarely of adults. We retain not only the anatomical stamp of childhood, but its mental flexibility as well" (p. 333).[23]

Childhood traumas leave scars and diminish the juvenile, but the *partial* reversibility of these conditions, be it marasmus, hospitalism, sexual abuse, or other insult, is not disputed. The resilience of the child, the depth of the damage, and its interaction with genetic variables of an idiographic character produce personality effects that are not wholly predictable. Neoteny provides backup for the families of children who have been harmed. A long childhood provides reparative potential. It is a period in which earlier psychological deformity can be mitigated, if not entirely undone. "In the wake of a traumatic infancy and childhood, prolonged immaturity provides humans with the affordances of redemption" (Dumont, 2000, p. 2090).

Other life-span perspectives

Carl Gustav Jung. Although most models of human development that have been proposed by researchers and theoreticians in the

[23] A cautionary note: S. J. Gould was a paleontologist, not a developmental psychologist.

past two generations must be passed over in this chapter, I would be remiss if I did not direct the student's attention to certain other theories that have been especially influential in personality psychology. Carl Gustav Jung, one of the great system builders of the twentieth century, developed a stage theory of personality development that is truly life span in character. The process of self-realization continues through the entire life course and culminates in the achievement of full *individuation*. This concept is quite different from the Freudian schema in which development is (conceptually) terminated at the end of the final psychosexual stage. As with other psychodynamic theorists, Jung developed a cartography of that domain of human mentation that is unconscious and that has received attention from a community of scholars in the neurosciences (for example, Damasio, 1999; Eysenck, 1967). The infant enters life with an undifferentiated unconscious and slowly develops a sense of personal agency/self that becomes increasingly conscious of itself and its relation to others. The individuality of the child emerges from the cocoon of the family, a transition that is usually memorialized by rites of passage.

Metamorphoses follow one another as the child passes into adulthood and assumes the civic and familial responsibilities inherent to that state. The major transition is from the youthful adult to the mature, middle-aged adult (in the Jungian paradigm, situated between the thirty-second and the thirty-eighth year of life). At this "turning of life," the deeper, unaddressed questions about identity, duty, life-meaning, and spiritual needs surface, torment, and press for attention and resolution. Men and women in this passage turn increasingly inward toward their unconscious, integrating such archetypes as the *animus* and the *anima* – complementary images of man and woman, respectively. Those who evade the questions pursue the pseudo-youth of many aging people in Western cultures. Those who accept the challenge repeat "shamans' journeys through the land of the spirits," and make the tortuous pilgrimage epitomized in the *Gilgamesh epic*, Homer's *Odyssey*, Virgil's *Aeneid*, and Dante's *Divine Comedy* (consult, for example, Ellenberger, 1970, pp. 707–13).

These Jungian principles appear to have percolated into the thought patterns (and been evidenced in the research) of later developmentalists, as we find parallel notions in the work of Bernice Neugarten (1977), George Vaillant (1977), and Daniel Levinson (1978, 1986). Vaillant participated in the Grant Study of Harvard University, which, beginning in the late 1930s, tracked a cohort of (White, middle-class) male college students into their sixties. He noted that at about the age of 40 these men tended to become "explorers of the world within" (p. 220). Neugarten, who participated at the University of Chicago in the much briefer Kansas City Study of Adult Life in the 1950s and 1960s (see, for example, Havighurst, 1973), noted the same tendency and characterized this "turning of life" as one of "interiority." The researchers in this project found evidence that developmental changes could be attributed in varying degrees to constitutional (nature-bound) and environmental (nurture-bound) conditions. Levinson echoed the theme by affirming middle-aged persons' need to reassess values and to integrate the feminine and masculine facets (read *anima* and *animus*, respectively) of their psyches. By this period of history, the arrested development model of classical psychoanalysis had been largely abandoned, and the research community had reached a consensus that not only did people sense that they were becoming "different persons" over time, but that this was a normative principle of development that terminated only at the end of life.

It should be noted that these are all stage theories, which bespeaks the human need to categorize and to impose order on the multitudinous arrays of data with which researchers flood us as they track individuals' progress over individual life histories and through longitudinal studies. Our penchant for chunking the life course into memorable stages overrides consideration of the vividness or faintness, or even, some would argue, the imaginary quality of these purported stages. Plato had long ago argued that in our studies of the world we must carve nature at the joints. We can suppose this need asserts itself even if the joints are wholly speculative.

Some of the more influential developmental theories are not stage theories at all. Examples include the ideas of John Broadus Watson (1925), Burrhus F. Skinner (1938), and Albert Bandura (1969), all of behaviorist inspiration. The researchers in the social learning tradition epitomized by Albert Bandura (for example, 1969, 1986) have argued that there are no "natural" stages in the human life course. The qualitative changes in human activity that one observes over the life course are at most a result of lifestyle fashions and fluctuating economic patterns of activity in various parts of the world. The stages observed in North America in the 1950s or the early years of the twentieth century are not only different from each other but different from those currently found in the "stone-age" societies of Papua-New Guinea. Two considerations flow from this. Briefly put, such life stages are not founded on a firm physiological understanding of the nature of the Human but on predominantly fluctuating and random variations in social conditions that can take myriad forms. Second, such stages are, therefore, less a "matter for the science of personality" than for sociology and cultural anthropology (*cf.* Featherman, 1983).

Aging

Although there has been strenuous resistance to the widespread conviction that human plasticity and adaptability suffer an inexorable decline as humans move into old age (for example, Mahoney, 1991 refers to this view as "a presumptuous insult to the life span potential for human psychological development," p. 12), the consensus among the world's foremost developmentalists is that there are serious and unavoidable losses particularly as one moves beyond the age of 80. Further, despite the ability of those who are in their sixties and seventies (the young-old, so to speak) to compensate for certain losses by proportionate gains enabled by better nutrition, exercise, culture-supplied "prostheses" and activities, and the intellectual stimulation that can be fostered by social expectations and opportunities (Birren and Schaie, 1996), the field of gerontology does not see clearly into what some (for example, Baltes, 1997) call "the "fourth age" (p. 375; the stage of the old-old). The architecture of the life course is, in fact, incomplete. We do not know what support and compensatory social prostheses culture can provide to ensure that losses are compensated by gains. This is the challenge that awaits society.

Humans endure losses in old age in large part because the evolutionary pressures that might have eliminated them do not appear until the reproductive phase of the life course is over. Consequently, plasticity and biological potential decline after the stage in which the human genome is transmitted to one's offspring. Baltes argues that for the growth potential of humans to be extended into the further reaches of the life course, we will need to rely on culture-based resources. One may argue that altering the human genome for these purposes is conceivable, but that option does not appear to be on the immediate horizon. Because of the age-related decline in biological potential, the remedies that society can offer as compensatory support become increasingly inefficient and, often, ultimately futile.

Baltes (1997) and his colleagues have developed a metatheory of human adaptation that in reality can be applied to all phases of human ontogenesis. The model, in brief, consists of the mechanisms of selection, optimization, and compensation (SOC). Emblematic of this pattern is his account of an interview with the concert pianist Arthur Rubinstein, then an 80-year-old:

> [When he] was asked in a television interview how he managed to maintain such a high level of expert piano playing, he hinted at the coordination of three strategies. First, Rubinstein said that he played fewer pieces (selection); second, he indicated that he now practiced these pieces more often (optimization); and third, he suggested that to counteract his loss in mechanical speed, he now used a kind of

impression management, such as introducing slower play before fast segments, so as to make the latter appear faster (compensation). Using such a SOC-related strategy and paraphrasing a quotation from Hesiod, "Half can be more than the whole." (p. 371)

As evolution and biology have not been, in Baltes' words, good friends of old age, it remains to medical engineering and other technological innovations, as well as socio-psychological and symbolic devices, to compensate for the inevitable decline in biological potential. Humans have been quite innovative in inventing cultural devices to compensate for biological deficits. Consider the heating and cooling systems we have developed, as well as various kinds of clothing to remedy our thermoregulatory deficits. The most effective prostheses for humans with dwindling capacity are, of course, other human beings who extend to the disabled the distributed cognition of the community, the neighborhood, and the family, as well as their physical support (Salomon, 1993).

Pure gain is not a realistic premise for conceptualizing human development. Any paradigm that does not model a complex pattern of losses as well as gains at all ages is manifestly false. The ratio of losses to gains increases as one passes the reproductive stage of life (Baltes and Graf, 1996). There are no simple, rectilinear pathways to development. On the contrary, the systemic and interactive nature of the complex biological and socio-psychological matrix in which each human is embedded requires an analysis that is multi-causal, multi-dimensional, multi-directional, and multi-functional (Baltes, 1997). As age-related losses begin to multiply and cascade into this complex mix, it remains for the ingenuity of humans to bring the gains into balance with these losses.

We close this chapter with the words of this distinguished psychologist:

The most pressing challenge for human developmentalists of the next century is to search for the conditions required to complete the biological and cultural architecture of the life span ... In this spirit, we need to keep in mind that the future is not something we simply enter; the future is also something we help create. (Baltes, 1997, p. 378)

4 The biology of personality

> Men ought to know that from nothing else but the brain come joys, delights, laughter and sports, and sorrows, griefs, despondency, and lamentations. And by this, in a special manner, we acquire wisdom and knowledge, and we see and hear and know what are foul and what are fair, what are bad and what are good, what are sweet and what are unsavory ... And by the same organ we become mad and delirious, and fears and terrors assail us ... all things we endure from the brain when it is not healthy ... In these ways I am of the opinion that the brain exercises the greatest power in the man. Hippocrates (fifth century BCE)

> We may say that personality is the central concept in our theory. The distal cause is DNA, the proximal cause is the physiological, hormonal, neurological set of intermediaries linking DNA to behaviour, and interacting with environmental factors. Hans J. Eysenck (1993)

Dismantling and conflating a dichotomy

Explanations of the construct personality that revolve about the relative contributions of "nature and nurture" to its development have served to polarize theoreticians and researchers in their approach to their science. This dichotomy has proved to be more mischievous than helpful for advancing our understanding of personality. Although there is broad consensus today that one cannot explain human behavior and personality without examining the interaction of genetic endowment and the geosocial environment in which all people are enmeshed, these two perspectives on human development are treated unevenly, and an examination of current textbooks on personality reveals that the preponderance of space is devoted to issues of nurture rather than nature. There are many reasons for this.

First, and most obvious, is the momentum of tradition. Personality psychologists have historically been grounded far more in the social sciences and mental health professions than in the basic sciences, although there has been a general shift in this tradition over the past generation. Second, reliable clinical and sociological data have historically been more accessible than have psychoneurological, physiological, or paleological data. Research tools have traditionally dictated the kind of research questions undertaken simply because scientists attempt to answer questions that current technology enables them to answer. New techniques of carbon dating in paleological studies and various tools of neuroscience are enabling a flood of research into determinants of human behavior that have not heretofore been possible. Third, political factors always influence the description and choice of research problems as well as funding for that research. Both psychologists and politicians will be attracted to the study and funding of those problems that are perceived to be more easily remediable by social than by biological engineering. Likewise ideological biases favoring environmental rather than genetic *explanations* of individual (as well as group) differences result in a premium being placed on discovering personality determinants that we can control. Fourth, our value system pervasively influences the character of the research we conduct. Democratic ideals of human equality suggest that with a properly structured environment individual humans, and all ethnic groups, can achieve an approximate equality with each other. John Broadus Watson and the behaviorist movement in the United States that he, and later Burrhus F. Skinner, epitomized promulgated this view (for example, Nash, 1978, p. 43). Western society, which values individuality, on the one hand, and mastery of one's environment (and consequently one's future), on the other hand, finds fluid environmental explanations of personality more appealing than fixed biological ones. *Biology as destiny* places a crippling constraint on our ability to make and remake ourselves as we wish. Even in science, we look for (and often find) what we prefer to find.

The good news, for those who share these views, is that evidence has poured in from all quarters that environmental factors are extremely important in shaping personality (Güntürkün, 2006; Hebb, 1949; Plomin and Caspi, 1999; Sulloway, 1996). Research has rarely demonstrated that, for any population, genes account for more than 50 percent of the variance in one or another personality trait. (The amount of variance attributable to epigenetic variables, considered below, has not to date been confidently assessed. The admixture of gene–expressive environmental events into this assessment greatly complicates the matter.) For example, birth order has been repeatedly demonstrated to shape, to some extent, individuals' attitudes and social values – and thus their behaviors. The peer groups and institutions in which individuals become cocooned, as well as the social circumstances into which they happen to drift, can alter behavior patterns. The question as to whether the drift into special social niches is genetically driven has yet to be resolved. Fortunately, barring serious pathology, there is plasticity in the human organism that supports both preventive and remedial measures for optimizing the development and alteration of personality.

Nature and nurture are not two discrete and disjoint constructs. The realities that nature and nurture refer to are interactive. Some of the forms that nurture assumes are themselves determined by a common human genetic inheritance so that, in a systemic perspective, to be nurtured is to be shaped by nature, at least in part. Some would argue, for example, that nature imposes constraints on the development by humans of their social environments. To cite one example, although the sex-differentiated division of labor that exists in practically all societies is socially constructed, biology has determined its parameters. Species-typical behaviors suggest the existence of genetic determinants of other such behaviors. That such traits as aggressiveness, nurturance, jealousy, impulsivity, or fearfulness appear in all societies and all historical periods suggests that there is a physiological and genetic substrate that enables their development. This argument is all the more compelling in that such traits are also found in the behavior of the great apes and other primates.

This chapter will examine various aspects of the biology of personality. First, I will briefly review evidence for the existence of a physiological substrate for temperament. Second, we will trace the development of an evolutionary psychology and the lessons it provides for the personality psychologist. Third, we will engage the issue of behavioral genetics and the most widely used methodology for exploring it. Finally we will briefly discuss the contribution of molecular genetic analysis to our understanding of personality.[1]

[1] Individualism, cloning, and environmentalism:

> That the psychologist as scientist has, in the past, inclined to a nomothetic and deterministic understanding of human beings is understandable. After all, it is difficult to build a science on the idiographic. Nevertheless, the concern of rank and file persons not to see themselves simply as one of billions of "knock-off's" from a universal template is also understandable. This may underlie the concern expressed by some about cloning. "If there are clones about, how will you know who people really are," ask some. Though there is a misconception about the phenomenon of cloning that underlies that question (after all, identical twins meet the definition of clones, and we come to know *who they are*), it indicates the concern about personal identity found in Western society. More specifically, the need to be an individual, and to be different (but not too different) from everyone else, is evident in many Western and Westernized cultures. This plays out in the theorizing of personologists who have developed personality theories that emphasize those environmental factors that impinge on individuals in all phases of their lives, and that individuate them. It is this diversity in personal histories that accounts to a large extent for the uniqueness of each person. That members of the same family who have similar histories still mature to adulthood with very different personality profiles attests to the interaction of genetic factors with education, not excluding stochastic events that may have powerful formative impact on development. (Dumont, 2000, p. 2084)

Human encephalization

The human mind evolved from the proliferating mechanisms of the human body, not least of which is the sub-set we refer to as the central nervous system (CNS). Many of these mechanisms, of course, are common to simpler vertebrates. There is a growing consensus on the gross anatomy of the human brain. The symmetrically bilateral development of advanced organisms allowed them to move themselves in a consistent way, placing, say, one foot before the other. The end of the body that leads its movement through space became the head (Polanyi, 1958). The evolution of the head, in a process that is labeled *encephalization*, eventuated in specialized mechanisms imbedded in a *hindbrain* (evolutionarily the most ancient), a *midbrain* (which evolved into the limbic system), and a *forebrain* (evolutionarily the most recent). The forebrain, referred to as the *neocortex*, is itself divided into two hemispheres, so that in reality humans can be considered to be in possession of four distinct brains, whose integration cannot yet be regarded as perfect (*cf.* Mahoney, 1991, pp. 423–35).

The infrastructure of the life support system, a *neural chassis* consisting of the spinal cord, brain stem, and hind brain, enables pulmonary, cardiovascular, digestive, and neuro-endocrinological functions, not to mention certain psychomotor and self-regulatory activities. This most ancient of the four brains is frequently referred to as the reptilian brain, for humans share this structure with reptiles, land-dwelling vertebrates in whom it evolved about 280,000,000 years ago. The personological functions of this brain (layered in through progressive encephalization) are the capacity for ritual behaviors (as in courtship), mobility and niche-seeking, territoriality, and aggressive and defensive patterns involved in predation and self-preservation. A segment of the hindbrain, the *cerebellum* (literally, "small brain") coordinates the complex psychomotor behaviors in which one engages when walking, playing the piano, or playing a computer game.

The second brain, the so-called *paleomammalian* brain,[2] encompasses the limbic system,[3] which is complexly developed in mammals. Much of human joy and sense of beauty, sorrow, and terror – the rich emotional tide of life – flows from this system, which is tightly enmeshed in humans

[2] *Paleo-*, of Greek origin, meaning old, is combined with other terms as in *paleomammalian* brain, thus referring to the primitive, "old mammalian" brain.

[3] The limbic system, which includes the hippocampi, the amygdalae, and other subsystems, was so designated by Paul Maclean. It has long been identified with the paleomammalian brain and presumed to comprise the components that govern emotional expression. Recent advances in neuroscience indicate so many disparate functions coexist in the limbic system that some researchers question whether the notion of a unitary limbic system is still useful (for example, LeDoux, 1996, pp. 98–102, 186).

with the cortex as well as with the reptilian complex. Psychotropic drugs, which humans use to reduce their terror or to heighten their pleasure, directly affect limbic receptors. The olfactory capabilities of the more primitive organisms serve to heighten the emotional life of the more advanced – all in the service of a more complex and adaptive interaction with a changing environment. Motivational complexity as well as the regulation of sensitivities, intense love of one's offspring, adaptive registration of fear, sorrow at the loss of those we most treasure, and the several emotions, such as jealousy, rage, terror, altruism, melancholy, and exhilaration, have their source in the limbic system. These emotions have adaptive functions in mammals. And though most of them certainly exist in all mammals with greater or lesser depth and sensitivity, it would be a mistake to assume that humans and even other primates experience them identically.

The evolution of the limbic system supported parenting functions, including vigilance for the safety of one's offspring allied to fierce protectiveness in the face of perceived threats. This human trait, also found in many other animals, is among the most important features of evolutionary biology. The willingness of parents to sacrifice their physical well-being and even their lives for the survival of their offspring, and the latter's reciprocal devotion to their parents, other kin, and even their remote forebears has inspired theater, cinema, and a wealth of art, song, and epic mythology. Rich emotional experiences are enabled by the limbic system. Intimacy, familial love and piety, and florid, diverse, and emotionally primed socialization patterns have taken root and flourished in that matrix.

The cortex, the uppermost "husk" of the brain, which emerges from, and shrouds, the older "brains" with which it is integrated, forms, in distinct hemispheres, the two cerebral portions of the central nervous system that support logic, abstract thinking, planning, creativity, and complex intentionality. But the human cortex, three times larger than one would predict if extrapolating from the ratio of cortex to body mass in other primates, is tightly interconnected with the (also) disproportionately large cerebellum. Granting that the cortex controls many motoric as well as ideational functions and the cerebellum many non-motoric, learning, and timing functions, their interdependence supports the insight of Pierre Janet that all ideation involves some motility. Thus, Mahoney (1991) stated that "movement and mentation are inseparable expressions of the same active, operational knowing processes" (p. 431). It can be reasonably assumed that as the limbic system also enables certain cognitive functions (as with the memory functions of the hippocampus), as well as profoundly influencing our cortical activities, reason can be regarded as the most rarefied and abstract of our emotions. Reason is a motile force that interacts powerfully with the various organs of the body.[4]

The complexity of the human body has over the course of its long developmental history ensured that humans can express themselves via numerous options afforded by the interplay of mind and environment. The reductionist schemas of some early twentieth-century personality theories that postulated only two or three motivational structures (for example, eros and thanatos) are clearly incompatible with evolutionary psychology. Buss (1999) states that "human behavioral flexibility comes not from having highly general psychological mechanisms, but rather from having a large number of specific psychological mechanisms that are activated and concatenated in varying complex sequences ... These implications render the study of human nature a difficult and daunting task ..." (pp. 41–42; also see Mahoney, 1991, pp. 427–35).

Temperament and biology

Temperament is difficult to define as it encompasses a wide assortment of behaviors. It is a *stylistic attribute* of behavior that is associated with mood and emotions and is relatively stable. This very stability, independent of environmental variables, is what militated against a strictly environmentalist view of personality development in the eyes of Alexander Thomas and Stella Chess (1980). Although "most features of personality are moderately heritable" (Sulloway, 1996, p. 172), the prevailing position after the Second World War (at least in North America if not in Europe) was heavily biased toward environmentalism. But, the biological basis of temperament had long been postulated in Western tradition. The Ancient Greeks proposed that the body secreted four different humors (fluids) that determined an individual's temperament depending on their volume, balance, and proportions (see Chapter 5 on traits in which we discuss this in some detail). Alcmaeon of Croton, a disciple of Pythagoras, distinguished "understanding" from temperament and situated intelligence in the human brain. Based on the dissections (indeed, vivisections) of the human brain conducted by Alcmaeon, Erasistratos, and Herophilos, later Pythagoreans such as Hippocrates and Plato endorsed this idea (Burnet, 1920). Galen (*c.* 130–200 CE), a Hippocratean and brilliant

[4] Right hemispheric activity involves varied emotional expression that the left hemisphere does not. The left hemisphere is largely the stage for reasoning, reading, etc. (Carlson, 1998, pp. 478–96).

Greek scientist/physician, vigorously continued this Hellenic tradition[5] (Kagan, 1994).[6]

The predominance of one or another of the four humors was hypothesized to give rise to four different personality types: sanguine, melancholic, choleric, and phlegmatic. What is significant about this schema, aside from the fact that it has percolated into contemporary formulations by Immanuel Kant, Pierre Janet, Carl Gustav Jung, and Hans Jürgen Eysenck, is that it affirms a biological principle to account for the variety of temperaments and derivative emotions that we witness in humans (Clark and Watson, 1999). Modern research has further substantiated the somatic origin of temperament by demonstrating that these personality types are innate, and that, therefore, they must be genetically influenced. Although not uncontested, the conflicting views are alluded to in Chapter 5 on traits. The early work of Thomas, Chess, and Birch (1968) marked a modern departure in this direction. Thomas and Chess (1980) stated:

As parents we were struck [in the 1950s] by the clearly evident individual differences in our children, even in the first few weeks of life. As clinicians, we were repeatedly impressed by our inability to make a direct one-to-one correlation

[5] Tracing contemporary European science to the Ancient Greeks has been labeled a Eurocentric exercise. This is true, and it is self-evident that other civilizations such as the Chinese, Arabic, and Indian have generated, over thousands of years, their own science, rooted in their own historical cultures. Each civilization has developed an indigenous science springing from its own favored cosmology and the fundamental questions it has posed about the nature of the world we all inhabit. There has been substantial cross-semination of ideas over the centuries, especially with the proliferation of mercantile structures and communicational technologies. However, it is also true that the Ancient Greeks had a singularly important influence on Western science and philosophy.

[6] *Mental illness as humoral imbalance.* During the Middle Ages, mental illness, like other maladies, was attributed to a lack of balance in the four humors: blood, yellow bile, black bile, and phlegm. Each specific humor was associated with a particular personality type, and each humor was associated with a specific body part. Individuals dominated by that part had specific properties (for example, melancholics [black bile] were characterized as cold and dry). Physicians of the day diagnosed and recommended treatments for their patients based on presumed imbalances in these four humors (not unlike contemporary theories about the effects of "chemical imbalances" of neurotransmitters like dopamine and serotonin). For example, in the "General Prologue" to *The Canterbury Tales*, Chaucer (*c.* [1386] 1990) described a capable physician as one who understands his humors:

> He often kept a patient from the pall
> By horoscopes and magic natural.
> Well could he tell the fortune ascendent
> Within the houses for his sick patient.
> He knew the cause of every malady,
> Were it of hot or cold, of moist or dry,
> And where engendered, and of what humour.
> He was a perfect practicing physician (p. 281)

between environmental influences, such as parental practices, and a child's psychological development. There was no question, of course, that these environmental influences played an important role in the child's life, and we, like other clinicians, devoted much effort to trying to persuade parents and others to provide a healthier environment for children. However, we saw many, many instances ... in which a child's development pursued a consistently healthy direction, even into adult life, in the face of severe parental disturbance, family disorganization, and social stress. (p. 69)

These views were supported not only by their own research, but by a generation of developmentalists who followed them (see, for example, Eysenck, 1990; Plomin and Daniels, 1987).[7] In 1956 they embarked on a systematic study of individual differences in human development. Titled the *New York Longitudinal Study*, it permitted Thomas, Chess, and Birch to extract from their data nine categories of temperament describing the *how* of behavior as distinguished from the *what*. They found that although the peculiar manner in which temperamental traits express themselves can be shaped by environment (pre- and post-natal), a large part of the variance in behavior is under genetic control. In brief, one is born with a particular temperament (which we now understand can be epigenetically modified) (Baltes, Reuter-Lorenz, and Rösler, 2006).

Thomas, Chess, and Birch (1968) postulated the following nine dimensions of temperament: (1) *activity level*; (2) *intensity of reaction* to stimuli; (3) *sensory threshold* to stimuli; (4) *approach/withdrawal* relative to the unfamiliar; (5) *adaptability* to novel situations; (6) *rhythmicity* or regularity in biological functions (as in sleep); (7) *quality of mood*; (8) *distractibility*; and (9) *attention span and persistence*. Some of these categories are highly related and, indeed, overlap. For example, (2) and (3) can intuitively be collapsed into one category as can (8) and (9). Later research has, in fact, reduced the number of dimensions to seven or even five (Martin, Wisenbaker, and Huttunen, 1994). Whether one finds fewer or more than nine categories of temperament depends on how microscopic or macroscopic one wishes to make one's perspective. Arguably, one's objectives and methodological preferences for establishing the number of trait factors will determine whether one finds, say, five as in the Big Five Model, or sixteen as in Cattell's (16PF) model (see Chapter 5, below). This work relates to that of Arnold Buss and Robert Plomin (1975), renowned for their bio-evolutionary approach to personality psychology, who developed the EASI (*E*motionality, *A*ctivity, *S*ociability, and *I*mpulsivity) Temperament Survey, which has been widely used for research purposes (Clark and Watson, 1999).

[7] Zuckerman (1987) stated: "All parents are environmentalists until they have their second child" (p. 42).

Comparative psychologists have demonstrated the (lower-order) temperamental analogues of aggressiveness, fearfulness, vigilance, excitability, gregariousness, and nervousness in animals, some strains of which have been bred to enhance certain of these traits (for example, Le Scolan, Hausberger, and Wolff, 1997; Suomi, Novak, and Well, 1996). The breeding of bulls, dogs, and horses is particularly instructive in this matter. Presuming neurophysiological continuity in the evolution of humans and other vertebrates, it makes sense to reason that human traits are comparably heritable. Draper (1995) draws a finer parallel by stating that canine and human personality similarities "have their origin in biogenetic factors stemming from common evolutionary sources and from canine breeding for human compatibility and assistance with human tasks" (p. 241). However, as I note elsewhere in this text there are epigenetic dynamics at the base of some non-genetic but heritable traits.

Temperament types

The two foundational dimensions of temperament according to Eysenck (Eysenck and Eysenck, 1975), referred to by other modern personality psychologists as *superfactors*, are extraversion–introversion and emotional stability–emotional lability. If we cross the two polar expressions of each of these dimensions with each other we get a matrix of four personological types: (1) extraversion and emotional stability (the sanguine); (2) extraversion with emotional lability (the choleric); (3) introversion and emotional stability (the phlegmatic); and (4) introversion and emotional lability (the melancholic). It can be argued that as these four temperament types can be found universally across cultures (multiculturally), among peoples thriving in the most diverse ecological niches, and across historical periods (polychronically), there are nomothetic principles at work here of a biogenetic character. It can be further argued that the social structures and conditions that promote the varied expressions of one or another temperament type are themselves a result of the neuroendocrinological potential encoded in our DNA – as well as random historical events. From this perspective, chance cannot account for the fact that all humans live in community – and are thus gregarious, political, and rule-governed. Of course, the particular forms that our gregariousness, political structures, and ethical codes assume are characterized by wide variability.

Environmental determinants of temperament. Abundant evidence exists that the *shared environment* of a stable home and a consistent pattern of childrearing exert little influence in shaping the temperament of the child (Beer, Arnold, and Loehlin, 1998; Eysenck, 1990b; see Goldsmith, Buss, and Lemery, 1997; Zuckerman, 1987). Although this is counter-intuitive and contradicts the "wisdom of the ages"[8] there is a good deal of evidence that *unshared* (and largely non-familial), environments account for virtually all of the non-genetic contributions to the formation of temperament (for example, Dunn and Plomin, 1990). Plomin and Caspi (1999) make the bold statement that research findings suggest "environmental influences that affect personality development operate to make children growing up in the same family no more similar than children growing up in different families" (p. 256). An explanatory theoretical framework for such findings has not been clearly developed. Children in the same family can lead very different, indeed separate, lives. The chance events, both constructive and harmful, that they variously experience in the different situations to which they may be genetically attracted can have epigenetic consequences as I point out below. Cultures (shared macro-environments) *have* been demonstrated to be an important variable in determining the divergent national temperaments of, say, the Swedes and the Japanese. Neapolitans generally have a temperament and character different from that of typical citizens of Bristol, Beijing, or Peoria, Illinois. Why a shared micro-environment, such as a familial culture, should have so little effect on individuals during the most formative years of their development has not been satisfactorily explained.

That "family culture" has a negligible influence on children's personality may be partially explained by birth order, which, argue some, differentially influences each child. Birth order effects may largely override and mask those enculturated effects that arise from an intimate, common social environment. The differing ages at which children in the same and different families are assessed can also dilute the evidence for the influence of a common culture. The Darwinian phenomenon of niche-seeking can contribute to the diversification of personality even among identical twins – differences that show up *especially* in a shared environment where sibling rivalry for parental investment becomes acute. This supports the hypothesis that innate but different temperamental features lead family members to seek different experiences both within and outside the family as I discuss in the section on shared and non-shared environments, below.

[8] As noted in Chapter 1, Plato argued that harsh discipline in the rearing of children would turn them into servile, spiritless, obsequious citizens unfit to assume their civic responsibilities.

The calculation of variance in personality variables results in considerable "measurement error" (rendered as the *error term* in the analysis) as it does in all social psychology research. As much as 15 to 30 percent of observed variance in any population sample is thus unattributable variance. To lump this error with variance due to the environment rather than heritability biases the conclusions we draw. Factoring out the genetic contribution to personality, we are left with only 20 to 35 percent of variance clearly due to environment. In the light of this consideration some researchers estimate that up to 70 percent of variance in the major superfactors is due to genetic determinants (Tellegen *et al.*, 1988; Riemann, Angleitner, and Strelau, 1997).

Shyness: an exemplar trait. Shyness is a component of introversion and "is one of the most heritable personality traits" (Sulloway, 1996, p. 174). As much as 60 percent of the variance in this trait may be directly attributable to genomic factors. Other biological factors, of course, enter the picture. Stressful pregnancies, during which the mother is traumatized, say, by a death in the family, the loss of family income, or grave marital conflict, can produce anxiety in the fetus, severe morphological anomalies, and impaired neurological development. This results from adrenocortical products and other stress hormones that enter the unborn child's bloodstream – hormones that, depending on the stage of embryonic development, are often inadequately metabolized. Such experiences can have long-term negative consequences and affect the personality development of the child.

Shy children are not just reserved, subdued, reticent, and withdrawn persons. They are *anxiously* so, and there are physiological markers for this trait. Shy persons have stable and elevated heart rates; indeed, children who reveal behavioral signs of shyness have shown higher than average pre-natal heart rates. While engaged in cognitive tasks they exhibit greater pupillary dilation (an effect, as well, of looking at pleasurable scenes).[9] They produce higher levels of salivary cortisol, have higher blood pressure, and evidence greater muscle tension in their vocal cords (Kagan and Reznick, 1986; Sulloway, 1996).

[9] The plant, Belladonna (literally *fair lady* in Italian), a species of Nightshade, has been used by courtesans since time immemorial as eye drops to enlarge pupil size. Enlarged pupils are regarded as a sign stimulus for sexual attraction (*cf.* Hess, 1965). When looking at a beautiful object, say a fine piece of jade, one's pupils dilate. A chemical found in Belladonna, atropine, is used by ophthalmologists to dilate pupils for purposes of conducting eye examinations.

The psychobiology of temperament

Many personologists are unsure as to whether personality types are simply mental constructs that humans devise to make sense of the world or "real" entities that map onto a neurological substrate and (even more reductionistically) onto the genome. As in all such dichotomous propositions, the answer is that both points are somewhat true. As long as scientists adopt a monistic rather than a dualistic model of the Human, they are obliged to postulate that *all behaviors* are influenced and partially determined by the CNS and the hormonal and visceral accessories to that system. A rejection of Cartesian dualism leads to the inevitable conclusion that there are *no* behaviors that are body-independent. Even behavior-shaping social environments have biological determinants, inasmuch as they are collectively created by intelligent, CNS-organized human beings. On the other hand, creating abstract, mathematical models of human behavior does not make such models into something that has more reality than a metaphor.[10] Their reality depends on the accuracy with which they can be mapped onto a causative (usually physiological) substrate – as well as the detail (or, contrariwise, the level of abstractness) with which they are described. In the final analysis, traits, as convenient lexical symbols of behavioral patterns, remain limited *analogs* of the neural mechanisms that influence them.[11]

Evolutionary psychology and personality

Ethological perspectives

When humans attempt to explain the origins of their behaviors they are prone to assess the influence of their social structures, the rich intellectual resources available to them, their powers of imagination, florid memories

[10] Building a model or labeling a construct does not confer reality on what it presumes to represent. Galileo polemicized against the position of a prominent and contemporary Jesuit astronomer, Orazio Grassi, and the virtuoso Danish star-gazer, Tycho Brahe, affirming that the comets they had studied were nothing more than will o' the wisps. "If their opinions and their voices have the power to call into existence the things they have considered and named," stated Galileo, "why then I beg them to do me the favor of considering and naming *gold* a lot of old hardware that I have about my house" (Sobel, 2000, p. 92).

[11] Clark and Watson (1999) discuss three factors that lead us to accept the psychobiological dimensions of temperament (p. 401). One of these factors is the growing consensus among researchers about the validity of a taxonomy of traits. They stated that "this development enabled researchers to focus more intensively on a relatively small number of consensually recognized traits, incorporating them into more complex and sophisticated conceptual schemes ..." (pp. 401–2). The very language of this supporting argument suggests "cognitive construction" more than "psychobiology."

of relevant experiences, and many other factors that are distinctively human. But when they try to explain the same pattern of behaviors in, say, chimpanzees, they are prone to resort to hormonal agents, spontaneous neural responses, innate release mechanisms, and other biological factors. There is a perennial temptation to create as broad a conceptual chasm between humans and lower organisms as possible, rather than to view all organisms on an evolutionary continuum that does not admit radical inter-species discontinuities. Like ethnocentricity, species-centricity promotes the sense that we are special among a throng of less well-endowed creatures. Unfortunately, this perspective diminishes our understanding of our nature and our awareness of the essential commonalities that bind all of us to a similar planetary destiny.

Bullock (1958) stated some 50 years ago that we know of no neurophysiological mechanisms in humans that do not exist among *invertebrates* as well. Much of the DNA that we find in the Human is identical to the DNA of unicellular protozoa. And though we pride ourselves on the spectacular complexity of the human nervous system, that system is constituted of *primordial mechanisms*, magnified in astronomic proportions (Sperry, 1958). It has long been known that embryological processes are essentially the same in all vertebrates, so much so that the embryos of fish and humans, both evidencing, for example, gill precursors, are to the eye of the lay person barely distinguishable at comparable stages of development.[12] The theory of evolution that these ideas presuppose presents a problem for the belief systems of some who are religious. On the other hand, many "spiritual" people have integrated these scientific notions into their religious schemas, as did the renowned Jesuit paleontologist, Teilhard de Chardin (1959).

The cell, which is common to all organisms, displays many varied and unique features. As one moves up the phylogenetic scale, one notes increasingly complex organic building blocks that are special or even unique to certain species. Evolution has taken myriad paths: some species can fly, while others can extract oxygen from water. Among the higher primates, humans can laugh and tell jokes, stage plays, and build complex homes. Whether the

[12] The discovery that the embryos of all vertebrates appear similar to scientists as well as to lay people partially explains the popularity of Ernst Haeckel's nineteenth-century aphorism *ontogeny recapitulates phylogeny*. The gist of the principle is that every vertebrate progresses through (or recapitulates) in its embryonic development all the stages of the many species from which it evolved. Those who endorsed this principle were called *recapitulationists*. Many extended the principle into post-natal life, asserting that each human repeats the *history* of the species in his or her development from infancy to adulthood. Freud, Jung, G. Stanley Hall, and other notable psychologists were numbered among them.

characteristic differences between humans and the great apes, as John Nash (1978) noted, are a matter of degree rather than of kind, is slowly being resolved. Language, the human instrument par excellence, is a capability that comparative psychologists demonstrate is shared to some extent with the higher primates, as well as with dolphins, whales, and other mammals.

Ethologists and comparative psychologists study the social behaviors of animals in their native habitats and, as much as possible, under normal conditions. They argue that their work has relevance to personality psychology, as the social organization and intra-species behaviors of many species and the Human are remarkably similar. The study of primate social behaviors and the emotional and apparently intentional character motivating them can teach us much about the biology of our own behaviors. Jane Goodall (1986)[13] studied chimpanzees in the Gombe National Park, Tanzania and discovered a natural shyness in their behavior when they were faced with the unknown. Shyness can be conceptualized as an aspect of an adaptive fearfulness in both humans and chimps. Although young chimps react to social novelty or to temporary abandonment with more florid displays of distress than do human infants – such as shrieking, wailing, and other tantrum-type behaviors – the parallels are stunning. Goodall's research reveals "mother chimps with attitude, rebellious kids, rivalry among females, bristling macho males and evil passion" (Glover and Gatos, 2001, p. 3). The sex-differentiated roles that revolve about the bearing, feeding, and rearing of infants, the fierce defense of their offspring, the development of dominance and social status, and the manifestation of aggressiveness have a striking similarity to personality traits evidenced in varying degrees in humans.

Evolution and human personality

Arnold Buss (1988) has delineated a number of trends in mammalian evolution that have implications for our understanding of human personality. First, although *anatomical adaptations* have been important in hominid evolution – note the importance of the freeing of the hands (enabled by our upright posture and the development of the human foot for balancing), and the advantage of an opposable thumb for the manipulation of tools – *behavioral adaptations* have been even more important. These behavioral adaptations have sprung from the very tool-making capability that the

[13] Jane Goodall, formerly known as Baroness Jane van Lawick-Goodall, has devoted most of her life to studying, and now protecting, chimpanzees whose numbers are drastically declining as a result of their shrinking habitat and the maraudings of bushmeat hunters. Some have argued that humans who eat chimpanzee meat are engaging in cannibalism (*cf.* Diamond, 1993).

anatomical adaptations made possible. Humans have come to rely less on brute strength and agility (outside the sports and entertainment arenas) and more on the development of prosthetic devices developed by their intellects. Sophisticated weaponry has replaced tooth and claw. Hydraulic hoists and motorized machinery have replaced highly muscular arms and torsos.

Second, many instinctive behaviors have receded among primates. The innate dispositions with which we have been endowed are less coercive, particularized, and stimulus controlled. Hormones continue, of course, to influence human behavior in both subtle and florid ways. Consider pubescence and witness the impact of hormones as they flood the bodies of young adolescents. A series of transformations takes place in Millon's words (1996) that creates "an element of discontinuity from prior experiences, confronting the youngster, not only with an internal 'revolution' of a physiological nature, but also with a series of psychological tasks that are prompted by emergent sexual feelings" (p. 104).[14] However, we are less influenced by them than are other species, particularly as we learn how to cope with these hormonal imperatives. New experiences, coupled with our maturing cerebral cortex, enable us to control our emotional impulses.

Third, all organisms, presumably, are endowed with *innate release mechanisms* (IRMs) that are triggered by special sign stimuli (say, steady gaze, wagging tail, or sexually suggestive gestures). Given such stimuli, these

[14] The phenomenon of "falling in love" has often been observed to resemble a mental disorder in that it distracts lovers from the routines of daily life, disorients them by force of passion away from the chores and imperatives of the "real world," underweights the long-term consequences of love-burdened decisions, crazes with jealousy, blinds, or distorts the exercise of reason, and, with loss, can reduce the will to live. It is the stuff of altruism, self-sacrifice, madness, and tragedy as Shakespeare and the medieval troubadours have brilliantly depicted it.

As interesting as a literary perspective is a physiological one. Marazziti *et al.* (1999) conducted a study that included twenty participants who had fallen in love within the past six months, twenty participants who had been diagnosed as obsessive-compulsive but as yet had not been medicated, and twenty participants who served as controls. They reasoned that:

> The evolutionary consequences of love are so important that there must be some long-established biological process regulating it. Recent findings suggest that the serotonin (5-HT) transporter might be linked to both neuroticism and sexual behaviour as well as to obsessive-compulsive disorder (OCD). The similarities between an overvalued idea, such as that typical of subjects in the early phase of a love relationship, and obsession prompted us to explore the possibility that the two conditions might share alterations at the level of the 5-HT transporter. (p. 741)

In their study their principal finding was that subjects who were in the early romantic phase of a love relationship were no different from OCD patients in terms of the density of the platelet 5-HT transporter, which proved to be significantly lower than in the normal controls. They concluded: "This would suggest common neurochemical changes involving the 5-HT system, linked to psychological dimensions shared by the two conditions, perhaps at an ideational level."

neural mechanisms are hypothesized to activate spontaneously when in an optimal state of readiness. We all live in environments in which there are multitudes of stimuli that impinge, but to many of which, of course, we cannot respond. We normally respond instinctively only to those that have a certain maximal adaptational value for us in terms of procreation, safety, sustenance, and comfort. Tinbergen stated (1951) that as a rule an organism "responds to only a very few stimuli, and the greater part of the environment has little or no influence, even though the animal may have the sensory equipment for receiving numerous details" (p. 27). Many of the most useful stimuli trigger spontaneous reactions.

Highly evolved species can exercise an inhibitory counter-response to these stimuli as when, for example, adolescent or mature primates find themselves in the presence of sexually attractive individuals who are "out of bounds." Further, the strength of the response is dependent on the degree of habituation to the stimuli and duration of relevant deprivation. Pubic bulges in males as well as the general configuration of their torso have sexual sign value for females as do women's breasts, buttocks, thighs, and ankles for men. Indeed, there have been historical periods when bustles and corsets, codpieces and padded shoulders, which enhanced the sexual appearance of women and men, respectively, were part of everyday apparel. Sign stimuli play a role in many other human behaviors, not least those involved in parenting activities. The shape of the infant's face, with its large forehead, protruding cheeks, and "kewpie doll" nose, not to mention its cooing and smiles, its rounded buttocks and dimples, have powerful sign value for the typical adult. It must be kept in mind, nevertheless, that there are wide individual differences in the responses of individuals to these sign stimuli.

As the strength and role of instinctive behaviors declined in the evolution of primates, stimuli increasingly assumed the role of environmental cues that *guided* rather than *coerced* behavior. Humans, through long training and socializing periods, learned to exercise self-control. Whatever vestiges of instinctive neural releasers still exist are, to a greater or lesser extent, brought under the control of conscience and reason.

Fourth, the prolonged childhood of humans relative to their total life span, during which so much learning occurs, ensures a longer formative period, and an apprenticeship in independent, adaptive living. This phenomenon, labeled *neoteny*[15] (Gould, 1977b, 1981; Montagu, 1981, 1986;

[15] Note our treatment of neoteny in Chapter 3 on developmental psychology and personality. Montagu (1986) defined it as a "process whereby the functional and structural features of the young (fetal or juvenile) of ancestral forms are retained ... from infancy to adulthood" (p. 53). Playfulness, love of fun, prolonged educability, and adult childlikeness are all characteristics of neoteny. See Montagu (1981) as well.

Nyborg, 1994) has several advantages. First, neoteny provides for greater cognitive control over the contingencies of daily life, for it promotes a creative rather than an instinctual response to the serious environmental challenges that face every human. The shorter the infancy and childhood of any species, the more that species must rely on innate dispositions, rather than learning to negotiate the challenges of life. Second, the neotenization of *Homo sapiens* permits individuals to become formed in highly individualized ways. As the opportunities for learning multiply over a lengthy childhood and adolescence, and its lasting plasticity, differences from child to child will occur. Such variation occurs not only as a result of socialized instruction but also from the random events that inevitably present themselves. The importance of chance events in the unfolding of individual lives has been amply demonstrated (for example, Bandura, 1998; Krantz, 1998).

Fifth, the explosive growth in the complexity, size, and capabilities of the human cerebrum enabled a rich cognitive life, with all the imagery, abstract thought, creativity, organizational power, planning, data storage, gratification delay, self-serving cunning, and altruism that characterize the Human. Whereas primates that are phylogenetically inferior must rely more on the mechanism of operant conditioning, humans do not have to rely as much on simple, habitual coping strategies. Their problem-solving in the face of novel challenges reveals ever more creative strategies to manage difficulties (which their very ingenuity has often generated). Prolonged childhood in combination with greater intelligence allows for multiple branching of interests and skill specialization. This in turn enhances the individuality of each member of the species, a potential that the larger community attempts to a greater or lesser extent to curb.

Sixth, in the reproductive realm there has been an evolutionary trend from large litters to single births. Single births and small families permit greater parental investment in the individual child. Closely spaced children, as in multiple births, result in a more uniform environment for such children. On the other hand, fewer and widely spaced children benefit from the characterological diversification that results from the impact of the evolving structure of the family, and the developmental changes in their maturing older siblings and parents. The advantages of greater parental investment per child would seem to outweigh certain individualizing advantages deriving from the large family. Although older children are often obliged to assume parental functions relative to their infant siblings for purposes of relieving some of the care-giving pressures on overburdened parents, this results in a shortening of childhood for the child providing care for his or her siblings. The advantages associated with this kind of early apprenticeship (in which they may uncritically adopt their own parents as models) may outweigh the advantages of more carefree, playful exploratory activities that are a source of much incidental learning.

Seventh, as Buss (1988) noted, "greater individuality appears to be the inevitable outcome of all the evolutionary trends" (p. 12) described above. Although there are no two protozoa that are identical in all respects, the scope for distinctiveness grows with the complexity of the organism. This trend toward complexity allows for an increasing number of adaptive (as well as maladaptive) departures from nomothetic behavior patterns for any species. Organisms as multifaceted as humans show virtually limitless variability in the expression of their genotype, most of which results from their interface with limitlessly variable environments. "Any evolutionary trend that extends the scope of acquired behavior inevitably increases the variation in behavior from one individual to the next" (Buss, 1988, p. 13). Sulloway (1996) concurs: "Evolution promotes adaptation. More than any [other] species, *Homo sapiens* possesses 'open' genetic programs that allow adaptation to occur during the course of individual development" (p. 119).

This evolutionary perspective provides us with a window on the idiographic development of individuals as well as on the nomothetic development of the species. Durham (1991) concluded from his analysis of the co-evolution of genetic and cultural systems that "genetic selection and cultural selection are generally expected to co-operate in the evolution of attributes that, from the point of view of their selectors, are adaptively advantageous" (p. 457). Thus, an understanding of evolutionary psychology and its interaction with the multitude of options that variegated cultures afford us is the richest "palette" for sketching a theory of human personality.

Natural selection

Theories of evolution have been circulating for millennia in both Asian and Western societies, and the ideas that have come to be associated with the Darwinian thesis "had been tossing around for many centuries before [Darwin] formalized their presentation and offered evidence for their accuracy" (Mahoney, 1991, p. 119). Considerable controversy still exists relative to this thesis, but Darwinism, in its evolved state (see, for example, Jones, 2001), is the currently dominant conceptual schema for understanding the emergence of various species of organisms, as well as intraspecies diversity. The most important variable assuring the propagation of traits of certain organisms is the *fitness criterion.* Simply put, if a trait can improve the survival of an individual or particular group of people who possess it, then that trait stands a good chance of being transmitted to the following generation. Those who do not possess this survival-enhancing trait have a lower probability of survival; those who do possess it have a higher probability. The organism can be seen as the vehicle for the conveyance through time of traits (and the genomic mechanisms that express them). Both Edward O. Wilson (1975, 1978, 1999), a *sociobiologist* (Wilson's coinage), and Richard Dawkins ([1976] 2006) take this view. Alternatively, genes – or, more broadly, traits – can be viewed as enabling organisms both individually and collectively to adapt, survive, and reproduce. These positions are two sides of the same coin, and we can flip from one to the other as our scientific objectives dictate.

The socio-biological perspective places the organism in an instrumental mode, viewing humans and other organisms as vehicles to ferry genes through time. The alternate perspective places the Human in an end-in-itself mode, and assumes that the organism will engage the environment, seek out hospitable niches for its kin and itself, and alter the world in all possible ways that enhance its personal survival. Genes exist for the sake of the organism rather than vice versa. The proactive, managerial, *adaptational* character of this latter view is evident. The organism picks niches that are most congenial to its survival.[16] The former perspective evidences its *selectional* character by postulating the power of the environment to determine which genes carry on and which perish.[17] As in many dichotomous paradigms, both paradigms are correct depending on the purposes of one's analysis. Organisms and genes exist for each other. This distinction is analogous to Karl Popper's (1982) distinction between *passive Darwinism* and *active Darwinism*. He characterized the former view as assuming that a coercive environment accounted for changes in *speciation*, the latter as assuming that organisms, especially the most highly evolved and complex ones, engage in a nonlinear, complex, systemic interaction with their habitats. They make do with what they genomically have – and scrabble about to that end, often unsuccessfully. The total organism thus participates in the selection and adaptational pressures that foster gradual, microscopic change.

Mutation, the mechanism of change, is blind. Mutations are random genetic events that usually reduce the functionality of the organism and, occasionally, the species. The typical human carries a genetic load of seven or eight sub-lethal recessive genes, most of which rarely get expressed. Expression can occur when both parents carry the same recessive gene. In

16 Adaptive humans negotiate adaptational problems (we will refer to humans as *adaptive* and to problems that require human adaptiveness as *adaptational*).

17 The evolution of a species may be termed macro-evolutionary, the evolution of individual lineages, as in Dawkins' schema ([1976] 2000), micro-evolutionary.

any event, whether because of the presence of maladaptive mutations, dominant predators, the inability to cope with the hazards of an inhospitable environment, catastrophic events, or a combination of these threats, it is estimated that over 99 percent of the species that have existed on earth have become extinct. The survivors are those that have successfully negotiated the hazards of a changing global ecology over hundreds of thousands of years. The most successful, such as humans and cockroaches, are either culturally adaptive and intelligent niche builders (the former) or superb adaptive niche pickers (the latter).

Evolutionary gradualism. Biological evolution may also occur in *abrupt* changes so as to effect functional alterations in organs and nervous systems within a single species. The changes may be so radical and far reaching that they result in the emergence of a new species. This view violates the ancient aphorism that "nature does not act by leaps" (*natura non agit per saltum*). The "leaps and bounds" thesis proposed by Goldschmidt (1933), Eldredge and Gould (1972), Gould and Eldredge (1977), Eldredge and Tattersall (1982), and Lewin (1988) suggests that a species appears abruptly at some point (*punctum*) in time and evolves very little thereafter; that is, it remains in a generally equilibrated state.[18] This would account for the long hiatuses that exist in the available fossil records.

Darwin stated that the gradualist principle of change through small increments (and decrements) was critical to his theory of evolution. The melding of Mendelian genetics with Darwinian evolutionism is consistent with the adaptational and gradualist thesis. The saltationist (or punctuational) position appears to be displacing it, although the fundamental Darwinian position remains strongly embedded in our science (and in popular culture).

Sexual selection

"Evolutionary psychology," stated David Buss (1995, p. 164), "predicts that males and females will be the same or similar in all those domains in which the sexes have faced the same or similar [adaptational] problems" (p. 164). Fitness criteria operate to ensure that those who are best adapted to the demands of an existing environment survive longer than those who are less well adapted. Evolutionary psychology puts forth the

[18] This theory is also referred to as the theory of *punctuated equilibria*, *punctuated* because it occurred at points in time, that is, occasionally; *equilibria* because the changes are stable over the long periods of time that separate the morphological leaps.

common-sense notion that those genic endowments that improve the probabilities of survival will get transmitted. But reproductive success is contingent on another criterion: the ability to compete for and to attract mates. *In vitro* fertilization and other reproductive technologies weaken the cogency of this evolutionary principle. This dynamic has often resulted in the development of psychophysical features that actually place the proud sexual competitor at risk. More brilliant plumage, more florid displays, louder chest-thumping, heavier horns, and more aggressive posturing all invite threats that diminish an organism's chances of survival.

Competition for mates becomes more cognitive as one ascends on the phylogenetic scale. Shrewdness and finesse in communicating to the opposite sex one's willingness to invest parentally in one's offspring counts for a good deal (Buss, 1995; Trivers, 1972). The advanced cognitive capability of humans is evidenced in a number of more or less subtle strategies that they employ in seeking out suitable mates: they frequent settings where they believe such potential mates can be found; they enhance their sexual attractiveness through dress, dieting, and demeanor; they indicate their interest by signaling their availability and directing their attention to one or another person of the opposite sex; and, in the more advanced stages of a relationship, they reveal in part the extent of their material and social assets.

Women are more discriminating in sexual selection than are men, though young men typically seek out women who appear to have reproductive potential. Parental investment by the male must clearly be evident to the female for she is more at risk from abandonment than the male. She needs assistance and nurturance during lengthy pregnancies and the early childhood of her children, all of which absorb much of her strength, dutiful attention, energy, and resources. An absent, philandering partner creates hardship for her and their children. In addition, evolutionists note that males have a greater inclination to seek multiple partners than do females, for the costs to them of multiple or simultaneous paternity are limited (Buss and Schmitt, 1993). Men who are seeking copulatory adventures can misrepresent an interest in parental investment to a potential partner, and then abandon her. Buss concludes what the journalistic press suggests: women who have failed to detect such ploys "failed to survive [and] imperiled the survival chances of their children" (1995, paragraph 4).

The mating strategies of men and women vary, as does the importance that they give to various qualities in potential mates. Women have traditionally been concerned about the social status and resources of a potential mate, and those who have not attained socio-economic independence still are. Shunning men who do not meet this economic litmus test marginalizes the poorest members of a society. (Thornhill and Thornhill, 1983 hypothesized that human rape is usually an act of last resort among those men who have neither the social skills, parental nesting motives, income, or status to competitively succeed in attracting women whose favors they covet.) In addition, women have sought partners who met virile fitness criteria, in part to protect them from the ravages of marauders and predators. Men, on the other hand, have assessed the reproductive prospects of potential mates and the human traits they would contribute to their children.

There are several implications of this psychology of mating. Humans (and other primates) are heavily endowed with certain motives, emotions, anxieties, and "drives" that plausibly explain mating selection pressures. For example, those species that form status hierarchies afford those with higher status greater opportunity for reproductive success (as well as resources needed for survival). As this will have occurred over vast periods of time in an "environment of evolutionary adaptedness" (Tooby and Cosmides, 1992) this condition would have favored the proliferation of individuals who shared the qualities that propelled their ancestors to the top of the hierarchy – most notably a striving for status.[19]

Another example is jealousy. When a third party threatens a prized relationship the partner does not react with a simple, calculated, and calmly reasoned countermeasure. A sense of distress occurs that precipitates a heightened sense of vigilance, hostility toward the presumed rival, and vigorous planned responses to preserve the status quo. The passion can often result in greater or lesser levels of violence. Jealousy is an example of an emotion found globally, which evolutionists speculate emanated from mate-retention strategies iterated through myriad generations. These theories about the ancestral conditions that influenced evolutionary aptitudes for successful mate selection suggest that human evolution has been driven exclusively by heterosexual individuals in their intrasexual rivalries and their intersexual strategies.

Dynamics of evolution

Evolutionary mate-selection pressures figure so prominently in evolutionary psychology that one must conclude that a personality psychology founded on it must grant greater importance to human sexuality than has heretofore been the case, and to the emotions with which it is larded.

[19] Simonov (1997) has provided a description of the motivatiogenic neural structures that indicate levels of probability that one behavior or another will achieve the results an individual seeks – not least, status.

Strangely, the formal systems that have evolved out of scientific American personality psychology do not give the same salience to sexual traits as did the nineteenth-century European psychiatric systems. Sexual pathology was intensively studied in that period (consider Krafft-Ebing's collection of case histories, *Pathologia Sexualis*, published in 1886) and found its most florid expression in the work of Freud, who postulated in *Studies in Hysteria* (Breuer and Freud, [1895] 1960) that it was the *only* drive. On the other hand, the neuroscientist Michael S. Gazzaniga wrote:

> What are brains for? Sex. Indeed I would argue that the cathedrals we build, the books we read and write, the science we create, the cars we drive, the stocks we buy and sell, all of the mergers, the politics, and the wars we wage – in short everything that constitutes the intricate web of life that we have constructed around ourselves with our amazingly large brains – serves a very simple purpose. Sex. (1997, p. 157)

Evolutionary psychologists, given the importance of mate-selection pressures in their theoretical bias, are moving the pendulum back to the center.

The term "design" is frequently used to characterize the evolution of a psychological mechanism. This can mistakenly connote an element of intentionality in the evolutionary process, for when a mechanism is *designed* one can presume an intelligence, an end state, and a planned, often creative, process. However, this theory of evolution is *not* teleological. Mutations in the structure of any organism are presumed to be random events.[20] Except in casinos they do not happen as part of a design, but for better or worse are simply exploited opportunistically. Those that promote reproductive success are happy aberrations; those that do not are unfortunate ones. Thus, evolutionary pressures do not cause variants; they only promote their propagation or their suppression. As soon as one introduces a vitalistic principle into the process, one compromises the parsimony of the model. Opponents regard that notion as a *deus ex machina* that is not only unnecessary but opens the door to theistic and unverifiable hypotheses.[21]

Related to the issue of process is that of product. Evolutionary adaptation, a product, can be distinguished, at least conceptually, from the process that effectuated it. Although few individuals list "maximization of gene replication" on their resumé as a personal goal, many have sought

[20] Random events are those for which we do not have a satisfactory causal explanation or, minimally, a method for quantifying their causes in individual cases. These events, which, of course, *are* caused, are considered to be noise in explanatory models. In statistical models they are lumped into the error term.

[21] See the Pomona College lectures of Nobel laureate Jacques Monod (published in 1971) for a review of this complex issue. A rudimentary knowledge of biology is required to understand his analysis of the "design" aspects of biological evolution.

"immortality" through procreation and the perpetuation of their group/colony. Even more have sought to proliferate for reasons of personal economic security or religious fulfillment. Obviously, the pursuit of personal satisfaction, whether spiritually or materially motivated, interacting with the evolutionary winnowing that sweeps away the debris of obsolete structures and hobbling mutations, effects marvelous adaptations over vast periods of time. Theoretically this is analogous to the process described in Bernard de Mandeville's *Fable of the Bees: or Private Vices, Publick Benefits* (1723) and Adam Smith's *An Inquiry into the Nature and Cause of the Wealth of Nations* (1776). When each person has sought his or her own prosperity in a free market, the collective end-result has normally been beyond the imagining or purpose of any single entrepreneur. Although evolution does not create adaptational variants, the fitness criterion operates to configure them. Some are selected and some are winnowed out, according to the constraints of evolving socio-geophysical niches in which individual species have chosen to seek their common good.

Behavioral genetics and personality

Nature is that aspect of the human (or any other object under study) that is presumably unchanging. (One can note here, incidentally, that "many environmental experiences produce enduring traits in brain structure, brain function, and behavior," Peterson, 2008.) Nature defines the essence rather than the existence of a species (hence, the essentialist rather than existentialist distinction in philosophy). Scientists are typically interested in defining the universal rather than the particular: flycatchers, sodium chloride, and amygdala, as examples, rather than the "flycatcher in my backyard," the salt on this kitchen table, the amygdalae of "Betty's CNS." Behavioral geneticists are no exception to this principle. They concern themselves with the lawfulness that can be established between the genome of certain species (typically, *Homo sapiens sapiens*) and the multiple behaviors manifested by the species. They are also interested in the interaction of the biological substrate of personality with environment, just as metallurgists and engineers study the tensile strength of various alloys and plastics under different climatic conditions. Nevertheless, it is the *regularities* in those interactions (or lack thereof) rather than the idiosyncrasies of any specimen that are the focus of their research. These principles present perils for the behavioral geneticist as there is more diversity in the human race and the way its capabilities are expressed than in any other species. The more complex an organism is, and the less it is constrained by instinctual processes, the more one runs the risk of mistaking that which is emic, local, or particular for that which is global, etic, and universal.

Twin studies and adoption studies

There are two broad areas of investigation for behavioral geneticists: the selective breeding of, say, rats, bulls, horses, and canines to arrive at purposeful breeds that consistently express certain levels of cognitive, athletic, or personality traits; and studies of humans using correlational (*ex post facto*) research techniques to determine the kinds and degrees of contribution that genes make to human behavior. The gold standard for establishing the genetic links to human behavior are twin studies. Identical twins (that is, monozygotic [MZ] siblings) enable us to study two human beings who have identical genotypes. We can reasonably conclude that differences in the behavioral patterns of MZ twins are attributable to environmental variables that trigger the expression of different genes. Comparable studies done with "fraternal" (that is, non-identical, dizygotic [DZ]) siblings, both twins and non-twins, half-siblings, parents and their children, and cousins in a diminishing order of consanguinity allow researchers to assess the relative contribution of genes to various personality traits and behavioral patterns. The most common model requires locating MZ twins who were separated at birth and raised by families who were unknown to each other. The variance that is found in the traits of such twin pairs yields estimates of the contribution of environment vis-à-vis genotype.

The classical twin method for studying the contribution of nature to personality was proposed by Francis Galton in 1875 (McClearn, 1962). Although his understanding of identical and non-identical twins was highly imperfect, Galton suggested that identical twins shared a common genetic inheritance. He postulated that behavioral differences between the individuals of a MZ twin-pair could reasonably be attributed to environmental factors. Similarly, differences between MZ twin-pairs, DZ twin-pairs, and other siblings would demonstrate the differential impact of genes on personality. Inferences made on the basis of that premise were limited due to the fact that researchers recognized that MZ twins and their siblings experienced rather different familial and extra-familial pressures. These largely uncontrolled variables confounded the outcomes in these studies.

The argument has often been made that the estimate of heritability of various traits is inflated if one does not account for the fact that MZ co-twins experience more similar environments than do DZ co-twins of the same sex. This has been disputed. As long ago as 1977, Lytton proposed that parents do not create a similar environment for MZ co-twins (that exceeds that for other children) but react to the evocative "demands" of the twins. He measured parental behaviors toward 30-month-old boys (seventeen MZ pairs and twenty-nine DZ pairs) and found that *parent-initiated* interactions were not systematically more similar with the identical twins than with the non-identical twins (*cf.* Plomin, Willerman, and Loehlin, 1976). A comparable result was earlier demonstrated by Scarr (1968), who studied the parental behaviors of mothers who were mistaken about the zygosity of their twins, that is, they thought their MZ twins were DZ or conversely that their DZ twins were MZ. Their interactions with their children were consistent with the twins' *actual* rather than *misperceived* zygosity. Children appear to elicit behaviors from others that are reflective of their innate temperament. In a sense children raise their parents as much as parents raise their children, for they subtly personalize the childrearing patterns that their parents evolve for them in the course of their development. And in any event, even though MZ twins should be treated more alike than DZ twins, there is little evidence that exposure to stereotyped parental treatment eventuates in correspondingly significant behavioral similarities in children, as research *on shared and non-shared environments*, which we will describe below, seems to reveal (for example, Pervin, 1996, pp. 150–59).

To circumvent the methodological difficulties presented by shared environments, behavioral geneticists have studied extensively the similarities and dissimilarities of MZ twins who have been raised apart (for example, Plomin and Caspi, 1999). The reasoning underpinning these studies is that the closer the genetic connection, the smaller the variance in genic inheritance that can interact with environmental variables – and the more evident the genetic influence.[22] In most cases, these children have been adopted by families that are unknown to each other. The greater the dissimilarity in the socio-cultural environment in which they have been raised, the greater the insight researchers can glean from the power of environment to shape personological features of humans. Comparable studies have been done studying the differential influence of adoptive parents on behavioral patterns and aptitudes of their adopted and biological children.

[22] Although siblings (and DZ co-twins) share, as a group, 50 percent of their genes in common, in any particular case, the genes identical by descent (IBD) can range from 25 to 75 percent. This level of genetic discordance in, say, nonMZ twin studies can be important in small-sample studies.

Shared and nonshared environments

The convention has arisen in behavioral genetics, for convenience sake, to label the family environment, which is shared by all the members of a single family, as a shared environment. The environment external to the family is referred to as the nonshared environment, even though it is obvious that much of it is shared by a family on an organizational, neighborhood, community, national, and cultural basis. A high correlation between the religion of individuals and that of their family of origin illustrates the influence of shared environment in shaping the belief systems, values, attitudes, and numerous social scripts that interact with and potentiate (or attenuate) inherited personality traits. However, after reviewing the available evidence, Plomin and Caspi (1999) stated that "environmental influences that affect personality development operate to make children growing up in the same family no more similar than children growing up in different families" (p. 256). In fact, the numerous adoption and twin studies that have been conducted in the past generation indicate that shared heredity accounts for a much greater proportion of family personality resemblance than does the influence of family ethos and childrearing patterns. Using self-report personality questionnaires (admittedly imperfect tools for research), researchers have found that shared environment accounts for a negligible proportion of the variance in the personality variables under study. If one estimates that 20 percent of the variance in twin studies is due to instrument measurement error (that is, to causes of undetermined origin), heredity and extra-familial influence account for virtually all other variance.

The social variables that are operative in the upbringing of children are intrinsically the same as those that are operative in the larger community (which are thought to play a major, if not preponderant, role in the shaping of personality). Both shared and nonshared environments in Western societies value the individual, both reward socially sanctioned behaviors, attitudes, and interpersonal styles, and both bestow a sense of belonging and importance. What is it about these two classes of causes that generically resemble each other that accounts for their differential effectiveness? The most plausible explanation is that individuals seek out nonshared environments that make the best fit with their genically determined traits. Even within the family, they seek niches that allow them the greatest scope to "feel good in their skins." The rivalry and competitive striving in which siblings engage to maximize their parents' investment in them individually result in niche cultivation and divergent lifestyle strategies. And the inherited platform on which children begin life launches them on a life course that potentiates innate dispositions and aptitudes.

Extra-familial opportunities are sought that give enlarging scope for such self-actualization. These nonshared environments account for over 40 percent of the variance that is found in measured personality traits. No doubt some of this can be attributed to the opportunities afforded by the world that beckons from beyond the family of origin.

Genetic causal pathways run in various directions and in complex patterns. Certain genes, for example, influence the phenotypic expression of other genes (a phenomenon labeled *epistasis*). Some mask others entirely (as when adult-onset baldness masks male widow's-peak hairline). On the other hand, novel experiences that abound *outside* the routine of family life can, epigenetically, trigger the expression of transient, "immediate-early genes" (IEGs) that may have enduring results, such as the repression or activation of further gene expression. This cascades into a series of CNS changes that have longer duration.[23] Some of these novel experiences are not sought out. Sadly, too often they are traumatic and leave permanent, disabling personality debilities. But human beings can be proactive in engaging their environment, and the influence of stochastic events is differentially moderated by genetic reactivity and by previous personal history and childhood experiences. We are *not* passive sensory receptors of environmental stimuli that impinge on us. We idiosyncratically shape our responses to self-selected as well as uncontrollable, changing environments. In many cases one can reciprocally alter one's environment. Witness how sunny, friendly, outgoing people, for example, can subtly alter a stiff, formal, sterile environment.

Molecular genetic analysis and personality

Molecular genetics of personality traces developmental pathways from genes to behavior. This is a daunting task as there are no single genes, thus far discovered, that determine in Mendelian fashion particular personality traits. Unlike eye color or other physical features, personality traits are influenced by multiple genes, many of which interact to bring about complex results. These analyses are further complicated by the fact that traits are complex constructs, which are difficult to calibrate and about which there is continuing discussion and controversy. Superfactors like extraversion–introversion or emotionality are themselves the conflation of many lower-order traits,

[23] A protagonist in Güntürkün's (2006) fictional exchange of letters between two scientists discussing gene-behavior causality writes: "First you have a specific thought, and then you induce [immediate-early gene] expressions, because your thought is, for your brain, equivalent to an (external) stimulus. Finally, your IEGs induce the production of gene products like hormones, for example. Finally, elevated levels of hormones can alter the shape of your body (possibly everything from synapses to hair growth)," p. 338.

which are no doubt multiply gene-linked. Cloninger's (2004a, pp. 44–60) Tridimensional Personality Questionnaire (TPQ) (see Cloninger, Svrakic, and Przybeck, 1993) labels a construct of exploratory, risky, fickle, impulsive, and quick-tempered behaviors as "Novelty Seeking." If, indeed, all of these different, though similar, behavior traits, which are grouped into one trait labeled "novelty seeking," can be reliably linked to a gene for dopamine receptor D4 (DRD4), a momentous advance will have been made in establishing the gene-behavior template for explaining human personality (Cloninger, Adolfsson, and Svrakic, 1996). However, studies investigating this linkage have been inconsistent in their results. As suggested above, the difficulty resides in the fact that any single gene accounts for a small percentage of the variance in the expression of this trait. A different specification of the behavioral patterns that can be gene-linked will no doubt bring about different results with different gene configurations.[24]

Whereas *molecular geneticists* trace lines of causality from genetic base-pairs, in a forward sequence, to genes, DNA markers, neural and endocrinological products, brain systems, and behavior patterns, *biopsychologists* begin with behavior patterns and trace lines of causality in a backward causal sequence to the genetic building blocks. This biopsychology strategy is a *reductionist* process.[25] *Neuropsychologists* are a third major group of scientists who study the relationship of behavior to biology. They focus on the brain and its sub-systems as they are functionally related to behavior, on the one hand (a constructivist orientation), and to gene determinants, on the other hand (a reductionist orientation).

Neuropsychologists also develop neural explanations for psychopathology. I have alluded to the work of Damasio and his colleagues, who have demonstrated the role of the amygdala in mediating social inhibition. Psychologists such as Pickering and Gray (1999) have similar interests. Gray (1982) has researched the contribution of mesolimbic brain structures

[24] The expression of genes is more complex than this might indicate. Deoxyribonucleic acid (DNA) is transcribed into ribonucleic acid (RNA). *Alternative splicing*, which enables a single stretch of DNA to produce various protein products, takes place in such transcripts of mRNA ("m" for messenger) by dropping out various segments of the transcript, called introns. This molecular enzyme-enabled process of transcribing DNA nucleotides into various combinations of RNA nucleotides determines the order in which different amino acids are linked to form a protein.

[25] Reductionist analyses, in which one proceeds from phenotypic units, such as one of the Five-Factor personality traits, to elementary units, such as genes, are contrasted with constructive analyses that proceed from first-order constructs, joviality, for example, to higher-order constructs, such as extraversion, which may integrate joviality with other same-family traits. Reductionism in psychology is a methodological tool that may or may not be useful, depending on one's research objectives, as distinct from reductionism as a philosophy of science, which involves a commitment to explaining the world in elemental terms.

to anxiety and sociopathy. His operative assumption is that persons with pathology are part of the same population as those who are clinically regarded as normal. Everyone experiences anxiety, but patients with trait anxiety suffer from a level of anxiety that incapacitates them. The anxiety continuum stretches between two poles, each of which is pathological. One pole is characteristic of the anti-social personality who is chillingly bereft of anxiety even in the most dangerous situations. The other pole is characteristic of highly emotional individuals whose chronic anxiety debilitates them.

Two constructs, impulsivity and inhibition, have been intensively studied (for example, Buss, 1988; Diamond, 1957). Not surprisingly, these constructs are essential components of Eysenck's original two superfactors: neuroticism (that is, high to low emotionality) and extraversion–introversion. The extreme poles of these factors, especially uncontrolled impulsivity and crippling inhibition, constitute serious anti-social and asocial disorders and are of particular interest to mental health professionals. As environments become increasingly urbanized, the need to control one's impulses in crowded, stimulus-intense settings becomes increasingly important.[26]

Two hypothesized brain systems, a Behavioral Inhibition System (BIS) that mediates anxiety and a Behavioral Activation System (BAS) that mediates novelty-, sensation-, and thrill-seeking, control levels of individual reactivity to inhibiting and activating stimuli (Pickering and Gray, 1999, pp. 278–86). The former is characterized as a "stop" system and the latter as a "go" system; not surprisingly, the two are often in conflict. Developmentally, these two systems are moderated by the constellation of punishments and rewards that society, natural environments, and the haphazard events of personal history impose on the template of heredity. Obviously, different societies in different historical periods reward and suppress varying types of behavior. For example, sedentary behaviors are widely rewarded in a post-industrial society, whereas they would have been punished in a hunter–gatherer society. The behavior patterns that characterize the construct *hyperactivity* are regarded as a disorder, but these same behaviors can be adaptive in some settings. Whether the BAS is adaptive or dysfunctional depends on the demand characteristics of a particular socio-economic setting. Football quarterbacks on the pitch or prizefighters in the ring are hampered by a penchant for reflective thought, slow reaction times, and methodical planning. Computer programmers,

[26] Consider, for example, the level of impulsivity that is tolerated on a large playground versus that which is tolerated in a classroom packed with adolescents. Clearly, different, changing environmental conditions alter the standards for classifying behaviors as pathological or normal. The shy student is considered more adaptive and reward-worthy in the classroom, whereas the impulsive student is considered more disruptive and punishable in the same setting.

in contrast, are hampered by restlessness, impulsive physical activity, distractibility, and spontaneity. Regardless of the psycho-physiological determinants of various patterns of behavior, being designated as a misfit in society depends on the activities on which society places a value – and the extreme (or moderate) level of inhibition or impulsivity that characterizes one's lifestyle. The particular configuration of traits that one exhibits reflects, in all probability, an interaction of both dimensions as well as the neural systems that mutually influence them.

Sensory receptors and personality

We can only speculate about the influence of sensory processes in personality development. Certainly, serious sensory deficits would seem to attenuate or enhance certain traits. The congenitally shy child, for example, would seem to be susceptible to becoming even shyer with the onset of visual or auditory impairment. That visual or olfactory acuity favors entry into certain career tracks, with the social rewards and subsequent self-esteem that flow therefrom, is self-evident. Chefs, and sommeliers, for example, need acute senses of smell and taste (as well as certain esthetic and cognitive abilities); combat pilots need exceptional vestibular and visual capabilities. The interaction of career success or failure and personality structure is a function, in part, of the value that a society places on the career that is pursued. The deeper question is whether there is any direct influence of perceptual capabilities on the structure of personality.

Phenylthiocarbamide. There are dramatic individual differences in sensory capabilities. A good example of such differences is people's perception of phenylthiocarbamide (PTC), a chemical compound with no industrial or commercial value other than its usefulness in genetics research. About 30 percent of the Caucasian population cannot taste PTC (Crandall and Spence, 1974). Of the 70 percent who can taste it, the majority regard it as vile and nauseating – in sufficient concentration it makes some retch (which is why, in experiments, it initially is placed on the tongue with an eyedropper in very dilute solutions). Others taste it as either bitter, salty, or sweet. The evidence for correlations between tasters and nontasters of PTC and personality variables is mixed. For example, Mascie-Taylor *et al.* (1983, p. 91) stated that "nontasters were significantly more placid ... than apprehensive, relaxed rather than tense, and practical rather than imaginative. Nontasters also scored higher on the more visuo-spatial component of an IQ test ... than their taster counterparts." Gentry, Polzine, and Wakefield (1985) found that the inability to taste PTC was related to high intelligence. Conversely, Kimmel and Lester (1987), using the Eysenck Personality Inventory (EPI), found no correlation between the ability to taste PTC and the EPI factors.

These studies suggest that there may be a correlation between sensory acuity and preferences, on the one hand, and personality traits, on the other hand. If there is, the question arises as to whether or not there are particular genes, whose causal pathways intersect, governing personological dispositions and sensory capabilities. Of course, it is also possible that the development of traits through life influences the development of sensory preferences and capabilities. Finally, culture-wide expressions of perceptual preferences may reflect either genetic variables inherent to particular ethnic groups, socio-historical influences, or the additive, possibly interactive, product of both.

Social cognitive neuroscience

Social cognitive neuroscience merges the methods and interests of social psychologists with those of cognitive neuropsychologists. This hybrid science builds on investigations by psychoneurologists, cognitive psychologists, and social psychologists to provide integrated, comprehensive, and socially relevant models of human functioning. For example, impaired amygdalae have been shown to compromise ability to evaluate, and appropriately respond to, risky social situations (Adolphs and Damasio, 1998).

All sciences proceed in the direction of greater specialization. Derek deSolla Price, an eminent sociologist of science, believed that science specialties break into sub-specialties when the total number of researchers in any one area become too numerous to be meaningfully engaged by any one of them.[27] A different, constructive progression occurs as studies in a maturing field bump up against the boundaries of other disciplines, and an interactive and integrative dynamic is triggered. Social cognitive science is an example of this.

Classic studies have documented the relationships between social events and proprioceptive, or neurohormonal, responses. Delgado (1963), for example, linked eye contact between monkeys and their keepers with specific brain wave patterns. This research area has been

[27] When the number of productive researchers in any specialty exceeds about 200 worldwide, the specialty tends to split in sub-specialties. The reader can imagine how difficult it is to follow the work of 200 laboratories, each producing findings relevant to one's own research.

catalyzed by neuro-imaging technology. Positron emission tomography (PET) and functional magnetic resonance imaging (fMRI) have been used, for example, to investigate and possibly account for the popular belief that people can remember the faces of those of their own race more easily than the faces of those of other races. Golby *et al.* (2001) used fMRIs to compare the activation of the neural fusiform face area in both Black and White participants exposed to unfamiliar Black and White faces. Their research documented a significantly diminished ability of both groups to remember other-race faces, and they identified clear neurological correlates for these differences. This same mechanism may be implicated in the visual discriminatory and memorial capabilities that appear to emerge from one's own personal history.

The clearest convergence between social psychologists and cognitive psychoneurologists is cognitive psychology. It is in the tradition-rich area of human mentation that psychoneurologists and social psychologists find a common language and merge the constructs, glossaries, and databases that allow the disciplines to engage one another (Ochsner and Lieberman, 2001). Whereas social psychologists have typically investigated broad (say, molar) issues, neuropsychologists have pursued narrower (say, molecular) issues, the former having much more tolerance philosophically for the grand and empirically nebulous constructs of psychologists in the tradition of Cooley, Mead, and Parsons.

Consider the vexing issue of "profiling" large groups of people, a need identified before September 11, 2001. Such phenomena as attribution of motives to others, assessing friendliness and threat on the basis of facial and other physical cues, categorizing people as "us" and "them," identifying and remembering faces, and making inferences and "maximum-likelihood" estimates of the intentions of others provide fertile ground for collaborative research. Social cognitive neuroscience holds rich promise for the study of social-orienting reflexes, behavioral activation and inhibition, tacit and overt mentation, scripts and schemas, stereotyping humans, attitude alteration, and the enmeshment and reciprocal engagement of emotion and reason.

Conclusion

"The human nervous system makes possible all that we can do, all that we can know, and all that we can experience. Its complexity is immense and the task of studying it and understanding it dwarfs all previous explorations our species has undertaken" (Carlson, 1998, p. 2). The evidence we have briefly reviewed in this chapter supports this statement and underscores the importance of research addressing the physiological and genetic determinants of personality and behavior. As we suggested in Chapter 1, psychology cannot be divorced from the human body. Though we have struggled to shed the Cartesian mantle that has shrouded our theory and research, its ubiquitous presence still bedevils the work of every personality psychologist. Any psychology that explains human cognition and behavior solely from a psychic viewpoint ignores and dismisses the myriad ways that physiology influences behavior. The research of those who have studied the *biology* of psychology will ensure that this discipline remains relevant to the student of human behavior.

There are profound ethical implications associated with our attempt to understand human behavior in terms of its biological underpinnings. Delgado (2000) wrote, "In the 21st century, psychophysiology will face the challenge of establishing ethical principles and practical means for the genetic and social influencing of the development of human beings" (p. 237). Delgado is one of the major primatologists of the past generation. He speaks with the authority of one who has devoted a lifetime to understanding how behaviors can be manipulated through the control of the CNS.[28]

Research findings themselves can be put to ideological purposes that are prejudicial to the interests of one or another segment of a larger (global) community. The work that is being done in the area of biological differences between the brains of men and those of women is a case in point. Kimura (1999), for example, has published an influential book demonstrating that these differences partially explain the average superiority of one sex over the other in various domains. As discussed in Chapter 8, those who do research in these fields need to take precautions that their findings do not prejudice the chances of visibly, socially distinct groups to grasp career opportunities for which they are fully qualified.

[28] Delgado (1963) found that a subordinate monkey in an established colony of monkeys could learn to press a lever to control the aggressive behavior of a boss monkey. The lever controlled radio signals to electrodes inserted into the boss's caudate nucleus.

5 Trait theories and the psychology of individual differences

> Why is the breed of lions violent, sullen or furious? Why are foxes sly, clever, astute? What makes the deer so swift, so timorous ? Why are all such traits and others consistent throughout all the generations? It must be that in every stock and seed the power of mind parallels growth of body ...
>
> Lucretius (55 BCE)

Traits as constructs

Traits, we must note from the outset, are not *per se* observables. Nor are they real entities. You will never be able to place them under a microscope. They are descriptive schemas that are the product of human reason and imagination. They serve a heuristic purpose, as do all other constructs about the world in which we live: namely, they give a conceptual order to our world and make it more comprehensible than it would be without them. That Allport ([1937] 1961), for example, stipulates that traits – or personality for that matter – have neuropsychic referents does not turn them into things (reify them so to speak).[1] Freud, for example, spoke of the Oedipal complex, "penis envy," and numerous other "dynamic" elements of the human psyche, affirming all the while that these descriptors of human psychical states and behavior could be mapped in principle to brain structures. Clearly, conceptualizing a psychical entity or a "complex," and then establishing a semantic explanatory link to a neurologic substrate, does not make the entity real or any less a hypothetical construct.

Traits are a way of thinking about personality. The recent scholar whose name is most often linked with this effort is Gordon Allport. His definition of trait needs to be placed under a bright light as his work is a point of

[1] The process whereby we take a label or an abstract concept and treat it as if it were a concrete thing is called reification. The verb, reify, derives from the Latin terms, *res* (a thing) and *facere* (to make). In scientific discourse, this is widely regarded as a logical pitfall.

reference for those who have continued to do research in this field. Allport ([1937] 1961) defined a trait as:

a generalized and focalized neuropsychic system (peculiar to the individual), with the capacity to render many stimuli functionally equivalent, and to initiate and guide consistent (equivalent) forms of adaptive and expressive behavior. (p. 295)

Why do we need to render many stimuli functionally equivalent? As we negotiate in our daily lives the hedgerows and byways of this world, we are faced with a constant flux of "images" and sensations that place invitations before us – or demands upon us. These stimuli can come from within or without, and they are ever-changing. As the philosopher Heraclitus reminded us, the experience of dipping our foot into a stream can never be perfectly replicated. Because we never encounter twice the identical and exact configuration of stimuli that cues us to a habitual response, it is usually adaptive for the organism to respond quickly to classes of stimuli that closely resemble each other. We constantly need to assess the functional equivalence of situations that we know from experience afford us reasons and opportunities to approach or avoid them. Likewise, categorization of functionally equivalent stimulus fields helps us to determine the quality and intensity of the response that we will make in any given situation.

Although the personality *types* that were described by earlier theorists, such as Carl Gustav Jung, Alfred Adler, and Ernst Kretschmer, had the properties of traits, they were limited in number and extremely broad in their conceptualization. These authors made no pretense at taxonomizing all the traits that characterize humans. Allport, on the other hand, believed that well over 4,000 distinguishable psychological traits are sprinkled throughout the human race – and he hoped to catalogue them all. We say sprinkled, as the traits he had in mind were *idiographic*, that is, descriptive of many *individuals* but not predicated of all humans by virtue of their human nature. Those that can be predicated of humans *qua* humans or, in a limiting case, of entire groups of people constituting a clan, tribe, or circumscribed culture, are characterized as *nomothetic*. That is, there is a certain lawfulness in the way they evidence themselves.

Allport and Odbert (1936), embarking on one of the more ambitious projects of recent psychology, proposed to create a comprehensive taxonomy of human traits.[2] They approached this task by first identifying all the

[2] Taxonomies are essentially lists of items with common properties. They express an order that exists in the list maker's mind. The order can be complex or simple, hierarchical or egalitarian, algorithmized, systemic, or linear. The way that items are aggregated and classified depends on the nature of the items and the degree of specificity and abstraction that the list maker wishes to build into the taxonomy. The most famous of taxonomists is Linnaeus (1707–78), a Swedish botanist, who devised a system for classifying plants into genus and species, among other levels of specificity.

terms in the English language that fit their criteria for a personality-relevant descriptor. They initially scanned an unabridged dictionary of the English language for such terms and were distressed by the large number they found. Their list, and it is not complete, numbered about 18,000. The problem was how to reduce these terms to some broad categories, call them traits, and make it easier for the human mind to grasp their commonalities and dimensions. At this point, it will be useful to present some ideas on the virtues and hazards of producing taxonomies.

Taxonomies

Lumpers

As Anne Roe (1961) and Stephen Jay Gould (1981) have reminded us, taxonomists fall into two camps: lumpers and splitters. (This classification is in itself an example of the simplest of taxonomies.) *Lumpers* take a large number of items that seem to be amenable to sub-grouping, say, apples, pomegranates, oranges, kiwis, peaches, bananas, among many others, and lump them all together to form a single category: fruit. Lumpers concentrate on the relevant similarities of items and aggregate them. Clearly, much information is lost in such a process, for much of the specificity of the constituent items falls by the wayside. Unless one is very clear about the rationale for a particular aggregation, this leads to confusion at best and error at worst. The mischief that can ensue from extreme lumping is that those who adopt such taxonomies may lose sight of the particularities of those items that they have lumped together. Where an analysis requires a high level of specificity, as in the mental health field of psychodiagnostics, it is troublesome to reduce personality disorders to two or three categories, as has been done historically. Those who generate such categories of events or individuals are what Paul Watzlawick (1974) has referred to as the *terribles simplificateurs* (pp. 40–46), a French term to denigrate those who lose sight of the trees as they attempt to describe the forest. Freud is an example of a radical lumper who reduced the multifarious impulsions, dynamisms, and motives of humans of all ages to two presumptive drives: the sexual and the aggressive. This (nomothetic) simplification had serious consequences for the kind of treatment that he and his followers provided to their patients.

Splitters

On the other hand, there are the *splitters*. Splitters are those who, as they begin the task of reducing a mass of data or an extremely long list of items to a small number of easily conceptualized and meaningful groupings, think it wise to give more attention to differences in the objects of their inquiry than do the lumpers. The particularities of the species seem more important than the generalities of the genus, and especially of the more abstract categories such as family, order, class, and phylum. Splitters believe the world around (and within) us is rife with more important and variegated differences than we can encompass in, say, three or five, or even sixteen categories of human personality traits.

Gould (1981) offered an example of extreme splitting in the work of the renowned zoologist and geologist, Louis Agassiz:[3] "He once named three genera of fossil fishes from isolated teeth that a later paleontologist found in the variable dentition of a single individual" (p. 44). The mischief created by extreme splitters is that those who study their taxonomies tend to lose sight of the forest for the trees. They may be prone to making distinctions where there are no differences, or see essential features where there are only superficial properties.[4] One of the consequences of this mischief occurs when we transfer this excessive "essentialism" to analysis of the human race. The splitter in this domain divides the human species

[3] It is interesting to note that although Professor Gould was the Agassiz Professor of Zoology at Harvard University, he felt free to harshly criticize his illustrious predecessor on a number of issues, as well as psychologists in whose domains he had been trained.

[4] Jared Diamond (1993), a professor of physiology at UCLA Medical School, relates a tale of a "red can" and a "blue can" and eight grades of Kung Fu. It conveys the extent to which even a biologist can feel free to ignore the simplest taxonomies:

> I was working in Indonesia with a wonderful young biologist named Ardy Irwanto. Ardy and I had come to like and admire each other, and to look out for each other's well-being. At one point, when we reached a troubled area and I expressed concern about dangerous people we might encounter, Ardy assured me, "No problem, Jared. I have kung fu grade eight." He explained that he practiced the Oriental martial art of kung fu and had reached a high level of proficiency, such that he could single-handedly fight off a group of eight attackers. To illustrate, Ardy showed me a scar in his back stemming from an attack by eight ruffians. One had knifed him, whereupon Ardy broke the arms of two and the skull of a third and the remainder fled. I had nothing to fear in Ardy's company, he told me.
>
> One evening at our campsite, Ardy walked with his drinking cup to our jerrycans. As usual, we had two cans: a blue one for water, and a red one for kerosene for our pressure lamp. To my horror, I watched Ardy pour from the red can and raise the cup to his lips. Remembering an awful moment during a mountaineering expedition when I had taken a sip of kerosene by mistake and spent all the next day coughing it back up, I screamed to Ardy to stop. But he raised his hand calmly, "No problem, Jared. I have kung fu grade eight."
>
> Ardy explained that kung fu gave him strength, which he and his kung fu masters tested each month by drinking a cup of kerosene. Without kung fu, of course, kerosene would make a weaker person sick: heaven forbid that I, Jared, for instance, should try it. But it did him, Ardy, no harm because he had kung fu. He calmly retired to his tent to sip his kerosene and emerged the next morning, happy and healthy as usual. (pp. 200–1)

Biologists come in a variety of personalities. We are counseled never to engage in this practice regardless of the grade we have achieved in the martial arts.

into two, three, or more races; the lumper reduces the physical diversity of humans to a single race, and culturally to a "global village." The radical splitter prefers to think in terms of dozens of groupings of traits; the radical lumper in terms of two or three. Neither approach is right or wrong, and both are more or less useful depending on one's purposes. However, these issues can rapidly take on a political character. Because every taxonomy is a product of the human imagination, not least its mathematical and political templates, it is a work of interpretation as well as observation, larded with implicit cultural assumptions and subject to bias.

The taxonomical project of Allport

The fundamental lexical hypothesis

There are several important assumptions that have underpinned the contemporary efforts of some trait psychologists. The first assumption is that natural language evolves in such a way that it can suitably describe a wide range of subjective experiences among those who use it. Language is also useful to observers for characterizing their own variably perceived behavior, as well as that of others. This assumption is a working hypothesis, and one that has a good deal of intuitive appeal. The everyday language that individuals use to describe what they are experiencing or what others are doing contains an intuitive vocabulary (for example, anxious, jovial, jealous, generous, deceitful, nurturant, reckless, cowardly) for defining the "structure" of their personality. Francis Galton hypothesized that the fundamental traits of human personality, which are by definition present in all human cultures, will be represented in the natural language of all groups. This concept has recently been rechristened the "fundamental lexical hypothesis" (Goldberg, 1990, p. 1216; see also Goldberg, 1993).

From an evolutionary and cultural point of view, a society develops a vocabulary that adequately allows it to conduct its business, and produce, harvest, and distribute the goods that sustain and improve its quality of life, on the one hand, and manage the social relations of its domestic, political, recreational, and religious life, on the other hand. The ability to use language to describe the qualities of individuals allows humans to lubricate those social mechanisms that permit them to live and work in close proximity to each other. Few skills are more important than those for describing the character of the individuals who will assume varied roles in society. The more complex the society and subtly differentiated these roles, the richer a practical thesaurus of descriptors (trait-names) becomes. It is useful to note that the expression of these traits, the more basic of which can be found in all languages, will be shaped, constrained, contextualized, and conflated in culture-specific ways.

A second assumption is that what a person experiences can be adequately expressed in human speech. As subtle, varied, semantically rich, and emotionally colored as human speech may be, it is still a limited instrument, even when flowing from the pen of a Shakespeare or a Proust, and we all experience a great deal that cannot be verbally described. Indeed, there is evidence that much, if not the preponderance, of the meaning in human communication is conveyed by auxiliary paralinguistic and paravocal expressions (Mehrabian, 1971). Speech is an analogical instrument that only remotely parallels the realities it is attempting to describe. One is reminded of the ancient affirmation, attributed to Anaxagoras, that we know more than we can tell, indeed, more than we know that we know. Damasio (1999) affirms this point when he writes "the brain knows more than the conscious mind reveals" (p. 42).

There is a more basic difficulty with the use of common words for describing human character. They were not developed by scientists, but arose out of the practical day-to-day needs of variously constituted societies. William James warned us some time ago that words are simply symbols that describe objects that are independent of our awareness of them, even when they express our own thoughts. The notion that these words represent, in any language, permanently existing ideas is a chimera. Indeed, it is impossible for any one of us to have the same thought twice, as the very condition for entertaining it (the physiology of our central nervous system) is in a constant state of change and modification. James ([1890] 1990) wrote: "*A permanently existing 'idea' or 'Vorstellung' which makes its appearance before the footlights of consciousness at periodic intervals is as mythological an entity as the Jack of Spades*" (original emphasis) (p. 153). These constructivist concerns did not dissuade Allport (and all those who succeeded him), whose work is fundamentally predicated on the lexical hypothesis.

It can be argued that those who have predicated their life work on the currently existing, and admittedly changing, lexica of many distinct human societies are engaged more in an applied science than a basic science. These researchers argue that they had to start with the raw information provided by the naive observations of myriads of humans and, thence, proceed to draw more or less probable inferences – and that it is precisely for this reason that inferential statistics were developed. Granted, no two people have the precisely same idea, nor does anyone have the same idea twice. But dictionaries are semantic alembics for the ideas of millions of people, and the correspondence between the words and their definitions is at least roughly the same for brief periods of time.

These words evoke different ideas – with the individual differences washed out. Be that as it may, the more cautious of the researchers in this realm have tapped into a variety of other data bases: psychobiographies; questionnaires; objective tests; and calibrated results of carefully constructed rating scales. It is precisely for the reasons adduced by James that researchers have used the responses of hundreds of research participants. If the psychological data generated in social science research were as unequivocal as those generated in the hard sciences, personality theorists (and agronomists, pollsters, and many medical specialties, for that matter) would not need to hedge their conclusions with probability statements and cautious reservations.

Despite the problematic assumptions associated with the lexical hypothesis, most contemporary researchers have based their work on the everyday language that individuals use to describe themselves (as well as others). Dictionaries have provided the vocabulary (for example, altruistic, dependent, aggressive, nurturant, inquisitive, venturesome, religious, witty, cunning) for defining the components of personality (*cf.*, for example, Cattell, 1950, 1965). Further, the equivalents of these terms, used to personologically distinguish individuals, can be found in most of the world's languages. It has been assumed that the more important a particular pattern of behavior is to the well-functioning of a society, the more it is apt to have developed a single word rather than a phrase for designating that pattern. That is simply more efficient. Researchers have recognized, however, that divergences between languages and cultures do exist.[5]

What did Allport contribute?

Although Gordon Allport is now largely relegated to the historical chapters in texts on personality, he was one of the first to propose a set of ideas that gave impetus to the study of personality traits, and one cannot think of trait theory without evoking his name and his life-long passion for this field of inquiry. The philosophy that animated his research in this field is captured in a series of lectures that he gave at Yale University (Allport, 1955) in which he proclaimed himself more Leibnizian than Lockean. Locke, he affirmed, endorsed the notion that the human intellect is a passive faculty, "acquiring content and structure only through the impact of sensation and the crisscross of associations, much as a pan of sweet dough acquires tracings through the impress of a cookie cutter" (p. 7).

[5] Students can find a succinct and useful review of cross-language and cross-cultural studies in John and Srivastava, 1999, pp. 106–9. Also M. Wertheimer recommends that we consult Karl Reuning's (1941) *Joy and Freude.*

Leibniz, on the other hand, proposed that the mind is an active agent, "perpetually active in its own right, addicted to rational problem solving, and bent on manipulating sensory data according to its own inherent nature" (p. 8). This orientation led Allport to inveigh against a raft of schools: the *associationism* of the British tradition, the related stimulus–response psychology and *behaviorism* that it generated, the molecular (laboratory-based) approach to the study of psychic and behavioral phenomena that lost sight of the Gestalt of human personality, mathematical models represented by the *factor analytic* studies of Raymond Cattell and Hans Eysenck, and comparative (animal) psychology with its adjunct principle of *species equivalence* (that is, the proposition that by studying, under rigorously controlled conditions, animals that are lower on the phylogenetic scale than we, one can arrive at an understanding of the basic motives and rules of behavior of humans [p. 10]).

Allport was a life-long critic of psychoanalysis and the trait psychology it characterized. He recounted an anecdote that sheds light on the negative attitude that he developed toward this system (cited in Hall, Lindzey, and Campbell, 1998). As a young man, Allport had the opportunity to pass through Vienna during some overseas travels, so he wrote to Freud requesting permission to pay him a visit. An invitation was extended. He described the visit as follows:

> With a callow forwardness characteristic of age twenty-two, I wrote to Freud announcing that I was in Vienna and implied that no doubt he would be glad to make my acquaintance. I received a kind reply in his own handwriting inviting me to come to his office at a certain time. Soon after I had entered the red burlap room with pictures of dreams on the wall, he summoned me to his inner office. He did not speak to me but sat in expectant silence for me to state my mission. I was not prepared for silence and had to think fast to find a suitable conversational gambit. I told him of an episode on the tram car on my way to his office. A small boy about four years of age had displayed a conspicuous dirt phobia. He kept saying to his mother, "I don't want to sit there ... don't let that dirty man sit beside me." To him everything was *schmutzig* [filthy]. His mother was a well-starched *Hausfrau*, so dominant and purposive looking that I thought the cause and effect apparent.
>
> When I finished my story Freud fixed his kindly therapeutic eyes upon me and said, "And was that little boy you?" Flabbergasted and feeling a bit guilty, I contrived to change the subject. While Freud's misunderstanding of my motivation was amusing, it also started a deep train of thought. I realized that he was accustomed to neurotic defenses and that my manifest motivation (a sort of rude curiosity and youthful ambition) escaped him. For therapeutic progress he would have to cut through my defenses, but it so happened that therapeutic progress was not here an issue. (Allport, 1968, pp. 383–84)

Whether or not Freud accurately characterized Allport with his question is immaterial. It was a gratuitous, unsolicited observation, and a serious misunderstanding of Allport's *reason* for seeking the interview. Allport always maintained that clients have a more accurate understanding of their own thoughts than their therapists have, a view now held by a large majority of psychotherapists in North America and Europe. He was firmly rooted in a humanist tradition, which seemed to thrive in the American Midwest. He was no less liberal in this regard than the theoretician and therapist, Carl Rogers, who never presumed, at least intentionally, to interpret for others what they might be thinking. This focus on the individual and his respect for the opinion held by that person, even of his or her unconscious motives, is a hallmark of Allport's life's work. He rejected the notion that the truth lies primarily in the unconscious and that everything else is suspect. He valued reason and believed that the dignity of individuals is compromised when one asserts that the ideas they hold in clear awareness are generally defensive, distorted, and duplicitous. It will pay us to quote him once again on this theme, for it will give the reader an insight into his approach to the study of traits:

> [In the psychoanalytic perspective] the individual's conscious report is rejected as untrustworthy, and the contemporary thrust of his motives is disregarded in favor of a backward tracing of his conduct to earlier formative stages. The individual loses his right to be believed. And while he is busy leading his life in the present with a forward thrust into the future, most psychologists have become busy tracing it backward into the past. (1960, p. 96)

Allport has not been a popular theorist among his fellows, as he pursued a path of inquiry that militated against the quantitative and positivist research ethos that had come to dominate the work of the most eminent current personality researchers. He preferred a qualitative approach to the study of personality, rather than an experimental or factor analytic one. (Some may suspect that this is because he had little aptitude – cognitive or temperamental – for such research, in the same way that Freud treated hypnotherapy with some contempt after he experienced difficulty in hypnotizing patients.[6]) In particular, Allport was always out of favor with those in the behavioral school of thought, who regarded "traits" as fictional constructs. Pratt (1939) had already affirmed that "neither traits nor types, as concepts, have any real existence. They are merely words, and words do not exist in the eye of the observer nor in the people observed" (p. 115). This debate continues.

Albert Bandura (1969), one of the foremost of American psychological empiricists, wrote the following parody of the way in which some trait and clinical theorists proceed in developing their constructs:

[6] This suggests a testable hypothesis: researchers tend to denigrate those methodologies for which they have little aptitude. Of course, they also may shun them because they have little value.

Let us designate behaviors in which persons violate social and legal codes of behavior and frequently engage in assaultive activities as the external expressions of an inferred **zoognick**. Based on prevailing clinical practices, the zoognick would come to represent an intrapsychically functioning agent. An honorific causative power would be conferred upon this hypothetical zoognick, whereas the observed behavior from which its existence is inferred would be depreciated as superficial behavioral manifestations. Before long, psychological tests would be constructed to measure zoognick strength on the basis of which diagnosticians would tautologically attribute clients' behavior to the action of the underlying zoognick. Proceeding on the assumption "that patient variables are not conceived to be behaviors, but constructs concerning internal constellations" (Wallerstein, 1963), psychotherapeutic goals would be stated in terms of removing the pernicious zoognick. On the other hand, direct modification of the deviant behavior would be considered not only superficial but potentially dangerous since elimination of the symptomatic expressions might force the zoognick to emerge in equally pernicious substitute forms. A sufficiently charismatic exponent of zoognick theory would undoubtedly develop a sizable following with the same extraordinary conviction in the vital importance and causative potency of zoognicks as that shown by the adherents of libidinal forces, Oedipal complexes, collective unconsciouses, and self dynamisms. Finally humanists would embrace zoognick theory as more befitting the complexities of human beings than those simplistic mechanistic doctrines that stubbornly insist that the zoognick is the deviant behavior rechristened. (original emphasis) (p. 72)

Allport's model of traits

Allport ([1937] 1961) wrote:

A specific act is always the product of many determinants, not only of lasting sets, but of momentary pressures in the person and in the situation. It is only the repeated occurrence of acts having the same significance (equivalence of response) following upon a definable range of stimuli having the same personal significance (equivalence of stimuli) that makes necessary the inference of traits and personal dispositions. (p. 374)

For a trait to qualify as such for any particular person, it is necessary for the behavior it characterizes to occur repeatedly in generally similar situations. Allport's thinking on this matter is subtle and effectively anticipates the criticisms of later researchers that people behave quite differently in some situations than in others (*cf.*, for example, James, [1890] 1990; Mischel, 1968). William James recognized with his characteristic clarity of thought that there is not cross-situational consistency in the varying personas that we present to the world. He wrote:

We may practically say that [a person] has as many different social selves as there are distinct *groups* of persons about whose opinion he cares. He generally shows a different side of himself to each of these different groups. Many a youth who is demure enough before his teachers and parents, swears and swaggers like a pirate among his "tough" young friends. We do not show ourselves to our children as to our club-companions, to our customers as to the laborers we employ, to our own masters and employers as to our intimate friends. (pp. 189–90)

Allport also recognized that one's extraverted behavior in one situation, say a fraternity party, is not grounds to predict that one will necessarily act that way in a gathering of worshippers, in a funeral parlor, or on some other solemn occasion where the inhibition of buoyant spirits is important. Likewise, it is well known that behaviors that are tolerated in one pub or bar may not be in another, and that patrons regulate their behavior in compliance with those unwritten rules. In one, for example, it is expected that one will drink beer, be boisterous, make passes at members of the other sex (or the same sex, as the case may be) and act aggressively. In another just down the street, where martinis and single-malt whiskeys are routinely served, deportment is more sedate, the noise level is low, conservative social manners are expected, and the personal space of patrons is respected. Allport believed the character of the situation as well as the character of the person, not to mention the cyclical moods and other momentary pressures that bear on one's state of mind, coalesce to evoke an habitual penchant to behave in a particular way.

Classic studies (for example, Hartshorne and May, 1928; Newcomb, 1929) had already demonstrated that cross-situational measures of traits such as honesty, extraversion, or aggressiveness are poorly correlated. For example, an adolescent who is disposed to shoplift a baseball in a department store is not likely, by virtue of that behavior, to steal money from his grandmother's purse. The situational determinants of behavior have long been recognized and were amply accommodated in Allport's theory. He always stipulated that the personal significance of the evocative features of the situation had to be roughly the same in order to elicit the same act. Our hypothetical youth had always respected his grandmother's property.

Nevertheless, the construal of widely divergent situations as essentially equivalent makes this formulation highly subjective rather than objective. It is the *personal significance* of a situation that is the operative variable in establishing equivalence in the actor's mind. For the youth who respects his grandmother but does not hesitate to park without placing money in a meter, there is probably a large range of significant others from whom he would not steal, extending possibly from his relatives to chance acquaintances.

Types of traits

The molar includes the molecular, and the nomothetic is revealed in the entire fabric of the individual's personality. Allport ([1937] 1961) cited, for example, Jean de la Bruyère (1645–96), whose satirical and pungent sketches of the airs and manners of certain personalities are as recognizable today as they were in the seventeenth century. The trait descriptors this satirist found for individuals of his society are universal, applicable as they are "to human beings in any age or land" (p. 58). The parameters of the nomothetic are found in the features of these individuals – indeed, of any individual.

Both philosophers and psychologists have wrestled with issues bearing on the relative importance to basic science of the nomothetic disciplines vis-à-vis the idiographic ones. Differences of opinion among personologists as to whether one should favor one approach over another may have led some to assume that these two approaches stand in opposition to each other. In fact, they complement rather than oppose, and the traits that one finds in an individual can be found in "human beings in any age or land," albeit in varying levels of vividness and intensity. How, then, does the nomothetic differ from the idiographic? One answer is: primarily in its perspective. The objectives of the researcher and theoretician determine the modalities they will choose in studying the Human. Dodge (1931) wrote: "Treating each individual as a special combination of capacities, accomplishments and tendencies, has been far more productive than treating individuals as though they were all alike or as though they belonged to mutually exclusive types" (p. 6). The Big Five Model factor analysts may justly ask: "Productive of what?" The answer to that question is the key to unraveling this knot. Nomotheticists and idiographicists, *given their research objectives*, have both been productive.

There are no two newly manufactured coke bottles or minted pennies that are identical down to their microscopic structures. With far greater assurance can we say this of humans. But if all individuals comprise in their make-up elements common to the entire human race, what then constitutes their individuality in terms of personality? It must be that they possess a unique configuration and assortment of polymorphic traits found "in any age or land." Even if humans were all constituted identically, the limitless varieties of experience that they enjoy and endure, prenatally and post-natally, would ensure that each is unique and speaks with a distinctive "voice."

In his later work, Allport labeled those traits that are subtly and idiosyncratically configured in an individual, "personal dispositions." Those qualities that appear throughout the human race were referred to simply as "traits." This may have been a distinction without a difference. The reader may choose to consider the personal disposition as a "personalized trait by a different name." In the matter of personal dispositions (and their configuration into a unique personality), however, we have left the realm of the nomothetic and, some would argue, the realm of science. This perspective can be applied to the domain of biography, descriptive art, and the visual properties of clouds, as well as to the science of personality.

Allport ([1937] 1961) quoted the post-Socratic Theophrastus in the latter's description of a prominent citizen of Athens. Which of the descriptors in this passage refer to a trait, a personal disposition, or a simple habit not necessarily revelatory of character?

> Giton has a fresh complexion, a full face and bulging cheeks, a fixed and assured gaze, broad shoulders, a projecting stomach, a firm and deliberate tread. He speaks with confidence; he makes those who converse with him repeat what they have said and he only moderately enjoys what is said. He unfolds an ample handkerchief and blows his nose noisily; he spits to a great distance and sneezes very loudly. He sleeps by day, he sleeps by night; he snores in company. At table and in walking he occupies more room than anybody else. He takes the center and walks with his equals; he stops and they stop; he walks on and they walk on; all regulate themselves by him; he is not interrupted, he is listened to as long as he likes to talk; his opinion is accepted, the rumors he spreads are believed. If he sits down you will see him settle into an armchair, cross his legs, frown, pull his hat over his eyes and see no one, or lift it up again and show his brow from pride and audacity. He is cheerful, a hearty laugher, impatient, presumptuous, quick to anger, irreligious, politic, mysterious about current affairs, he believes he has talents and wit. He is rich. (p. 59)

Allport notes that we are likely to recognize this person among our own acquaintances. But Giton also is different from our acquaintances, and we know no one who resembles him exactly. He fits Allport's notion of the unique individual. There is no central tendency that we can discern, unless it be to dominate. Boorish habits abound. The sketch reveals personal dispositions and, certainly, some traits. The most that can be said with assurance of this parsimonious description of an individual who lived 2,400 years ago is that it is a portrait of a lifestyle. It is an example, states Allport, of "character writing," and is of limited interest to the scientist for that very reason.

Cardinal, central, and secondary traits

Cardinal traits in Allport's ([1937] 1961) terminology are units of personality that are pervasive and highly influential in the life of the individual, so much so that much of the emotional life, the cognitions, self-image, interests, life goals, and the behavior, both private and public, of the individual are imbued with this feature. He states that "such a master quality has sometimes been called *the eminent trait, the ruling passion, the master sentiment*, or *the radix* of a life" (original emphasis) (p. 338). Often someone's personality is characterized by such a global trait, as for example, when we comment that he's a Machiavel, she's a recluse, he's a party animal, he's a dominator, she's hypervigilant, he's paranoid, she's work-obsessed. Such cardinal traits are generally not flattering by reason of the fact that they influence the individual's presentation of self to the rest of the world in such a radical way that they preclude the balance that is essential to make her or him a pleasure to be around. We like our friends and associates to be of several if not many parts; otherwise, they become tiresome, if not obnoxious. It is not by chance that the *Diagnostic and Statistical Manual* of the American Psychiatric Association (in all its editions) characterizes personality disorders in terms of a cardinal personality trait that has largely taken over a person's life. In this connection, Allport ([1937] 1961) stated that it is difficult "to achieve a psychological vocabulary of non-censorial trait-names" (p. 340).

Central traits are like cardinal traits except that several can coexist in the same individual. They give balance and richness to personality (unlike the cardinal trait that so dramatically shapes the behavior of the individual who possesses it). There are normally five to ten central traits; they are the personality descriptors that are usually found in nicely crafted letters of reference. But they lack the specificity of *secondary traits* or dispositions. These latter are those that are found in "thick descriptions" of people, that appear in some situations but not in others, admit of greater or lesser vividness in the behavior of the same individual, that are more subtle, varied, and (perhaps) clinical, and that correspond to Allport's notion of the idiographic.

Correlational approaches to trait formulation

As important as Allport is in the history of research on traits and trait theory, Raymond B. Cattell, Hans Eysenck, and a large number of other influential theoreticians who have used correlational approaches to arrive at an understanding of traits have overshadowed him. This approach, one of the most important developments that have occurred in personality theory, is typified by the systematic and logical rigor of the procedures used. The contributions of the great psychiatric system builders of the past were clearly important, but they lacked the parsimonious theoretical foundation and the systematic, empirically controlled procedures that one finds in the work of Cattell and Eysenck.

The search for elements

Every scientific discipline must define the elements (that is, the basic units) that constitute its building blocks and on which it is going to do its research. This is no less true for the science of personality than for chemistry, physics, and biology. For example, pathophysiologists needed to determine that the microbe was a basic unit of their discipline and that in large numbers contributed to disease. Only then could applied researchers begin to discover and classify them, and determine the disorders they generated. The field was relatively impotent before that, as the history of Ignaz Semmelweiss, a nineteenth-century epidemiologist, attests.[7] It is this same motivation that has impelled personality researchers to assume that the basic unit of personality is the "trait," and to search for evidence that it exists. We say "assume that it exists," not because we doubt that there are neurological correlates of all human behavior, but only because there is no proof that there are "neuropsychic mechanisms" on which traits are tightly, isomorphically mapped. We simply postulate that trait theory is a plausible basis on which to build a system that describes human behavior as well as poses hypothetical explanations of the determinants of that behavior. The typical approach to this task is to examine behavior that shows patterns of regularity.

It has been alleged that those who state that traits cause behavior are engaging in circular reasoning. For example, Cloninger (1996) writes: "Mike is aggressive. How do we know? We have seen him beating up people. Why does he do it? His trait of aggressiveness causes him to beat up on people. The trait explains the behavior, and the behavior is the reason that we infer the trait. That is circular reasoning" (pp. 76–77). On the one hand, they affirm that traits cause regular and consistent types of behavior. On the other hand, they state that behavior is the reason that we infer the trait.

Others argue that there is no circularity at all, and causality is always directional, connecting a pre-existing disposition to a consequent act. The line of inferential reasoning is always from a behavior back to an explanatory or descriptive principle. This is no different from the reasoning that good detectives use. They study the scene of a crime and make inferences

[7] Semmelweiss was a physician in mid-nineteenth-century Vienna who investigated the causes of the high mortality rate among mothers in the maternity wards of the Vienna General Hospital. He noticed that it was the patients of physicians who had returned from autopsy rooms and sick wards who contracted "puerperal" fever and subsequently died. As many as 12 percent of these child-bearing women succumbed to these iatrogenic infections. Semmelweiss' solution was to insist that physicians wash their hands before approaching women in labor. The suggestion was derided by the medical profession, which could not clearly envision the causal relation, in this pre-Pasteur era, of the work of obstetricians in disease-ridden wards of the hospital with the labor and maternity ward deaths of previously healthy women and their neonates. The tragedies continued. Semmelweiss eventually died in an "asylum," where he had been committed for treatment of a mental disorder.

about the character and motives of those who perpetrated the crime. This latter line of argument is simply a restatement of the philosophical principle that every effect participates in the nature of its causes. Unfortunately, this line of reasoning, related to the *representativeness heuristic*,[8] can result in false conclusions.

Factor analysis

There are literally thousands of distinct labels in the English language describing the characterological features of humans. Though each of us can make sense of these labels as we read them, one after another, we cannot make sense of them all at once, and in their totality. It is like looking at a jumble of numbers that state, for example, the annual income of randomly selected individuals in a defined population. Each individual income figure could make some sense to us, but we cannot make sense of all of them viewed at once, in terms of their median, mean, range, or pattern of distribution. Whenever humans are faced with a blur of information, they seek to reduce it to meaningful composites.[9] The human mind craves simplicity in important matters, and this is the motivation behind most reductionist systems. As Aldous Huxley once said, the human brain is essentially a reducing valve. We have developed ingenious techniques for making these reductions, and factor analysis is one such technique.

Although the technique of factor analysis has evolved into a sophisticated statistical method, now algorithmically coded into the software used by our powerful desk computers, the logic that supports it is straightforward and simple. We will content ourselves with sketching the logic rather than providing the methodology for doing factor analysis by hand (which is not easy).[10] It should be noted beforehand that the use of this method in the social sciences has its critics (Block, 1995a, 1995b, for example). Gould (1981), among others, criticizes scientists like Raymond Cattell who think "that abstract measures summarizing large tables of data must express something more real and fundamental than the data themselves"

[8] See, for example, Nisbett and Ross, 1980 for an analysis of this heuristic.

[9] One is reminded of the lines from T. S. Eliot's poem, *The Rock*, in which he plaintively asks "Where is the life / we have lost in living? / Where is the wisdom / we have lost in knowledge? / Where is the knowledge / we have lost in information?"

[10] The renowned zoologist, Stephen Jay Gould, wrote in *The Mismeasure of Man*, that the reason that little of a popular nature has been written about factor analysis (as it bears on mental testing) is due to the difficulty in explaining it. He continued: "Factor analysis, rooted in abstract statistical theory and based on the attempt to discover 'underlying' structure in large matrices of data, is, to put it bluntly, a bitch" (1981, p. 238).

(p. 239). This criticism would be justified if it were only this single technique, utilizing only one kind of data, that provided evidence of "underlying structures." However, researchers have used many kinds of data, and their techniques have varied. This has allowed them to converge on a number of "source traits" that are presumed to govern all behavior. The findings are more than simple statistical artefacts; however, it is still not entirely clear whether they are any more real than any other mental construct.

The critical statistic that is used by all factor analysts is the correlation coefficient. As we have seen, Galton devised this technique in the nineteenth century in order to assess the relationship between individuals' performances on two or more sets of tasks. His hypothesis was that measures of performance on tests of various mental abilities would be highly correlated, and that we might therefore infer that intelligence is a single general ability, which each person, at any moment in time, possesses in greater or lesser degree. His view, shared by his student, Karl Pearson, was that much if not most of the variance found in various individuals' scores on tests of intellectual ability could be accounted for by hereditary factors. They used data from family histories, biographies, and measures of twin resemblances, among other sources as grist for their mill. The technique of choice for establishing these relationships was a process through which they arrived at an "index of co-relation." Pearson perfected what is now called the *Pearson product–moment correlation*, designated by the symbol *r*, to more effectively test their hypotheses. Factor analysis evolved from these humble beginnings.[11]

Factor theorists studying personality theory have gathered extensive personality data, arranged in no particular order, and exhaustively computed their intercorrelations. These data are then aggregated into tables called matrices – matrices, presumably, because they serve as the *mother* lode for the further computations that will eventually yield the factors from which the theorists build their systems. It is not unusual for a matrix to contain hundreds if not thousands of correlations. What factor analysis does with these correlations is sort them into groups whose variables have a certain affinity with each other: that is, they correlate highly with one another but not substantially with variables from other groups. This data-reduction process is repeated until one arrives at the smallest number of

[11] Charles Spearman, Louis Leon Thurstone, and other talented statisticians elaborated the simple correlational coefficient into a method for assessing relationships between more than two variables at a time. Such multiple-factor analysis set the stage for the ambitious factor analyses that have been conducted in recent decades by personality and educational psychologists, among others.

coherent groupings possible. Each group is regarded as a factor, a core, nomothetic dimension of personality. The practical task that then remains is to find a term that is descriptive of this factor – in other words, a label. Ideally, the label selected should describe the commonalities from which the factor was extracted, on the one hand, and not exclude any of the sub-traits from which the factor was extracted, on the other hand.

Psychometrists have developed inventories, questionnaires, and tests that can be used to categorize people. These instruments are directly linked to various factorial systems, and they raise vexing questions. Heidbreder (1933, p. 111) asked just such a question: Are the melancholic, for example, a different population than the jovial and the light-hearted, even the manic, or are they the same population but with varying degrees of depression or hyperactivity, say, in their demeanor and state of mind? Are they simply at different ends of the same continuum on which a single comparable population of individuals is placed for purposes of understanding how intensely or thinly they share a common trait? The very fact that we have a label for depression and that that trait is the dominant, pervasive, and defining feature of certain people's lives would seem to indicate that these individuals constitute a different population than the one that is largely defined by consistent joviality, optimism, and even mania – those at the other end of the continuum. If we accept that there is a radical discontinuity between these two populations (as discussed in Chapter 2), must we then postulate a third distinct population to categorize those who are, alternately, both depressive and manic? Or do they alternately shift from one population to another?

This argument recapitulates a similar quarrel among cognitive psychologists as to whether those who on reliable measures of intelligence fall beyond the conventional cut-offs for normality – plus or minus one standard deviation from the mean – are the same or a different population than those who fall within the normal range. Are the intellectually gifted a different population than the intellectually delayed? In a very different field, is the person who is low on a neuroticism scale a member of a different population than is the cyclothymic? Gordon Allport was inclined to say, in the theoretical way that was his wont, there is a real if not radical discontinuity between the "normal" and the "abnormal." Others have denied this (see Hall, Lindzey, and Campbell, 1998, pp. 307–8). The answer lies, it would seem, in the objectives of the theoretician and system builder. For applied scientists, the value of our categories does not reside in their "ultimate reality," as if they were Platonic forms that existed independently of our ability to conceptualize them. It is enough that they have a foundation in reality and that they serve a reasonable objective, say, to improve students' opportunities to develop adaptive skills, or to create environments that reduce pitfalls for those who have a penchant for falling into them.

Raymond B. Cattell

Many thinkers and researchers have studied human character and personality over the centuries, but none has done so as thoroughly, intensively, and systematically as Raymond B. Cattell. His acute intellect, innovative spirit, and prodigious energy allowed him to create a body of scholarship that must be given special attention in any work on personality. His awesome scholarly output can, indeed, intimidate some students. To give a sense of Cattell's stature in the 1960s, we quote from the *Annual Review of Psychology* (Wiggins, 1968):

> Cattell occupies such a unique position in the field of personality structure that his work demands separate consideration. In the three years under review (May 1964–May 1967) Cattell has published four books, 12 chapters and 40 articles, a total of almost four thousand pages that must somehow be summarized. In addition, he has found time to launch a new journal (*Multivariate Behavioral Research*) and edit a massive handbook (*Handbook of Multivariate Experimental Psychology*). This alone would warrant separate consideration but there is more. The appearance of so many major works and especially the publication of his *Collected Papers (Personality and Social Psychology)* has once again forced an evaluation of a body of literature so vast, uneven, and demanding that many American workers have simply tended to ignore it. (p. 313)

There may be a cautionary tale embedded in this account. The production of voluminous monographs, books, journal articles, and research reports, which constitute the scaffolding of a complex, detailed, architectonic system, causes approach–avoidance conflicts in those whose time is severely limited and who have their own productivity schedules to meet.

In the front matter of his 1952 book, *Factor Analysis: An Introduction and Manual for the Psychologist and Social Scientist*, Cattell dedicated that work: "To the Pioneers: Charles Spearman *and* Louis L. Thurstone" (p. v). (To those two pioneers we might add his frequent allusions to Sir Francis Galton and Karl Pearson.) He then quoted Robert Browning who, after alluding to four giants of Ancient Greek letters and arts, wrote:

> I am not great as they are, point by point:
> But I have entered into sympathy
> With these four, running these into one soul,
> Who, separate, ignored each other's arts. (p. vi)

Cattell (1952) saw himself as an heir of Spearman's and Thurstone's legacy; they developed, as we noted above, what for him was a key for unlocking the "structure" of personality – factor analysis. Cattell deplored the rarity of those who combined the mathematical skills of the statistician and the "artistic, intuitive, empathic skills of the good clinical or educational practitioner" (p. x). The very reason why this combination mattered so much was that the researcher of that era had to do all the computations by hand. Though the mathematical reasoning underpinning the procedures still needs to be understood, today we have powerful desk computers that can accomplish in a matter of seconds what took Cattell months to compute. As John and Srivastava (1999) have noted (p. 104), this created data-analytic challenges for Cattell that obliged him to take short-cuts that compromised the reliability of his findings. They also noted that Digman and Takemoto-Chock (1981) re-analyzed Cattell's own correlation matrices and found "unfortunate clerical errors."

In spite of the limitations that a later generation of personologists found in Cattell's earlier work, it is useful to briefly review his methodology and findings as their influence has had a profound impact on personality theory. It must first be noted that the analyses that any researcher does cannot be more reliable than the data in which they are grounded. There is almost limitless information about innumerable individuals that could be sampled for factor analytic purposes. *One needs a reason* for choosing one sample of behavior rather than any other or one sample of individuals rather than any other. A totally random choice of either type will in all probability leave one with results that are almost impossible to interpret.

Rychlak (1968) demonstrated that even if we do not have an explicit theory that underlies our choice of a data source, there will, at a minimum, be a subtle and implicit theory that is all the more dangerous because it is covert and misunderstood. He wrote: "Suppose we hypothesize that in autumn there is a significant positive relationship between the number of leaves falling from the trees [and left on the lawns] of homeowners in a given area and independent measures of the homeowners' personality dominance" (p. 44). The impracticality of testing this hypothesis is only the least of the problems that a methodologist might raise. Any good project needs to be built on a sound research base. If someone comes up with a novel research approach that reveals "no understanding or appreciation of the scientific matrix in which his work is embedded," one has *prima facie* evidence that one is dealing with a "crank" (Bernstein, 1993, p. 20). On the other hand, information that would lead any serious personality theorist to look twice at this research project would be an explicit and plausible theory underlying the choice of variables – a theory, moreover, that was not in conflict with his or her own deeply held theoretical biases.

Suppose we pressed this (presumably male) researcher to explain the reasons underlying his choice of variables. He might write something like this:

> Dominant personality types are highly competitive. [There is an assumption here that these two traits are highly correlated.] They generally like their lawns to look as tidy, or even more so, than those of their neighbors. [There is an additional assumption that these two traits are correlated with this preference.] The number of leaves left on the lawn, as a proportion of the total number that have fallen, is an indicant of the level of competitiveness, hence, dominance.

Although funding agencies would not find this project competitive, at least theoretical grounds for judging its soundness have been furnished. If judged to have plausibility, it would have to pass muster on a large number of methodological grounds as well. Beyond that it would have to convince those who adjudicated the request for support that it was worth doing at all. Cattell founded his research on a copiously articulated theoretical base. As to method, unlike Allport and Henry Murray (for example, [1938] 2007), Cattell was a fierce partisan of the multivariate approach, and he put virtually all his eggs in the factor analytic basket. In 1965, he argued: "Whereas the univariate method follows the older sciences in bringing the man into the laboratory chair, surrounded by brass instruments, the multivariate method says that with sufficient analytic subtlety we can tease out the connexions from the behaviour of the man in his actual situation ... " (p. 20). Two principles leap out from that statement. One, he preferred to study the whole person in an ecologically valid and natural environment rather than only part of the person in a laboratory. Two, he preferred to use multivariate approaches rather than the univariate approaches generally preferred in his time by laboratory-based researchers.[12] It is obvious that Cattell is embedded solidly in the correlational, and not experimental, research tradition.

Data sources

Although it is generally easy to find flaws in Cattell's work, he cannot be faulted relative to his concern for sampling a carefully chosen yet broad

[12] Cattell was embedded in the British psychometric tradition. In 1965 he wrote:

> Sir Francis Galton gathered statistics on personality and ability wherever it was available ... while Karl Pearson was especially interested in what the *average* man is like, anthropologically and psychologically. In its break from the classical univariate experiment theirs was a more bold and imaginative step in studying human psychology than Wundt's. Wundt's clinging more closely to established sciences may have been oriented to giving to the new science of psychology more standing, whereas Galton had the sublime indifference to appearances which one sees in a fox terrier with his nose in a rabbit burrow. (pp. 20–21)

assortment of data sources, and he always built a strong theoretical justification for choosing the data sources that he used. He used three principal sources for arriving at a description of the totality of human behavior, which he called *personality sphere*. They were, first, ratings that observers made of individuals' various and specific kinds of behavior, in terms of both their frequency and intensity. Although these data were typically the behavior exhibited by subjects over a sample 24-hour period, they also included long-term records of such events as "number of automobile accidents over twenty years," and "number of societies to which the person belongs." He designated them "life-record data" (or *L-data*).

A second source of data was questionnaires that were administered to participants in his studies. He designated these *Q-data*. Hundreds of carefully selected items were presented to large groups of normal people. An example is true or false responses to statements such as "I could stand being a hermit" or "I am careful to turn up when someone expects me." Such questionnaires were the least reliable source of information as they depended on subjective and inevitably biased assessments by individuals of their own behavior. The questionnaire, Cattell stated (1965), "depends on introspection and is liable to distortion by imperfect self knowledge, delusions about the self, or an intention deliberately to 'fake'" (p. 61).

A third source of data for Cattell was objective tests, which yielded what he called *T-data*. These tests are different from those typically given in academic courses or found in the Scholastic Aptitude Tests (for example, those in a multiple-choice format). Test takers are given a series of standardized tasks to perform and their level of proficiency is measured. For example, to assess suggestibility, one of six source traits defining the so-called neuroticism factor, individuals are asked to place their back against a wall, keep their eyes closed, and listen to a recording of a voice that repeatedly suggests "You are falling slowly forward. You are falling forward ..., etc." A mechanism is used to measure the degree of forward sway. Writing about this test, Cattell noted, "A few neurotics actually have to be caught before they fall!" (p. 112).

Cattell's views on traits

Cattell (1950, p. 2) was primarily a nomotheticist, that is, he was concerned with discovering the personological laws governing patterns of human behavior. In that respect his theories were a departure from the idiographic emphasis that one finds in Allport's work. Cattell often wrote that he wished to formulate a system that accounted for the totality of the human personality – the personality sphere to use his terminology. Though he multiplied constructs (always in a reasoned way) and postulated a number of source (that is, foundational) traits beyond those his successors were inclined to generate, he nevertheless integrated them into a holistic perspective. Cattell did not make theoretical parsimony one of his intellectual virtues, and he delighted in inventing new terms for the factors he extracted from the multitudinous data that he gathered. Students may possibly be amused by his use of terms such as "harria" and "premsia," among many others, which they will not find in any unabridged English-language dictionary. Their meanings, which you may find defined by trait descriptions in Cattell's principal psychometric creation, the *16PF*, are not intuitively evident; however, they are not entirely arbitrary. Harria, for example, is roughly an acronym for "hard realism." Premsia, its contrary, is based on an acronym for "protected emotional sensitivity."

Cattell began with approximately 4,500 personality descriptors that Allport had garnered by perusing dictionaries. Through a succession of clustering procedures he ultimately reduced these to under 200. Using the various data sources described above, Cattell then gathered grist for his factor analytic mills, conducted numerous analyses, conflated and reduced various lists of traits, and at length presented the world with sixteen *source traits*. Cattell later developed the *Sixteen Personality Factor Questionnaire (16PF)* to assess these core personality variables. These traits were proposed by Cattell (1950) as "the structural influences underlying personality, which it is necessary for us to deal with in developmental problems, psychosomatics, and problems of dynamic integration" (p. 27). These source traits are analogous to Allport's central traits.

Surface traits are the epiphenomena of the source traits. They are the observable cluster of variables that correspond to an underlying factor. The reader will note that they are analogous to a psychiatric syndrome that reveals the presence of an underlying and specific disorder. The essentially descriptive character of surface traits does not provide us with an explanation of behavior; it allows us only to begin to make sense of them in terms of an organizing principle, and it provides us with the elements necessary to conclude that we have a significant pattern of behaviors. For example, testing may document a rich vocabulary, some knowledge of mathematics, and a broad and accurate, if not detailed, understanding of world geography. All of this indicates a higher-order pattern of high intelligence and studiousness. In the last analysis surface traits provided the numerous elements that, when factor analyzed, allowed Cattell to generate the source factors. The most popular and widely recognized

Table 5.1 *Primary factor scale descriptions*

Factor	Left meaning	Right meaning
A Warmth	Reserved, impersonal, distant	Warm, outgoing, attentive to others
B Reasoning	Concrete	Abstract
C Emotional stability	Reactive, emotionally changeable	Emotionally stable, adaptive, mature
E Dominance	Deferential, cooperative, avoids conflict	Dominant, forceful, assertive
F Liveliness	Serious, restrained, careful	Lively, animated, spontaneous
G Rule-consciousness	Expedient, nonconforming	Rule-conscious, dutiful
H Social boldness	Shy, threat-sensitive, timid	Socially bold, venturesome, thick-skinned
I Sensitivity	Utilitarian, objective, unsentimental	Sensitive, aesthetic, sentimental
L Vigilance	Trusting, unsuspecting, accepting	Vigilant, suspicious, skeptical, wary
M Abstractedness	Grounded, practical, solution-oriented	Abstracted, imaginative, idea-oriented
N Privateness	Forthright, genuine, artless	Private, discreet, non-disclosing
O Apprehension	Self-assured, unworried, complacent	Apprehensive, self-doubting, worried
Q1 Openness to change	Traditional, attached to familiar	Open to change, experimenting
Q2 Self-reliance	Group-oriented, affiliative	Self-reliant, solitary, individualistic
Q3 Perfectionism	Tolerates disorder, unexacting, flexible	Perfectionistic, organized, self-disciplined
Q4 Tension	Relaxed, placid, patient	Tense, high energy, impatient, driven

Source: M. T. Russell and D. L. Karol 1994. *The 16PF Fifth Edition Administrator's Manual.* Champaign, IL: Institute for Personality and Ability Testing.

aggregation of these factors is the *16PF Questionnaire* associated with his name. Table 5.1, above, lists these factors.

Cattell wished to reduce this aggregation of sixteen factors even further. This moved his taxonomy up the Linnaean ladder to a higher level of abstraction and generality. He eventually lumped them into five global constructs, which he called "second-order factors" (see Table 5.2 for this reduced list). As Cattell (1965) explained, correlations among the sixteen primary factors can be factor analyzed to give us a smaller number of higher-order factors that organize the primary factors in the same way that the primary factors organize the surface traits (pp. 117–18). The student who is interested in gaining a more detailed understanding of (a) the instrument that measures these traits, (b) interpretive approaches to using the results, and (c) the way it compares with other psychometric

Table 5.2 *Global factor scale descriptors*

Factor	Left meaning	Right meaning
EX Extraversion	Introverted, socially inhibited	Extraverted, socially participating
AX Anxiety	Low anxiety, unperturbed	High anxiety, perturbable
TM Tough-mindedness	Receptive, open-minded, intuitive	Tough-minded, resolute, unempathic
IN Independence,	Accommodating, agreeable, selfless	Independent, persuasive, willful
SC Self-control	Unrestrained, follows urges	Self-controlled, inhibits urges

Source: M. T. Russell and D. L. Karol 1994. *The 16PF Fifth Edition Administrator's Manual.* Champaign, IL: Institute for Personality and Ability Testing.

instruments of a similar character will find *The 16PF Fifth Edition Administrator's Manual* (Russell and Karol, 1994) helpful.

Environmental and constitutional traits

Cattell made another taxonomical distinction. Although all traits represent a disposition, that is, an habitual inclination on the part of an individual to act in a specific way in the presence of a certain stimulus field, the origins of that disposition can be primarily hereditary or primarily experiential, that is, environmental. He called the former *constitutional traits* and the latter *environmental-mold traits* (Cattell, 1950). As with most categories in the social sciences, these two are fuzzy sets. Traits are never the product of hereditary only or environmental factors only. Every behavior is thoroughly infused with both kinds of influence, but in different degrees. Even congenital dispositions, such as those demonstrated by the long-term New York Longitudinal Study of Thomas and Chess (for example, 1980) may (as noted in Chapter 2) be the result in part of stress that gravid females experienced at specific points in the gestation of their embryo and fetus. Stress hormones flowing into the embryo, ill-equipped as it may be to metabolize them, can result in physiological damage. (Such stress could be caused by financial hardship, domestic violence, death of a family member, or other traumatizing events.) It is not unreasonable to affirm that some infants are born anxious – as a result of the anxiety of their mothers. In addition, tobacco and other drugs ingested during a pregnancy can profoundly affect the neurological development of the unborn and the subsequent behaviors related to that impairment. Cattell (1950) would consider traits resulting from such events as constitutional, as the behaviors expressive of such traits were less the result of learning than of internal psychophysiological conditions.

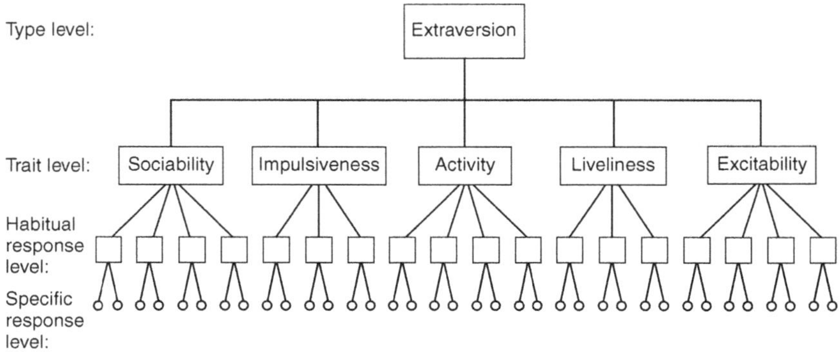

Source: H. J. Eysenck, *The Structure of Personality*, London: Metheun, 1970.

Figure 5.1 Diagrammatic representation of hierarchical organization of personality

Hans J. Eysenck

Eysenck (1916–97) was a brilliant and indefatigable researcher. His extensive interests included psychopedagogy, criminology, behavior genetics, psychopathology, and the science of personality.[13] He devoted much of his prodigiously productive life to dimensionalizing personality and developing measures for assessing those dimensions. His two-factor structure of personality (extraversion–introversion and neuroticism), later expanded to three factors with the addition of *psychoticism*, formed an initial scaffolding for a theory of personality that stands as one of the signal achievements of twentieth-century personality science (Farley, 2000). These factors, which Eysenck referred to as types rather than traits, correspond to the second-order factors of Raymond Cattell. Such constructs are situated at rarefied levels of abstraction (see Figure 5.1), extracted by factor analyses – unlike Cattell, he preferred orthogonal to oblique rotations – from voluminous data that were meticulously gathered. (He averred that oblique rotations introduced a bias into Cattell's bloated inventory of traits.)

What is unusual about Eysenck's model is that he initially reduced the overall theoretical structure of personality to two dimensions, *introversion–extraversion* (derivative of Jung's concept [1921] 1971), on the one hand, and *neuroticism*, on the other hand. This is as reductionist a model as one can find. Its utility is demonstrated by the fact that its two dimensions are

[13] In my estimation Eysenck had a choleric temperament. He was an inveterate polemicist as well as an estimable scientist.

integrated into "all major models of temperament and personality" (Clark and Watson, 1999, p. 400). If the model seems unduly parsimonious, it must be noted that Eysenck acknowledged that it only partially captured the true variability of personality.[14] Nevertheless, these two personality types have proven to be useful in highlighting two venerable notions of human variability and for providing the scaffolding necessary for the construction of instruments to assess that variability.

The characteristics of the extravert are, quite plausibly, spontaneity, expressiveness, impulsivity, optimism, gregariousness, assertiveness, and dominance. Introversion, which characterizes those at the other end of the continuum, is evidenced by shyness, pessimism, unobtrusive social behavior, a tendency to solitude and quietude, and inhibitedness. Those who press their extraversion or introversion to the extreme, and who by definition fall outside the normal range, are characterized as pathological if this extreme expression of the trait compromises their ability to function in their home society. Extremely high scores on Eysenck's factor of neuroticism suggest that the subjects are, indeed, very likely to be neurotic. Those who are emotionally stable, with extreme scores on the other end of the scale, are considered to be mentally healthy. The question that needs to be raised is whether or not one can be so extremely stable that the trait becomes dysfunctional.

Clark and Watson (1999) have observed that the Ancient Greeks preceded us in this taxonomy – as in so many other matters. As was noted in Chapter 1, the Greeks divided human temperament into four categories: sanguine; choleric; melancholic; and phlegmatic. If one crosses Eysenck's extraversion–introversion dimension with the emotional stability–emotional lability dimension, four cells emerge that correspond to the Greek model of Hippocrates and Galen. "Thus, the [emotionally] stable extravert is sanguine, the unstable extravert is choleric, the unstable introvert is melancholic, and the stable introvert is phlegmatic" (Clark and Watson, 1999, p. 400). It is fascinating that the psychobiological formulation of the Greeks is so nicely replicated in the work of Eysenck. The drawback of such a molar approach to conceptualizing human variability is that it can mask the complexity that underlies our constructs.

Eysenck later added a third dimension of personality to his schema, to which he gave the infelicitous label, *psychoticism*. Like the term neuroticism, it connotes an orientation to personality theory founded on a psychopathological template – a subject treated in Chapter 2, which examines the pathognomonic bias in theory building. The Maudsley

[14] Readers can speculate about how a one-factor model of personality might be conceptualized and labeled. Such a factor would be situated at the pinnacle of the trait hierarchy and theoretically include the universe of personality traits.

Hospital, London was Eysenck's professional and research base for much of his life, and it is not surprising that his theorizing bears the stamp of his medical surroundings. Eysenck might resent this environmentalist insinuation, were he alive to read it, but the fact remains that two of his three constructs bear "negative" labels and give prominence as much to what goes wrong as to what goes right in personality development. In any event, this particular personality dimension is less a measure of psychoticism than of impulsivity. Individuals who are impulsive are less constrained by conventions, monitor themselves less carefully, are prone to speak their mind, are often flamboyant in the expression of their emotions, most notably their anger, and are reactive to perceived slights. Those who score high on this dimension appear to be more extraverted than introverted, and more emotionally labile and theatrical than stable. Indeed, they manifest many of the more problematic behaviors of the emotionally unstable (high-neuroticism) individual.

The three trait types that Eysenck postulated, often referred to as the *three-factor model*, are truly dimensional as each is a bipolar, continuous variable. For example, the *extraversion–introversion* dimension ranges from the extremes of extraversion, characterized as highly sociable, animated, active, assertive, impulsive, and excitement-seeking (among other qualifiers) at one pole to reclusive, submissive, placid, apathetic, hypervigilant, at the other. Similarly, the *neuroticism* factor ranges from extreme emotional lability, moodiness and chronic anxiety, and depressed affect at one pole to high levels of self-esteem, self-confidence, emotional stability, and calm, reasoned approaches to problem-solving at the other. The so-called psychoticism factor ranges from the impulsive, sociopathic, and self-absorbed personality at one pole to high levels of self-inhibition, risk-avoidance, and excessive planfulness and gratification deferral at the other pole. Clearly these latter descriptors do not correspond with the term psychoticism as it is used by either the "man-in-the-street" or the professional psychotherapist.

Eysenck, like every other serious psychologist, postulated that there are both environmental and genetic determinants of personality development. He differed from other psychologists in his heavy emphasis on the genetic contribution to individual temperament and, most controversially, to intelligence. Like Raymond Cattell, Eysenck was trained in the British tradition of Galton and Spearman. He revealed this background in the emphasis that he gave to genetic determinants of a hypothesized intelligence factor, "*g*," and the constitutional conditions that limit it. Clark and Watson (1999) assert that Eysenck, beginning in the 1960s, provided the initial "seminal modern work on the biological bases of temperament ..." (p. 413). The heritability of temperament, as well as cognitive potential, is now widely accepted. What is still shrouded in mystery are the neuropsychic mechanisms and the enabling neural transmitters and endocrinological products that determine temperament.

The Five-Factor Model

The Five-Factor Model (FFM) of human personality, informally referred to as the Big Five, has a long history. Thurstone came up with one model as early as the 1930s (1934). Winter and Barenbaum (1999) affirm that it was fathered by Allport in that same time frame and discovered somewhat later, at least implicitly, by Cattell (p. 12). Tupes and Christal (1992) offered evidence that its discovery dates back to 1949. Donald Fiske's (1949) analysis, using data gathered and analyzed by Cattell, revealed five factors. Other investigators, most notably Digman and Takemoto-Chock (1981), re-analyzed Cattell's data and they also could only extract five robust factors. Cattell is now, according to many notables, the putative father of the FFM, but Goldberg (1993) has observed that Cattell "has consistently denied his paternity" (p. 27).[15]

The five higher-order factors that constitute the FFM have been labeled: *neuroticism*; *extraversion*; *openness to experience*; *agreeableness*; and *conscientiousness* (see Table 5.3). These are the labels used by McCrae and Costa (1999), but other researchers have chosen other terms to characterize this configuration of superfactors. Because each higher-order factor is extracted from representations of numerous traits that have more specificity than the factor does, we are faced with a problem of nomenclature. It is not a simple matter to choose a name for a factor that neither over-specifies that factor nor privileges one aspect of the dimension more than other more important aspects of it. Take neuroticism, for example. As it embraces such characteristics as emotional stability, calmness, hardiness, and self-satisfaction, as well as emotional lability, nervousness, high anxiety, and hypochondriasis, one may justly ask why the designation neuroticism was chosen instead of, let's say, *wellness*. And do they, either singly or together, give an adequate representation of the multifaceted constituents of the factor? Cattell's penchant for creating

[15] Is it important to know who was the first person to suggest that there were five, and only five, factors that could reasonably encompass all the basic dimensions of personality? It may be of greater interest to know who the researchers are who have made the major contributions to the most widely accepted taxonomy of superfactors that is used today. The serious student is urged to consult several chapters in *Handbook of Personality: Theory and Research*, 3rd edn., edited by John, Robins, and Pervin (2008), which address this issue.

Table 5.3 *The Big Five trait factors and illustrative scales*

Characteristics of the high scorer	Trait scales	Characteristics of the low scorer
Neuroticism (N)		
Worrying, nervous, emotional, insecure, inadequate, hypochondriacal	Assesses adjustment versus emotional instability. Identifies individuals prone to psychological distress, unrealistic ideas, excessive cravings or urges, and maladaptive coping responses	Calm, relaxed, unemotional, hardy, secure, self-satisfied
Extraversion (E)		
Sociable, active, talkative, person-oriented, optimistic, fun-loving, affectionate	Assesses quantity and intensity of interpersonal interaction, activity level, need for stimulation, and capacity for joy	Reserved, sober, unexuberant, aloof, task-oriented, retiring, quiet
Openness to Experience (O)		
Curious, broad interests, creative, original, imaginative, untraditional	Assesses proactive seeking and appreciation of experience for its own sake, toleration for and exploration of the unfamiliar	Conventional, down-to-earth, narrow interests, unartistic, unanalytical
Agreeableness (A)		
Soft-hearted, good-natured, trusting, helpful, forgiving, gullible, straightforward	Assesses the quality of one's interpersonal orientation along a continuum from compassion to antagonism in thoughts, feelings, and actions	Cynical, rude, suspicious, uncooperative, vengeful, ruthless, irritable, manipulative
Conscientiousness (C)		
Organized, reliable, hard-working, self-disciplined, punctual, scrupulous, neat, ambitious, persevering	Assesses the individual's degree of organization, persistence, and motivation in goal-directed behavior. Contrasts dependable, fastidious people with those who are lackadaisical and sloppy	Aimless, unreliable, lazy, careless, lax, negligent, weak-willed, hedonistic

NEO-PI-R facet scales associated with the Big Five trait factors:
Neuroticism: anxiety, angry hostility, depression, self-consciousness, impulsiveness, vulnerability
Extraversion: warmth, gregariousness, assertiveness, activity, excitement seeking, positive emotions
Openness to experience: fantasy, aesthetics, feelings, actions, ideas, values
Agreeableness: trust, straightforwardness, altruism, compliance, modesty, tendermindedness
Conscientiousness: competence, order, dutifulness, achievement, striving, self-discipline, deliberation
Source: P. T. Costa, Jr. and R. R. McCrae 1985. *The NEO Personality Inventory Manual.* Odessa, FL: Psychological Assessment Resources, p. 2; P. T. Costa, Jr. and R. R. McCrae 1992. *NEO-PI-R, Professional Manual.* Odessa, FL: Psychological Assessment Resources, p. 3.

relatively neutral, neological labels for his factors has some merit. On the one hand, it addressed the problem that is posed by choosing vernacular terms for scientific concepts, for the former inevitably engage all the emotional overtones that we each bring, to some extent idiosyncratically, to our understanding of the factor label. On the other hand, there was an underlying rationale for the neologisms that Cattell created, as we noted above. Of greater importance than its origins or its labels is the purpose this model serves today in guiding theory building and research, and in developing instruments for personality assessment.

Paul Costa and Robert McCrae (1992), who have been very active in FFM research, have developed one of the better known questionnaires for measuring the strength and direction of these factors in individuals (see their *Revised NEO Personality Inventory* [NEO-PI-R] and the companion manual). The value of a highly valid and reliable instrument is that it allows one to do basic as well as applied research. Of course, any research makes assumptions that others can question or dispute, but once one has accepted both the philosophical and the methodological assumptions the researchers make, the results of rigorously conducted and replicated studies can give us further insight into the Human.

McCrae and Costa (1999) make four assumptions about human nature. First, they state that it is knowable. There is a certain lawfulness in the way humans experience and conduct themselves; this implies that the nomothetic is not overwhelmed by the personologically idiographic. We do have a basis for developing a science of personality, classifying behaviors and experiences, and generating a metric for assessing people according to meaningful personality dimensions. Second, they assume that humans are capable of rendering a relatively objective judgment about their own character and that of others. This is important given that their research has utilized self-reports and peer and spousal ratings of their subjects. They would seem to have little time for post-modernist constructivists who cast doubt on the veridicality of any individual's assessment of the world in which he or she lives. Third, they affirm that individuals can vary widely on the numerous traits that constitute the Five Factors – "an obvious premise for differential psychology" (p. 141). This accommodates the cherished views of humanist psychologists about the individuality and uniqueness that are the basis for human dignity – at least in part. Fourth, they hold that in spite of the dispositional and longstanding character of our orientation to the myriad stimulus fields with which we are presented, there is a spontaneous element to much of human behavior. The source of our behavior is, in significant ways, within us, and we are not simply prisoners of our personal history. This is clearly a rejection of hard determinism.

Costa and McCrae (1988) examined data from a six-year longitudinal study and concluded that there is relative stability in the typical behavior we exhibit in the face of environmental demands. In other words, we *are* programmed in large part by our histories of reinforcement. We are pushed from behind as much as pulled from in front. Some back-pedaling on this issue is evident in a recent study in which they and colleagues sampled five populations that differed linguistically and culturally. They found that older men and women measured lower on extraversion and openness than did younger adults, but higher on agreeableness and conscientiousness. The fact that these populations were so divergent largely obviated the threat to the validity of these findings from generational (cohort) effects. This latter finding corroborates what developmental psychologists have long affirmed: that is, human development is a lifelong process that does not get arrested at any stage of life (Baltes, 1987).

Reservations about Three- and Five-Factor Models

We have already signaled some of the troublesome aspects of the lexical approach to trait formulation and addressed cross-situational inconsistency in people's behavior. William James cogently addressed this problem, as have later behaviorally oriented theorists such as Bandura. There are criticisms that are more specific to the FFM. Parsimonious models run the danger of excluding many important personality traits and of raising categories to such a level of abstraction that they have only limited practical use. This is as true of Digman's (1997) two-factor model as of any other (Blackburn *et al.*, 2004). This model is a reduction of conscientiousness, agreeableness, and neuroticism into the singular Factor Alpha, and openness and extraversion into Factor Beta. In the vast canvas of personality traits, Eysenck's three-factor model integrates important dimensions of human variability, but much of importance has been left out. The Five-Factor Model, proposed by Costa and McCrae, purportedly embraces the universe of human dispositions and the traits that are subsumed by them. Out of this model they have elaborated the Five-Factor theory, which is a highly interactive system of dynamic elements predicated on postulates that have variously compelling rationales (McCrae and Costa, 1996).

It is not the highly abstract character of the Five-Factor schema that provokes the most criticism. After all, any descriptive science necessarily employs abstractions. What is most troubling is that the FFM does not comprise many important facets of personality, at least in a manner that does not look forced and contrived. Where is wit and humor, whether dry or moist, generated in a party or in the midnight solitude of one's study?

Where is hedonism, the Falstaffian self-indulgence that rejoices in the tavern or the boudoir? Does it get shoehorned into openness? Under which of the basic tendencies do we place stubbornness and tenacity? Within conscientiousness? Even when it seems more tinged with obsessiveness than with norm-compliance? Likewise, how should we conceptualize spirituality and religiosity? Sexuality or lubricity? Craftiness and cunning? Playfulness and teasing? Creativity and imaginativeness? Simplificationism and astuteness of judgment? How does the FFM help us to grasp the developmental and motivational nature of responses made to myriad situations that evoke complex responses (McAdams, 1996)? Creative readers, functioning as lay scientists, can do an "eyeball factor analysis" and place these and many other traits in one of the five clusters of human tendencies – and defend their decisions. But the critic will ask, now that you have shoehorned these traits into one or another of the categories that are called factors, how have you thereby advanced our understanding of the complexity of the human personality? More importantly what have you taught us about the causes of these traits? "Where in this system is there a place for the concepts of the grand theories of the past" (Digman, 1997, p. 1246)?

Competent personality researchers have proposed many schemas for understanding personality. The constructs they have developed have different nomenclatures and are, moreover, predicated on somewhat different arrays of raw data. As a consequence, objections have been raised to this "jingle-jangle of labeling" of the five factors. Over and above the jingle-jangle, Block (1995b) stated that "there are good conceptual and empirical reasons for being unhappy with the labels and meanings that have been variously ascribed to the five absolute factors" (p. 226). Of course, unhappiness over the matter of nomenclature depends in large measure on the purpose served by the label. Names for things do not necessarily have a meaning. They may simply be indicants: for example, labels such as Roman numerals. If one's purpose is to clothe a personality type in a meaningful label, say, "paranoid," one needs to be able to justify the accuracy of the label. If one's purpose is to be vague and nondescriptive, for example, designating a personality as "borderline," one escapes criticism for being inaccurate, but not for being vague and noninformative, which is a different problem. Likewise, using umbrella terms, and lay persons' terms at that, with all their surplus meanings, to designate factors that are fuzzy and overlapping sets, necessarily entails some confusion.

Summing up

Heidbreder (1933, p. 8) wrote, "Knowledge sometimes stimulates thought, but the right amount of [knowledge] must be present – neither too much nor too little." The 18,000 descriptors for human personality that Allport found in the English lexicon is clearly too much for the human mind to grasp at one time, and it befuddles our comprehension. If we wish to grasp it all at once, we have to have some organizing principles that allow us to wrap our minds around these myriad data. Those who have factor analyzed these data sets have reduced and sorted the descriptors of personality into categories that simplify the cognitive task for the student of personality. Of course, in simplifying the task, researchers have had to collapse their descriptors and blur their specificity. When this occurs, much of the precision, detail, and vividness found in these descriptors is lost. The big picture which the Big-Five factorialists have given us has beclouded the multitude of little pictures from which it was generated. Have we paid too high a price in specificity and detail in order to achieve the level of simplicity that makes "personality" comprehensible? Is there not a middle ground?

The answer must be framed, again, in terms of scientists' objectives. What is their human concern, and what are they trying to achieve? What advantage is there to reducing the number of personality units from 1,000 to 16, from 16 to 5, or from 5 to 3? From 5 to 2? Why not reduce the 2 to 1 as a classical metaphysician would? Is this a scientific issue or is it simply a matter of "taste" and temperament? Are some of us characterologically splitters and others of us lumpers? What are the lumpers in this scientific domain attempting to achieve that the splitters are not? The researchers in the Big-Five Model or the Big-Three tradition are lumpers *for a reason.* They have chosen to move up the Linnaean scale rather than down. This tip-of-the-hierarchy focus reveals a greater interest in the commonalities of personality descriptors than in the complex, richly detailed, subtle molecular differences. The objectives of the factorialists have been more nomothetic than idiographic. But even this generalization needs to be qualified, as there have been excellent personality measures that have been developed within the framework of the FFM and the sixteen factors of Cattell.

If we are set upon reducing the complex to the simple, the multifariously phenotypic to the universally ideal form, we are in a neo-Platonic world where we explain the multiplicity of phenomena by resorting to the idealist forms that underlie them all, using abstract, simple Platonic schemas. In its ideal form, this produces a theory of everything that can describe – and possibly explain – the totality of the human personality, but little of its humanly meaningful diversity.

6 The puzzle of the self

> Motile microorganisms of the same species are like solitary eccentrics in their swimming behavior. When they are searching for food, some tumble in one direction for precisely so many seconds before quitting, while others tumble differently and for different, but characteristic, periods of time. If you watch them closely ... you can tell them from each other by the way they twirl, as accurately as though they had different names.
>
> (Thomas, 1979, p. 2)

The search for *self*

William James ([1980] 1990) wrestled with the problem of "selfhood" and "personal identity." He stated that determining what these "entities" are was the "most puzzling puzzle" of the psychology of his time (p. 213). No wonder; there was no agreement (nor is there still) as to just what these constructs comprise. Cushman (1995) stated that "each era has a predominant configuration of the self, a particular foundational set of beliefs about what it means to be human" (p. 3). In the same vein, Leahey (2000) commented on the numerous historical notions of *self* that have been the legacy of Western culture. He gave several examples:

> Augustine looked inside himself and found God. Descartes looked inside himself and seemed to find himself. However the leading British philosopher of the next century, David Hume, looked inside himself and found sensations and images from the world and the body, but no self. Rejecting this alarming conclusion, leading German philosopher [Immanuel] Kant could not find a self either, but posited it as a logical necessity. He then imbued it with enormous powers, and his followers, the German idealists, said that the invisible Self is so powerful, it creates the universe. (p. 154)

Given the historical diversity of views on selfhood, one is entitled to ask that if one does not know what one is looking for, the forms in which it manifests itself, what its constituent parts are, where it is to be found, or even if it exists, how can one subject it to a reasoned examination? The broader question, in fact, that has been posed by contemporary empiricists is whether such constructs as "self" (or "soul," "will," "consciousness," "mind," and "personality" for that matter) fall into a category of constructs that are more objects of belief than of scientific scrutiny. This is the problem this chapter will address.

Toward a definition of self

Needless to emphasize, the roots of these questions can be traced to the ancients, most notably to such Stoics as Chrysippus (*c.* 280–*c.* 207 BCE), Marcus Aurelius (*c.* 121–*c.* 180 CE), Porphyry (*c.* 234–*c.* 305 CE), and Seneca (*c.* 1 BCE–*c.* 207 CE).[1] But whereas the matter of character, individuation, and identity was of great interest to both the Stoics and the nascent Christian Church (especially for the latter in view of the problems posed by the doctrine of the resurrection of *individuals*) it did not beget the ethos of individualism that emerged in the late Middle Ages.

William James, one of the most eminent of modern psychologists, thought and wrote extensively on the subject of the self, puzzle though he thought it was.[2] We need to understand that although James had been schooled in a German laboratory-based tradition, he had a flair for the philosophical that bespoke his reservations about a strictly empiricist approach to the more elusive realities of the human person.[3] For this reason, William James had many ideological enemies, principally the structuralists such as Titchener who was also associated with the laboratory-based institutes of German science. But James, always the pragmatist, never let himself drift into the rarefied realms of metaphysics. He was one of the founders of the American school of pragmatism and

[1] Consult Richard Sorabji's (2006), *Self*, ch. 8, among others; the book is a mother lode of references and insights on this subject.

[2] Students who have the time and interest will find James' ([1890] 1990) chapter, "The consciousness of self," in his work *The Principles of Psychology* both instructive and entertaining. James will, furthermore, introduce his readers to some great eighteenth- and nineteenth-century thinkers (for example, Bain, Carlyle, Herbart, Hume, Kant, Lotze, T. Huxley, and J. S. Mill) to whom they might otherwise have scant exposure. Of interest is the fact that his *The Principles of Psychology* is the only volume in the *Great Books* series of Encyclopaedia Britannica that qualifies as a tract on psychology *qua* psychology.

[3] James anticipated a large number of eminent psychologists who distanced themselves from, if they did not actually repudiate, a laboratory-based, natural sciences approach to the study of the human psyche. In the 1920s, Lev Vygotsky (1997), for example, took the position that a methodology used to investigate the subject of one science is "blind, and leads nowhere when transposed to another science. Such blind transpositions of the biogenetic principle, the experiment, the mathematical method from the natural sciences created the appearance of science in psychology, which in reality concealed a total impotence in the face of the studied facts" (p. 280).

retained a philosophical commitment to a positivism[4] that postulates that in the final analysis we must ground our science in the consensually validated evidence of our senses. This ambivalence continued well into the twentieth century. Cosmologists like Antony Hewish like to refer to themselves as essentially observers, distinguishing themselves from others, such as Sir Martin Rees or Stephen Hawking, who are modelers. Sir Arthur Stanley Eddington, however, is reputed to have said that one should not trust an observation unless it is grounded in sound theory.

As the positivist current of behaviorist and structuralist[5] psychology took possession of American academia in the first half of the twentieth century, the study of such "hypothetical constructs" as self and consciousness, difficult as they were to manipulate and measure experimentally, was largely ignored or abandoned to the speculative and anecdotal writings of philosophers and others outside the vaunted towers of academic psychology. Radical behaviorists considered it more useful to engage in a rigorously empirical, and preferably experimental, study of behavior than to develop ambitious theoretical systems for explaining constructs that were only tenuously linked to "phenomena,"[6] that is, to objects and events that are observable. One should not conclude, however, that behaviorists denied the existence of mental processes. They did not. Their position was simply that science is founded on the analysis of variables that can be rigorously manipulated, contaminant-controlled, and precisely measured. When these qualities are present, then such studies can be empirically replicated. Successful replication of a well-structured study that results in comparable findings increases the probability that psychologists have advanced their science (see Marx, 1963).

There has been a retreat from this position over the past half century on a number of fronts. First, a new cluster of psychologies has emerged that was formerly characterized as the Third Force in contemporary

[4] Positivism is the term coined by Auguste Comte (1798–1857), as well as the doctrine he developed stipulating that knowledge must be based on empirically verifiable data rather than on theological or metaphysical speculations.

[5] Recall that structuralists are interested in the architecture of the psyche more than in how it functions. Just as anatomists study the structure of the *soma*, the structuralists proposed that psychologists should study the structure of the *mind*. And one needs to understand what the structures are before one studies what they do; this latter task is the domain of the functionalist.

[6] The term *phenomenon* derives from the Greek noun that means "appearance." Phenomena are people's perceptions and awareness of objects within their sensory (visual, auditory, etc.) field. There can be no assurance that *how an event or object is perceived* and *how it exists* independent of the knower are isomorphic. When we speak of *phenomenal field*, for example, we are often referring to the private and subjective experience individuals have of their world (both internal and external). As James would say, reality and the knowledge of it are two different facts.

North American psychology. This movement, which falls under the comprehensive rubric, Humanistic Psychology, is associated with the work of Abraham Maslow (1954, 1968), Carl Rogers (1947, 1955, 1961), Fritz Perls (1969; Perls, Hefferline, and Goodman, 1951), Gordon Allport ([1937] 1961, 1955, 1968), and a gathering of thinkers who wished to distinguish themselves from the behaviorist tradition, on the one hand, and the psychoanalytic tradition, on the other hand. Further, they wished to distance the study of the Human from the methodologies that are so successful in the realms of physics, biology, and other nonhuman sciences. We will return to an analysis of this broad movement in Chapter 11, but the reader may wish to see contemporary statements on this issue by Kurt Danziger (1997), Michael Mahoney (1991), and Martin and Sugarman (2000).

Second, a remarkable burst of activity in cognitive psychology occurred in the 1940s and 1950s: Noam Chomsky (1957), Warren McCulloch (1965), Allen Newell, J. C. Shaw, and Herbert Simon (1958), Lev Vygotsky ([1925] 1997),[7] and Norbert Wiener (1954), among others, placed cognitive activities under academic klieg lights.[8] All of these researchers gained added respectability for inquiry into the processes of our "interior life." With the development of information science and the new field of cybernetics, psychologists were given powerful new metaphors for their ideas as well as new heuristics from which they could develop novel theories of cognition. As Gigerenzer (1991) has demonstrated, new tools and technologies are fertile ground for spawning new psychological theories. The computer and other servo-mechanisms rapidly became metaphors for conceptualizing the structures of the mind, just as scientists in the Newtonian era used the clockworks of the Renaissance as a metaphor for conceptualizing a celestial mechanics (created by a "Divine Watchmaker").

Third, advances in the neurosciences have revealed much about the neural maps that we make of our internal body states and the state that we refer to as consciousness – and eventually as self. Prominent among the pioneers of this domain were Herbert Jasper and Wilder Penfield of the Montreal Neurological Institute who, beginning in the 1930s,

[7] Lev Vygotsky's work, largely done in the former Soviet Union, dates from the 1920s and early 1930s; he died in 1934. Although a committed Marxist, he did not find favor with the ruling regime as he was flagrantly critical of the "reflexological" (Pavlovian) psychology that fit so well with the purposes of that regime. A newly translated six-volume collection of his work has recently been published by Plenum Press.

[8] Consult table 4.1 in Mahoney, 1991, for a list of many of the signal events that mark the history of the cognitive sciences and their ascendancy in the post-Second World War era (pp. 67–76).

collaborated on studies of consciousness and learning in both healthy and neurologically impaired individuals. More recently, critically important work in neuropsychology has been conducted by prominent researchers such as Patricia Churchland (1996, and 1998 with Paul Churchland), C. Robert Cloninger (2004; see especially pp. 50–60), Antonio Damasio (for example, 1994, 1999), Daniel Dennett (for example, 1991), Gerald Edelman (1992; for example, with Giulio Tononi, 1998), Eric Kandel (2007), Roger Penrose (for example, 1994), and John Searle (1992). The work of these neuropsychologists has given an added impetus and respectability to work in this field. This is most notable in the work of Damasio, who provides a neural explanation for a psychology of *self.*

The self: a semantic distinction. The term self occurs frequently in speech, in English as well as in other languages. But as the philologist Nietzsche has warned us, language is a blunt instrument, and humans often fail to make proper distinctions when using it. This is complicated by the fact that most words that are not proper nouns, or that do not specify some highly specific – often technical – operation, have more than one meaning. Any unabridged dictionary will provide ample evidence of the numerous definitions associated with the most ordinary words. This problem is further compounded by the subtle alterations that occur in the meanings of words as contexts change and paravocal cues overlie the words spoken. Because we habitually use expressions such as himself or herself, self-esteem, self-consciousness, or self-enhancement does not mean that we are free to attribute the psychological state of selfhood and personal identity to their referents. After all, we say that bacteria have the capacity for *self*-replication and *self*-preservation without implying that they have a self, and still less a sense of self.

The use of the term self in a purely recursive way, as in "she dressed her*self* for the occasion" or "he scalded him*self* with the broth" is quite different from what James calls "the nuclear Self" ([1890] 1990, pp. 195–96), what we intend to convey when we say that humans can sense at their core the agency and the ownership of their unique experiences and acts. Their sense of self springs from the conviction that the thought they are mulling over in their minds or the pain or pleasure they are experiencing belongs uniquely to them. When we say, "I have a headache," we are saying, "this pain belongs to *me* and not to *you*, as empathic as you may be." The question we must ask ourselves is whether or not such terms as "himself," "herself," or "itself" are just colloquial terms that, when introduced into scientific discourse, carry surplus and irrelevant meanings that tend to obfuscate any genuine attempts to identify and describe a true psychological self.

Two divergent views of self

Two divergent explanations of self, one from the nineteenth century and another from the cusp of the twenty-first century, will bracket a larger discussion of the issues involved in understanding what constitutes selfhood and personal identity. Let us begin with the classic position of William James ([1890] 1990, chapter X, "The Consciousness of Self") and then proceed to the vision of some contemporary neuropsychologists, focusing specifically on the work of Antonio Damasio (1994, 1999, 2000, 2003).

William James' view

James asked profound, searching questions about the self and made many distinctions in his discussion of this construct. The first distinction bore on what each of us considers "the me" and "the not-me." This distinction first occurs when humans develop a sense of self-awareness within the first 15 to 18 months of life. The elegant studies of Lewis and Brooks-Gunn (1979), for example, revealed that infants demonstrate a mirror-recognition of themselves at this age (*cf.* Robins, Norem, and Cheek, 1999, pp. 454–56).[9] But for James, this distinction is not as simple as one might initially think. He stated that at certain times one may consider the same object, say, one's big toe,[10] "part of me, at other times as simply mine, and then again as if I had nothing to do with it at all" ([1892] 1963a, p. 188). Some people think of their body as simply theirs; others think of it as *them*. Some people who consider their life a religious pilgrimage to a celestial hereafter treat their body "as prisons of clay from which they should some day be glad to escape" (p. 188). James made further

[9] Mirror-recognition is not the exclusive capability of humans. Gallup (1970, 1977) experimentally demonstrated that chimpanzees easily develop the ability to recognize themselves in a mirror (MSR), suggesting that self-awareness is a characteristic of the great apes. Students may ask themselves how far one can descend on the phylogenetic scale before the capability of "self-recognition" no longer occurs. The answers to such questions fall within the domain of comparative psychology. A 2008 article by Helmut Prior, Ariane Schwarz, and Onur Güntürkün, "Mirror-induced behavior in the magpie (*Pica pica*): evidence of self-recognition," is a rich source of references to such research.

[10] After I wrote this reference to a toe, I came upon the same example in McAdams' (1994) gloss on James' concept of the Me (p. 536). As Goethe is reputed to have said, a big problem for all writers is to discover who anticipated them in articulating a good idea (indeed, even a bad idea).

distinctions and wrote, "In its widest possible sense, however, *a man's self is the sum total of all that he* can *call his*, not only his body and his psychic powers, but his clothes and his house, his wife and children, his ancestors and friends, his reputation and works, his lands and horses, and yacht and bank-account" (original emphases) (p. 188).

Among the constituents of the *self* in James' view are a biological self that constitutes our physical identity; the social self that reflects the skein of ever-changing relationships by which we are linked to our fellow humans; a private, psychological self to which only we are privy; and a spiritual, largely unconscious, self which is only imperfectly accessed by individuals themselves. Our self-concept embraces with greater or lesser specificity all of these selves, albeit in a highly subjective and imperfect way. If our view of ourselves in a normal state is necessarily limited and imperfect, it is nonetheless usually more reliable than the assessment of us that can be made even by astute observers.

The material, empirical "me" that is the self is first of all one's body. Without the body, the science of psychology would have to acknowledge the disappearance of the self, for there would no longer be an *empirical* me. However, our sense of self is much more than that. When our mother and father die (or when, possibly, our spouse and children predecease us), we experience a diminution of our self. Should we lose all of our material possessions, or if the product of a lifetime of our labor should get swept away, we would even sense the shrinkage of our personality, "a partial conversion of ourselves to nothingness" (James, [1890] 1990, pp. 213ff.).

The Jamesian view of "the social self" is that it is the recognition that we get from our fellow humans. That self blossoms under the benign, nurturant, perhaps admiring, regard of at least several significant others. Like Willy Loman in Arthur Miller's *Death of a Salesman*, or the Black protagonist in Ralph Ellison's *The Invisible Man*, each of us has a desperate need to be seen, recognized, and acknowledged. But as we are regarded by others in a variety of social contexts and multiply diverse relationships, the social aspect of our self takes on various characteristics depending on whom we are with and where we are. Different social situations have different demand characteristics with which, in varying degrees, we idiosyncratically conform. In James' view, this calls forth a plurality of selves. Our self-concept is necessarily a composite of these selves, albeit highly subjective and imperfect. If our view of ourselves is necessarily limited and imperfect, it is still typically more reliable than the assessment others have made of us.

As I noted briefly in a different context (Chapter 5), James wrote:

Properly speaking, *a man has as many social selves as there are individuals who recognize him* and carry an image of him in their mind. To wound any one of these images is to wound him. But as the individuals who carry the images fall naturally into classes, we may practically say that he has as many different social selves as there are distinct *groups* of persons about whose opinion he cares. He generally shows a different side of himself to each of these different groups. Many a youth who is demure enough before his teachers and parents, swears and swaggers like a pirate among his "tough" young friends. We do not show ourselves to our children as to our club-companions, to our customers as to the laborers we employ, to our own masters and employers as to our intimate friends. From this there results what practically is a division of the man into several selves; and this may be a discordant splitting, as where one is afraid to let one set of his acquaintances know him as he is elsewhere; or it may be a perfectly harmonious division of labor, as where one tender to his children is stern to the soldiers or prisoners under his command. ([1890]1990, pp. 189–90)

If you find this has a modern ring to it, that may be because so many of James' ideas have entered, and to an extent shaped, the *Zeitgeist* of contemporary thought. Hartshorne and May (1928, 1929) demonstrated that the traits that we manifest in one social situation might not be expressed in another. The person who acts with bravado and aggressiveness in a school yard or in a bar may act with meekness and passivity in a police station. The youth who acts like a "hooligan" at a sports event may be a paragon of discreet, even demure manners at the wake of a close relative. Walter Mischel (1968), anticipating the formulation of the Fundamental Attribution Error (Ross, 1977), provoked an academic tempest when he demonstrated that scholars underestimated the influence of social situations on people's behavior and exaggerated the influence that character traits have relative to the character of the social situation in which they find themselves. (The exact proportions of the variances in behavior that can be attributed to enduring personality traits, contextual situations, and, say, biological or drug-induced variables are vigorously debated in the 2000s.)

This theory of the self is both relativistic and relational. The relativistic issue raises two difficulties. First, there is situational relativism. Observers cannot distinguish which behaviors that they have just witnessed arise from the demands of the situation in which they have occurred, and which arise from "the true self" of the actor. This is all the more problematic if the only sampling of someone's behavior we have is that which has occurred in coercive and ambiguous circumstances. This is why judging character and aptitude by an applicant's studied "performance" in a job interview or an interview for admission into a graduate program is a dicey project. The façade, that is, the *persona*,[11] that individuals will adopt in order to create an impression that will be advantageous to them, raises vexing issues. To what extent can we attribute to the *authentic self* the various personas that individuals adopt in various situations to meet (possibly flout) the social expectations of different sets of observers? Are we acting inauthentically when we act differently in different situations? How habitual must the adoption of a particular persona for a specific situation become before it is authentic? And to what extent are these behaviors under our conscious control? Further, how can one factor into one's self-image the richness and subtlety of one's repertory of behaviors and personages? Some do this role-changing and persona-switching as easily as slipping out of dress shoes and into loafers. If, indeed, some heretofore self-consciously adopted personas have become "second nature," how can one deny that they are truly part of their owners' personality, no matter how inconsistent they may be with other personas adopted by them for other venues and occasions?[12]

Second, actors (those who engage in a behavior) have a different take on their thoughts than do those who are external observers. They have an intimacy with their own thoughts that others can never share; they are generally more aware of how they want to come across to others than are external observers. They make evaluative judgments of the situation in which they have to perform, the importance of making a good impression, the extent to which they may have to tailor their behavior to achieve their

[11] The term *persona* derives from the Latin term for the individualized mask that was worn by an actor in the Roman and Greek theater to signify the character he or she was portraying. Carl Gustav Jung referred to the public façade that individuals present in various social situations as their persona. Its function was to conceal those aspects of their personality that they did not want the world to view, and to enhance those aspects that they did. In brief, the persona is an edited version of one's personality. This Jungian usage has by and large been endorsed by social psychologists.

[12] Sigmund Koch (1981) asked profound questions about the self and how we get mired in its ambiguities and bottomlessness. Am I, he asked:

> a success or a failure? If a failure, is it by virtue of having successfully sought failure, or by ineptitude, the fault of others, or just bad luck?
>
> Do I understand this equation, this line of poetry or prose, this view, theory, subject, person, event? Really understand, or merely think that I do? Do I really like X (e.g., any object of taste) or only think that I do because I should? As a matter of fact, should I really? Should I wear X or Y today, or are they both inappropriate? Am I showing favoritism toward one of my children, or is that child the one who needs special attention?
>
> Can I sustain this performance? Am I doing brilliantly or did I lose it somewhere? Does the audience resonate to me or does it loathe me? If the latter, does it loathe my person, my ideas, or merely my words?
>
> Sometimes I am convinced that they are about to fire me, but five minutes later the same evidence seems to mean that I am on the verge of being promoted …

objectives (which may only be vaguely formulated), and their level of "comfort" with the persona the situation calls for. Observers can view them only through inferential lenses and then make conjectures about what is *really* going on. When William James speaks of multiple "social selves" as one aspect of the human self, these are some of the operative dynamics that constitute them.

The relational aspect of James' theory lies in the relation of personality to self. One can be considered the flip side of the other. When Ethel Merman sang "Personality" in the Broadway musical, "Anything Goes," she exalted those aspects of the social self that are evident to observers. When James wrote of the mutations of the self, he was not only alluding to how we construe our autobiographical record in attempting to understand ourselves and embellish our self-image, but also asserting that we are often engaged in presenting that embellished record to others. Memories are never exactly veridical, and our memories of our past are under constant revision, even on a daily basis. James ([1890] 1990) stated that "false memories are by no means rare occurrences in most of us, and, whenever they occur, they distort the consciousness of the me ... The most frequent source of false memory is the accounts we give to others of our experiences" (p. 241). What is most sobering about this view of the self (should one give credence to it) is that even when we are intent on giving true testimony to others about episodes in our personal history, significant distortions creep in.

Consistent with the Jamesian view is the recent work of Laurence, Day, and Gaston (1998), who demonstrated that one's characterization of oneself or of others can be profoundly shaped by the situational as well as self-imposed demands that one experiences when one is called upon to testify under oath in court. Elizabeth Loftus (2003) makes a comparable observation: "Who we are may be shaped by our memories, but our memories are shaped by who we are and what we have been led to believe" (p. 872). She quotes Sally Satel, a psychiatrist: "We are always angling the prism of memory" (p. 31). Loftus continues: "We seem to reinvent our memories and in doing so, we become the person of our own imagination" (p. 872).

Beyond the phenomenal self (the self, that is, that one can observe), there is the "self" as a pure principle of personal identity. James refers to this latter "self" as "the pure ego," that constituent of the self that since the time of David Hume has been "the most puzzling puzzle." It is not the phenomenal *me*, the me that can be observed by others as well as myself that is a puzzle. After all, that can be subjected to examination. It is the "I" from which we get a sense of personal sameness from day to day and year to year that is difficult to nail down. Should *you* remember yourself as a young child, as a pubescent teenager, as a young adult, as an older adult (should such be the case), as you were yesterday, or even 10 minutes ago, you will note you are not truly the same in all respects *at this very instant* as you were at any of those times. What is it, then, that permits us to say that we have the same identity or are the same self as we were yesterday? When so much else changes – your maturing body structure, your ideas and affections, your tastes in clothes, politics, and diet, your values and career aspirations, perhaps your legal name, even the face that peers vacuously at you from the bathroom mirror in the morning – what is the enduring residue in this sea of changes that you can label "self" and that allows you to say, I am the same *person* that I always have been?

Following a set of finely reasoned passages, James concluded that it is the (unbroken) continuum of bodily feelings, the remembrance of events past that happened to you, the recurring pulse of consciousness as thought succeeds thought, the interiority and "warmth" of your ideas that generate a sense that they belong to you and not to someone else, the recognition of your body in the morning as the same one that drifted off to sleep the night before – it is this aggregate of variables that constitutes the self. When Peter or Paul wake up in the morning, each says, "Here's the same old bed, the same old room, the same old world" just as he says, at least implicitly, "here's the same old self." It is as if at each moment that we are conscious we implicitly affirm our ownership not only of our present state of consciousness, and the body that underpins it, but of all those states that preceded this one – our autobiographical record in other words – that give us a sense of continuity. If we can remember nothing that has occurred before this present instant we have lost our sense of self and personal identity. This self endures as long as our memory is not destroyed by disease or, presumably, by death.

A neuropsychological view

Numerous researchers in the domain of cognitive psychology and neuropsychology (some of whom we cite elsewhere in this book, for example, Chapter 6) have attacked the problem of human consciousness and the "puzzle" of the self. Their work is enormously important to students of personality as advances in this field will increasingly become enmeshed with the science of the mind and its neurobiological foundations. Antonio Damasio, based on his clinical research, has written two books (1994, 1999) that are accessible to students without a neurological background. I have made the difficult choice to focus on his work, rather than that of so many comparable others, as it is as lucid, cogent, firmly grounded in research, and as centrally situated at the leading edge of this field as any.

In the early pages of his book, *The Feeling of What Happens*, Damasio (1999) describes an incident in which a patient, while remaining awake, made an abrupt passage from a state of consciousness with a sense of self, to a state in which he was deprived of both. "The loss of consciousness was radical, something like a complete power outage" (p. 6). We can let Damasio tell the story in his own words:

> A man sat across from me in a strange, entirely circular, gray-painted examining room. The afternoon sun was shining on us through a skylight as we talked quietly. Suddenly the man stopped, in midsentence, and his face lost animation; his mouth froze, still open, and his eyes became vacuously fixed on some point on the wall behind me. For a few seconds he remained motionless. I spoke his name but there was no reply. Then he began to move a little. He smacked his lips, his eyes shifted to the table between us, he seemed to see a cup of coffee and a small metal vase of flowers; he must have because he picked up the cup and drank from it. I spoke to him again and again he did not reply. He touched the vase. I asked him what was going on, and he did not reply, his face had no expression. He did not look at me. Now he rose to his feet and I was nervous; I did not know what to expect. I called his name and he did not reply. When would this end? Now he turned around and walked slowly to the door. I got up and called him again. He stopped, he looked at me, and some expression returned to his face – he looked perplexed. I called him again, and he said "What?" (pp. 5–6)

This patient suffered from an absence seizure followed by an absence automatism. But what else did he suffer from? What was his identity, as he perceived it? Psychologically, who was he? What had happened to his emotional processes – and the feelings induced by his emotions? If these feelings were compromised, then his knowledge of these feelings, his awareness that he was experiencing these feelings, was also compromised. What had happened to "the pure ego" as James called this self? And what exactly is this sense of self, *his self*, which apparently was absent for several minutes?

The sense of self is always compromised when consciousness is impaired. There are two kinds of consciousness however. First, the basic kind is core consciousness which provides us with a sense of self at this moment but not in the many moments that preceded it – nor anticipated moments that may follow it. Core consciousness allows us to sense the here-and-now in which we exist without reference to either "a before" or "an after." Core consciousness is a biological process that does not require reflection, reason, working memory, or, of course, language. Second, the consciousness from which the *fully* human self emerges is called "extended consciousness."[13] It comprises the rich storehouses of memories of the past – and the "memories of the future," those cherished scenarios that we nurture and hope to actualize. Our autobiographical memory is the shifting, ever-eroding, ever-expanding, self-renewing foundation of our elaborate, multi-layered sense of *self*. To the extent that this extended consciousness is lost through normal or pathological processes, *self* is diminished.

Damasio (1999) believes these two kinds of consciousness generate two kinds of self: "The sense of self which emerges in core consciousness is the *core self*, a transient entity, ceaselessly re-created for each and every object with which the brain interacts" (p. 17). The self in its conventional and popular sense, the self that establishes our identity, is "the nontransient collection of unique facts and ways of being which characterize a person" (p. 17). Damasio calls this entity the *autobiographical self*. This historical record includes our idiosyncratic tastes, character traits, typical problem-solving and coping styles, and biographical data such as our name and address, date and place of birth, memories of our family of origin, friends and neighborhoods, signal events in our personal history, and so forth. Without extended consciousness there is no autobiographical memory; and without autobiographical memory, there is no autobiographical self, that self that we commonly understand as our "self": a repeatedly reconstructed, biologically grounded, and protected sense of who we are, have been, and hope to become (Damasio, 1994, pp. 226–29). At varying levels of consciousness one must simultaneously live in the past, present, and future. Living in the present without reference to the future or the past, one loses one's self.

Consciousness, an essentially private experience, occurs as part of the process that we conceptualize as mind. It is an I-based (first-person) phenomenon that can be distinguished from wakefulness, as the vignette of the patient in Damasio's case history illustrates. It can also be distinguished from low-level attention that precedes consciousness; note how you can vaguely recall having attended to something before you became truly conscious of it. For example, one can be daydreaming about an interesting experience, say a party of the previous evening, as an object comes hurtling toward you, like a frisbee or a car. In the instant before the rumination is disrupted by the new image, a low-level attention is paid, that is just before *red alert* flashes and conscious and semi-reflexive responses erupt. On the other hand, separable as it may be from some

[13] This notion of extended consciousness approximates "self-aware consciousness" that C. Robert Cloninger (2004) developed throughout his architectonic treatise on "feeling good."

attentional processes, consciousness cannot be separated from emotion, for whenever there is conscious ideation there is emotion. A corollary of this is that normally when consciousness is impaired so is emotion. Biological motility (and all the hormonal and neural constituents of emotion-generation) always accompanies consciousness. Indeed, any experience of the normally conscious person evokes a sense of recognition or surprise, at times a combination of both – and an emotion or mood state gets played out in the theater of the body. In contrast the person in a comatose or vegetative state does not experience emotions. This is a subject to which we will return in Chapter 9, dealing with emotions.

The search for self. The primitive brain's first responsibility is to regulate the body of which it is a part. Even the simplest organisms, those bereft of a brain, have a need to regulate their internal chemistry. As the eukaryotic cell evolved into the multi-celled organism, and the regulation of its life-sustaining processes grew more complex, brains evolved to monitor and regulate those processes. The human brain, a master switching station for the organism of which it is an integral part, monitors and controls the functioning of the rest-of-the-body (ROB), including the organ systems encased in the head. Through a complex network of communication channels, neural and hormonal, the brain is informed of what is transpiring in the remote frontiers of the visceral self. At each instant it forms an "image," that is, a neural representation of the biological state of the organism. The information flows in, the directives flow out – we are not aware of them – instant by instant. This constantly reconstructed biological state is what Damasio called the *proto-self.*

The proto-self is the antecedent of the *core self*, which is the conscious, cerebral representation of the changes that from moment to moment are occurring not only in the brain itself but in the ROB as well. These changes arise from the many events that impinge on the total organism, events that originate outside the boundaries of the organism as well as within it. The brain "captures" these knowable events in sensory and motor terms and represents them in neural patterns that are distributed in various regions of the brain. These regions work in a coordinated way to exhibit the event to the core self, that "something-to-which-knowing-is-attributed." The core self is generated by higher-order (but nonverbal) symbolizations of changes in the proto-self induced by a continuous stream of events that "happen" to the organism and that trigger responses. Like the frames in a film, it is generated anew by each pulse of consciousness. This self exists only in the here-and-now, but as it is continuously renewed, moment by moment, it appears to be continuous.

Core consciousness leaves a trail of images, some of which are permanently stored. These records of past experiences, which bear not only on the past but also on anticipated scenarios for the future, constitute the *autobiographical self.* Of course, as life goes on, the forward edge of this biographical record advances, adding new, vivid memories, and the memories of the distant past get nibbled away at the edges or (if they are intimately tied into one's self-definition and identity) reconstructed or confabulated.[14] This autobiographical self is the self of tradition. It is the museum of our archaic selves, the long-lost lives of our infancy and adolescence, as well as the workshop of our present. This self is the self of the person constantly renewed, ever recreating herself or himself, using dispositions and tools from the past, and improving, if not newly designing, skills for the future. This is the self of which we are conscious and to which we attribute ownership of our knowledge and actions.

There is a complex neural architecture that supports and can be correlated with consciousness and the selves that Damasio and others posit. Although those structures are being scientifically investigated and their functions hotly debated, the neuropsychology of the twenty-first century has definitively jettisoned a spiritualist, Cartesian dualism that is more closely aligned with the rational psychology of Thomas Aquinas than the neuropsychological science of Damasio, Churchland, Dennett, and others committed to this approach. Neither the human mind nor the self is regarded as a spiritual entity except by those who choose to accommodate their psychology to the structures of a transcendent and transpersonal belief system.

Bridging from James to Damasio

The nomothetic self. The self that Damasio has described is one that, in large part, defines our human nature. This self is not simply a mental construct that humans have devised as a conceptual and rhetorical device for describing themselves. It is an ontological self for he has presented it as an entity, a structure, a reality that is grounded in the total organism – brain and the ROB. It involves all the organ systems, the neural and hormonal pathways to the brain, all the viscera, not excluding that most extensive of the organs, the envelope that we call the skin, and,

[14] Confabulation is the process through which plausible "new memories" are created to fill in the gaps in recent memory that result from pathological processes such as Wernicke–Korsakoff syndrome (a disease that results primarily from the thiamine deficiency associated with long-term alcoholism).

of course, pre-eminently, the various regions of the brain that are implicated in constituting our evolving identity and sense of who we are. Because the nature of the self derives from the nature of the human organism, all selves *qua*[15] selves are identically constituted. This is the self of the normal human being, a universal property of people of all times and places, independent of their social status, ethnic background, or life history. Among the universal properties of this self is that in its fully evolved state it is grounded in the individual's autobiography.

The self that all humans have in common, by virtue of their belonging to the same species, becomes individualized by the unique history of its interactions with its environment. In terms of the nature–nurture dichotomy, this self that is constituted by nature is differentiated along a life-long continuum by the influence of its experiences. It is a conscious, idioreflective, emotional, feeling, reasoning, environment-interactive entity that is socially defined as a person. The identity of this self? It is constituted by this life-long biopsychosocial history – which brings us to the idiographic self.

The idiographic self. Needless to emphasize, boundaries are essential to the constitution of any individual whatever, and not least, the notion of the self. If one cannot say *here* is where I end and *there* is where you begin, one cannot speak about *oneself*. The selves of various individuals become quite different from each other depending on their gender, educational background, culture, psycholinguistic training, critical life events, physical endowments, and other biographical features. Barring serious pathology, however, none of this alters their essential features. Once we begin to describe the various ways in which the self can be idiographically configured by experience, by context, and by intention, we have moved out of the ontological and into the phenomenological[16] realm, from the domain of the nomothetic to that of the

[15] The Latin conjunction, *qua*, is used as a neat, succinct way of telling readers that the term it modifies should be considered only in terms of its essential characteristics. In this instance consider selves only in terms of their nomothetic rather than their idiosyncratic aspects.

[16] The term phenomenological refers to the immediate experience of individuals, implying that their perception of reality is personally constructed and, therefore, largely idiosyncratic. For example, everyone, by and large, would agree that a Granny Smith apple is green. That is, the substance, apple, has the property (or accidental quality as Aristotle termed it) "greenness." However, we have no grounds for affirming that each of us sees the color green with the same precise shades of hue, saturation, and lightness. (The term phenomenon, and such cognates as epiphany, derive from the Greek verb, *phainein*, which means "to show.")

idiographic. It is this highly differentiated and individualized self that is the subject matter of personality psychology.

When people characterize themselves, that is, present a private portrait of themselves, it differs in certain respects from the portrait that objective observers might draw of them. Private selves and public selves can diverge dramatically depending on how self-disclosing or secretive the individuals in question are, on the one hand, or how well the observers know them, on the other hand. Robins, Norem, and Cheek (1999) have refined the perspectives from which we can analyze the self (pp. 449–51), by layering it on the following psychosocial levels: the *personal* (that is, the private); the *relational* (that is, the intimate); the *social* (that is, the interpersonal); and the *collective* (that is, the communal).

The first layer embodies the individual self with its traits, values, private aspirations, and regrets. An emphasis on this level of self-regard is characteristic of cultures that value independence, autonomy, and individualism. The self has experiences that are private and inaccessible to anyone else but the one who owns them. As one gazes into the Grand Canyon, listens to a song by Barenaked Ladies, or attends the funeral of a loved one, one is swept by feelings and ideations that no one else can know – others can only imagine or suspect what one is feeling or thinking at any given time. This is the phenomenological self. From another perspective, one can study oneself as one would any other object in the universe. One can try to get to know oneself better – a process that normally transpires in psychotherapy or growth counselling. This represents the epiphany of the "me" – the self as known, to the "I" – the self as knower. One outcome of such a process is to represent oneself in terms of personality variables.[17] One can think, "I see myself as an ambitious,

[17] Students may be confused as to the boundaries that exist between self psychology and trait psychology. This is understandable as the boundaries have never been drawn in a convincing way. The two fields exist in a symbiotic relationship, each feeding off the research and theorizing that is done in the other. This is not a bad thing, and it can be argued that they set a good example for other psycho-social disciplines. Related to this, Lewis Thomas wrote an engaging essay, *The Medusa and the Snail* (1979, pp. 1–5), in which he described how one marine creature attaches itself to another, thereby allowing each to benefit – for a time. The medusa, a jelly fish, affixes itself to the ventral surface of the snail, near its mouth, thereby making a comfortable and easy living and generating lots of offspring who eventually wander off. As the snail produces its larvae they are engulfed, trapped one might say, by the medusa. Being hungry, these little snails munch on the medusa, reducing it to its original size, as they grow into mature, healthy adult snails. And the dramatic cycle is repeated, again and again, to the mutual benefit of both the medusa and the snail. We leave it to the reader to speculate as to whether this story is an apt metaphor for the relationship of self and trait psychologists. It would certainly seem aptly to describe the relationship of psychotherapists and self/trait psychologists.

cautious, planful, fearful, but fun-loving and lazy person." Or one can study oneself in terms of the Five Factor or 16 Personality-Factor models presented in Chapter 5 dealing with personality traits.

The other layers, in a descending order, delineate the self in ever increasing involvement with its surroundings. At the second level is the *relational* self that involves one's personal interactions and intimacy with family and close friends. An emphasis on this level implies a self-regard that is qualified by one's relation with these significant others, and the esteem in which one is held by them. This entails a certain interdependence that encroaches on personal autonomy. We measure the impact of our behavior on important others, and shape our presentation of self accordingly. The third level involves the larger *social* self which is implicated in the institutional life of one's immediate social world. This level involves all our professional, occupational, recreational, and legal roles. This level places still stronger constraints on the self, and to the extent that one allows this to influence one's life, moves the individual into a more collectivist and interdependent lifestyle. The fourth level defines the *collective* self that implicates the ethnic, religious, cultural, and racial group(s) with which one identifies, and in which one takes pride. The reader may wish to add further dimensions to this multi-layered schema, for example, the racial self which allows one to identify with the species, *Homo sapiens*, the terrestrial self in which one self-identifies as an earthling, and a galactic self, in which one identifies with the denizens of the Milky Way, described by *sci-fi auteurs*.

Relative to a social perspective, the self is never regarded in isolation from the social and biological world in which it is enmeshed. The social self psychologist specializes in the study of the evolving, shifting, adaptive, and chameleonesque personas that humans assume – sometimes in spite of themselves – in the face of environmental challenges and demands. Partisans of this approach often reproach those who have adopted a more individualistic approach for neglecting the pre-eminently social character of the Human. Certainly, the Zeitgeist of the past generation has favored a more sociologically-oriented approach to an understanding of the human (see Tooby and Cosmides, 1990, relative to the *Standard Social Science Model* of human nature). Plato and his student, Aristotle, both referred to humans as essentially gregarious[18] animals and recognized the influence of the family and community on the development of the Human from infancy to adulthood.

[18] Gregarious derives from the Latin terms, *grex* (a flock or a crowd) and *gregarius* (inclined to assemble or live in groups).

Back to the future: James, Baldwin, Cooley, Mead, and Sullivan

The "I" and the "Me" of James. James ([1892] 1963a), as noted above, made a distinction between the self as "I" and the self as "me." He wrote:

> Whatever I may be thinking of, I am always at the same time more or less aware of *myself*, of my *personal existence*. At the same time it is *I* who am aware; so that the total self of me, being as it were duplex, partly known and partly knower, partly object and partly subject, must have two aspects discriminated in it, of which for shortness we may call one the *Me* and the other the *I*. I call these "discriminated aspects" and not separate things, because the identity of *I* with *me*, even in the very act of their discrimination, is perhaps the most ineradicable dictum of common-sense, and must not be undermined by our terminology here at the outset, whatever we may come to think of its validity at our inquiry's end. (p. 166)

This twofold conception of the self proposed by James, but anticipated by Kant and other earlier thinkers, had "legs." It entered the lexicon of Western self and personality theory and still has a profound influence. The "I" of the self is the observer, the transient, thinking part of the self, and the "Me" is the object known, the empirical self. This twofold perspective is intuitively appealing. For example, as one contemplates one's mortality the *I* of the self fills one's psyche; one is the quintessential "knower" pondering, evaluating, and ruminating, Hamlet-like, "to be or not to be." On the other hand, as you try to pull a pair of socks over your slightly damp feet after coming out of the pool or shower, it is your self as "object," the self as "known," that fills your awareness. Although you have thoroughly objectified your feet and socks, you have not lost sight of the fact that they are you and yours – at least for the moment.

Before leaving the "empirical me" of James, that is, the self (as observed by others as well as by oneself), it may be useful to point out that he had a profound influence on other psychologists, including his contemporary, James Mark Baldwin. Baldwin ([1897] 1906), like James, saw the self as a socially mirrored construction of its owner: dynamic, evolving, adaptive, and in continuous interaction with its fellows. Baldwin noted that a child's "wants are a function of the social situation as a whole" (p. 31). He cogently demonstrated that it is futile and ultimately unfair to attach a trait label such as "generous" or "selfish" to a nonpathological population of individuals who respond with varying levels of both selfishness and generosity according to the social situations in which they find themselves. James would have endorsed this view. He observed that gamblers pay their gambling debts though they often pay no other debts in the world, and he noted the relativity of our virtues: "One must not lie in general, but you may lie as much as you please if asked about your relations with a lady" (p. 191). He continues on the issue of selfishness: "We know how little it matters to us whether *some* man, a man taken at large and in the abstract, prove a failure or succeed in life, – he may be hanged for all we care – but we know the utter momentousness and terribleness of the alternative when the man is the one whose name we ourselves bear" (p. 205; see pp. 204–13). The empirical "me" is a complex, "psychic" organism, and there is little of the static trait psychologist in either James or Baldwin.

James Mark Baldwin: social psychologist. Baldwin, whose star has begun to rise in the twenty-first century after a century of neglect, can be numbered with William James as one of the major creative theoreticians of the nature of the self. Like James, he had not "lashed himself to Wundt's lab-bench," which Titchener and other eminent psychologists had imported into American academia. Although, like James, trained in bench work, Baldwin became noted for his theoretical work. Nevertheless, Baldwin was influenced by the findings of the great empiricists of his times, most notably Darwin.

Baldwin is important historically because of the primacy he gave to the social determinants of behavior. He knew that the developing self is shaped by the social and cultural institutions that provide it a psychological home. This is initially the family of origin with its mores and values, a social unit that is itself a microcosm of the larger society in which each person is embedded. The forces that expand, mold, and socialize individuals into persons who are recognizably *members of their society* are the beliefs and value systems in which they are indoctrinated. The media, schools and universities, and even organized recreational activities, all propagate their specific ethos.

The genetic substrate for social development has been provided by an ancient (and continuing) evolutionary process. The potential for a personal self is built, in Baldwin's view, into the ontology of the organism. The infant has an innate need to respond to (and be shaped by) whatever social stimuli it encounters, and social adaptation to these stimuli takes place in an interactive process. All children are born with a proclivity for imitating the adults who surround them, and they rapidly develop a repertoire of effective responses to changing environmental conditions. They learn the power of a social smile, as well as the power inherent in the expression of negative emotions and personal distress – and the variable limits to this power. As children physically mature to adulthood their socialization develops at a proportionate pace. Although there are individual differences in the acquisition and expression of the scripts and social schemas that are valued in a culture, there is a recognizable similarity in the dominant patterns that occur across cultures. Baldwin's basic thesis was that humans are largely a product of the social institutions in which they are enmeshed by the accidental coordinates of birth.

If all of this sounds rather modern, pedestrian, and unexceptional, that is because it is. Social constructionists, as we note below, have promoted this Baldwinian perspective in the recent past. Although many social scientists do not confer the same importance on the social determinants of the "personal self" as Baldwin (and Charles Horton Cooley, George Herbert Mead, Harry Stack Sullivan, and others in that tradition), they attend respectfully to the notion, in Rom Harré's (1984) words, "that the self owes its form and perhaps its very existence to the circumambient social order" (p. 256).

Charles Horton Cooley and the "looking-glass self." Cooley was a no-nonsense, cut-to-the-chase pragmatist who had little patience with the fine philosophical distinctions and nuanced hair-splitting in which previous investigators of the self construct engaged. His writings resemble his personality. He was, and is, appreciated for the terseness, lucidity, and simplicity of his prose, and this may be one reason why he is more often cited, and has remained more visible than his contemporary, Baldwin. Baldwin's texts, should you wish to sample his work directly (rather than read others' interpretations of him), will appear obscure, complex, and turgid by comparison.

Although Cooley paid token respect to his contemporary, "Professor James," he was dismissive of the notion of a "pure ego – whatever that may be" characterizing it as interesting only to metaphysicians and moralists. His interest was in the empirical *me that could be observed by one's fellows as well as by oneself.* The self, for him, was what is represented in common parlance by *I, me, myself, and mine.* The first-person pronouns have a self-evident and substantial meaning. Why else could they be in use the world over for the intelligible communication of needs and feelings? The feelings that are excited in each of us when we use first-person pronouns are clear evidence of the substantiality of the self – and, indeed, of its instinctual basis. This self that we sense as we address the challenges of daily existence is the engine that powers the coordination, unification, and excitation of the very behaviors that ensure our survival.

If you want to understand the unvarnished meaning of *I,* Cooley counseled, listen to children at play, especially when they are engaged in a heated contest of wills or passionate disagreements. Alternatively, one can get a sense of the feeling of the self that is expressed in the words of those consumed by a cause and a mission. Cooley ([1902] 1922, p. 173) asks us to consider the following lines of Lowell's poem, *A Glance Behind the Curtain*, in which the poet describes Cromwell's sense of personal destiny:

> I, perchance
> Am one raised up by the Almighty arm
> To witness some great truth to all the world.

If one does not doubt the consciousness of self in Cromwell as he contemplated the exalted mission to which he felt summoned, then one cannot doubt the substance of the self in whose name every other human plots his or her life course.

In his description of the manner in which the selves of all humans are molded by the society in which they live, Cooley coined the metaphor for which he is best remembered today. He referred to the human self as a "looking-glass self." That is, we perceive ourselves as we imagine others perceive us; we shape our behaviors and social schemas to accord with the values that they wish to find in us. And he cited an illustrative couplet:

> Each to each a looking glass
> Reflects the other that doth pass.

Cooley ([1902] 1922) amplified this theme by stating that "we perceive in another's mind some thought of our appearance, manners, aims, deeds, character, friends, and so on and are variously affected by it" (p. 184).

The group self, or *we*, is simply an *I* that includes other people. The group is a surrogate *I* in which we take personal pride or feel personal disappointment. "We beat Chelsea on Saturday" or "We lost the Conference title by a missed field goal" may be spoken by thousands who never attended the sports event. "We fought honorably in Vietnam" or "We were defeated in Vietnam" is stated or admitted by millions of Americans who may not have been born when the war was fought. Our personal identity is intimately linked to the institutions and social groups whose values we have embraced. Each person must have an identity that is founded on his or her membership in a society, even if it be imaginary. Cooley stated, "[This identity] is more necessary to him than bread; and if he does not find scope for it within existing institutions he will be likely to make trouble."

Looking-glass is a powerful metaphor for personologists. The looking-glass is also a powerful artefact, for it allows each individual, in a variety of situations, to view himself or herself appareled in a variety of footwear, suits, gowns, scarves, and hats, adorned with wristwatches, rings, brooches, pendants, and earrings. One may ask if the growth of individualism is not correlated with the prevalence of the looking-glass. The mythical Narcissus became obsessed with his image reflected in a pool of water. How much more is an entire society in thrall to the self images with which individuals view themselves in photographs, and in mirrors daily. Whereas primitive man had only rare glimpses of his own face, contemporary humans in the West look into a mirror each time they visit a washroom, which they do on average every three hours, not to mention the numerous times they view their reflection in dressing rooms, store windows, and other large plate-glass surfaces. Salman Rushdie (2005) astutely observed that "it is in our faces that we face ourselves, and in this regard the invention of the looking-glass is an event of some importance, making possible as never before the prolonged daily study of the self, the self-as-face, the self as reflected self from which that further reflection, the self-portrait, can be born" (p. 12). The ubiquitous mirror has in recent centuries allowed people to study themselves in all their individuality. One can legitimately ask how this artefact has systemically permitted us to shift some focus from the other onto the self. Among all the tools and devices that have shaped our psyches, the mirror must rank as one of the most formative.

George Herbert Mead and the social self. James, Baldwin, and Cooley laid the foundations for an understanding of the self that would find major expression in the work of George Herbert Mead. These four, among a host of others, were members of the Pragmatist tradition, a typically American orientation affirming that the truthfulness of an idea depended on its practical consequences. The interested student can find their influence embedded in the work of many social psychologists, psychiatrists, and personality psychologists who followed in their wake, not least Harry Stack Sullivan and George Kelly. Mead, like James and Titchener, had been schooled in the physiological psychology institutes of nineteenth-century Germany. Although he studied in Leipzig and Berlin (for a period of time, indeed, with Wilhelm Wundt himself) his later studies were increasingly sociological. He eventually found a lasting academic home at the University of Chicago with John Dewey and others who are historically associated with the Symbolic Interactionist Movement (see below). After Dewey moved to Columbia University, Mead became the pre-eminent representative of this group.

Much of Mead's 40-year academic career was spent doing research in the field of social psychology, and he is important to personality theory because of his interest and work in the development of the self. All of his work was originally published in the form of monographs and journal articles, rather than in book form. We can thank his students for having compiled his copious writings into several books, one of the more notable of which is *Mind, Self, and Society* (1934). This book makes clear that fundamental to all of his work is the idea that individual psychology can only be truly understood in terms of the social life of each individual. Humans are by nature social animals, and any attempt to understand them independent of their enmeshment in the ambient society into which they have been born and in which they were educated will leave one with a seriously truncated understanding of them. Humans function within the group largely by way of a *symbolic interactional process*. This expression simply means that they use gestures and other symbols such as language to communicate with each other. From the moment that they are born, it is this interactive process that constitutes the matrix of their sense of identity, their mind, and, in the last analysis, their sense of self.

Communication is a "conversation of significant gestures," and of meaningful linguistic symbols. Gestures are nonverbal and often unconscious transmissions of feelings and sentiments. Linguistic symbols, on the other hand, are usually consciously selected and articulated; yet even these can be given shadings of meaning by inflection, tone, and other paravocal properties that we do not consciously intend. Through these conversations – the symbolic interaction process that characterizes human social life – the mind emerges. Clearly, the mind is not a substance distinct from the body, nor a construct explainable entirely by physiology or its neurological components, nor even the product of a Pavlovian conditioning process. It is the result of the organism's continuing and proactive *interaction* with fellow humans in meaningful discourse.

The self, no less than the mind, emerges in response to the organism's interaction with its human surround. However, there is no self, in Mead's view, before one reflects upon who one is. Self emerges from this objectification of the *me*, very much as William James theorized. Further, it is this objectification of the self that permits self-consciousness. One becomes aware of oneself, as a person, as one faces an audience[19] – of one or more others – with a sensitivity as to how one is being perceived by them. Additionally, the self that we develop is a reflection of the perceptions that others communicate to us about ourselves. The more positive their attitudes toward us, the more positive the sense of self that we harbor;

[19] In the first sentence of the first chapter of Damasio's book (1999) he writes: "I have always been intrigued by the specific moment when, as we sit waiting in the audience, the door to the stage opens and a performer steps into the light; or to take the other perspective, the moment when a performer who waits in semidarkness sees the same door open, revealing the lights, the stage, and the audience" (p. 3). He continues, observing that the moving quality of this moment is the passage from darkness into light, from oblivion into consciousness.

conversely, the more negative the views of us that they communicate, the poorer our view of self. Our self-concept and our self-esteem are largely products of this interaction. But we are not wholly in a passive mode, according to Mead. We shape ourselves by the quality of the responses and adjustments we choose to make in various social contexts. We make further adjustments by the selective attention to, and construal of, what others tell us, as well as by the even more selective process of choosing those to whom we wish to listen or by whom we wish to be valued. It is obvious that our choice of a group to belong to and identify with is a powerfully reactive determinant of our qualities of self and personality.

Gordon Allport's proprium. Allport's term for self is proprium. Like the attempts made by Henry Murray, Raymond Cattell, and many others to introduce new terminology to their field, this attempt to substitute a term unburdened with surplus meanings (and their potential for evoking irrelevant feelings) has not truly succeeded.[20] Nevertheless, it will have currency as long as the psychology of Allport is taken seriously by personality psychologists. Allport defined the proprium as the unifying principle in the personhood of each human.

The proprium has seven distinct functions that emerge in seven sequential developmental stages, which can be distinguished as the individual matures to adulthood. First, a sense of bodily self develops in the first one to two years of life. Humans acutely distinguish that which is part of their body and its various fluid and solid products from all that which is not. Humans readily swallow saliva and snuff back mucosal products, but once they have left one's body they normally cause disgust. Despite what James might have thought, they are no longer considered part of the self, unlike, say, one's clothes. Second, self-identity can emerge only as our autobiographical self comes into focus. This second function of the self is also awakened in the second year of life, as we noted above in speaking of mirror recognition. The child (like many higher primates) can recognize herself when she peers into a mirror. She also learns to respond to her name during this same period. There are few labels in our social world to which individuals respond as warmly and viscerally as they do to their own name. It is an affirmation of the self and identity that each of us cherishes and wishes to nurture.

[20] The reader will recognize the many cognates of this Latin word: proper, property, proprietary, proprietor, proprioceptive, *amour-propre* (self-love). They all have reference to "what is one's own."

Third, self-esteem begins to flower in these early years. As it develops through childhood, it normally receives setbacks only as children enter school situations and get buffeted by the rivalries and tensions that arise in struggles for social recognition and scarce resources (for example, preferred toys). The family optimally provides a safe refuge and nurturing haven for children who have begun to learn that the larger social arena provides occasions for blows to the ego as well as resources for acquiring psycho-motor, cognitive, and social skills. Fourth, self-extension occurs with the extension of the material and social world with which the child identifies. This overlaps to a great extent with the material and social aspects of the self that James defined as its constituents ([1890] 1990, pp. 188–89). Individuals' extension of self includes their bureau and closet, bedroom and family room, their parents and close relatives, cuddly toys and blankets. As mature adults, this self-extension embraces the gang, the crowd, the village, the favorite hockey, baseball, or football team, and, not least, their occupation – the jobs by which they support themselves and their loved ones. Fifth, a functioning *self-image* emerges between the ages of four and six. This is largely the "looking-glass self" of Charles Cooley, a reflection of the esteem and love that is showered on the child by its parents. This self-image, which includes one's sexual orientation, one's self-perceived status in an affiliative group of choice, and one's sense of competence and mastery, gets bruised and restored, diminished and enhanced by the normal transactions of life.

Sixth, *rational coping* begins to mature during the elementary school years. This is the function that allows the child to address problems rather than escape from them (which, indeed, is a reasonable response of last resort). Children develop rudimentary strategies to cope with threats in these years, using the nascent rational skills that will flower in adolescence. Much of this revolves around building friendships and alliances, using various affiliative skills, and consulting with adults. Seventh, *propriate striving* places the finishing touches on the proprium. In later adolescence the self strives to develop life objectives and career goals. During this stage the person takes ownership of his or her feelings, aspirations, values, and critical decisions for the short and long term. It is in young adulthood that one's sense of identity fully evolves.

These seven functions of the proprium are consistent with the principles of William James. In their aggregate they define the self. Allport's system is derivative of a tradition that began with Hume and Herbart in the eighteenth century,[21] yet it is highly pragmatic and, in that latter

[21] James ([1890] 1990) wrote, "It is to the imperishable glory of Hume and Herbart and their successors to have taken so much of the meaning of personal identity out of the clouds and made of the Self an empirical and verifiable thing" (p. 217).

respect, not only Scottish and German but also very American. Whether acknowledged or not, contemporary personality theory has been profoundly shaped by Allport's ideas.

Post-modernism and the self

The twentieth century was ushered in by a robust faith in our ability to access the secrets of nature. That faith has begun to erode at the edges. Modernists believe that there is in principle a knowable reality, which diligent research, meticulous experimentation, and objective and dispassionate observation will reveal to us. Though we may never attain a perfect comprehension of the truths that define the reality of which we are a part, our studies will allow us to formulate closer and closer approximations of that reality. This is one part of a positivist creed, and its partisans point to the advances in medicine, engineering, astrophysics, and microbiology, among other sciences, to vindicate their claims. As soon as we leave the domain of the "hard" sciences, however, and enter the domain of the psycho-social (and, indeed, theoretical physics), we note a weakening in this epistemological view. The confidence of the brass-instrument psycho-methodologists of the early decades of the past century is less evident among contemporary students of the human mind and human society. The modernist philosophy of science has been superseded by that of post-modernism, which casts doubt on many of the psycho-social postulates of the past.

Post-modernists in recent decades have shaken the pillars of the scientific establishment (for example, Bruner, 1986; Foucault, 1980, 1988a, 1988b; Neimeyer and Raskin, 2000). They have (some think convincingly) demonstrated that the knowledge bases of the social sciences of the early twentieth century have been largely invalidated. For post-modernists, facts are simply constructions of the human mind, and our acceptance of them depends on fashion as much as proof. Science is alleged to be a highly politicized industry whose findings gain acceptance by society only to the extent that they serve currently valued and pragmatic ends. Various cultures express divergent worldviews and conceptions of truth, and individuals within the same neighborhood and even the same family have idiosyncratic perceptions of the world they inhabit. Thus, the central postulate of post-modernism is that the social organization of life is less shaped by objective science than science is shaped by the political objectives of society. Consequently, there is no stable, canonical knowledge base detailing how society, the crowd, the family, the human psyche, or personal self functions.

Post-modernists have been in large measure sociologists, historians, philosophers of science, hermeneuticists, social psychologists, and psychotherapists. Their concern has veered from consideration of "ageless facts" to transient meanings – and the meaning-making in which we all engage. Their investigations have focused principally on the human mind and the meanings that we give to the torrent of events that sweep over us daily. Post-modernists ask themselves – and speculate that every one else does as well – "Which world is this? What is to be done in it? Which of my selves is to do it?" (Allen and Allen, 1995, p. 328). The post-modernists who designate themselves as constructivists (for example, Mahoney, 1991) give particular salience to the experiential nature of self-construction. Because each person construes the world in which he or she lives from a unique position in the universe, there are a multitude of "realities" rather than one. The corollary of this is that just as each self is uniquely constructed and positioned in the world, its construal of that ambient world is unique – and we are consequently all encapsulated in our private epistemological worlds.[22]

There is another cadre of post-modernists, the social constructionists, who de-emphasize the way we privately give meaning to the institutions in which we live and work and stress the way the dominant culture has inculcated in its citizens its beliefs, values, and normative behavioral patterns. In other words, reality is socially constructed. To be born and raised in a society is to be indoctrinated from infancy into socially sanctioned patterns of behavior and value systems. These systems are imbued with all the socio-political tenets that are most prized by vested interests and the power structures of the society.

Narratives, life stories, and the self

Jerome Bruner (1986, 1987) has argued that narrative thinking is a basic way of explaining and giving meaning to our lives. Stories give order to our experience and are a primary process for "constructing" reality. The sequencing of events in our lives establishes, to our satisfaction, the causal connections between decision points we reached in the long past, on the one hand, and our long-term objectives, on the other hand. Of course, the memories that constitute our autobiographical self cannot be considered veridical, and our personally constructed life stories inevitably contain a tangle of distorted if not false memories. Of the many matters that we ascribe to our life history, we may in fact have been engaged in them, seen them, or only have dreamed them. We may have so integrated false and embellished memories into our sense of self that they are as influential in

[22] This results in an existential isolation that imprisons each of us. A fascinating treatment of this matter can be found in Yalom's (1980) *Existential Psychotherapy*.

shaping our sense of who we are now as the most veridical of them (Adler, [1931] 1958; James, [1890] 1990, p. 241; see also Parry, 1997). How often have we heard one adult sibling tell the story of a childhood misadventure in the presence of another sibling, only to have the second interject something as follows, "Hey, that didn't happen to you, that happened to me. Then Mom grounded both of us." It is clear that in constructing our autobiography we have difficulty sifting in and out what we ardently wish had or had not happened from what actually did happen. We "self-report" our own story so often that the errors, glosses, self-attributed intentions, excuses, and burnished self-images take on a reality more vivid and plausible than the historical truth. The organization of memory, it appears, is a process in the service of self-esteem – and perhaps of mental health itself.

The burgeoning field of narrative therapy is founded on the principle that humans are by nature story tellers (see McAdams, 1999). When people go to therapists they couch their problems and the meaning of their lives in the stories they tell about themselves (Drewery, Winslade, and Monk, 2000; Monk *et al.*, 1997).[23] Social constructionists who have endorsed that belief system are those most inclined to favor narrative therapy as a way of re-ordering their lives and those of others. Their thesis is that if one can help clients to refurbish, even confabulate, their life stories, one can help them delete the dysfunctional self-attributions and personal styles that impair their social functioning (White and Epston, 1990). Helping people to re-author and re-interpret their life stories creates a pathway to enhanced self-esteem[24] and a modified and healthier personal identity.

Our thoughts about ourselves are saturated by dominant cultural beliefs and values that we rarely attend to, and they form a part of the psychological ground that rarely comes into focus. We bathe in this ethos the way fish swim in the sea, often as inattentive to the former as fish presumably are to the latter. This becomes relevant in psychotherapy when the story that clients tell is bleak, dark, pessimistic, and, worse, incoherent. A skilled therapist can deconstruct a patient's story, pick apart the disabling cultural assumptions that operate in the background, and expand those chapters that provide a basis for personal growth and hope.

[23] This was a technique used by Alfred Adler, who began therapy by asking his clients to relate to him their early recollections. From these memories, which he acknowledged were often flawed, he was able to piece together portraits of his clients' current style of life – and the context for, if not the cause of, the difficulties they were experiencing.

[24] Roy F. Baumeister and his colleagues suggest that from a dimensional perspective high self-esteem can mask sub-clinical narcissism (for example, Krueger, Vohs, and Baumeister, 2008; see also Baumeister and Twenge, 2003).

Distinguishing self psychology from trait psychology

Self psychologists usually analyze such constructs as self-esteem, self-concept, self efficacy, and self-image in exploring the many ways in which the self evaluates and defines itself. More specifically, they speak of self-consciousness, self-importance, self-disparagement, self-effacement, and a host of other descriptors of the way the self is disposed to observe and characterize itself and its acts. In the questionnaires that Raymond B. Cattell, a trait psychologist, used to ferret out the basic traits that constitute human personality, he asked questions about how the respondents evaluated themselves in terms, for example, of self-confidence, self-assurance, and shyness. He was not alone in this as Chapter 5 on traits demonstrates. One may then be tempted to ask at what point one crosses the boundary separating trait psychology from self psychology. What precisely is the conceptual difference between the components of self-concept or self-esteem, which influence the way we interact with our fellows, and human traits, which dispose us to respond in more or less consistent ways when faced by demands that we either impose on ourselves or have imposed on us by our social and physical environment?[25]

In the same chapter in which James ([1892] 1963a) distinguished between the *self* as partly known and partly knower, partly object and partly subject, he warned us not to think of these two aspects of the self as "separate things." Although *I* and *Me* are identical ontologically, that is, they are the same *thing*, they are also discriminable aspects of that thing. James alerted readers that they must not let their common sense be undermined by the terminology employed. Caution is indicated here for the same reason. Taking a personological trait such as indulgence (of others) and prefixing the term "self" to it – hence, self-indulgence – does not move that construct, *ipso facto*, out of trait psychology and into self psychology. As with many other constructs, the meaning of a term and how we wish to classify it is determined by our research and clinical objectives. That traits of an individual as evaluated and applied to himself or herself are often examined under the rubric of self psychology does not make them any the less traits. Self-respect and self-esteem can

[25] The overlap between trait psychology and personality psychology is as problematic as that between personality and social psychology. Should one Venn-diagram the subject matter of these disciplines, one would find that they are largely intermeshed. Baumeister (1999) insisted "it is becoming obsolete to speak of the 'interface'" (p. 369) as if one were treating distinct disciplines. The Platonic dictum "carve nature at the joints" is futile advice when one cannot find the joints.

appropriately be treated in a chapter on the "self." They can just as well be treated in a chapter on "traits."

An evolutionary perspective

A *sense* of self depends on consciousness. If one is unconscious the sense of self has dissipated. In a deep sleep, a coma, or a syncopal episode, one is as unaware of one's identity as if one had never existed. Once awakened, however, all the neural structures that support the *self* are reactivated. Having slept, we rediscover ourselves on awakening, normally without reflection, pause, or a shadow of a doubt as to who we really are. But what is it in the human that assures the continuity of self in our unconscious as well as conscious states? According to Damasio (1999, p. 229), it is the complex apparatus of "body–brain linkages, both neural and hormonal" as well as the continually reactivated neural circuits that archive our autobiographical data.

Although the organism's environment has played an essential role in shaping the features of the self, it is the genome that has put the mechanisms in place to achieve this end. These mechanisms have not always existed. Geneticists as well as psychologists are certain that our eukaryotic (unicellular) ancestors did not possess the neurohormonal infrastructure that enables the development of a self. Indeed, this capability is regarded as a late evolutionary novelty (Dobzhansky, 1964). There is much that is remarkable about this infrastructure. It is stunningly complex and well engineered, integrating a number of different organismic sub-systems. This ultimately enabled the sublime consciousness that generated the sophisticated moral, ethical, philosophical, scientific, and religious achievements of the human race. Could we have achieved them without a sense of self and extended consciousness? At what point will an advanced, non-organic robotic system be such a close *cognitive* simulacrum of the Human, that for practical purposes it could achieve comparable results – and thus merit the label human? Would it need a *sense of self* allied to the kind of consciousness that "the knowing self" enables in order to accomplish what the human species has achieved?

This is a somewhat different question than "Does the *self* contribute to the fitness of the human for survival and self-reproduction? How necessary is consciousness for these purposes?" (Robins, Norem, and Cheek, 1999, p. 460). Pinker (1997) observes, "If consciousness is useless – if a creature without it could navigate the world as well as a creature with it – why would natural selection have favored the conscious one?" (p. 132). The force of the implicit argument in this question is diminished by the fact that there are many species whose evolutionary fitness exceeds that of humans – cockroaches, to name but one – which are presumed to lack a sense of self and an extended consciousness. The question would need to be addressed in terms of the ecological niches and the intrinsic vulnerabilities and strengths that condition the struggle for survival for *any* species.

The struggle for survival goes on in a number of contexts. The adaptive function of the self is most demonstrable for the regulation of conflict within the species, conflict that can threaten the survival of the species from the enemy within, rather than the enemy without. The argument made by Sedikides and Skowronski (1997), that survival and reproduction depended on intra-group harmony and cooperation and other refined social skills that the self enables, is plausible if not compelling. Arguably the great wars of religion and national aggrandizement, pitting humans against humans, wreaked greater devastation among humans than, say, the Black Plague in medieval Europe, that pitted vermin and microbes against humans. Nevertheless, self-monitoring and the level of introspection that allows one to predict (and control) how one will behave in novel situations enables adaptive and prosocial behavior (for example, Krosnick and Sedikides, 1990). This self-regulatory behavior may well be the most important function of the self. In the words of Robins, Norem, and Cheek (1999), the self "helps control, regulate, and organize the mind, and in so doing, it guides our behaviors, our emotions, our thoughts, and our goals" (p. 461).

A correlated question is whether higher primates and other vertebrates have a sense of self. If the self is a late evolutionary development, and an entity deeply rooted within the central nervous system, particularly the brain, it is interesting to speculate about the level of the phylogenetic scale at which the sense of self first developed. Hart and Karmel (1996) listed three classes of evidence they called markers that suggest a sense of self exists in infra-human species. They are linguistic markers, cognitive–behavioral markers, and emotional markers. These aptly named markers refer to behaviors that suggest the existence of a "knowing self" not only in humans and the great apes, but in dolphins, chimpanzees, whales, and other species. Admittedly there is much that is inferential in the conclusions one may draw about, say, the expression of self-consciousness, embarrassment, and pride in an orangutan, but there is a convergence of evidence that emotions analogous to those in humans exist in less evolved species.

The self in old age

The rhetorical question, "Can personality change?" is often asked by psychologists (*cf.* Heatherton and Weinberger, 1994). The writings of major thinkers from Heraclitus (fifth century BCE) to Baltes (1983, 1987, 1997) and other contemporary developmentalists suggest it does. Indeed, criminologists, psychotherapists, drug rehabilitation counselors, educators, milieu therapy staff, and geriatricians are often surprised if not offended by the question. They have predicated their professional *raison d'être* on an affirmative answer to that question. And if personality does indeed change over time, surely the character of the self,[26] which is its substrate, does as well.

Self-concept, the portrait that one draws of oneself, is, in the perspective of symbolic interactionists, influenced by the social roles that one exercises. As one shifts from one role to another during the course of a day or of a year, the concept that one has of oneself requires multiple transient adjustments. These multiple roles are integrated into a complex core concept of who one is. As individuals move, say, from dancing in an avant garde discotheque to worshipping in a church or temple, they shed one persona and don another. Various aspects of the self-concept recede and emerge as the need arises. One may call them *working self concepts* (Markus and Kunda, 1986), but they are nonetheless, like alternating figure and ground, part of a broad canvas of personal capabilities on which the self shines a shifting limelight.

Over the years and decades, our roles change and our various capabilities grow and decline. This alters the way our colleagues, neighbors, and family regard us, and it necessarily influences the way we regard ourselves. People with multiple sexual partners typically reach a point in their lives when the activities associated with this lifestyle no longer appeal to them. Parents with adult children view themselves quite differently than they did in their youthful, single period. Declining physical powers and increasing social responsibilities eventuate in altered self-perceptions. There normally occur corresponding changes in their core self-concept. Carl Gustav Jung maintained that as individuals grow into middle age profound changes occur in their existential concerns. Several of these are provoked by intimations of mortality – the inexorable movement into the unknown realms of disability and death. They search for a transcendent meaning in their lives, and religious issues take on salience and vividness. If the swinging mode subsides, the worshipping mode waxes.[27]

[26] We refer here to those aspects of the idiographic self that vary from individual to individual, including those that vary within individuals as a function of the various stressors, challenges, and situational demands that they face at various periods of their lives. Of course, the universal, ontological self that is the natural inheritance of every human being differs from the selves of other highly evolved species.

[27] In Chapters 3 and 5, on human development and on psychological traits, respectively, I have argued that there is long-term stability in temperament types and personality traits (for example, Caspi and Silva, 1995; Costa and McCrae, 1989). These constitute only one aspect of the self and the self-concept. In any event, both subtle and dramatic changes can occur in self-concept as individuals move into old age.

THERE was a time when meadow, grove, and stream,
The earth, and every common sight,
To me did seem
Apparelled in celestial light,
The glory and the freshness of a dream.
It is not now as it hath been of yore; –
Turn wheresoe'er I may,
By night or day,
The things which I have seen I now can see no more.

William Wordsworth, *Intimations of Immortality from Recollections of Early Childhood*

As we age day by day, changes are made to our autobiographical self, which entail changes in our self-concept and self-esteem as well as the expression of our personality traits. Earlier memories fade; new ones are added. Friendships are strengthened, others are weakened. This social network is constantly changing, not only in terms of the intimacy of the bonds that link us but also in the type of arenas in which the social interactions are played out. We experience new successes, sometimes heightened performance in old skills, or modest achievement in newly acquired skills. This is accompanied by declines in competence in activities that used to thrill us but now no longer interest us – due simply to neglect, to self-recognized hints of genomic limitations in the physical endowment required to make, say, the national diving team or a roster of concert-standard vocalists or pianists, or to the passage of the sensitive period when we had the strength, cognitive acumen, and aptitude to acquire a complex skill.

Although these changes do not necessarily require a daily, conscious re-evaluation of our self-concept, the neural circuits involved in sustaining our sense of self undergo constant, albeit imperceptible, alterations. Many psychometricians have built sophisticated paper-and-pencil instruments to measure changes in traits and self-evaluations. These have generally revealed high levels of stability in self-report measures of temperament, traits, and self-concept (for example, Costa and McCrae, 1988; see Caspi and Roberts, 1999, for a broad but succinct review of the issues operative in the assessment of personality continuity across the life span). These findings may simply indicate that the changes in the variables measured are not substantial or significant enough to be detected, given the current state of our psychometry. The evidence for the stability of traits, moreover, has largely been gathered over *relatively* short periods of time from those in the first "half" of life. If one examines these variables as people progress from age 30 to age 85,[28] one finds

[28] Needless to emphasize, few longitudinal studies of this scale are currently available. Costa and McCrea (1989) did an interesting cross-sectional study of individuals ranging from age 21 to 91.

that self-image changes in significant ways. These changes pervade one's personality structure and have profound implications for the way individuals interact with the larger society.

The impact of aging on self-concept. There are numerous myths (and verities) about the effects of aging, most of them negative. Unfortunately, people's self-concept may be shaped as much by the dominant myths and belief systems of the society in which they live, whether or not they are favorable to them, as by less publicized, scientific evidence that would give a more solid basis for self-evaluation. However, there *are* significant changes that take place in the bodies of humans over the decades that impair in significant ways their ability to do the things they were capable of in their youth. Decline in all the sensory modalities has begun by the time people reach their forties (Hoffman *et al.*, 1988). Although the loss of sensory capacity in taste and smell has less of an immediate impact on people's social and professional life (unless they are in such occupations as wine tasting and cooking, or other professions in which odor detection is important), the decline in tactile sensitivity, and visual and auditory acuity can diminish the scope of one's activities, and this will inevitably alter one's self-image.

As individuals who have engaged in vigorous recreational activities move into middle age, they often attempt to recapture the experiences and robust lifestyle of their youth. Even a well-tuned middle-aged body does not support the strenuous lifestyle that youths are capable of. Unfortunately, 40-year-olds suffer more grievously from "all-nighters" than they did as 20-year-olds. Aging gracefully and healthily requires the consistent assimilation of proprioceptive data about our changing body and a proportioned accommodation of that information (*cf.* Markus and Herzog, 1992). The decline in physical robustness is paralleled by a correlative (albeit not always proportional) decline in cognitive functioning. This is not surprising as the mind is a totally embodied entity. As physical stamina declines and sensory processing and psychomotor reflexes significantly slow, one has to be in denial not to perceive it. The efficiency with which memory and information processing occur also declines *pari passu*. B. F. Skinner (1983), approaching his eightieth birthday, published an article entitled, "Intellectual self-management in old age" in which he acknowledged these declines and described his personal strategies for accommodating them. This is reminiscent of the process developed by Baltes (see Chapter 3, above, on human development) to compensate for the inevitable physiological limitations that emerge in old age, including sensory and motor deficiencies and memory loss.

Health, home, and social connections. Serious illness, like old age, can drastically alter the body one calls *self* (see Charmaz, 1995). The effect of muscular dystrophy on self-concept is vividly noted by the anthropologist Robert F. Murphy (1987):

> In my middle age, I had become a changeling, the lot of all disabled people. They are afflicted with a malady of the body that is translated into a cancer within the self and a disease of social relationships. They have experienced a transformation of the essential condition of their being in the world. They have become aliens, even exiles in their own lands. (p. 111)

Rather more severe challenges to the sense of self arrive in later adulthood. Although old age is not a disease, it can masquerade as such, bringing in its train all the stresses, fatigue, and weakness that are symptoms of serious illness and disability. Among the old-old (those over 90), illness and disability are often considered virtually normative. The assault on the sense of self that extreme old age and weakening cognitive functioning occasion is comparable to suffering a stroke, which can generate the experience of a diminishment of self.

A loss of continuity with the past is also precipitated by elderly persons' loss of their home, which has become over the decades so much a part of their existence that it can almost be regarded as their extended body, in effect, a second skin. To be precipitously, often involuntarily, moved out of a home in which they have lived for 40 or more years can be traumatic and a severe rupture of continuity with their past. Most grievous and destructive to continuity and sense of self, however, is outliving the friends and acquaintances of one's own generation – and in some cases one's own children.

Despite the changes that occur with aging, personal *identity* transcends changes in personality and sense of self. Barring severe pathology, such as advanced Alzheimer's disease, the assaults on one's self-concept do not erase one's identity. The continuity provided by an autobiographical record, tattered and raveled as it can become, authenticates the conviction that one is the child of one's parents, born at a moment in time, at a single location, and often, early on, with a social security tag that anchors one in the body politic. A compassionate community of family and strangers provides for the rest – and for the end – of life in these most vulnerable of circumstances.

By way of conclusion

It is curious that this construct, the self, which some of the most astute observers and thinkers of history could not find, and whose existence some have even disputed, should be the subject of so many studies in the past century. It is all the more extraordinary that so much labor has been expended in trying to explain the self in the twentieth, and now the twenty-first, century. However, all of these investigations and the sophisticated instrumentation that the modern scientist uses to study the elements of self have still not provided compelling, and universally accepted, evidence of its existence. Even though it is more than a substance, a way of being, empirically "the self" is still found as a term of convenience in all the languages of the world. If everyone talks about it, one may be excused for thinking that it must exist. Although it has been fashionable to rail against "the Cartesian Error" and to blame the body–mind problem on Descartes and his dualistic predecessors, the fact is that we have not been prepared to abandon the concept of mind – nor the other broad constructs that have been such a convenience to us. This is certainly true of the construct of self. We are not prepared, nor do we yet need, to embrace a self-less psychology.

7 Culture and personality

> Persons in power should be very careful about how they deal with a man who cares nothing for sensual pleasure, nothing for riches, nothing for comfort or praise, or promotion, but is simply determined to do what he believes to be right. He is a dangerous and uncomfortable enemy, because his body, which you can always conquer, gives you so little purchase upon his soul.
>
> Gilbert Murray on Mahatma Gandhi

Toward a working definition of culture

Evolving cultures, however construed, have an impact on the personalities of those who identify with them and are enmeshed in their expectations, directives, standards, political rituals, and values. This commonplace notion merits analysis. To what extent can culture shape personality and even override genetic trait variances in a population? Can humans become more fully integrated personalities by embracing the belief system of a society distinct from the one in which they were educated? Can members of one culture, thoroughly imbued with its values, enter into a meaningful dialogue with members of societies that propagate values incompatible with their own? These questions are of importance to the personality psychologist, and this chapter will attempt to address them.

A definition of culture

We define *culture* as the distinctive belief systems, patterns of behavior, and significant moral, artistic, familial, vocational, and political structures that are valued by a society, members of which usually speak the same language.[1] It

[1] We often speak, for example, of a Finnish culture, or a Canadian culture, even though there are two official languages in Finland and Canada – Finnish and Swedish in the former and English and French in the latter. Other examples of multilingual societies are Switzerland and India. Nietzsche famously affirmed that language subtly reflects and shapes the distinctive cognitive structures of its speakers. "The Sapir–Whorf hypothesis goes even further and states that the very structure of the language that a cultural subset of the human race speaks affects the character of their ideation and the consequent conduct of the group's affairs" (Dumont, 2000, p. 2092).

includes such aspects of life as members' preferred cuisine, religious practices, dominant sports, and recreational patterns. Defining a national society as a distinctive cultural entity is not a simple matter. Among the sub-cultures of a *single modern nation*, there are cultural variances that are as striking as those among the macro-cultures that define nations themselves. Even the assumption that cultures are co-terminous with the political states in which they exist is problematic. Provence, for example, a region of southern France, has more in common with its northern Italian neighbors than it does with Brittany or Alsace-Lorraine; the state of Maine has more in common, culturally, with its Canadian neighbors in New Brunswick and Québec than with Mississippi or southern California.

The question that the personality psychologist needs to ask is to what extent do the culture and sub-cultures in which people are immersed shape their personality? If one concludes that patterns of culture significantly shape personality, and it would appear that they do to some extent, one must ask if certain cultures *do* this more effectively than others, say, those that value social conformity more than self-individuation. To the extent that the various expressions of culture confer "national" traits on those raised and "indoctrinated" within it, citizens of the same nation-state will tend to share the same personality traits.

This line of reasoning is compelling, of course, only to the degree that we accept that environment influences values, attitudes, and behavior. We also need to accept the premise that the dominant cognitive and behavioral patterns present in a society define the culture, and not vice versa. After all, "culture" is a mental construct designed to create a readily comprehensible model of human activities and the institutions humans have created to perpetuate their values. Although we often speak of culture as if it had an agentic function, the concept of culture must not, in the last analysis, be reified, for culture does not *do* anything. Instead, culture is a mental device that helps us to understand what *people do*. When *their* activities and institutions change, and the values that undergird them are replaced by new ones, the cultural model should be redefined correspondingly.

Herein lies a challenge. It is now axiomatic that societies are *always* in flux. That *mores* shift with the times as well as with the societies through which we travel has been commonplace wisdom – though often ignored – throughout the ages.[2] Definitions of various *cultures* in our contemporary

[2] The classic expression of the passing vogue of customs is Cicero's lament, *O tempora, O mores*, loosely translated from the Latin, "How the times are changing!" This temporal view of varying customs is complemented by the French cross-national take on it, *Autre pays, autre mœurs*, loosely translated with a shrug of the shoulders, "Another country, other customs" with an implicit "what do you expect?"

world need to be adjusted on an annual basis, unless we are speaking of, say, isolated and disappearing "stone-age" societies of Papua-New Guinea or the headwaters of the Amazon Basin – disappearing, it must be added, to the chagrin of those who study them and prize the diversity we have heretofore found in the world. The vast swirling tides of people that are migrating from their ancestral homelands to other regions and continents of the globe are accelerating these changes. Further, the industrialization of the "developing world," on the one hand, and the increasing interconnectedness of people by an information technology (IT) that is "catching" them in its web, on the other hand, are creating change even where vested political, economic, and religious interests are fiercely resisting it.

Changing our stereotypical definitions of various cultures is even more imperative when we shift our perspective from one generation to another, or one century to another. For example, models of Asian cultures that are static (or perceived to be changing only under the impact of Western influence) are antiquated and dysfunctional. Likewise, there never was a neatly defined European culture. Compare the personality of the typical Spaniard with the typical Dane, the Englishman with the Pole, the Frenchman with the Finn, the Italian with the German. One may be impressed more by the differences in their social conventions and politics than in the commonalities of their religions and complexion. Enlarging the model to include North Americans (as in Euro-American culture) is even more problematical.

Zuriff (1999) deplores the gross categories that certain psychologists use to characterize large heterogeneous populations: "Following a tradition that has become politically correct in contemporary psychology ... they treat 'mainstream' or 'European American' culture as if it were homogeneous – as if Greek Americans, Irish Americans, Jewish Americans, Bible-Belt Americans, and so on, all share one culture" (p. 71). This appears all the more incongruent when we add African-Americans, Hispanic Americans, Amerindians, Japanese-, Chinese-, South Asian (say, Indian), Filipino- and Vietnamese-Americans to the larger construct, American culture. On the other hand, should we move from the molar end of this continuum of culture definition to the finer, more idiographic end, we need to ask ourselves what are the objectives of our analysis, for there is no *proper* level of analysis independent of those objectives.

Questions to be addressed. I use the term "culture" in this chapter to apply to a dispersed ethnic group as well as to a localized community, to a neighborhood as well as to a family, to a corporation as well as to a classroom, to a singles bar as well as to a bowling alley. Whatever the social entity that is the focus of one's analysis, the question that needs to be asked is: does culture as expressed by the people and institutions that constitute our environment shape personality? If yes, then *how* do the prevailing "rules" shape the behaviors and the values of those enmeshed by them? What components of culture are most influential in shaping personality? What levels of analysis are most appropriate for which purposes? When should we prefer an emic and idiographic approach rather than an etic or nomothetic approach? Those readers will be amply rewarded who address (still more, answer) these questions.

The etic and the emic

Emics. The terms *etic* and *emic*, coined by the linguist Kenneth Pike (1967), are found extensively in anthropological and linguistic literature. They are now used in social psychology as well. Emics refers to the domain of behaviors that are characteristic of a single society or cluster of related societies. The term is important for delineating cultural patterns that are peculiar to one group of individuals but not to all human beings. Ruth Benedict's emic approach (evident, for example, in her classic work, *Patterns of Culture*, [1932] 1949) allowed her to study the Zuñi Pueblo from within and to describe their collective character as austere and Apollonian as distinguished from the character of the Plains Indians, given, she claimed, to frenzied, Dionysian excesses. Clearly the emic behaviors of these two cultures could not be generalized universally to humans of all times and places. The enforcement of such cultural norms has clear implications for shaping a national character and the personality of its members. That Benedict saw a correspondence between national character, as she construed it, and individual personality led her to affirm that culture is "personality writ large." However, there are pitfalls in this approach.

Benedict's "national character" studies have been discredited in part for their static and stereotypical descriptions of "national personalities." Her study of the Japanese, *The Chrysanthemum and the Sword* (1946), has been criticized for this reason, as well as for the theoretical assumptions and data-gathering procedures that she used. For example, most of her data came from interviews with Japanese-Americans interned in North American camps. Her purported objective, clothed in academic language, was to "see how the enemy looks at life through his own eyes" (p. 5). This American "construction" of a foreign culture was not only based on an archaic "model" of a nation toward which there was at the time hostility in the West, but one that represented a crystallized model of a particular, numerically small, elite Japanese class. This was not unusual, for as Marsella *et al.* (2000) stated, "many cultural anthropologists and cultural psychiatrists became involved in the allied ... [Second World War] effort by delineating the national personality structures of various countries including Japan, Germany, Russia, Thailand, Romania, and China" (p. 41). These studies became suspect by reason of the ideological and rhetorical biases to which they were vulnerable.

Etics. Etics, in contrast to emics, refers to behavior patterns that are found in all, or almost all, societies (for example, Brown, 1991; Moghaddam, 1998, pp. 8–16). They characterize humans *qua* humans, independent of the diverse cultural contexts in which they live. For example, all societies have a *norm of trust* for the generality of people who belong to their society, a *norm of truth-telling* in the routine transactions of daily life, and a *norm of turn-taking in dialogue* in which individuals take turns listening to one another. Promoting these norms exercises a civilizing influence on social intercourse in all societies. To the extent that these conventions are violated, breakdowns in the health and well being of the society as a whole become evident.

The nomothetic and the idiographic

The analogues of *emic* and *etic* in the psychological literature are *idiographic* and *nomothetic.*[3] In Chapter 5 on traits I contrasted the work of personologists, such as Henry Murray and Gordon Allport, with the research of the factor analytically oriented personality psychologists who followed them. The former not only characterized their work as idiographic, but also insisted that the individual was *the* proper focus of work in this domain. This view denies validity to the notion of a universal or even national personality, such as proposed by Ruth Benedict. If the "proper study of mankind is man," the proper study of the personologist is the personality of the individual. Their view is that as we expand our study beyond the individual, we move into the fuzzy realm of average behavior – that of types and factors – herding our ideas into procrustean beds. The fewer the beds the more gross and constraining are our characterizations.

[3] Corsini's (1999) *Dictionary of Psychology* defines *idiographic approach* as "The study of individuality, the uniqueness of an individual's behavior and adjustment, as contrasted with the study of the universal, or nomothetic, aspects of experience or behavior." Consult the definitions of *idiographic* and *idiographic psychology* as well. *A nomothetic approach* is defined as "Pertaining to the formulation of general laws as the goal of scientific method, as opposed to the study of the individual case ..."

Some have argued that a focus on the individual and individual differences is an expression of a Euro-American bias. The cogency of this argument is weakened by the fact that Europe and North America have also been home to theoreticians, philosophers, and scientists who have predominantly proposed universal notions of human psychology. The United States in particular has to some degree been a hothouse of research on the nomothetic principles of human behavior, principles tinged by American traditions. Two prominent "comparative" psychologists in this tradition were Harry F. Harlow and Burrhus F. Skinner.

B. F. Skinner. Extrapolations from animal studies to human behavior are by their very nature nomothetic. It could not be otherwise as one is generalizing from one species to another. There is nothing idiographic in the work of Skinner with Norwegian white rats, except the creative, personal contribution of Skinner himself (see his seminal work, *The Behavior of Organisms*, 1938). He worked with one species of animal – one rat at a time – with a view to making qualified inferences about the human race as a whole. If one assumes that the individual differences among rats of the same species are negligible, at least relative to the variables that are under consideration, the results of Skinnerian operant-conditioning experiments should be generalizable to all rats – and, by extension, not just to some humans but to all humans. (His wartime experiments with pigeons, which had other research objectives, are equally interesting.)

The central focus of Skinner's research, and of those who followed him in the tradition he established, was the relationship of varying schedules and conditions of reward to resulting patterns of behavior. These patterns, they would argue, with the "expectations," strivings, "concerns," and social skills they imply, are integral to a comprehensive model of personality – suitably defined.[4] The learning principles that this formidable body of research generated are of universal application.

Harry F. Harlow. The survival of the human infant depends to a great extent on the bond that forms between that infant and a primary caregiver – traditionally, the mother. This poignant fact has spawned a body of research literature generally found under the rubric *Attachment*

4 Radical behaviorists have not denied the existence of an interior, mental life with its rich and complex cognition, imagery, feeling, willfulness, creativity, and consciousness. What they doubted was the usefulness of the constructs developed for understanding those interior activities and the correspondence of those constructs to their operational definitions that purportedly allowed us to measure them. They had scientific, not ontological, reservations about studying the "mind," given the technology of the time.

Theory. The child has multiple needs – emotional, tactile, nutritional – which must be satisfied in the earliest post-natal periods. A failure to meet those needs over a sustained period of time will result in developmental setbacks to the infant. These setbacks can sometimes be remediated, depending on their severity. In addition, there are wide individual differences among infants in their resilience to deprivation (for example, Belsky and Nezworski, 1988; Cicchetti and Barnett, 1991; Egeland, Carlson, and Sroufe, 1993). The implications of such findings for personality development are manifold and are addressed in more detail in Chapter 4 on the evolutionary and biological determinants of personality.

Harry and Margaret Harlow (for example, 1962, 1966) conducted longitudinal studies of infant Rhesus monkeys, separating them from their mothers several hours after birth. They replaced the mothers with mechanical surrogates. The substitute mothers were essentially monkey-size cylinders made of wire mesh, fitted with crude heads. One of the wire mesh cylinders was fitted with a nipple and milk container that the infant could easily access. The other "mother" was upholstered with simple terry cloth, but had no feeding mechanism. Over a period of time, the experimenters were able to determine that infant monkeys engaged in more attachment behaviors with the terry cloth-covered mother that provided them with some tactile comfort but no milk than the surrogate that furnished them only with milk but no tactile comfort (Harlow, 1959). They would cling to the cloth surrogate when frightened, and they would use it as a safe haven after their exploratory forays into an alien world. In effect, the terry-cloth mother became *the* mother. The Harlows concluded that the tactile and emotional sustenance mothers furnish is more important in creating a maternal–infant bond than the food they provide. Other studies demonstrated that the Rhesus infants that were deprived of maternal care were seriously and permanently damaged in terms of their adult social skills and sexual relations. They showed little affection for one another and, when artificially inseminated, bore infants that they mistreated and otherwise neglected.

The seminal studies emanating from Harlow's primate laboratory at the University of Wisconsin have implications for humans as well as lower primates (Montagu, 1986, pp. 38–44). The conclusions that many have drawn from these studies are that *human* infants need continuous nurturance from one or more caregivers to whom they bond, and that any disruption of the attachment that they have formed with their parents compromises their personality development. These deformations in their character are partially if not entirely reversible. Moreover, there are individual differences in the vulnerability and resilience of individuals to such insult to their psychological (and by extension physical) integrity

(Garmezy, 1993). This theory is essentially nomothetic. It applies to all societies of humans in all times and places and, presumably, to all primates.

Universal human traits

The temptation to define the *universal* Human can blur our vision of the emic and the individual. The culturally encapsulated philosopher-clinicians of nineteenth-century Europe fell into this trap. The other pitfall to avoid is overcorrecting for this past error and neglecting that which is truly universal in human behavior. Some aberrant behaviors, such as homicide or child molestation[5] occur in every society (Triandis, 1978), but are not thereby characterized as nomothetic or normative. In fact, there are taboos against such behavior, and in literate societies such proscriptive norms are usually embedded in legal codes. Prescriptive norms are given distinctive meaning in each society, and their expression takes on the character and coloring of the dominant culture (Moghaddam, 1998). No matter how different the children may be in other respects, family and society largely succeed in inculcating such values in their children – values that are seamlessly knitted into their personalities.[6]

The norm of trust. As suggested above, a society cannot function without an implicit faith that its citizens will *follow the rules* that have been consensually agreed upon to govern the routine transactions of daily life. When we are told that a high-speed train will leave King's Cross at 17:26 and arrive in Cambridge 75 minutes later, we do not cross-examine the ticket agents to determine whether they are deliberately deceiving us relative to the published timetable (or the fare). We allow for a certain margin of error both on our part and on the corporation's part and present ourselves on the train platform several minutes before scheduled departure. Of course, ordinary citizens use due diligence in accepting information, if for no other reason than to protect themselves against lapses of memory or simple distractions by those who are serving them.

[5] An aversion for sexual relations with close family members (*contra* Freud) appears to be universal (Wolf and Huang, 1980). See a discussion of the *Westermarck Effect* in Chapter 12 on disordered personalities.

[6] For a somewhat contrarian view read Sulloway's (1996) *Born to Rebel*, the fruit of 25 years of research into the effects of birth order.

The norm of truth-telling. Functional societies are noted for the reliability of the contracts, explicit and implicit, which individuals enter into with each other. A society will slide into dysfunctionality to the extent that a norm of trust is violated. The lubricant for the social mechanisms that provide the complex assortment of services and products that an industrialized or agrarian society needs to maintain its quality of life is simple honesty – individuals keeping their word. "Flaky" behavior and intentional deception, like systemically unreliable service, is like a handful of sand thrown into the gears.[7] Thus, every functional society has a norm of truth-telling governing the interactions of multitudes of its citizens, as well as those private and public bureaucracies that promote specialized, ideological ends. Moghaddam (1998) states, "whereas the norm of trust places the burden on the receiver of information, the norm of truth places the burden on the provider" (p. 9). These norms are two sides of the same coin, and violations of one erode the other. When members of a society cannot trust each other to tell the truth, to honor their contracts, and to meet reasonable expectations, then the social fabric begins to shred and unravel. Indeed, serious violations of these two norms are symptoms of the *sociopathic personality* – hence its name. An economy would grind to a standstill if the majority of buyers routinely defrauded vendors, or vendors buyers. The healthiest cultures are those in which a handshake alone, or simply one's word, can confidently seal a contract.[8]

Second-tier norms

Moralistic attitudes. Every society has a more or less complex set of moral and legal imperatives. Although the specific character of these activities varies from culture to culture, each has some implicit standards

[7] Although a norm of trust exists in every society, a reading of history and the daily newspapers leads us to understand that (a) the norm is never uniformly respected by all the members of a society, and (b) there is between-culture variance in the general levels of trust governing daily social and business intercourse. Margaret Mead's study of the Mundugumor, a people of the Sepik region of New Guinea, affirmed that their social norms revealed a high level of paranoia by Western standards, and fierce hostility between members of the same sex (her studies have been vigorously disputed by fellow anthropologists; *cf.* McDowell, 1991). She endorsed Franz Boas' position that culture shaped the personalities of its members (see Mead, [1935] 2001).

[8] Lore has it that a million dollars worth of diamonds can change hands in Antwerp with no more than a handshake to seal the transaction. Gems worth thousands of dollars are routinely left by one dealer with another for appraisal, or to have them cut, polished, or mounted, without asking, or expecting, a receipt. Interacting like members of a large family, they pick the jewels up several days later and walk out with them, with the promise to remit payment within the week. The sub-culture functions smoothly among individuals who keep their word – and who know each other very well.

of what is virtuous and what is evil – and possibly what is neither. Good acts are those that advance collective as well as individual well-being, however it is construed within their moral universe. Bad acts are those that militate against the achievement of social ends for which there exists a broad social consensus. Among the virtuous acts that are valued in every society is that of "helping" one's neighbor. The more closely related those in a position to help are to those in need of help, the more psychologically compelling is the appeal. Individual generosity, however, ranges from the donation of one's organs or the sacrifice of one's life for the benefit of others, to total disregard for their value as "humans-in-need." Every society has concepts of fairness, equity, justice, and "due process," though the modalities of their expression vary widely as does the precision of the language they use for encoding them. All the above are constituents of personality, and they produce relatively enduring attitudes about what is permissible within a particular set of circumstances. The consistency with which people act in accordance with their principles and the social sanctions they invoke when these principles are violated are other aspects of their personality.

Curiosity. The prediction and control of forces that affect our well being is an objective for all societies. Underpinning all goal-seeking is the presumption that one's actions (or someone else's) can be instrumental in achieving goals. For that reason, in order to understand events (and hence to be able to predict and possibly control them), humans attribute causal influence to the things that happen to them – indeed, they posit complex networks of causal events. This can be faith-based or science-based. In a faith-based system, for example, people believe they can influence the future by their prayers, thoughts, or certain rituals. In a science-based system, they believe they can influence the future by understanding and manipulating natural objects and forces. These two systems can, and do, exist side by side. Of course, cause–effect paradigms differ from culture to culture, from the very simple to the highly complex.

The search for causes is at the root of much of the *curiosity* that appears to be an innate aspect of human personality (see Chapter 13, where this is explored). Although much of this curiosity is directed to solving practical problems and acquiring utilitarian knowledge, humans also engage in exploratory behavior that brings no reward other than the simple pleasure of gaining information or solving a puzzle. Certainly this served an evolutionary purpose, for survival depended on gaining knowledge and establishing a mastery of one's environment (this presumes that one has causal schemas in mind that are appropriate to the effects one wants to achieve).

Advancing science for its own sake is another aspect of curiosity. Subrahmanyan Chandrasekhar (1985),[9] for example, devoted seven years of his life to exploring the mathematical mysteries of rotating ellipsoids, admitting that he had no practical end in view other than to bring order into a disorderly domain of physics. Rhesus monkeys were observed by Harlow (1950) to "work" diligently at solving puzzles such as opening a latch that neither appeared nor actually did give access to a tangible reward. Even the white rat, renowned for its laboratory exploits, engages in exploratory behavior by poking around in corners of its habitat. This suggests that all organisms, not just humans, are curious and seek out comfortable levels of arousal in exploring novel aspects of their environment.

The norm of turn-taking in dialogue. In every society when two people enter into an exchange of views and ideas, they take turns listening to one another. When it is perceived that one or the other is not listening or not allowing the other to speak, there is a breakdown in the dialogue and in the relationship (Moghaddam, 1998, p. 10). The individual who fails to respect this norm in any society is regarded as uncouth and is shunned. The secure, respectful person allows others to speak and be heard; the self-absorbed and narcissistic individual listens only to his or her own voice. Respect for one's neighbors (and their right to be heard) is a personality trait valued around the world and encoded in most religious and secular moral systems. There are wide individual differences, however, in the level of respect individuals give to this norm. The right to be heard is restricted, of course, by societal judgments about what constitutes harmful communication.

One body, one self. A human self is always associated with one body.[10] Multiple personality disorder has been voluminously discussed over the centuries in both the psychiatric and religious pastoral literature.[11] In this disorder different personal identities usually manifest

[9] Chandrasekhar was awarded the Nobel Prize in 1983 for his work in astrophysics. This genius, born in Lahore (in what is modern Pakistan), created the mathematics that later enabled our international missile programs.

[10] Rarely do people claim to have two distinct bodies (see Damasio, 1999, relative to the "one body, one self" principle). On the other hand, there are people who claim to be several different persons sharing the same body in a condominium arrangement. In some cases they manifest themselves simultaneously, more often sequentially. If they are aware of each other's existence, they may not remember what the other has done (that is, they are mutually amnesic). In other cases, the amnesia exists only in one or the other.

[11] Consult Ellenberger (1970, pp. 126–47) for a brief, scholarly, historical overview of this condition, beginning with Saint Augustine's self-analysis.

themselves sequentially, and each of them (singly or collectively) is considered in most developed societies to be mentally ill. This resembles a "time-share" rather than a condominium arrangement. When an original "multiple" periodically goes "into hiding," one of the alter egos takes over the premises. This condition has long been known as Multiple Personality Disorder (MPD) and is still called by that name in the *International Code of Diseases* (ICD-10). This disorder has been recently retagged Dissociative Identity Disorder (DID, and coded *300.14*) by the architects of the *Diagnostic and Statistical Manual of Mental Disorders* (DSM-IV-TR). As a syndrome it has been amply documented since the end of the eighteenth century, but was more frequently diagnosed in nineteenth-century Europe than it is now. Students of religious history will recognize similarities between this syndrome and the numerous reports of "demonic possession" that preceded our era. The ebb and surge of these personality syndromes in various periods of history strongly suggest that they are in part socio-culturally constructed.

Birds of a feather. Individuals have a cross-cultural penchant for preferring the company of those who are similar to them to those who are dissimilar. For this reason, social groups with a collective self-consciousness exist around the globe. A corollary of this is that populations, as well as their constituent sub-groups, categorize the world into "us" and "them." Olsak, Perreault, and Moghaddam (1997) examined the *similarity-attraction hypothesis* (SAH) that posits that there is a strong, persistent tendency for people to surround themselves with people who resemble them rather than not. Their study of Chinese, English-Canadians, French-Canadians, Greeks, Indians, and Jews in Canada supported this hypothesis and corroborated the earlier findings by Levine and Campbell (1972), who found confirmatory evidence of this hypothesis in a "vast range of world cultures." Needless to say, there are wide individual differences within any culture relative to these findings – whether or not they prove to be durable. Moreover, there are moderator variables that must be considered, three of which are the level of personal and "ethnic" security that individuals feel in their environment, the climate of tolerance, if not acceptance, that exists in the society in which they live, and the nature of the relationship, personal, social, and physical, in which attraction is being assessed.

Although there is mixed evidence in idiographic research for the obverse of SAH, that is, the *opposites-attract hypothesis*, Berscheid (1985) and Buss (1985) found evidence that individuals tend to marry partners whose personalities are similar to their own. It must be noted, however, that societal imperatives can override individual preferences, for there are cultures in which individuals who marry have never even met – unions that have been arranged by their families. Hindu Indians, for example, routinely select partners for their children, and researchers report that the "success" of these marriages equals and may be superior to marriages grounded in the romantic ideals of the West. Gupta and Singh (1982) did a longitudinal study of twenty-five arranged marriages and twenty-five love-inspired marriages and found that the arranged marriages were superior in terms of their durability and the levels of affection between the spouses to the marriages that were love-based. This finding is not surprising, as the hormonal floods that initially potentiate and sustain the attraction in the love-struck group subside within a matter of months, whereas the rational factors that enter into the arranged unions have greater staying power.

Research in cross-cultural psychology

Methodological concerns. The personality of psychologists inevitably influences the research they conduct. Indeed, because so many of our assumptions operate outside our awareness, it is fair to ask whether it is *ever* possible for researchers to bracket (that is, isolate) their assumptions – as carefully and conscientiously as they may try. The same cognitive structures that we use to analyze the world about us, with all of the subtle norms, values, and schemas that seep into and undergird the meanings that we attribute to phenomena, are derived from the culture in which we were born and raised. Consequently, it can be argued that the cognitive tools we use for "objectively" appraising other cultures inevitably contaminate the process and the conclusions of our study. It is inherently a boot-strapping endeavor. The distortions stemming from these biasing assumptions can be attenuated by steeping ourselves in the cross-cultural paradigms of an international community of psychologists. Pedersen (2000) proposed that psychologists, journal editors, academics, and other scholars integrate a multicultural perspective into their psychologies. Laudable as that is, historically there have been failures in this kind of research for the simple reason that one cannot escape entirely from the bias that is intrinsic to the methodologies developed in one or another culture, as well as the content of the materials used (Marsella *et al.*, 2000). Marsella and his colleagues stated that some "psychologists seemed oblivious to the possibility of cultural bias in the very nature of their concepts, scales, and norms" (p. 44).[12]

[12] The problems that Marsella and colleagues allude to will be addressed in Chapter 11, on measuring the Human.

Despite this concern, today there are hundreds of books and scores of journals dedicated, in whole or in part, to topics in cross-cultural psychology. Advances in our knowledge about equivalences in cross-cultural concepts, scales, and norms have provided new insights into the universals and cultural specifics of personality. In addition, some writers have suggested that psychology cannot be separated from a study of the culture in which it is rooted. Shweder and Sullivan (1993) state, "culture and psychology ... make up each other" (p. 339). For this reason Shweder (1991) suggested that a new discipline needed to be developed whose purpose would be to understand the interaction of culture and psychology. Shweder dubbed this new science "cultural psychology," an analogue to cultural anthropology.

At least two approaches can be discerned relative to this problem. First, Marsella (1998), consistent with Pedersen's approach, has proposed that we develop a "metapsychology," that is, a superordinate set of principles "concerned with understanding, assessing, and addressing the individual and collective psychological consequences of global events and forces ... [which] are occurring with alarming frequency, intensity, and acceleration, impacting virtually every aspect of our lives" (p. 1284). The prophetic tone of this article is evident in the light of the cataclysmic events of recent years. From a different perspective, but consistent with Marsella's objectives, Jessie Bernard (1950) advanced the still popular argument that safeguards can be put in place to protect the researcher from subjectivity and bias. He responded to the challenge placed by students of culture, who affirmed that science is itself a cultural construction and cannot be trusted outside its culture of origin: "Although the direction of scientific inquiry may be culturally determined, the methods of pursuing scientific inquiry – scientific techniques – contain their own protection against cultural (and even personal, idiosyncratic, and perceptual) biases" (p. 268).

Whether the psychological constructs deriving from one culture can be transposed to a study of a different culture is still hotly debated. Those who have significantly contributed to the development of the Five-Factor Model (FFM) of personality (described in Chapter 5) argue that their research has found evidence for the same Five-Factor structure in collectivist cultures as in individualist cultures (for example, McCrae *et al.*, 1998). Critics of this research argue that "these studies will likely only map a portion of the life space that can be characterized by traits" (Cross and Markus, 1999, p. 391), omitting other important culture-specific constructs of the target populations. Equally problematic, the inventories used in these studies are translations from American English, and are derived at least in part from lexical norms and cognitive structures of an alien culture. Marsella *et al.* (2000), among others, have raised doubts about the conceptual and linguistic equivalences of the instruments used in such research. When there is great disparity in the connotative meanings of terms used in two different cultures, critics understandably ask what the reliability will be of behavioral predictions based on responses to such terms, especially when the testing situation as well as the test itself may seem strange to the research participants.

Indigenization of personality theory. A newer position has surfaced: accept the futility of trying to escape one's ineradicable ethnocentricity – indeed, embrace it while conducting one's research. For example, Yang (1999) has proposed that the Chinese develop a Chinese personality psychology in which they can synthesize the paradoxes and apparent contradictions that non-Chinese have difficulty understanding (Liu and Liu, 1999). Chinese psychologists can understand the philosophical roots of their relational and familial collectivism in a way that strangers cannot, no matter how hard they try. The corollary of this is that Americans should develop an indigenous American psychology – which, indeed, they have. To force it into an international procrustean bed may not only plug the creative wellsprings that generated Euro-American psychology in the first place, but make it of less value to others who wish to make use of it.

In this vein, Yang (1999) stated, "the detrimental over-dominance of Western social sciences in the development of social sciences in non-Western societies is the outcome of a worldwide academic hegemony of Western learning" (p. 182). The problem then is not whether there can be a legitimately indigenous American or European psychology, one that authentically reflects the kaleidoscopic nature of those multicultured collectivities, but whether we will refrain from forcibly exporting it to those peoples for whom it simply does not work. Consistent with this is the view of Cross and Markus (1999), who stated "that the articulation of a truly universal understanding of human nature and personality … requires the development of theories of behavior *originating* in the indigenous psychologies of Asian, Latin American, African, and other non-Western societies" (p. 381).

Other methodological dilemmas. Doing descriptive research in cultural psychology forces us to make difficult choices in methodology, and these choices often have profound philosophical implications. For example, one may adopt an essentialist, Platonic model of what a member of a particular culture looks like (see Scarr, 1993). The approach proceeds, states Scarr (1993), from the vantage point of a more static, Platonic model. It defines what a German, Malaysian, or Sicilian *typically* looks like and then only selects participants that match that model.[13] Of course, there may be so many dramatically disparate sub-groups within a larger society (say, the sub-Saharan in Seville, Albanian in Genoa, Algerian in Perpignan) that a composite portrait may bear little resemblance to any single citizen in the larger group.

Highly competent researchers in this domain use this model. In this method only those who match a specific cultural profile will be chosen as research participants, and documented members of the society who do not fit this template will be eliminated from consideration (*cf.* Marsella *et al.*, 2000). If we do not make this presumptive judgment before using our research participants, state Marsella *et al.*, "we cannot know if we are measuring their culture or their [distinct] ethnic ancestry. The reason for this is quite simple, it is not whether a person is Japanese or Swedish, it is how much they are Japanese or Swedish in practice" that meets our research objectives (p. 50).

Selecting participants for any study that purports to define the character of, say, traditional Swedes or Japanese requires that researchers use prior, defining assumptions about the character of these ethnic groups (rather than that of others who have recently immigrated). A great advantage of this approach is that, with precautions in place for proper sampling, it fits with the indigenization of one rather than the blending of several distinct cultural psychologies. As in other domains, one's research objectives determine the assumptions that one makes respecting the population under study.

A distinctly different approach proceeds from an *existentialist* point of view: no matter who legitimately moves in or out of a society, they contribute in a systemic way to the evolving culture that characterizes that society. If one is a legal member of the society, one contributes to the complex Gestalt that is its culture. Every individual's ethnocultural identity is thereby defined in an *existential* way. The accuracy of the assessment will be partially assured by the size and representativeness of the sample used. Thus Algerians, for example, contribute to the evolving culture of France, Jamaicans to England, and Puerto Ricans to America. Every culture is permeable and to some degree receptive to the technical skills and arts that "new arrivals" contribute. One cannot understand any culture without factoring in those contributions. Obviously tourists and undocumented aliens would not qualify, except in the most liberal

13 Sandra Scarr (1993) stated that "the study of the typical human is philosophically Platonic: Although individual differences within species are observed, they are considered merely unimportant variations on the ideal type" (p. 47).

construal of culture. The assumptions one makes about a population at issue is, again, determined by the research questions one is asking.

Entering a society and enlisting the participation of individuals, say, in a questionnaire-based research project is problematic if (a) the research instruments reflect the mentality of the researchers and their school of origin, and (b) we have not defined the contours of the culture that we intend to study from the standpoint of its members. Further researchers may choose to define a culture by simple reference to those who are legitimate members, no matter where they may have been born – in other words, those who have been lawfully accepted into the polity of interest and have all the rights appertaining to membership.

With the accelerating globalization of immigration patterns, traditional cultures are becoming transformed by the influx of exogenous "personality" patterns, much as has happened in the United States. With the influence of telecommunications and CNN-type international news broadcasts, cultures are changing both subtly and dramatically in many ways – and on a daily basis. It is literally impossible to find a culture, be it Norwegian, French, or Czech, which has not been dramatically altered by innumerable influences vectored in part by those who have recently arrived from foreign backgrounds.

Monocultural approaches to personality psychology

If one assumes that culture has an influence on the development of personality, it follows that personality psychology must take into account socially constructed values, scripts, and models of behavior when examining the development of personality. A personality psychology deriving from the assumptions of a singular culture runs the risk, of course, of (a) reflecting more the peculiarities of that culture than the scientific exploration of individual differences, and (b) distorting our vision of personality development in mosaicized societies and blended cultures. On the other hand, as soon as one begins to distinguish variant patterns of behavior among different ethnic groups (say, marital arrangements or clitoridectomies), it becomes difficult to separate out the value implications of those distinctions. This is a minefield for social scientists.

We recognize that the modern expansion of a panoply of "human rights" is being driven not so much by one culture or another as by international politics. The values inherent in these rights reside in the mind of the observer. They cannot, however, be presumed to have the same level of validity to all observers, as a cursory glance at the geopolitical conflicts embroiling the planet indicates. Even if the value judgments of observers remain tacit or out of awareness, they inevitably shape the tenor of their conclusions. One cannot separate out the *meanings* that "observers" impute to an activity from the processes of personality development described by them. Further, how a researcher construes a pattern of behavior becomes a concern for the "consumers" of these scientific products as well as a methodological problem for the researcher. In that light, cross-cultural and cross-ethnic studies make the researcher vulnerable to allegations of bias – of ethnicism and even racism. In addition, insofar as personality is a product of genetic as well as cultural factors, any research that demonstrates salient group differences leaves unresolved the question of biological contributions to those differences, should they exist. We simply do not know what proportion of between-group variance in various personality variables should be attributed to biological factors. Perhaps none.

What do race and ethnicity mean? The imprecise vocabulary used to describe groups is often confusing. The term *race*, for example, usually designates a population that has "typical" physiognomy, complexion, or blood-group characteristics. This is a controversial construct and is generally regarded as not scientifically definable. The term *ethnicity*, on the other hand, emphasizes the cultural aspects of a distinct population. Populations that were originally closely related in terms of, say, blood-group characteristics have evolved quite distinct cultures, characterized by different languages, religions, political structures, and family systems. For example, the Semitic people who evolved the Phoenician culture that spread throughout the Mediterranean Basin and founded thriving cities such as Carthage, Sidon, and Tyre are related to the Semitic people who evolved the Judaic culture, whose influence radiated primarily north, northwest, and northeast into Europe and Russia. Even if we could trace the myriad ethnic groups that have evolved from the common *Ur*-culture[14] that gave birth to all Semitic peoples, we would find many historical forces including migration, conquest, acculturation, and intermarriage that confound the genomic basis for distinguishing them from the Anglo-Saxon, Nordic, Hispanic, or Russian peoples of recent history. This has occurred to a greater or lesser extent among all the populations of the world with few, isolated exceptions. If we burrow back far enough in history, we would discover that we all have the same ancestors and are all "cousins" to one another (*cf.* Manrubia, Derrida, and Zanette, 2003; Rohde, Olson, and Chang, 2004). And if we progress, say, 2,000 years into the future each of us will have become an ancestor, directly or

[14] The prefix *Ur* is a handy German term meaning *original* or *primeval* (as in *Urpsychologie*, *Urvolk*, or *Urwelt*).

collaterally, of every human alive at that time. For this reason, the terms "ethnicity" and "ethnic group" are more appropriate than a term such as "race,"[15] even for populations that have a high degree of genomic relatedness.

The risk that biological explanations of cultural differences can be used to perpetuate harmful stereotypes is an additional reason for focusing on the cultural rather than biological basis of population diversity. Thus, many psychologists, and sometimes anthropologists, are reluctant to study group differences in personality, especially when trying to understand the genic determinants of these differences. However, although there are risks of perpetuating ethnic and "racial" stereotyping by this kind of research, some researchers argue that it is worth doing. Cloninger (1996) wrote, "Why study such a risky area? Besides a general hope that knowledge will bring a better world, it seems unwise to leave all descriptions of group differences to those whose aims are divisive." There are devices for minimizing these risks. One set of safeguards, briefly reviewed here, has been proposed by Paul Pedersen.

Covert assumptions in research. Experts in the field of multicultural psychology have argued that cultural bias is often unintentionally introduced into otherwise fair-minded research (for example, Pedersen, 2000; Ridley, 1995). Pedersen states, "if mainstream psychology is to become more global and less culturally encapsulated several changes will be required" (p. 20). He then proceeds to make a number of recommendations. First, textbooks need to present cross-cultural perspectives on the various topics that are treated in the text, and core principles should be illustrated by examples drawn from a variety of societies.

Second, psychological theories, wherever proposed, need to be explicated in terms of their culture of origin. For example, we must ideally know not only who the theoreticians were, but also the cities in which they were born and bred, their religious, political, and socio-economic background, the dominant worldviews in their cultural strata, and the critical events in their personal lives. Readers need to be helped in understanding the generalizability of the proposed principles, given the profound impact that cultural variables can have on theories that purport to be universal descriptions of human behavior. In general, behavioral norms for one society should not be presumed to apply to all societies.

Third, psychological theories need to be understood in their historical context. The specific worldviews that prevailed in the era and societies in

[15] A corollary of this principle is that the term *racist* gives respectability to the construct *race*, purportedly a construct more rhetorical than scientific.

which theoreticians were educated when they developed their theories give an informed perspective on their usefulness and applicability in other times and places. Of course, this hermeneutic principle applies to the entire field of cultural psychology.

Fourth, journals of psychology need to internationalize their editorial boards as well as invite contributions from a variety of cultures. A corollary of this is that American, European, and Canadian researchers should read publications appearing in the journals of other countries. The utility of this recommendation is limited by the fact that many appear in languages we may not know, nor are all cultures at the same level of sophistication in the various disciplines to which different journals are dedicated.

Fifth, the co-authoring of textbooks by persons from different cultures allows varying perspectives to influence their development. This practice precludes monoculturalism (but not, if there are only two authors, biculturalism). Finally, "institutional support and funding for collaborative research across cultural and national boundaries needs to become more available" (p. 21).

To a great extent, these recommendations assume that a blended, global psychology has a value that various indigenized psychologies do not.

Cross-disciplinary issues

Distributed cognitions

Each group to which we belong is an extension of ourselves. The communities within which we live are social prostheses that help us to achieve personal objectives that would otherwise be out of reach. Division of labor is one socio-economic mechanism that makes this possible. Common interests place the power of the group at our disposal to achieve goals ardently sought both by the group and by any individual within it. Writing a good journal article is a challenging task for a scholar, but it would be impossible without the support of numerous other individuals. This includes, among many others, the contributions of (a) those researchers who contributed to the archival resources for the article and were generous enough to vet the manuscript, and (b) other scholars who managed the laboratory, had the technical skills to conduct one or several experiments, and then gathered and treated the data. Bernard de Mandeville and Adam Smith are the emblematic spokesmen for this socio-economic phenomenon, which reigns even in academia.

Socio-cultural and educational psychologists have proposed a theory of "distributed cognition" in which they dispute the idea that knowledge resides simply in an individual's mind (Resnick, Levine, and Teasley, 1991; Salomon, 1997). These theorists maintain that knowledge is distributed through communities of scholars as well as in the books, journals, and electronic devices that store our thoughts and harbor our data banks. Salomon has argued that "people think in conjunction and partnership with others and with the help of culturally provided tools and implements" (p. xiii). An individualist perspective on the fabrication of knowledge may be useful in understanding the creative achievements of Galileo, Copernicus, Brahe, or Kepler, although even they (excepting Brahe) relied on the optical technology of the day as well as the mathematics and (not entirely flawed) Ptolemaic schemas of antiquity.[16] But normal science (Kuhn, 1970, pp. 10–34)[17] develops through the interchange of ideas that scholars generate by way of thousands of experiments and journal articles – which they avidly share with each other. They are nothing if not a mutually enmeshed community of competent, industrious, talented laborers advancing projects toward the "Hill."[18] These legions of contributors to the accumulation of human intelligence are too numerous to remain in public consciousness. However, they are represented by those luminaries who serve as historical markers for later generations, who in their turn continue to advance science.

Resnick, Levine, and Teasley (1991) enlisted the collaboration of cultural anthropologists in arguing that cognition does not simply transcend the mind of individuals, it transcends the bounds of single cultures (see, for example, Hatano and Inagaki, 1991). The challenge of understanding this principle in its many dimensions requires the collaboration not only of cognitive scientists, but of social psychologists, developmental psychologists, psycholinguists, anthropologists, and sociologists. Bhattacharya and Chatterjee (2000) stretched the envelope further by stating that the distributed cognitions of people and artefacts are integrated during the process of collaborative innovation. The availability of the World Wide Web makes it possible for us to work with

16 Those who tend to be dismissive of the work of Ptolemy should consult his masterwork, *The Almagest*. A convenient source is *Great Books of the Western World*. Chicago: Encyclopaedia Britannica, Inc., vol. 15.

17 For Kuhn (1970), "normal science means research firmly based upon one or more past scientific achievements that some particular scientific community acknowledges for a time as supplying the foundation for its further practice. Today such achievements are recounted ... by science textbooks ... These textbooks expound the body of accepted theory, illustrate many or all of its successful applications and compare these applications with exemplary observations and observations" (p. 10).

18 See the section, "The hive and the Hill," below.

groups of cross-culturally compatible colleagues to conduct research and effect innovation in select fields. "This, in turn, enriches cognition, metacognitive abilities, techniques of interpersonal communication and reflective thinking skills" (p. 295). Researchers, however, do not agree about the properties of the paradigm that guides theory-building and experimentation in distributed cognition. There are still definitional issues that create ambiguity among those working in these areas, including the field of educational psychology (for example, Moore and Rocklin, 1998).

Individualism and collectivism

Distributed cognition has its analogue in distributed patterns of behavior and volition. If cognition is not bounded by one's body, then the wellsprings of human behavior are not bounded by "the individual" but are distributed throughout the complex social network of group and national norms that enmesh the individual. An example of this principle, presented by Cross and Markus (1999), is that "in many Asian cultural contexts, the source of individual action and behavior is distributed throughout a configuration of one's relations to others; it can be found in the net of expectations and role obligations that binds one to a particular set of others" (p. 379).

Every society generates normative behavioral scripts governing the interaction of its members with each other. The degree to which a society is successful in prescribing and enforcing those norms depends on the coercive means the group has at its disposal. The value a society places on individualism or collectivism partially determines the level of compliance with social norms. If we conceive of this variable as a continuum ranging from the most laissez-faire to the most totalitarian, we will find cultural differences that vary as a function of this (permissive–inhibitory) variable. The more collectivist the society the more homogeneity will we find in the behavioral patterns manifested in its population. Less homogeneity and greater diversity will characterize individualistic societies. Hence, we can meaningfully speak of collective personality to the extent that a society is compliant and law-abiding. The society that is not governed by uniform and internalized religious, social, and political rules cannot be characterized as having a collective personality except with a multitude of reservations.

The Euro-American view of what constitutes optimal development of the individual is distinct from that of the preponderance of the world's populations. This Western model (even this term, like the term "Orient," is an ethnocentric creation of Europeans to distinguish them from peoples living to the east of them)[19] is a minority and non-dominant paradigm in that respect. Although there are innumerable distinct cultures within this North Atlantic entity, certain transcendent values have shaped the indigenous psychologies of Europe and North America. For example, the pivotal construct by which one can distinguish Western cultures from others is that of *individualism* (Triandis *et al.*, 1988). It is a superordinate[20] "way of being" that shapes all the other values of the society. The revolutions that brought social upheavals to Europe and North America in the past few centuries characteristically had "freedom" as a central objective. (The terrible but short-lived totalitarian regimes of the twentieth century are notable exceptions.) When Patrick Henry proclaimed, "Give me liberty or give me death," he was exalting the right of the individual (presumably a White male) to shape his or her own life projects, with inalienable rights to "life, liberty, and the pursuit of property,"[21] rather than have them shaped by foreign or domestic government. This apotheosis of personal freedom was built on a creed of individualism and the subordination of governing bodies to the will of its citizens. For this reason, modern collectivist regimes in the West have had difficulty in establishing and maintaining themselves for long.

Every society has its culture-heroes, its myths and preferred histories – the narratives that recount with great conceptual power and vision their military and political origins, the founding of their religions, their beliefs about their class structures, and the virtues that make them a distinct society. These beliefs and values pervade every dimension of national life. Cultural psychologists affirm that they pervade all the social sciences as well, including, of course, psychology – and more specifically, personality theory. In the opinion of some, personality theory is not only an invention of Western societies, it is one that in its language, conceptual categories, and axiomatic principles has little correspondence with the reality of Asian, Latin American, or African cultures.

[19] If the Japanese had sectored the world as "the West" has, one can speculate what might have constituted the *West*, the *Near West*, the *Middle West* and the *Far West*.

[20] The polar opposite of subordinate. Those students who are interested may read the classic study of Muzafer Sherif and his colleagues (Sherif, [1961] 1988; Sherif, White, and Harvey, 1955) in which they demonstrated that groups that were hostile toward each other were motivated to collaborate when a higher, superordinate goal was imposed on them. The common goal had the result of diminishing their antagonism toward each other.

[21] This Lockean formula of human rights was altered by Jefferson in the American Declaration of Independence to read "life, liberty, and the pursuit of happiness."

Introversion and culture

People in North America who score "high" on introversion scales are considered to be less adaptive and psychologically well positioned to succeed than those who are more extraverted. The North American culture can be said to value extraversion. This is not the case with other cultures. The Japanese, noted for their social reserve and emotional control, have an aphorism: "The nail that sticks up will be hammered down." They value conformity and are intolerant of the exuberant individualism that is considered normative, say, on college campuses in North America. Among some Caucasian cultures, the Swedish, for example, shyness is valued. It does not as a rule compromise career success in Swedish society. Shaffer (2000) stated, "[Swedes] view shyness more positively than Americans do and prefer shy, reserved behaviors to bold assertive or attention-seeking antics. Consequently, shyness is not really a disadvantage for Swedish men, though it possibly is for Swedish women who tend to marry men who have lower incomes" (p. 117).

Shyness is only one aspect of a broader typological dimension: introversion. Personality types, as noted in Chapter 5 on trait psychology, are unlikely to be simply artefacts of culture. An important contribution of Hans Eysenck (1990) to the field of trait psychology was his demonstration – it was not the first – that such types are largely determined by biological factors. In this view certain ethnic societies are as *peoples* genetically disposed to develop an introverted lifestyle (which would then explain why they have historically developed such a social pattern of behavior). This genetic disposition becomes randomly assorted among humans of all ethnic backgrounds, but in varying proportions. Factoring out the variance attributable to genes, we can conclude that culture is also a result of geo-political, spontaneous, creative, and accidental forces.[22] "Birds of a feather" in sufficient numbers eventually establish dominant norms.[23]

[22] In a commencement address Richard Feynman (1974) alluded to the brief historical epoch in the Second World War when aboriginal people in New Guinea developed a Cargo Cult Science modeled on the logistical procedures employed by Western forces. Eric Lippert (2008) has described the term: "During the Second World War, the Americans set up airstrips on various tiny islands in the Pacific. After the war was over and the Americans went home, the natives did a perfectly sensible thing – they dressed themselves up as ground traffic controllers and waved those sticks around. They mistook cause and effect – they assumed that the guys waving the sticks were the ones making the planes full of supplies appear, and that if only they could get it right, they could pull the same trick."

[23] A friend of mine recounts the experience of teaching a class of native English-speakers at the University of Hawaii, half of whom were of native Hawaiian and other Polynesian stock, the other half Caucasians ("haolies"). The taciturnity of the former and the garrulous, argumentative style of the latter were in striking contrast.

Behavioral plasticity, on the other hand, is remarkable in the face of changing cultural demands. The power of institutions to alter behavior is readily evident among the products, say, of military boot camps, élite prep schools, seminaries, medical schools, and trades unions. Even if one grants that self-selection into these organizations is not the sole determinant of membership, plasticity at a late adolescent, early adult age largely accounts for the influence of social situations and institutions in shaping the various social selves that individuals appear to adopt.

The implications of such a notion are self-evident for immigrants to Western nations from cultures that differ not only in the value they place on shyness and a self-effacing demeanor but also in the numerous scenarios and scripts that govern day-to-day social life. Immigrants tend to buffer themselves against culture shock by initially moving to a "ghetto," from which they can, as from a safe haven, gradually venture deeper into the new society that (often on its own terms) is welcoming them. If introverted immigrants are intent on succeeding in a new, extroverted culture they *may* choose to pursue a career that does not require a lively, outgoing, talkative, sociable presentation of self (as, for example, doing bench work in a microbiology or physics laboratory). Alternatively, they may choose to cultivate a more assertive, gregarious persona, often required in the business world. (Of course, people can only forget; they cannot unlearn.[24])

Inner-directed and outer-directed persons

That a culture tilts more to an individualist or a collectivist value system is itself an aspect of its collective personality. This dimension serves as a superordinate principle affecting the evolution of the personalities of the many individuals who compose the society. It prompts the question: how do collectivist and individualist societies inculcate in their members fundamental attitudes of responsibility for self and family, as well as the larger community? Clearly, it is done through all the educational media and institutions that the society has at its disposal. For example, millions of school children in the English-speaking West have been required to memorize W. E. Henley's poem, *Invictus*. It presents the individual, defiant and decontextualized before the universe, proclaiming:

[24] Piaget gave as an example, once we have the solution to a puzzle we cannot unlearn the solution (Polanyi, 1965, p. 780) – but one can forget it, which may be desirable should it be incorrect.

Under the bludgeonings of chance
My head is bloody, but unbowed.
I am the master of my fate;
I am the captain of my soul.[25]

Rudyard Kipling's poem *If* is also a grandiloquent exaltation of the individual, redolent of the self-reliance that was the characterological armor of people building free enterprise economies and girdling the globe with their mercantile, corporate structures. He writes: "If you can keep your head when all about you are losing theirs and blaming it on you … If neither foes nor loving friends can hurt you … [then] yours is the earth and everything that's in it." This is the song of the empire builders who clothed themselves with heroic, Homeric attributes. And though few college students now read Ralph Waldo Emerson's essay, *On Self Reliance*, its message has entered into Western culture like a subtle ether that we are not aware of but that we breathe daily. Adult Americans, for example, are taught that they should be autonomous and inner-directed, making their own decisions rather than having them made for them by their families or their communities; that they should rely first of all on themselves and only at a later moment on others; that they must take personal responsibility for their acts; and that they may accept credit but should also assume blame for their successes and failures in life.

Allied to this spirit is Max Weber's ([1904] 1930) notion that traditional ways of life in pre-industrial, undeveloped societies needed to be abandoned in order for rational, technocratic thinking to assert itself (Elwell, 1996). A transformative Protestantism asserted that it was *individuals* who were numbered among the *elect*, and that their success in this world was evidence of their eternal salvation. Modern *Existentialism*, a philosophical psychology of northern and middle European inspiration, affirmed that one of the ultimate concerns of *every* human is his or her *aloneness*. We must die as we were born – by ourselves. No other can make the journey with us, even through life. This spirit continues to influence the arts, literature, cinema, television, and theater. The exploits of the individual are celebrated in American films such as *Unforgiven*, *Top Gun*, *Erin Brockovich*, *Hud*, *Cool Hand Luke*, and *Hombre*. The psyches of the past several generations of Americans were shaped in the movie halls and in front of the TV screens of Middle America, as well as in its schools and churches. The family ideal in the United States is not a harmonious, self-subordinating collective in which the children adopt the views of their

[25] A new generation was introduced to this poem when Oklahoma City bomber, Timothy McVeigh, included it in the statement he released just before his execution by lethal injection.

parents by adapting and "fitting in"; rather it is exemplified in the Bunker family portrayed in the sit-com *All in the Family*, which enjoyed exceptional popularity for the latter decades of the twentieth century. The Bunker family was rancorous, quarrelsome, generationally conflicted, and individually defiant on almost every significant social issue, underscoring and illustrating the individualist nature of American society.

Self-actualization

The humanist school of psychology, epitomized by such clinicians and theoreticians as Adler, Goldstein, Maslow, and Rogers, supposes that there is an intrinsic dynamic in humans to make the most of themselves. Steele states that those who are faithful to their nature will find that motivation "to affirm the self ... [is] an inherent and important part of the self-system" (Horenczyk, 1996). As noted in Chapter 6 on self psychology, a host of psychologists researched the hypothesis that individuals seek a "self-completion," an ultimate fulfillment, and the exploitation of their potential as individuals. If you refer to the pyramidal figure depicting Maslow's hierarchy of needs you will note that the social need for group affiliation and belongingness is at the third level of the hierarchy, two levels below the need for self-actualization, which is at the pinnacle.

Relevant to the individualist–collectivist issue is Julian Rotter's (1966, 1975, 1990) construct of "internal and external locus of control." Rotter (1966) developed an inventory to measure the extent to which individuals believed they could influence their environment and the rewards that society made available to them. Implicit in the hundreds of research articles that this theory inspired is the assumption that the individual who thinks that social groups, or fate, chance, or powerful others determine how well he or she will fare in life is at a disadvantage in comparison with individuals who think that success is largely a product of their own efforts. Consistent with this ideology is the work of Shelley Taylor and her colleagues (1983; Taylor and Brown, 1988), who found that the mental health of her subjects (who had suffered personal tragedy in their lives) was correlated with re-establishing a *personal* mastery of their environment – a mastery that was in many instances largely illusory, but still influential.[26] Locus of control and self-actualization relate to a panoply

[26] It should be noted that this thesis has evoked a storm of controversy and many research articles, not least the contrarian studies of Colvin and his colleagues (for example, 1994; 1995).

of other value-laden Western "virtues," such as inner-directedness or autonomy (see, for example, Balog, 1997; Hyman, 1988), optimism, extraversion, self-affirmation, creativity, independence, and gregariousness.

Collectivist societies do not by and large resonate to these views. Pedersen (2000) cites Rigney, who has highlighted the fact that in many traditional cultures the Maslow hierarchy of needs, as an example, is totally inappropriate: "Placing self-actualization at the top of the so-called hierarchy is simply incorrect for my people. We are who we are because of our relations with the group." To the peoples of Southeast Asia, Japan, China, and Korea, these personality traits bespeak a self-centeredness that group-centered cultures find uncongenial if not repugnant.

The hive and the Hill

What is meant by the term *environment* when we speak of person–environment interaction? Environment encompasses all the external stimuli that impinge, or are capable of impinging on us, but more specifically and importantly it refers to people who are in contact with us, either directly or indirectly. For better or for worse, every person-to-person interaction shapes, if only in a minuscule way, the self-perceptions of the interacting individuals and the behavioral consequences of these changes. Of course, it is not just persons, as bundles of raw data, who impinge on us, but all the prosthetic devices they use to extend and amplify their influence. Most significant among these are institutions – political, educational, social, cultural, economic, and religious – whose power is enhanced by print and broadcast media (Resnick, Levine, and Teasley, 1991). To understand individuals is to understand not only the temperament with which they were endowed at conception, but to implicitly understand the significant people and the institutions in their lives that have shaped their life goals.

Some scholars have attributed greater importance to the influence of institutions in shaping personality than to random experiences. A different school has affirmed the pre-eminent importance of personal history, sown with accidents, tears and rejoicing, disappointments and successes, loving nurturance and rejection, chance friendships, and ambivalent group affiliations, sickness, frailties, and periods of good health (*cf.* Albert Bandura who has written extensively on the importance of chance in life-courses).[27] An exemplar of the former model is the

[27] An illustration of the power of chance illness on the life course of an individual's personality is revealed in the work of Milton H. Erickson (for example, Zeig, 1980). Arguably one of the foremost psychotherapists of the twentieth century, Erickson affirmed that the best teacher he ever had was poliomyelitis. It taught him to be patient, to accommodate to

anthropologist, Ruth Benedict, who believed that individuals are stamped with the personality of the dominant culture within which they were raised. Sicilian culture or Parisian culture is fundamentally the personality of a "typical" Sicilian or Parisian, writ large. The concept of Dutch culture, for example, results from the fact that modal Dutch citizens resemble their respective countrymen more than they resemble typical members of other societies.

A society of organisms, whether of humans or ants, multiplies the power of its members by several magnitudes. These multitudes seem to "accumulate intelligence" as they collaborate. Lewis Thomas, the biologist, wrote a typically insightful essay, "Societies as Organisms" (1990), in which he vaguely related the "Hill" and the hive to human society, which resembles a thinking organism more than it does an agglomeration of individuals. He wrote:

> A solitary ant, afield, cannot be considered to have much of anything on his mind; indeed, with only a few neurons strung together by fibers, he can't be imagined to have a mind at all, much less a thought. He is more like a ganglion on legs. Four ants together, or ten encircling a dead moth on a path, begin to look more like an idea. They fumble and shove, gradually moving the food toward the Hill, but as though by blind chance. It is only when you watch the dense mass of thousands of ants, crowded together around the Hill, blackening the ground, that you begin to see the whole beast, and now you observe it thinking, planning, calculating. It is an intelligence, a kind of live computer, with crawling bits for its wits. (p. 11)

Shunning a dichotomous perspective. Human variables are intrinsically continuous in nature. However, we explain behavior using dichotomous constructs that can lend a false sense of precision to our thinking and induce simplistic paradigms for understanding the world around us. Although creating "boxes" in which to store data has its advantages, it generally leads many to believe that the boxes exist in reality rather than only in our minds and later on paper. I have done this, but recognize that there are clumsy limitations to dividing the human population of this planet into collectivist and individualist societies. There are no absolute criteria for categorizing cultures as one or the other. In the language of factor analysis, collectivism and individualism are not orthogonal factors, nor are they totally independent of one another. In fact, they are constructs that have an assortment of constituent sub-factors that overlap.

a sedentary lifestyle, to observe the minute paravocal and nonverbal cues in others, to probe the tacit, intuitive resources of his mind, and otherwise exploit such talents as had not been destroyed by disease.

Many Americans, say of Italian extraction, live in an extended family that extends over three or even four generations. They buy duplexes or triplexes, distributed by related family units, which allow them to enjoy a close-knit communal life. A grandmother is often available to care for the children while the parents are at work, and she often welcomes them home with a casserole for their evening meal. The children run errands, help in emergencies, and do repairs and maintenance for those members of the family who have "lost capacity." All of this suggests a collectivist orientation. However, the members make individual forays into a competitive labor market, a highly individualist socio-economic theater, and they succeed there. The shared property and domestic responsibilities are complemented by personal independence, self-reliance in the public domain, and individual pursuits. On a collectivist–individualist continuum such a family cannot be equated with families of the Kibbutz, where property is owned in common and child care is distributed among nonrelated members of the commune. Other Americans, say, those of Jewish background, promote respect for family norms, congregate for religious ceremonies with due regularity, *tend* to marry members of their own tradition, and are strongly affiliated with Jewish institutions. But these same individuals promote self-expression in their children, value individual thinking, and promote entrepreneurial self-reliance in the socio-economic arena. How can we calibrate movement on that continuum? Where on an individualist–collectivist continuum should these individuals be placed? Likewise, British society is perceived as traditional and stratified still, with constraints on social mobility. A hereditary peerage system that is intrinsically nonegalitarian is deeply embedded in the British culture (although it is slowly being liberalized). Yet who would argue that British society, a society that prizes the individual liberty of its citizens as highly as any other,[28] and has fostered nontradition-bound science like no other, is a collectivist society? Collectivism and individualism appear to be two different dimensions of every culture and personality.

Science and collectivism

Twenty-five centuries ago, Anaxagoras was placed on trial in Athens on charges of impiety and atheism – notwithstanding his status as an instructor of Pericles, the pre-eminent statesman of that era.

[28] It has long been recognized that all science is politicized and serves social and even nationalist ends (for example, Mannheim, 1936). As with all other judgments, this is a matter of less rather than more, not a matter of none rather than some.

Anaxagoras' crime was to speculate that the sun was simply a glowing rock, possibly half as big as Greece. Relative to that event, Frye (1992) stated, "What is important for us ... [to realize] is that some of these early speculations collided with the anxieties of [Greek] society" (p. 32). The scientific work of the Greek thinkers that preceded the Common Era and set the stage for a scientism that was only to flourish fully many centuries later was not significantly influenced by the values, comfort levels, mythic beliefs, and culturally sanctioned schemas of the day. That the geometers, architects, astronomers, mathematicians, logicians, biologists, and cosmologists, not to mention the playwrights, poets, and philosophers of that civilization, were individuals who were willing to propose ideas that violated the *communitarian* mores of their society set the standard for the individualist European culture that was its cultural heir. The people who bequeathed to the world the highly individualist and competitive ethos of the Olympic games did not evolve into a classically collectivist society. Instead, they created the paradigm and the standards for the individualist society. And the principle that they exalted "was the right of individuals to make their own speculations, a right that few societies have been willing to concede" (Frye, 1992, p. 32). To this day in the West, the espoused ideology, though it is often violated in the rough-and-tumble of factional and international politics, is that the pursuit of scientific truth must not be compromised by religious, regional, tribal, or ethnocentric ends. "Science is one faith and has one great commandment. The faith is ... that man can control his future in his own interest" (Becker, 1952); the commandment is to pursue truth in the service of that end, wherever it may lead.

There is a vast literature on the subject of "individualism" (often characterized as pertaining to the "developed" world) as opposed to "collectivism" (characterized as pertaining to undeveloped and developing nations).[29] Triandis, McCusker, and Hui (1990) have illuminated the factors underlying individualism and collectivism with a fifteen-nation, multi-probed analysis of the schemas that undergird these constructs. In this study collectivism was best represented by the constructs family integrity ("children should live with their families until they get married") and interdependence ("I like to live close to my good friends"). Individualism was best represented by the constructs

[29] Implicit in this categorization of cultures, linking individualism with economies that are more advanced and collectivism with economies that are less developed, is a subtle way of promoting one ethos over another. Placing these issues in an historical context is necessary to understand the factors that have enabled some groups to prosper more than others.

self-reliance ("it is best to work alone rather than in a group")[30] and separation from in-groups (for example, one should not try to accept credit for the achievement of other members of your group).

Clearly one can distinguish collectivist attitudes and values from individualist ones. Triandis found that within any population individuals are distributed along a continuum of interdependence, on the one hand, and self-reliance, on the other hand. The factors that best capture these orientations are "family integrity" and "emotional detachment from the in-group," respectively. The more self-reliant are called *idiocentric* (from the root word for self: idio-); the more other-dependent are called *allocentric* (from the root word for other: allo-). When distinguishing cultures along these dimensions, Triandis recommended that we use the term "idiocentrism" to characterize those who are individualist, and allocentrism to characterize those who are collectivist. Self-reliant idiocentrics pop up in certain numbers in collectivist societies. These are the individuals who find it attractive, if not imperative, to migrate to more individualist societies. Correspondingly, group-attached allocentric individuals in a highly individualist society do not suffer the same culture shock when migrating or posted to a collectivist economy.

Language and personality

Language arguably is the most important aspect of culture – a slowly evolving instrument that mirrors the values, activities, belief systems, and artefacts of a society. The meanings that events or achievements take on for individuals in a culture are shaped by the words habitually associated with them. But language perpetuates as well as mirrors a culture (Shweder and Sullivan, 1993). To be raised speaking a particular language teaches us not only to think with the cognitive structures and ideas that the language makes possible, but steeps us in the literature of

[30] As with all lexically based research, this study is far removed from the "reality" being studied. One is dealing with words that only imperfectly capture the constructs being assessed; the constructs only imperfectly capture the meaning research "subjects" place on those constructs (which, in turn, only imperfectly correspond with the meanings that the researchers give to these linear assortments of words); and the meanings, structure, and dimensionality the subjects place on the external reality they are asked to assess differs widely from culture to culture, and within each culture, from person to person. Take the dimension "self-reliance," for example. There are self-reliant people who have a deep appreciation of teamwork. They take varying levels of responsibility for their work, believe that in the long run success favors the individually dedicated person, and are variably committed to civic causes. The paper and pencil inventories only allow us to look at the subject of inquiry darkly as through smoked glass. Marsella *et al.* (2000) present an excellent review of these issues.

that language. Europeans of the Dark Ages, largely unlettered, received most of their intellectual sustenance from listening to sermons phrased in a Biblical language. They reasoned using dualistic schemas of body and mind, polarities of allegiance (as in "Whose side are you on?" or "Those who are not with me are against me"), political dichotomies (as in the sacred and the profane, the Church and the Crown), moralistic categories (as in good and evil, "state of grace" and "state of sin," believers and infidels), of finalism and teleological schemas, free will and final causes. When the Bible was translated from Latin into the vernacular, these rich phrases and prosodic cadences entered the language of the common people.[31] The best example of this is the King James version of the Bible (largely the work of the religious martyr, William Tyndale [1492–1536]) that significantly reflected and shaped spoken English of the post-sixteenth century. The words and phrases of that translation permeated the speech of the common people and for centuries subtly shaped the mores, customs, and values of the societies in which it was used.

If English (or German, Chinese, or Tagalog) shapes the thought and the cognitive categories of the people who use it as a native language – and indirectly their character – one may ask how native speakers of any language escape the constraints this puts upon their thinking. Researchers who specialize in social linguistics and cognitive anthropology have been obliged to address this question. It is of special interest to those engaged in cross-cultural studies, for many of their studies involve the use of self-report inventories conducted in the varying languages of the populations being studied. An unavoidable dilemma presents itself. Researchers may develop a scale or other instrument in one language, say standard American English, normed on performance of educated White North American adults, and then translate this instrument into a foreign language, presumed to convey equivalent meanings to the speakers of that other language. This presents one set of problems – not least, the one that Nietzsche refers to as "looking into the world differently" as a function of the idiom one uses. Or researchers may start with similar constructs and research questions and independently develop instruments in several languages, which they then attempt to cross-validate in the various populations at issue.[32] This presents a

[31] The influence of the King James version still resonates in the language of English-speaking nations as we note in the film *Pulp Fiction*. Verses from the Prophet Ezekiel are vividly proclaimed by Samuel L. Jackson, who played the role of a common thug living in, but eventually emancipating himself from, the corrupt underbelly of late twentieth-century American society.

[32] There are some nettlesome difficulties in asking self-report questions of populations speaking different languages. Marsella *et al.* (2000) have detailed some of them. Many

different set of problems – not least, the relative incomparablility of the very constructs and cognitions that are imperfectly encoded in words. Addressing the challenge has given birth to a new discipline labeled *psychosemantics*.

The constraints are not absolute as language is constantly being reinvented precisely to allow the expression of novel ideas. Shifting worldviews enlarge and enrich the idiom needed to symbolize them adequately. English speakers have traditionally been among the most flexible in importing words from foreign languages to match precisely the paradigms and schemas that they invented or imported. What would we do without such terms as *machismo*, *junta*, *apparatchik*, *lumpenproletariat*, *esprit de corps*, *imbroglio*, *jungle*, *chutzpa*, *shlep*, *intelligentsia*, or even *ciao* and *so long*?[33] Assimilating such words is to assimilate, at least partially, their meanings, which in turn reverberate throughout the host culture.

Metaphorical language. Metaphors abound in our language. Though they are valued in poetry, they must be suspect in scientific discourse, for they both reveal similarities and conceal differences. This is particularly problematic where they blend rhetoric (for purposes of persuasion) with scientific discourse (for purposes of explanation). The extent to which we festoon our language with metaphors to create a tone, image, or feeling about an event is well demonstrated by Lakoff and Johnson (1980) in *Metaphors We Live By*.[34] Popular metaphors permeate all aspects of human thought. They ultimately shape a nation's culture and collective personality.

The argument needs to be balanced, however, by alluding to the usefulness of metaphor in scientific discourse. Light, for example, is said to have a particle nature as well as a wave nature. Minsky (1986) pointed out that Volta and Ampère "discovered how to represent electricity in terms of pressures and flows of fluid ... [thus] they could transport

of these difficulties are similar to those that affect any self-report instrument (such as the desire to please the researcher by "giving the right answer," complicated by ethnocultural differences in construing what "the right answer" may be). For this and many other reasons, formal, fixed, paper and pencil-based quantitative approaches to studying complex, nuanced socio-cultural questions are yielding to qualitative approaches that are relatively unstructured, more *phenomenological*, less number-based, and highly nuanced in their results as well as their conclusions.

33 The student who is interested in language may enjoy Eugene Ehrlich's (2000) book, *You've Got Ketchup on your Muumuu*, that explores words from around the world that have become current in English. Another rich and pleasurable source is *The New Joys of Yiddish* by Leo Rosten (1999) that explores hundreds of Yiddish words that have passed into common American parlance.

34 The use of the word festoon both as a verb and as a metaphor is used as much to describe a fact as to create an impression, as above.

much of what they already knew about fluids to the domain of electricity." He continued: "such cross-realm correspondences can enable us to transport entire families of problems into other realms, in which we can apply to them some already well-developed skills" (p. 299). These are the metaphors of the natural sciences,[35] which have become a global language, generating a global culture of science. In fact, these scientific metaphors have become so widespread that one would be hard pressed to identify different metaphors used by Euro-American, Australian, African, or Asian scientists.

Divergent uses of language. Martha Borgmann Crago (1988) did a study of the communicative interactions among the Inuit of northern Québec.[36] There are striking differences between the cultural patterns of the Inuit and those of Caucasian Canadians, and aside from their socio-economic organization there are few differences more unusual, even startling, than in the manner in which they educate their children to communicate with them.

Inuit mothers keep their infants continually pressed against their own bodies and are acutely sensitive to all the physical cues by which their infants communicate their needs. Visitors among this ethnic group are surprised at the absence of crying among the children. Crying can result from their experience of pain, but not of hunger, fear, or inattention. As a consequence children rarely have to ask for what they need, nor are they encouraged to. Taciturnity is a value inculcated from the earliest days of life. This Inuit virtue is layered into all the activities of the growing child. Although the "southern" classroom has been imported into their social structure, the mode of instructing their children has not. Children learn by watching and listening. Questions are discouraged; students are encouraged simply to *attend* to what is happening in the classroom setting. Teachers in that society evaluate a child's mastery of language not by the production of speech, but rather by their ability to respond to directives of an increasingly specific character.

[35] John Dewey, George Herbert Mead, and Lev Vygotsky repudiated the division of science into the natural sciences and the human sciences, and rejected the idea that the "natural" sciences were aseptic, context-free enterprises, while the "social" or "human" sciences were subjective, interpretive, historically bounded, and inherently compromised by changing subject matter (Cole, 1991).

[36] The label *Eskimo* is no longer used to designate the Inuit people of the Arctic regions of the globe, as it is a pejorative term used by other Amerindian peoples for "outsider." The Inuit refer to themselves as Inuit (in the plural). A single member of this group is called an *Inuk*. Inuit refer to the Caucasian as a *Qallunaat*.

Crago relates an illuminating classroom event. A recently arrived non-Inuk (that is, Qallunaat) teacher was impressed by one of her students, Joseph, whom she found to be attentive, curious, interactive, questioning, and obedient. A model child in her estimation. When she met with the indigenous teachers of that school she enthusiastically brought up Joseph's name. We know about him, they quickly responded. "Do you think he might have a learning problem? Some of these children who don't have such high intelligence don't know when to stop talking"[37] (p. 219). Joseph's classroom behavior had been perceived as evidence of giftedness by the "foreigner," when in fact he had a personality that made him stand out among the Inuit as a deviant and troublesome child.[38] Numerous missionaries, educators, government officials, and researchers have visited Inuit communities with a view to helping them in one respect or another. Insufficiently prepared for their work, in many instances, they have projected unrealistic expectations on the natives they wished to help. The locals have insistently reproached them for their hard-headed resistance to the principle that well-raised Inuit children should *not* talk.

The implications of these cultural norms for the formation of the personality of Inuit children are clear. In addition, the interaction of nature and nurture among the Inuit as they grow to adulthood in a harsh geo-economic environment generates personalities that, for all the individual differences that one finds among them, makes them resemble each other much more than they resemble those who come to visit them from the south. Or so it appears to the southerners.

Personality and the sensorium

Proxemics and haptics. A significant aspect of personality is the preferred pattern of sensory approaches to the world that individuals, and the cultures in which they are embedded, utilize for negotiating the demands and the opportunities that their environment provides. *Proxemicists* are social scientists who study the way we structure space

[37] A comparable example is given by Tobin (1989), who studied pre-school education in China, Japan, and the United States. American researchers observed a Japanese child throwing objects about and otherwise disrupting the educational process in the classroom. They suggested to the teacher that perhaps he was just a more gifted, but bored, student who was not sufficiently stimulated by his environment. This took the Japanese teacher by surprise; she countered that if the child were intelligent he would know how to behave properly in this setting.

[38] Actually, as the child's father was a Qallunaat who had married an Inuk, the child was being indoctrinated with social scripts that were maladaptive in the context of the dominant culture in which he was being educated.

and our preferred modalities for interacting with others within such space. Edward T. Hall (for example, 1966, 1969), an anthropologist by profession, was an early pioneer in the exploration of cultural differences in the use individuals make of their senses for modulating, monitoring, and interpreting their social environment. *Haptics* refers to patterns of bodily contact that are socially approved or condoned. The rules for touching others change as a function of the situation people are in as well as the social relationship that exists between them (Lyman *et al.*, 1999). There are wide disparities as well between the patterns of touch that are prescribed (and proscribed) in various cultures. The rules that govern touching others vary greatly from culture to culture as well as within sub-sectors of each culture. The interaction of haptics with proxemics is highly complex and space will not allow an extended treatment of this subject. However, these are fascinating fields with significant cross-cultural implications.

Even the different ways in which cultures structure their perceptual worlds determine how individuals will interact. Hall (1966) found that Arabic cultures typically place emphasis on the sense of smell, and respond to the rich, varied, seductive (or repellent) odors of the bazaar and people. These are high-context, contact-inclined ways of being that require cognitive processing that is more existential, immediate, intuitive, and subjective. Northern Europeans, on the other hand, give primacy to distance receptors like vision and hearing, and they are culturally inclined to less personal, more abstract and objective cognitive processing (see Montagu, 1986, pp. 293–392). As the world becomes smaller and cultures become enmeshed, we are witnessing the softening and blurring of cultural distinctions. This continental drifting is deplored by some, applauded by others.

Conclusion

The study of "culture and personality" is, by its very nature, emic. There is an assumption that individuals who are born and bred in a particular culture will show commonalities in their personality traits that behaviorally distinguish them from individuals of a different culture. Although theorists and researchers in human development have traditionally sought to discover *universal* laws governing the development of humans, that is, laws governed by our species-typical genic inheritance, there has been a shift in recent decades to give equal attention to individual, that is, *idiographic* variability in personality traits (discussed in Chapter 5, on trait psychology). Cultural psychology, on the other hand, focuses neither on nomothetic laws of human development nor on individual differences in personality profiles of humans, independent of the larger social context in which they have developed. Emics deals with social groups, attempting to describe the typical patterns of behavior of distinct societies. This places it in an intermediate position bridging the nomothetic and the idiographic.

It is not by chance that people born and raised in Naples speak a dialect of Italian, have a typically Neapolitan set of coping strategies and socializing patterns, and profess to be Roman Catholic. Comparable statements can be made of those born and bred in a Viennese, Niçoise, "Saigonian," Amish, Hassidic, or Tamil community. The power of the social environment to shape humans is formidable, and all the more so when its contact with alien cultures is limited. But one has to be careful not to exaggerate or simplify cultural differences. There are innumerable universal principles of human development that are present in the behaviors of humans of all times and every environment (Brown, 1991). There are, of course, wide individual differences in personality in every culture, although even the range and statistical distribution of those differences for any given population are culturally constrained by local tolerance for deviance.

8 Gendered personality

> Nothing can be more absurd than the practice which prevails in our own country of men and women not following the same pursuits with all their strength and with one mind, for thus the state, instead of being a whole, is reduced to a half, but has the same imposts to pay and the same toils to undergo. (Plato, *c.* 350 BCE, *Laws* [804], p. 722)[1]

Social norms and gender

Readers have various expectations when they approach a chapter on gender and personality. First, they can expect to find an investigation of the ways in which males and females differ universally: that is, of the nomothetic principles grounded in biology and evolutionary psychology that govern sex-differentiated human development. Second, however, as the term *gender* refers primarily to social and societal aspects of sex[2] (Corsini, 1999) they can expect an exploration of the emic dimensions of gender-differentiated personality development, for different cultures assign different sex-appropriate roles and behaviors to those living within them. Third, even within any single culture, there is going to be wide variability in the gender-linked personality traits that are expressed by females and males. Idiographic principles will come into play, all the

[1] Emilie du Châtelet, an ardent feminist as well as renowned physicist and mathematician of the early 1700s, wrote a phrase that is reminiscent of this Platonic quotation: "If I were king ... I would redress an abuse which cuts back, as it were, one half of human kind. I would have women participate in all human rights, especially those of the mind" (cited in Sulloway, 1996, p. 165).

[2] The simple division of the human race into two sexes has begun to be contested. Fausto-Sterling (2000), a professor of biology at Brown University, avers that there are five sexes. To the traditional dimorphic model of sexes, she adds three intersexuals: the true hermaphrodite, male pseudo-hermaphrodites who are born with testes and some aspect of female genitalia, and female pseudo-hermaphrodites who have ovaries and some aspects of male genitalia. It is estimated that between 1.7 and 4 percent of all live births are intersexual. Maccoby's (1998) statement that "the sex identity assigned to an individual by society, as well as the sex identity adopted by the individual, corresponds to the person's biological sex" (p. viii) can be qualified accordingly.

more so in that humans, more than other primates, are exceedingly plastic in their development. The degree of variability in this third instance will depend on how much freedom one society or another permits individuals within it to deviate from societal norms.

Historical shifts to gender-equal paradigms

The theoreticians of the nineteenth and early twentieth centuries fashioned templates of human nature that were largely masculine in character as noted in Chapter 3 (developmental psychology and personality). The historical record attests that a belief in the natural superiority of men has long suffused the *Weltanschauung* of European and Abrahamic cultures (see Bohan, 1992).[3] It is not surprising that it permeated much of the psychological theory building of that period. By the turn of the twentieth century three distinct views on women had gained currency. The *dominant* one was that women were naturally inferior to men, not only by reason of their lesser physical strength, but also due to weaker will power, less robust character, and inferior moral intelligence and creativity. This view is epitomized by Freudian personality theory. The *second*, epitomized by Adlerian personality theory, and espoused by feminists and activist socialists as well as other liberal publicists, theoreticians, and journalists, was that there is a natural equality of the sexes. Whatever inferiority was observable in the generality of women was the result of centuries of male domination and oppression in every walk of life. The *third* template, epitomized in Jungian psychology, presumed important qualitative differences in the psychic and physical make-up of men and women, differences that made them complementary to one another rather than inferior or superior to one another. This view was related to the Classical Greek myth of *Androgyne*, whose life combined the characteristics of both genders. With respect to this view, Ellenberger (1970) quoted Michelet who wrote in 1860 that "man and woman are two incomplete and relative beings, since they are two halves of the same whole" (p. 292).[4]

[3] The various Judeo-Christian religions have affirmed that the "second-created" sex is subordinate to the "first-created." Agonito (1977) showed that Thomas Aquinas followed in this tradition: males supply the essentially human features of "man" in procreation; females simply supply the fertile matrix in which the human seed takes root and develops. The biblical story of Adam and Eve is near the source of many Western attitudes relating to the roles of the two sexes.

[4] Ellenberger (1970) wrote that each of the three system builders, Freud, Adler, and Jung, later adopted one of these three positions: "Freud seems to have taken the natural inferiority of women for granted, since in one of his early writings, he assumed that the stronger sexual repression in woman is the cause of her intellectual inferiority. Later he

There is no doubt that the first position was dominant in turn-of-the-century Europe. The universities, let alone the armed forces, did not admit women. Women were excluded from political life and deprived of the vote in civic elections. They occupied no positions of power, executive or judicial, in the general society; the sole exceptions were those societies in which there were no men, as in cloistered religious communities. They found themselves segregated in the social world of women's salons, train compartments, schools, and the private household. The power of the husband and father over the entire family was unquestioned and it was exercised with authoritarian rigor. This ethos was like the air that nineteenth-century Europeans breathed. As Ellenberger (1970) observed: "All this must be considered [the backdrop for] the genesis of Freud's Oedipus Complex" (p. 255). A sound hermeneutics will not only explain the origin of much of the theorizing of the day, but excuse the distortions that appear to be outrageous in the light of later political and social developments.

Alfred Adler was the first major voice raised in favor of a gender-equal model of human nature. He was married to an ardent feminist activist, Raïssa Epstein, who implemented her views of women's equality within the home as well as in the political arena. Phyllis Bottome (1939), Adler's biographer, noted that Adler found it easier to declaim a feminist rhetoric than to live with a feminist activist. His feminist views began to be evident as early as 1910 when he proposed the notion of "the masculine protest." This protest is women's reaction to the roles that have been imposed on them by a society dominated by men. He did not regard it as a healthy reaction as it represents an attempt by women to model themselves on the least desirable traits of the European male, stereotypically "virile," aggressive, coldly rational, and emotionally subdued. They are prone to dress, talk, groom, and conduct themselves interpersonally in a "masculine" manner. Men caught in the toils of their own inferiority feelings are also vulnerable to this lure, often manifesting their insecurities by callous and abusive behaviors toward women. Masculine protest, which would be regarded today as sheer *machismo*, bespeaks an insecurity in the male and a cry for significance that he can achieve only through the passivity and submissiveness of the female. In the female it is a servile mimicry of

came to speak of the natural masochism of woman. Adler on the other hand was a staunch defender of the natural equality of the sexes. As for Jung, it is obvious that his theory of the Anima in man and the Animus in woman is related to the third contention" (p. 292). The myth of Androgyne is the root for the androgynous template for humans that have pronounced masculine and feminine qualities (see Sandra Bem's schemas for androgyny, 1974, 1977). Psychological androgyny was termed psychological hermaphroditism by Adler.

some of the least adaptive features of a masculine, culturally constructed model of social competency that serves to mask the male's vulnerability.

Karen Horney was another strong voice that was raised against the prevailing notions of the sex differences in development that were part and parcel of the psychoanalytic system in which she had been trained. Through the 1920s and 1930s she slowly drifted from a Freudian to an Adlerian position on the psychology of women. She challenged the doctrine of "penis envy" in the 1920s and later repudiated the notion that biology, to the exclusion of cultural factors, determined woman's destiny (1939). In 1937 Horney wrote that Freud "concludes from his observations that woman is more jealous than man, and then tries to account for this presumably general phenomenon on biological grounds" (p. 17). She presented anthropological evidence that "in making statements like these Freud was yielding to the temptation of his time: to make generalizations about human nature ... though his generalization grows from the observation of only one culture zone" (p. 17). She noted that therapists who "have to deal day after day with neurotic persons lose sight of the fact that normal persons, too, exist in our culture" (p. 17). With the publication of *New Ways in Psychoanalysis* in 1939, Horney continued to fashion a feminist response to some stereotypes of *woman* that were popularized by her discipline. So influential were her publications (see Horney, 1967) that she is widely credited today as a founder of the field broadly designated as *psychology of women* (Shaffer, 2000, p. 43).

Post-Second World War trends

In a succession of post-Second World War decades, interest in sex-differentiated traits and capabilities grew. All science is politicized, of course, and this is especially true of the social sciences, where the results of scientific advances are seen as differentially advancing the interests of various groups of people both in the homeland and abroad. The war itself had obliged women to assume traditional masculine roles as entire societies were wrenched onto a war footing. Males left in the millions for military and logistical overseas assignments, and countless women were "conscripted" to replace them in factories and nontraditional commercial settings. This demographic shift had long-lasting attitudinal effects that ramified throughout politics, academia, science, the professions, and other traditional male preserves. The feminist *politique* that had long been simmering came to a full boil. A large segment of an American population of women never looked back.

The literature of the 1950s evidenced an increasing interest in the psychology of sex typing. Articles appeared presenting demonstrations of the power of social modeling and indoctrination in shaping the dominant sex-appropriate standards for apparel, toys, play, speech, gait, and posture – standing and seated (for example, Emmerich, 1959; Hartley, 1959, 1960; Pope, 1953; Rabban, 1950). The stability of these "habits" from childhood through adulthood was demonstrated in the influential, longitudinal Fels Institute study (Kagan and Moss, 1962). A critical issue in the debates these studies excited was the extent to which sex typing was due to both genic and social variables. As Alice Eagly has noted, although lip service is always paid to the fact that both nature and nurture (in itself a tiresome cliché) contribute to the pattern of behaviors that are typical of either sex, ideological biases subtly, at times flagrantly, influence the importance given to one or the other of these sources of behavior.

Two points need to be made relative to "nature vis-à-vis nurture" issues. Nature exerts profound influence on the cultural expression of human traits, for it places severe constraints and exercises pervasive guidance on cultural variability. There are limits as to how much the heritable factors can be overridden by collective human volition.[5] Behavioral geneticists have found, as noted in Chapter 4, that events, culture-specific or not, can influence the expression of genes that might otherwise remain dormant. By and large cultural behavioral variables have virtually no influence in shaping genic determinants of temperament. In a Darwinian perspective, as cultural adaptational patterns change over a long span of human evolutionary history in the face of powerful geo-biological challenges, certain traits are selected in and out of human (and other species') genomes. Although there is some truth to the aphorism that biology is destiny, gene-splicing technology, quintessentially a cultural artefact, certainly can alter biology.

Maccoby and Jacklin: issues of stereotypes

A landmark book authored by Eleanor Maccoby and Carol Jacklin in 1974 created a flurry of critiques and endorsements in the developmental literature. It gave empirical support to the position that the majority of gender-role stereotypes were without biologic foundation. They did not

[5] The United Society of Believers (popularly known as the Shakers), a religious sect that originated in England and flourished for a time in the northeast of the United States, was a celibate sect, whose members did not marry or bear children. This communal society has virtually disappeared. This is an extreme example of how the nature of the organism has established fitness criteria that cannot be altered on a society-wide basis; disregarding them simply works to the detriment of its collective survival. On the other hand, analogous sects (say the Cathars and Albigensians of southern Europe) were hounded to extinction by the Church and its secular arm, which pre-empted the course of nature.

deny that different patterns of interests and behaviors *de facto* exist. What they did attempt to show was that few gender stereotypes are uniquely grounded in biology rather than in culture. As descriptors of human behaviors, the gender-based stereotypes had a core of truth in them, for society had foisted onto both men and women roles for which certain personality traits were more adaptive than others. Often these social conventions worked to the detriment of women. The Maccoby and Jacklin book, appearing as it did when a feminist grass roots movement was in full efflorescence, took on a politico-cultural significance that radiated throughout the culture wars of the next decade. Although many of the book's conclusions have been partially invalidated by later research and sounder historiographic methods, it still has historical importance for the impetus it gave to inquiries into the sex-differences issue.

The book underplayed the influence that biology has in shaping human behavior. Although clinical evidence had long been presented showing that there were biological determinants of feminine and masculine traits – psychodynamic theories of the previous century had been predicated on the assumption that sex-linked traits were biologic in origin – there were fewer rigorous, hypothesis-testing studies at the time exploring these questions. Western psychologists had been aware following the work of James, Cooley, Mead, and Vygotsky that individuals are shaped by the *mores* in which they have been indoctrinated from infancy. Moreover, behaviorists in the Skinnerian tradition and social learning theorists in the Banduran tradition (see Bandura, 1969) affirmed the power of reward systems and social modeling to shape human behavior. These views had not entered popular, mainstream thinking, which was largely of the men-are-from-Mars, women-are-from-Venus kind. Maccoby and Jacklin's book, grafted onto a populist feminism, was a powerful statement that nature is not at the root of these stereotypes. Historical, random, often capricious, socio-religious, and journalistic variables are.

Maccoby and Jacklin (1974) reviewed over 1,500 research studies on gender differences. This is a significant number of studies and should dispel the notion that this field of inquiry was neglected and barren until the Woodstock generation moved into influential academic positions. From that trove of studies they concluded, not that individuals are free of indoctrination into sex-appropriate roles of the *Ozzie and Harriet* (1950s) or *Archie and Edith Bunker* (1970s) sit-com variety, but that there are few sex-role differences that are predominantly determined by biology. Their evaluation allowed them to conclude that males and females differed in cognitive functioning in only three abilities: verbal, mathematical, and visual/spatial. These gender differences have been largely confirmed by a generation of later researchers. Girls acquire linguistic and other language skills at an earlier age than do boys, a finding that can be partially explained by the fact that girls, pre-natally and (for many years) post-natally, are neurologically more advanced than boys. Female superiority is evidenced in "rapid access to and use of phonological and semantic information in long-term memory, production and comprehension of complex prose" (Halpern, 1997, p. 1091). Male neural circuitry is more or less compromised by the androgenization of the embryo. The higher proportion of males than females who fall within the autistic spectrum as well as suffer from clinical disorders of speech makes this vividly evident. Girls' advantage continues throughout childhood and adolescence, both in terms of reading comprehension, speech fluency, and, indeed, related social skills.

On the other hand, boys showed evidence of superior performance in certain spatio-temporal tasks. This advantage has been found in later studies, in such tasks as spatial perception, mental rotation, and spatial visualization (Voyer, Voyer, and Bryden, 1995).[6] Boys gave evidence of superior performance in mathematical reasoning beginning in early adolescence. Their advantage in this domain was small but consistent and these findings have been replicated in subsequent studies (Halpern, 1997; *cf.* Helson, 1990).

Maccoby and Jacklin also found that boys are more aggressive than girls. That this applies to verbal as well as physical aggression is not surprising, as physical aggression is usually accompanied by intensely hostile verbal and paravocal communication. Consistent with this is the fact that the preponderance of violent crime in the United States is committed by males (Shaffer, 2000; United States Department of Justice, 1995). This pattern is evident in cultures throughout the world, despite the fact that aggressive behavior, like all other behaviors, has environmental as well as biological determinants. The unresolved question to this day is: to *what extent* is aggressivity or any other personological trait a function of biology and to *what extent* culture?

Recent research in gender psychology

Maccoby and Jacklin's 1974 review of previous research in gender psychology was largely of an informal, narrative, eye-balling kind. Those who continued to use those methods to synthesize findings from numerous

[6] Mental rotation tasks require that by using one's imagination two- or three-dimensional figures are rotated as viewed, say, in printed form. In the Voyer, Voyer, and Bryden study the mean effect sizes in the spatial perception and mental rotation tasks were medium ($d = 0.44$) and large ($d = 0.73$).

research projects allowed a larger degree of subjectivity to enter into their analysis than did those who attempted to use more quantitative approaches. As Alice Eagly (1995) has pointed out, it is not enough simply to do a count of studies that support or disconfirm a particular hypothesis and then make subjective, somewhat intuitive judgments about the trends the studies *seem* to reveal. Those who continued in this "impressionist" mode contested even the limited number of gender divergences that Maccoby and Jacklin had conceded, for as Lipsey and Wilson (1993) report, there is a tendency for such narrative and intuitive approaches to underestimate effect sizes. In the absence of a more objective method for quantifying sex differences, subjective bias was given more room to operate in the impressionist approach used by Maccoby and Jacklin. An example is Frodi, Macaulay, and Thorne (1977), who wrote in their review of the literature, "commonly held hypotheses that men are almost always more physically aggressive than women and that women display more indirect or displaced aggression were not supported" (p. 634). The evidence in this study that would support the authors' conclusions came largely from what males and females reported in general terms *about themselves* – data that are intrinsically suspect. Circumstances that can provoke aggression – say, prosocial behavior such as defending one's children – would seem to generate as much physical aggression in women as in men. The critical question is not whether *some* women can act physically aggressively (they obviously can), but how women have been acculturated to express their aggressivity – and how males and females have been conditioned to physically react differentially to certain cues and provocations.

Meta-analytic approaches to gender research

In the latter part of the 1970s psychologists began to use meta-analytic (statistical) procedures for grasping whether a large number of aggregated studies revealed substantial differences in performance by distinct groups. This often involved combining, evaluating, and synthesizing the results of hundreds of studies. Roughly comparable performances of various groups – in some cases experimental, in other cases simply correlational – are simultaneously analyzed. The resulting, distilled effect sizes, often from hundreds of studies, are group differences calculated in terms of a basic unit – the standard deviation. Some critics of this statistical procedure feel that there is insufficient control for the uneven quality of the studies that are lumped together, the specific populations sampled, the subjective bias that excludes and includes studies in the analysis, and the comparability of the research hypotheses and objectives of those studies that "made the cut."[7] Nevertheless, meta-analytic techniques have gained wide support from eminent statisticians and researchers and are widely used (see, for example, Glass, McGraw, and Smith, 1981; Lipsey and Wilson, 1993; Rosenthal, 1991).

When researchers conduct a study in the social sciences they are always faced with the challenge of interpreting the results. If their studies are soundly designed and executed, their first decision is to determine if the results are attributable to simple chance or to the influence of the independent variables under consideration (in this case the sex of the subjects).[8] Second, once they decide to accept the results, they need to determine whether they are trivial and meaningless in context, of moderate significance, or of great importance. Now the difference between males and females in their performance on a variety of measures has often been very small, so small say some that they can effectively be ignored. But small effect sizes have often been a problem in social science research where there are many "uncontrolled" moderator variables (that is, extraneous variables not directly relevant to the study) that can wash out differences in the main effects under consideration. Eagly (1995) has stated that those who allege that the sex differences found in large numbers of individual studies investigating male and female personological differences are trivially small seem not to realize that many of these differences are, on the whole, comparable to other findings they have reviewed in their textbooks, findings not accompanied by any warnings about their small magnitude or about overlapping distributions.

Debate about the theoretical and scientific soundness of meta-analytic *procedures* and of the *interpretation* we may place on their results continues (see Henson, 2006, for an analysis of these issues; also Leaper and Ayres, 2007, relative to the inconsistent, small, and inconclusive results that some meta-analyses present us). That meta-analyses, such as those for the solid review article by Tamres, Janicki, and Helgeson (2002), are not superior for review articles in general is a difficult case to make.

Variant gender-role theory

Parsonian theory

Talcott Parsons (1902–79), the eminent sociologist, proposed that the roles that men and women assumed in society could be broadly divided

[7] An extreme criticism of this statistical procedure was voiced by Hans Eysenck (1978), who referred to meta-analysis as "an exercise in meta-silliness."

[8] Scholars understand that it is virtually impossible for two groups ever to perform at identical levels on sufficiently sensitive measures. Paul Meehl (1978) in his classic article on psychological research and probability measures asserts that one must try very hard *not* to find differences between two groups of subjects, no matter the size of one's sample.

into instrumental roles (largely engaged in by men) and expressive roles (largely engaged in by women). As a consequence, men are temperamentally suited to engage in tasks outside the family, such as foraging, hunting, laboring, and generally assuming responsibility to provide for the family's material needs. Women, on the other hand, are personologically suited to deal with the many needs that arise within the family, such as reducing tensions, assuaging emotional distress, socializing the children, and generally managing the household (*cf.* Parsons, 1955).

Traditionally, women have remained at home, bearing and nurturing infants, mediating intra-familial squabbles, providing for the comfort and quality of life of the entire family including the elderly and other near relatives (given the means at their disposal), educating and training the young, and entertaining visitors and home-dwellers alike. These tasks are as "instrumental" as those of prehistoric man, but require an additional interpersonal, expressive dimension. Because of men's greater physical strength and related psycho-motor aptitudes, they have historically assumed the responsibility not only physically to defend their home and community from the predation of other men and beasts, but have been generally required to behave in a more instrumental, tool-dependent, physical, often aggressive mode.

Social role theory

Related to this paradigm is social role theory (for example, Eagly, 1987), which postulates that differences in the social behavior of women and men originate from the differential distribution of tasks that men and women are expected and required to perform in society. At the most basic level the fact that women have traditionally been homemakers and men have been the hunters and providers has shaped the stereotypic traits that they manifest. This is reminiscent of the collectivist–individualist distinction made in cultural psychology. If women represent a communal sub-*culture* and men a more self-reliant, individualist sub-*culture*, the correspondence of these divisions with the economic roles and political fortunes of these two "cultures" becomes apparent. In the view of those who have promoted this theory (for example, Deaux and LaFrance, 1998; Eagly, 1987, 1995, 1997; Eagly and Wood, 1999), sex-differentiated dispositions are largely the result of this division of labor – a division of labor that is less the product of evolutionary forces than the vagaries of historical cultural constructions. They propose that it is the dynamic nature of gender, as socially constructed, that powerfully influences the individual, interpersonal, corporate, and cultural activities of any society. This particular pattern of tasks, which enables a society to achieve its economic ends effectively, has been extrapolated into the mercantile and manufacturing sectors of the society. For women to seek employment utilizing the aptitudes and skills that most closely resembled, in its various aspects, the work they had traditionally done in the home seems to have been a path of least resistance, and one presumably imposed on them in part by males using male-formulated employment criteria.

Women are concentrated in different economic sectors than men, doing different kinds of work. Whether in service, clerical, or professional occupations, women and men have segregated themselves into domains that require signally different, task-specific aptitudes and skills. We find more women than men, for example, in social work, dental hygienics, school counseling, nursery through primary school teaching positions, nursing, and, more recently, mental health services, not to mention the textile and other relatively sedentary manufacturing sectors.[9] It would seem plausible that social expectations about, and the demand requirements for, performance in these positions would exercise a pervasive effect on women's social behavior, their self-concept, and their attitudes toward socio-political institutions and problems. Social role theory postulates that the occupational landscape is the "primary origin" of women's social behavior (Eagly, 1997). When a large segment of the population has been conditioned to present, as properties of its economic function, a cordial, welcoming, accommodating, and comforting persona to all who come in contact with it, and those numbered in that population come to understand that continuing in their employment is contingent on exhibiting these traits, it is not surprising that they should behave accordingly. It is primarily the coercive influence of socio-economic expectations that shapes that behavior. The structure of society organizes the activities of men and women, and, consequently, it is the differential structural demands placed upon men and women in any political economy that is at the root of their gender-specific roles and personalities (Eagly and Wood, 1999).

Historical, secular, political variables have profoundly affected this distribution of occupations. If men have in the past been more numerous in the teaching professions in the West than have women, the moot question is whether this is because they were congenitally better suited for this work – an unlikely hypothesis – or because of a socially coercive norm that kept the woman in the home, caring for a large number of

[9] Brabeck and Brown (1997) have noted that feminists have established beachheads in "psychotherapy, supervision and consultation, forensic psychology, psychological assessment, research, training, social action, teaching, community organization, organizational consultation, and political practice in formal structures, psychological organizations, universities, legislatures" (p. 17) among many others.

children, many of whom often became sick. Society simply did not free them up to work outside the home.

When the occupational sluice gates were opened to women during the Second World War, as noted above, they opted in the millions to take work outside the home. As machines replaced brute force in the manufacturing sector, the rationale for employing only men in that sector began to evaporate. After all, a crane or stamping machine exerts the same force and achieves the same ends whether operated by a man or a woman. And the further that modern societies have advanced into a "post-industrial" age, relying on information technology and robotics for wealth production, whatever advantages the human male of previous epochs enjoyed over women in the more valued occupations have evanesced. Male monopoly rapidly eroded in such related occupations as war-making, diplomacy, politics, personnel and corporate management, medicine, and their related white-collar professions. If this theory proves to be correct, it remains to be seen whether women will adopt the personality traits traditionally thought to be adaptive for the exercise of these professions. An alternative hypothesis is that "evolved" (read evolutionary) feminine dispositions will shape the dominant style in which professions are practiced.

Role assignments

Corporate institutions require a certain level of reliability in the performance of the varied tasks and roles that enable them to function well. Members of a society are not permitted to do *whatever* they please with impunity, for the simple reason that no one would be able to make plans based on reasonable expectations of others' behavior. If the large social arena in which we wish to operate lacks rules, then planning becomes impossible, as there will be a limitless number of contingencies for which we would need to prepare. Our best-laid plans would be undone by the actions of others which we had little ability to predict. Further, the more rationalized and complex a society is, the more rigid are the norms governing individual behaviors, for even small mistakes will quickly ramify throughout the entire system. For this reason, modern industrialized societies have developed "bureaucracies," rigid, large-scale structures that coordinate the activities of multitudes of individuals who are unknown to each other.

Gender-based roles constitute one domain in which every society has seen fit to establish norms. Although the evolution of norms appears to be an organic process, it would be a mistake to believe that collective consent was always freely given to these norms. Patriarchal societies, that is, those governed (typically) by elderly males, have set the rules for the appropriate conduct of members of both sexes. These rules are first and foremost economic[10] in character: that is, they relate to the management of the household. The assertion that patriarchal societies have fashioned rules that economically and socially favor males has been largely substantiated. This is the position of those within the women's movements, which have primed a vast socio-political campaign to refashion the norms governing the status, roles, economic rewards, political influence, socio-religious activities, and sexual mores of women.

Whether there are universal psychological differences between men and women that arise from genetic determinants of gender-specific behavior is an empirical question that has largely been answered but is still being studied by behavioral geneticists. If, indeed, there were no significant difference in the psychological development of the two sexes independent of fluid, relativistic cultural variables, it would not seem to matter which sex one used as the nomothetic template for healthy human development. The focus of analysis would necessarily shift to the socially constructed roles that are imposed on females and males. Assessing women as inferior by virtue of possessing personality traits they have internalized in a society that has prescribed them would seem patently unjust.

Evolutionary psychology

Evolutionary drivers

Darwinian psychologists have searched through the evolutionary record for the origins of those enduring dispositions that guide human sex-linked behavior (for example, Buss, 1995). If a particular behavioral trait improved the fitness of the individual within a geo-social niche – a fitness validated by mate-selection pressures and social integration – then that trait genomically self-propagated. For example, a female courted by an array of males selected the *one* she judged to be endowed with behavioral skills and aptitudes that were most promising to secure her future and that of her children. If a behavioral trait jeopardized (or did not improve) the social group's welfare, the individuals who possessed it languished and fell off the tree. According to Darwin the mechanism for this transmission of gene-determined dispositions was sexual selection, which he "viewed as a result of inter-male competition for nubile females and reciprocally the female selection from the array of suitors" (Archer, 1996, p. 912).

[10] The terms, ecology, economy, and ecumenical all derive ultimately from the Greek term, *oikos*, meaning house, household, habitat. Economy relates to the rules for managing a household and, by extension, the larger community.

Females contribute more of their personal resources to their offspring than do males. They are more discriminating in their choice of a mate, as they have a greater need for someone with the ability to defend and nourish them during gestation, lactation, and the early child-rearing years. They need a mate, in brief, who is dependable and shows promise of being a responsible parent and partner. Future mothers seek a reliable partner more earnestly than does the single male who can sow his seed and then "split" with relative impunity. The implications for this in terms of competitiveness, aggressivity, and patterns of courtship are drawn out by evolutionary psychology. These theorists take a small leap to a theory of sex-linked behaviors such as we witness in our work-a-day world: women specialized in a constellation of behaviors that we have labeled as communal, nurturant, and socially oriented; men specialized in a comparable set of behaviors that we have labeled as instrumental, agentic, and task-oriented. The conclusion is drawn that men are able to dominate women, not simply because of their greater strength and size, but through this selection mechanism, and the subsequent parenting cycle that constrains a woman's freedom more than it does a man's (Smuts, 1995). In this perspective, we can reason that men's different bonding styles, reactions to loss, and relative lack of expressiveness, aside from the historical prevalence of patriarchies, is less the result of a societal division of labor than of evolutionary selection pressures.

The cogency of this line of argumentation hinges on the acceptance of a classical social Darwinism. Eagly (1997) has emphasized that this line of reasoning is highly speculative. It is a system of inferences that build upon each other; it could not be otherwise as the empirical evidence for these hypotheses has disappeared into the mists of pre-history. But the arguments have an intrinsic plausibility. Although it is difficult for the proponents of an evolutionary psychology approach explaining gender differences to demonstrate that they are right, it is equally difficult for anyone to demonstrate that they are wrong. The burden of proof lies with those who affirm a proposition, however, not with those who question it.

An example of their line of reasoning is that the genically programmed behaviors and sex-differentiated roles that exist, with some variation, in contemporary societies had their origin in the mid-Pleistocene era.[11] Groups of hominids began to propagate under very challenging conditions, including glacial periods when ice sheets covered large parts of the

[11] The Pleistocene era began about 1,800,000 years ago and ended about 12,000 years ago. It is in the earlier reaches of this period that it is believed that the species *Homo sapiens* first made its appearance. This subject is dealt with more extensively in Chapter 4, on biology and personality.

globe. Under these conditions evolutionary pressures selectively shaped the physical qualities and behavioral dispositions that improved the chances of "human" survival. These successive and unforgiving *environments of evolutionary adaptedness* (EEAs) were the matrix of the hunter-gatherer tribes. Patterns of behavior that were expertly, if not effortlessly, learned and which enhanced survival in these EEAs resulted in them being reproductively continued. Those that were weak or contra-indicated withered in the face of brutal predation and competition for scarce resources and adaptive mates. That procreating and rearing children placed differential demands of an economic character on men and women, demands that became translated into genomically favored, sex-based traits, seems plausible. Behaviors, of course, are not, by and large, genomically programmed, but a predisposition and aptitude to acquire them seem to be.

The generality of evolutionary psychologists as well as those in a social constructionist orientation attempt to avoid a strictly dichotomous view of the determinants of human personality. Buss (1996) stated that "human behavior cannot be explained without articulating evolved psychological mechanisms combined with the social and cultural input to those mechanisms. Through this interactionist framework, we can move beyond the false dichotomies of genetic versus learned, nature versus nurture, and biology versus culture." Nevertheless, he immediately continued, "Evolutionary psychology provides the best account we have of where men and women show psychological differences, where they share the same psychology, and the social contexts in which each is expressed" (p. 162). Adherents of social role theory likewise place themselves on a theoretical nature–nurture continuum. The difference between the two camps is clear-cut, however. The "causal arrow" in the latter flies predominantly from social determinants to personological effects; the "causal arrow" among the evolutionists flies predominantly from Darwinian fitness-criteria in mate selection to personological effects. The debate continues and will be resolved by those who can provide the most solidly grounded, empirical validation of their position.

Biology-based sex differences

Reprising the nomothetic issues raised in the opening paragraphs of this chapter, we look at the ecological and evolutionary pressures that have shaped *Homo sapiens*. They explain some of the sex differences that are apparent in the social behaviors of men and women. These pressures have shaped the genomic constitution of the race and the neurological and hormonal characteristics, as well as the more obvious musculo-skeletal features, of men and women (Geary, 1998). That there would be behavioral correlates of these genomic and physiological correlates is intuitively plausible. However, the power of socialization, coupled with the characterological plasticity of a species whose juvenile and formative period is prolonged by comparison with other mammals, is evidenced by the diversity one finds in disparate cultures in male and female behaviors. As in all other phenomena, there is a multitude of causal paths leading to the idiosyncratic behavior of an individual: genomic; cultural; family-historical; contextual; medical; and, of course, purely haphazard and accidental.

There is a presumption that the moment the child is born he or she is relatively unshaped by social mores and role assignments. Although the Western family knows the symbolic meaning of pink and blue, for example, newborns do not. They need some time to learn it. And the longer they are exposed to explicit as well as covert expectations of parents, family, and community relative to gender-appropriate behaviors, the more they internalize such attitudes; these finally become "second nature" to them. Developmentalists have reasoned, therefore, that the earlier in life that one can distinguish typically masculine and feminine patterns of behavior – especially if they show up in a cross-cultural perspective – the more confidence one can have in the heritable, genomic influence that purportedly accounts for them (Geary, 1998).

Evidence exists that behavioral sex differences appear within the first few days of neonatal life, clearly pointing to genetic determinants of such differences. The fact that the expression of certain of these sex traits is timed for later developmental periods (say the onset of puberty) does not weaken the position that there is a significant heritability factor shaping human behavioral gender-based traits that are manifest from day one. David C. Geary (1998), in his book *Male, Female: The Evolution of Human Sex Differences*, has brought together a formidable array of empirical evidence to support this position, one that he has in large measure based theoretically on a Darwinian fitness schema for mate selection.[12] Newborn girls, for example, engage in more eye contact than do boys, even on their first day after birth. They spend more time looking at faces, and in later infancy demonstrate a superior ability to recognize faces – which suggests also that they have a better memory for faces. This bespeaks a superior awareness of, and sensitivity to, social cues. Boys, on the other hand, evidence more gaze aversion at age 6 months than do girls. After stating that boys are generally more aggressive than girls, Carlson (1998) asserted that "prenatal androgenization increases aggressive behavior in all species that we have studied, including primates …

[12] Given its basic (though tendentious) premises, this well-regarded book serves as an excellent review of the literature on its subject as of the date of its publication.

After puberty androgens also begin to have activational effects. Boys' testosterone levels begin to increase during the early teens at which time aggressive behavior and intermale fighting also increase" (p. 351).

In this biosocial perspective, studies indicate that in toddlerhood, girls evidence more nurturant, comforting, prosocial behaviors than boys, and empathize more when witnessing the distress and suffering of either humans or animals. This compares markedly with the *relative* emotional indifference of boys to dramatic social events. Boys also appear more willing in infancy to approach unfamiliar objects, a penchant that presages their risk-taking behaviors in adolescence and young adulthood – particularly evident in their high-speed, sensation-seeking driving behavior (see supporting evidence by Christopherson, 1989). Perceptually, boys attend to the physical properties of objects more than do girls. On the other hand, girls are more sensitive to the responses of their social environment when they have oriented themselves to it. Their investment in parenting is also apparent from an early age as they reveal their knowledge of the variety and subtleties of family roles in their play. The play of boys on the other hand centers more on motifs of power, dominance, and aggression.

All of these traits are normally distributed in any population. That means that, although there is a significant statistical difference in the *averages* of boys and girls on any of the behaviors addressed above, there are innumerable females who would score more like typical males than like other girls in their population. Conversely, there are innumerable boys who would score more like typical girls than like other boys in their population. The extent to which certain personality traits vary as a function of fluctuations in levels of endocrine products is one of many psychobiological issues under investigation. Adler *et al.* (1997), for example, have demonstrated that among fifteen male experimental participants, the nocturnal excretion of cortisol and testosterone fluctuates much more among those who had tested emotionally labile than among those who had tested emotionally stable. Such results did not appear in the amount of adrenaline or noradrenaline. Thus, it was still not clear how the levels and the fluctuations in endocrine products explained the causal pathways between these variables.

Sex-linked traits and the brain

The study of sex differences is less the study of individual differences than of group differences (indeed, of the largest distinct sub-groups of humans – women and men). Plomin (2003) nuances the point: "Life scientists are more interested in the generalities of the genome," but as "we each have a unique genome ... behavioral scientists are more interested in the variations in the genome" (p. 7). Brain sciences, *per se*, are more nomothetic than idiographic in their perspectives on the agentic functions of brain organs. This is not to minimize the many individual differences in the structure of human brains, which are as distinctive as the differences in human faces. Neuroscience has some of the characteristics of all three approaches to human behavior – genomic, neuro-endocrinological, and behavioral – including, of course, the approach discussed in Chapter 7, on culture and personality.

The study of gender and personality is not, then, simply about idiographic differences; it is also about within-gender similarities – *similarities* that exist within and across cultures. All human embryos, which are typical mammalian, begin life as female organisms. This trajectory is the default setting for both female (XX-coded) and male (XY-coded) zygotes. When the nascent male brain gets bathed in androgens at about 8 weeks of age, it gets shunted onto a male track. This entails far-reaching changes in the structure of its central nervous system and a significant alteration of its developmental path. This, of course, not only generates neurological setbacks for the male fetus, but for the future boy. Males are consequently born with a 3- to 5-week neurological handicap relative to females, a maturational gap that does not get closed until late adolescence, if then.

A more serious consideration is that boys are more prone to language as well as socialization disorders than girls. Primary-grade boys are referred to child-guidance clinics many times more frequently than girls. The incidence of autism, in its severe and attenuated forms, is eight times greater for males than for females. Males are weaker in their ability to empathize and stronger in their penchant for systemizing (see Baron-Cohen's studies, for example, Baron-Cohen, Knickmeyer, and Belmonte, 2005; also Baron-Cohen, 2005).[13] This suggests that they are also less prone to telling the "white lies" that can shield the feelings of their companions and, on another plane, more inclined to engage in object-focused natural sciences. Among other long-term sequelae, boys' risk-taking and failures in prudential judgment are legendary. Empathizing is different. It is about being able to imagine what someone else is thinking and feeling, and having an emotional response to that person's feelings.

[13] Baron-Cohen (2005) states: "Systemizing is the drive to analyze or construct a system. It could be any kind of system – a mechanical system such as a computer, or a natural system such as the circulatory system of the body. Or it could be an abstract system like mathematics. The key thing is that when you systemize, you identify the rules or the laws that govern that system in order to predict how it will behave" (p. 1).

Girls, on the other hand, appear to be hardwired for scanning the facial behavior of others to assess fleeting, tell-tale signs of concern, interest, annoyance, joyfulness, or fear. Louann Brizendine (2006) presents in a breezy, popular style a scholarly and solidly documented book on these neurological differences. She avers (p. 191) that women not only have 11 percent more neurons in brain sectors dedicated to language, hearing, and relational skills, but they have larger hippocampi, CNS organs "designed" for certain memory and emotional functions.[14] It is not an accident of history that women are now flooding through the gates of the professional mental health services. Given their neuro-hormonal endowment for empathy, human connecting, and fine linguistic skills, they make exceptionally fine psychotherapists, counselors, and social workers. Excellent psychotherapists have many of the trait behaviors associated with typical female personality profiles – sensitivity to social cues, capacity for empathy and nurturance, ability to "read" faces and paravocal cues, and ease in making eye contact, among others. That women are now entering the mental and physical "helping" professions in increasing numbers and currently constitute the (growing) majority of psychotherapists in North America is consistent with this personality profile. Needless perhaps to say, there is a large number of extraordinarily effective male psychotherapists who manifest these human qualities as well.

In this light, trivializing the issue of gender differences is no longer a scientifically sustainable position.

Small differences and large fallout

When one compares the (estimated) averages for two relatively large samplings of various populations (such as *all* women and *all* men) on any measured personality trait or some reliably measured task performance, it is still virtually impossible for there not to be differences in those averages – if the tests or instruments one is using are sensitive enough (Meehl, 1978). One will always find differences between men and women (of course, given the same mean differences, the larger the samples one uses, the more one approximates true population values and, therefore, the greater the level of

[14] Cognitive neuroscientists of the twenty-first century have available to them brain-imaging technology which enables them not only to measure the "organs" of the brain, their configuration, and the severity of lesions to them, but to determine their cognitive functions. These so-called haemodynamic methods, for example, functional magnetic resonance imaging (fMRI) and positron emission tomography (PET), measure the relative flow of blood to various regions of the brain (consult, for example, Ward, 2006, especially pp. 48–59).

confidence one can have in such studies).[15] This may obscure the fact that the personological traits that all men and women share in any society or subculture may be more meaningful than their inter-sex differences.

How significant is a small difference?

That kingdoms can be lost for want of a horseshoe nail is the moral we have all drawn as children from the nursery rhyme about trivial deficits. Prizes in track and field events or downhill skiing are often won and lost on the basis of several hundredths of a second. There are no major consequences of this except for the individuals involved. A tiny but not so trivial difference in performance separates those who climb to the winner's podium and those who fade into a crowd of also-rans. However, if any *group* consistently bests another group by "millimeters and microseconds" in the realm of the socio-economic, then the consequences are not trivial; they are momentous, affirms Eagly. Those who argue that the differences that may exist between women and men are small enough that they need not be taken seriously underestimate their consequences in the competition for the more valued positions and jobs in a modern political economy.

Alice Eagly (1995, 1996, 1997) has cogently demonstrated that small differences in measures of male and female behaviors can have significant consequences in the socio-political world where the pragmatics of social status and economic equity are played out. By conventional standards of what constitutes small, medium, or large differences in statistical results[16] recent research has demonstrated that a number of domains exist in which small sex differences are important, meaningful, and laden with socio-economic and political consequences, a subject that will be addressed below. Eagly (1996) cited Martell, Lane, and Emrich (1996), who demonstrated "that a difference in performance evaluations favoring men but accounting for only 1% of the variability in scores can produce an organizational structure in which only 35% of the highest level positions are filled by women" (p. 158). The tendency to dismiss statistically small differences as inconsequential is a serious misreading of the variables that

[15] Some readers may not yet have been introduced to the science of inferential statistics. The mathematical principles are fixed and unforgiving, but the criteria for reaching reasonable conclusions about one's findings can fluctuate from the most rigorous to the most relaxed. These standards are established by consensus within a community of experts and take fully into consideration the probability of error that may enter into any research design and its execution. It is important to note, however, that not all *statistically* significant findings are of significant *import and meaning*.

[16] Using a statistical metric, *d*, which represents the proportion of a standard deviation by which two groups differ from one another.

impinge not only on the sexes but on other groups within society as well. "In practical terms, the importance of a difference depends on the consequences of the behavior in natural settings" (Eagly, 1995, p. 152).

There are innumerable variables that influence the outcome of personal efforts to achieve particular career goals. Aside from the genetic contribution to variance in performance, whether psycho-motor, cognitive, or socio-emotional, there are systemic variables such as ethnicity and socio-economic class that enhance or diminish the probabilities of success. If gender, ethnicity, social class, "family connections," membership in fraternal orders, and entry into the best law schools accounted, individually, for only a small fraction of the variance in the task performances that qualified one for movement up corporate and political ladders, the edge that would give those who shared that background would still ensure that a disproportionate number of them would rise to the upper echelons. And those who did not share that background, no matter their native endowments, would be disproportionately (under)represented.

Abelson (1985) demonstrated that, cumulatively, team batting averages that were only a few points higher than those of an opposing team had dramatic consequences for the final standings of the teams. Rosenthal (1990) noted that in a study of the effect of consistent use of aspirin on the incidence of heart attacks in a sample of 22,000 physicians, the use of aspirin accounted for 0.1 percent of the variance in mortality rates between those who took aspirin and the control group that did not. This minuscule percentage of the variance was so momentous in its consequences that the experiment was called off and all participants were counseled to begin taking aspirin (actually 104 of the aspirin-takers suffered a heart attack as distinguished from 189 among the placebo group). Though the amount of variance in specific task performances accounted for by sex is small, one needs to consider the consequences of this discrepancy especially in the upper reaches of the socio-economic hierarchies where it spells the difference between working in a lower-echelon office and sitting among the board of directors. In the field of sex-differences research, the effect sizes may be small, but their consequences, as in this example, are much larger than has traditionally been suggested.

"Not all gender differences are the same size"

Hyde and Plant (1995), on the other hand, made clear that there is wide variation in the magnitude of the gender differences[17] that research has

[17] It is noteworthy that whereas Eagly writes of sex differences, Hyde and Plant write of gender differences.

brought to light. Some are minuscule; others are very large. They further demonstrate that "the difference between the distributions of gender effects and other [psychological] effect sizes is highly significant" (p. 160). Gender differences that are extremely small show up in the literature much more frequently than do differences in other psychological studies; gender differences that are large or *very* large show up much less frequently than do contrasts of other psycho-social variables. This, of course, is understandable as psychology research is often of an experimental treatment kind, in which the effects of intervention strategies are assessed. Whenever one group of subjects is subjected to a powerful treatment to which another group is not, there is a high probability that the two groups will differ in the criterion variables[18] the treatment is hypothesized to affect. This argument does not invalidate Eagly's argument that even small differences in effect sizes in performance for individuals as well as groups can have significant personal and social consequences.

Averages of effect sizes for task performances that are a function of sex is only one consideration in assessing gender differences. The *variability* in the effect sizes is also important. Although there may be virtually no differences in *average* effect sizes for one variable or another, the tails of the distributions, representing the high and low scores for males and females, may show disproportionate representation of these two groups. Males, for example, of a certain age may be overrepresented in the *giftedness* and in the highly cognitively *impaired* tails of the distribution for, say, math ability. That the dispersion of scores (standard deviations) is different for males and females has to be taken into consideration when examining averages, that is, the central tendencies of measures (see Deary *et al.*, 2003; Feingold, 1992, 1994).

There are other limitations that one must consider when assessing the results of sex-differences research. That there are personality and cognitive changes that occur over the course of the life span has been amply demonstrated (see Chapter 3 in which life-span issues are discussed). One cannot assume that such changes occur at the same rate and in the same magnitudes for both sexes. American college students have long been samples of convenience for much of the research that has been done in social, clinical, and counseling psychology. That this segment of the human population can represent vast tracts of humanity that are neither

[18] A criterion (or dependent) variable is a measurable response to a treatment or to an independent condition. In a well-controlled study, changes in the criterion variable are presumed to be a function of changes in the independent variable.

collegiate nor American is easily disputed. Finally, there are secular[19] trends that are emerging which as yet it has not been possible to investigate. Such research requires long-term institutional support, large data bases, and rolling membership in a cadre of investigators.

Mating selections

Clearly there are biological differences between men and women, the most obvious of which are celebrated in the arts as well as in commercial adverts. Due to the fact that these characteristics serve as sex-signaling stimuli for the attraction of sexual partners, they are recognized by everyone, even in those cultures where by reason of modesty, weather, or religious scruple they are largely concealed. Brizendine (2006) sketches in copious detail the powerful floods of neurotransmitters that catalyze the mating behaviors of both males and females. Following a review of the literature in 1994, Sherwin concluded that "it would now seem that testosterone is the hormone most crucially implicated in the maintenance of ... sexual desire, in women just as it is in men" (p. 423), but estrogen, oxytocin, dopamine, and serotonin are implicated as well. The evidence continues to accumulate that a panoply of hormones in complex interaction with the central nervous system (for example, Aron *et al.*, 2005) drives much of the non-rational behavior in romantic love, sexual pursuit, courtship, pairing, coupling, and nesting. In other domains, however, mechanisms for the regulation of body temperature (the sweat glands), metabolism (for example, the thyroid gland), and blood sugar levels (the pancreas) appear similar for both sexes.[20]

Clearly, the competition for mates goes beyond issues of sexual aesthetics. The survival of men and women depends on *adaptive* skills (some of which are identical for both sexes), to the extent the adaptational problems that they must deal with are comparable. Reproductive activities implicate many of the same physiological responses. This is particularly important in a Darwinian (see Darwin, [1872] 1972; Jones, 1999) perspective, for evolution has favored those who are most

19 *Secular trends* are evolving and measurable patterns or directions of variables that appear over *long periods*. For example, the increasingly early onset of *menarche* (that has declined in the past century from about age 19 to 13 among certain populations) is an example of a secular trend. *Secular*, derived from the Latin term *saeculum*, is defined as an epoch, or age, or long period of time.

20 Even in these standard physiological mechanisms, there are sex-differentiated expressions of these capabilities. For example, women are anabolic, men are katabolic. Anabolics have a slower metabolism and tend to build tissue reserves; katabolics have a more rapid metabolism and tend to consume their fatty reserves.

successful in attracting partners who appear to be able to cope with the adaptational problems that a competitive, often hostile and inhospitable, environment presents – and who have pleasing social and physical features as well. Buss *et al.* (2000) studied college age Korean, Dutch, and American men and women to determine the level of distress they experienced when faced by rivals for a prospective mate who surpassed them in various attributes. They found that for the men in all three cultures, superior financial prospects, job prospects, and physical strength in their rivals intimidated them the most. The Korean, Dutch, and American women, on the other hand, reported greater distress when their rivals surpassed them in facial and bodily attractiveness.

Simply because the same pattern of behavior appears in quite disparate cultures in (apparently) comparable contexts does *not* mean that they *are not* socially constructed or that they *are* largely grounded in biology. After all, there are some social conventions that have been globally propagated. And, of course, coincidences do happen, all the more improbably as their number increases – as anyone who engages in games of chance can attest.[21] Generalizing on the basis of cross-cultural research becomes more hazardous as the swirl of ideas, attitudes, tastes in music, art, literature, and socialization and communicational patterns sweeps around the world. There is a certain homogenization of culture happening worldwide. Entropy is breaking down the cultural distinctions between one people and another – diminishing the surprise as well as the pleasure that comes from visiting places that are not as exotic as they used to be. The similarities that one finds among people around the world may be the result of culture-based assimilation that is increasing at an exponential rate. In any event, both nature *and* nurture account for the differences between men and women, and it is the differences that are the focus of interest in this chapter.

Birth order and sex differences

Children born of the same parents share many of the same genes, many more in fact than they do with unrelated children from a community of the

[21] It is this logic that underpins David Hume's argument that we cannot infer causation just because we have observed a large number of coincidences. Such reasoning is at the source of much superstition as well as scientific theorizing of the *post hoc, ergo propter hoc* variety. To put a finer point on it, because we observe that one phenomenon precedes or coincides with another does not justify the conclusion that the former, in a chain of events, *caused* the latter. This is why Karl Popper affirmed that even a perfectly designed experiment that resulted in statistically significant results could not demonstrate that the variable that we manipulated explained changes that followed in another variable of interest. Students who are interested in this subject may explore the scientific subject of *disconfirmationism*. A useful place to start can be the work of Imre Lakatos (1970) or Michael Mahoney (1976).

same blood group. But it is possible for siblings to have in common relatively few *polymorphic* genes (that is, those that confer on us our individuality, namely, within-species differences, rather than our human nature).[22] The reason is that the assortment of genes that occurs in *meiosis*[23] can give vastly disproportionate numbers of genes from different maternal and paternal grandparents to the children of the same four grandparents. This assortment no doubt accounts for a great deal of the diversity one sees among siblings. As noted above, on a population-wide basis males seem consistently to surpass females in expression of assertiveness and aggression. Females, on the other hand, surpass males in the expression of caring and nurturant attitudes and tender-mindedness. Yet studies of personality *phenotypes*[24] within families demonstrate that birth order often and systematically reverses these gender-linked expectations.

In the 1950s Helen Koch (for example, 1995, 1957) devised sophisticated and detailed studies of birth order effects, measuring their interaction with sex of subjects, sex of siblings, inter-sibling age spacing, and number of siblings in the family. Sulloway (1996, pp. 75–79) assessed the impact of her studies on subsequent work in this field. After demonstrating that birth order accounts for *five to ten times as much variance* in personality traits as it does in academic achievement and IQ, he noted that birth order reversed expected gender-linked traits in females as well as males. Whereas females are stereotypically less aggressive and quarrelsome than males, females who are first-borns show an aggressiveness comparable to that of males. First-born males and females use the same range of strategies for achieving their goals, namely, status enhancement and domination (rather than cooperation).

Women who become "alpha males"

Orville Brim (1958) took Koch's database and subjected it to new and different analyses. He placed it into the sociological paradigm of the family that Talcott Parsons had devised, dividing Koch's measures into agentic, instrumental measures (presumably masculine) and expressive,

[22] The preponderance of our genome consists in monomorphic genes that all humans share and that confer on them their distinctive human morphology and capabilities. We share most of these genes with chimpanzees, and many others even with lowly bacteria and other eukaryotic cells.

[23] In the production of germ cells, *meiosis*, that is, cell division, takes place in which there is a random assortment of chromosomes, one from each original pair of chromosomes found in the diploid set produced in each parent's gonads.

[24] *Phenotype* refers to the appearance of a genetic trait in the ontology of a human being. It is distinguished from *genotype* which refers to the total genetic inheritance of an individual, whether the genes are expressed or remain latent.

affectional measures (presumably feminine). Sulloway interpreted both Koch's and Brim's findings as demonstrating that birth order was more influential in shaping personality than was gender. He wrote, "Although sex differences were numerous in Koch's study, birth order frequently overrode these differences" (Sulloway, 1996, p. 77). The first born of *both* sexes reveal themselves as "alpha males." Later-born males more readily adopted expressive and female-appropriate behaviors than first-born females. First-born females were typically more assertive, competitive, aggressive, status-conscious, and dominant than the later-born of either gender. In this connection it must be noted that birth order also interacted strongly with the sex of the second-born. First-born girls were actually *more masculine* on average than their second-born brothers. They tend to behave like alpha males.

Niche picking

Sulloway (1996) did an extended historical analysis (pp. 148–71) of nonconforming behavior among women, in his book, *Born to Rebel*. Citing Bragg and Allen (1970), he first demonstrated that in two-child siblingships, there is a complex interaction of gender and birth order determining the level of conformity. First-born females with a sister are low in conformity, whereas first-born females with a brother are high in conformity. These covariations are reversed with later-borns. The later-born females with an older sister are high on conformity, whereas the later-born females with an older brother are low on conformity. The explanations for these relationships are speculative. Obviously, being a girl is a very different experience for her when she is an only child than when she is one in a small crowd of children. But what is the mechanism that shapes the divergent patterns of rebelliousness and conformity? Sulloway has amassed evidence to indicate that children in any family pick niches to cultivate in which they can excel. All children look for special ways to obtain and increase parental approval and affection. Similarly they are very sensitive to signs of discrimination in favor of their siblings. "Niche picking" is the preferred strategy for gaining the advantage in parental investment. To paraphrase Sulloway, what resourceful siblings do best is what the most creative thinkers do best: they exploit unoccupied niches. And this is what immigrants do in host societies. They seek out the unoccupied niches. Generation after generation, they seek, contest, and occupy increasingly secure, rewarding, and socially valued niches.

The Darwinian explanation for this is the "principle of divergence." Adaptive species faced with adaptational problems self-modify to capture niche-wise advantages in the economy of nature. Not only do various species engage in such diversification in their competition for scarce resources, individuals do so as well. In the economy of the household, children "radiate" (to use Darwin's expression) into unoccupied niches to increase parents' emotional as well as material investment in them. Radiation into these family niches offers opportunities to establish their primacy in domains where their siblings have *not* established themselves. It is this mechanism, Sulloway hypothesizes (p. 86), that accounts for the fact that history as well as personality tests show later-borns to be more open to experience. Sulloway asks, "*But why?* Openness helps younger siblings in their quest to find an unoccupied family niche. This psychological capacity is the engine that drives Darwin's principle of divergence" (p. 86).

Women who are first-born have been generally highly conformist, endorsing the status quo, and closed to reform movements unless they were underpinned by prevailing moral standards. Examples of first-borns, long prominent in the public eye, that Sulloway parades through his engaging chapter on gender are Anita Bryant, the vocal critic of homosexuality, who later became a born-again Christian. Phyllis Schafly, who saw her campaign against women's rights as a religious commitment, is also a first-born. That there have been rebels among first-born women is not surprising when one considers that their parents have also been rebels or members of oppressed minorities, founding their socio-political activities on a traditional socio-ethical bedrock. Charlotte Forten Grimké, an influential and dynamic African-American abolitionist, was a first-born in a family that had been abolitionist for three generations. As first-borns tend to identify with the authority structures of their parents and elders, they assimilate the social values implicit in those structures. Their rebellion is grounded on a lofty but traditional moral conscience, which is why, like so many alpha males, they are overrepresented in the most militant women's rights groups.[25] If it is difficult to understand these and other psychological dynamics without consideration of the political context in which they are played out, it is even more difficult to develop a psychological science without placing it in its historical context.

[25] Jean Hollands' (2001) book, *Same Game, Different Rules: How to Get Ahead Without Being a Bully Broad, Ice Queen or Ms. Understood*, is an expression of the need for powerful, high-profile female executives with steel in their spines to maximize their influence by assuming the suave manners of successful diplomats rather than professional wrestlers.

The intersection of science and politics

Whenever scientists compare two or more groups of humans on personality traits, especially those that implicate levels of cognitive functioning, they stir up ideological and political issues. The reason is that the belief systems and moral structures of any society invest various personality traits with values – values embedded in the cognitive structures we in turn use for judging people's worth. Should you assert that males or females (or any other demographic group) score higher or lower on measures of certain human traits (say, mathematical intelligence, psychomotor skills, or verbal aptitude), you are simultaneously asserting that they score higher or lower on the corresponding values society invests in such traits. Such comparisons, usually termed invidious, are fraught with political implications, often inflammatory, and all the more so if the valued attributes are viewed as products of genetic inheritance rather than social advantage. As Stephen Jay Gould (1981) has stated, because the term *inherited* is equated with *inevitable* or not remediable, enormous political and social mischief can be perpetrated by asserting that an entire group of humans is genically disadvantaged by comparison with another. Stereotypes are thus a vehicle for propagating the view that certain groups are of intrinsically less value than others in a society. For this reason, invidious comparisons, often encapsulated in harmful stereotypes, have been the target of advocacy groups and political factions.

For historical reasons, trait attributions placing women at a social disadvantage relative to men have been readily tolerated, whereas prejudicial attributions to other groups of humans have incurred relentless recrimination and accusations of bigotry. No ethnic group would have tolerated academic judgments of their innate intellectual inferiority, say, in mathematics or spatio-temporal tasks, such as have been researched and published about women – by women as well as by men. Jeremy Bernstein (1993) treated Albert Einstein very gently for his bigoted observations about women, when he commented on the career of Marie Curie, a Nobel laureate. In 1931 Einstein wrote "In Mme. Curie I can see no more than a brilliant exception. Even if there were more women scientists of like caliber they would serve as no argument against the fundamental weakness of the feminine organization" (Bernstein, 1993, p. 177). On the other hand, Bernstein vituperated in the strongest language an eminent American psychologist, Robert M. Yerkes,[26] for making comparable

[26] Robert Yerkes was an eminent psychologist, among whose accomplishments was the coauthorship in 1908 of the Yerkes–Dodson Law. The Law states that moderate levels of arousal promote optimal performance in a variety of tasks. Too little arousal leaves one

statements in 1908 about specific ethnic groups, and concluded that "The spirit of Yerkes and his ilk persists" in 1993.[27] The disparity in the judgments passed on Einstein and Yerkes for their denigrating statements about women and ethnic groups, respectively, is remarkable but not surprising.

Obviously, male and female trait attributions are still a major field of inquiry, arousing partisanship among gifted female researchers as well as among their male colleagues. Alice Eagly (1995) acknowledged the tendentious and sentiment-based character of much of this work: "Much feminist research on sex differences was (and still is) intended to shatter stereotypes of women's characteristics and change people's attitudes by proving that women and men are essentially equivalent in their personalities, behavioral tendencies, and intellectual abilities" (p. 149). In the decade following Maccoby and Jacklin's (1974) work, the Zeitgeist favored the discovery of personological similarities rather than differences between males and females. Where indisputable differences were shown in quantitative studies to exist, there was a tendency to view them as trivial, on the one hand, and socially constructed, on the other hand. A feminist political agenda appears to have been operative here. Women's struggle to achieve parity with men in all domains of civic and socio-economic life would be all the more difficult as long as stereotypes that disparaged women remained unchallenged. "Caught up in the passions of a burgeoning social movement" researchers pursued research ends they hoped would yield null results. The usefulness of continuing research in sex differences was questioned by some and disparaged by others (for example, Baumeister, 1988).

A countermovement emerged, linked in the public mind with Carol Gilligan's (1982) influential book, *In a Different Voice*, in which she affirmed that women's moral reasoning was, in part, of a different character than men's. This was a reaction to the stage-based maturational schema of Lawrence Kohlberg (Kohlberg and Kramer, 1969; *cf.* Kohlberg, Levine, and Hewer, 1983), according to which women were prone to stall at a lower maturational level than men. Gilligan's position is reminiscent of the Parsonian notion that men are more instrumental than women and women are more expressive and nurturant than men. She affirmed that in the moral realm women are

unmotivated. Too much arousal disorganizes the performer by reason of stress and anxiety. An inverted U-shaped curve describes the relationship, for example, between anxiety and performance.

27 Interestingly, Bernstein does not refer to "Einstein and his ilk," although he was as defamatory of women as Yerkes was of African-Americans. Women, moreover, have had as long a history of oppression and are as large a component of the human race as any other demographic group.

more family- and neighbor-oriented, attending to human needs as they are presented, whereas men are more task-oriented, given to strict observance of the rules. Of course, if one ranks the "principled, law-and-order stage" of moral development as superior to the "good interpersonal relations stage," where empathy, nurturance, mutual concern, and trust are privileged, then one arrives at the conclusion that women tend to a lower standard of morality than men. The arguments supporting the superiority of either approach to morality are not compelling.

Other studies proposing that there were important personological differences between women and men appeared in studies that were largely qualitative in nature (for example, Chodorow, 1978; Helgesen, 1990; Ruddick, 1989). Rachel Hare-Mustin and Jeanne Maracek (1988, 1989) highlighted the issue when they wrote that there is a tendency for subjective bias to manifest in either of two directions: to exaggerate sex differences, which they labeled *alpha bias*, or to minimize such differences, labeled *beta bias*. Their approach is one that they characterize as postmodern and *deconstructivist*. They note that the meaning they attach to *deconstruction* is that it is an operation that demythologizes "received wisdom," particularly those concepts that maintain the status quo and weaken challenges to vested interests. In a feminist perspective, it unmasks the covert ideological agenda and culturally constructed nature of concepts that emphasize, perhaps exaggerate, those sex differences that endow males with *privilege*.

Beta bias is evident among some feminist theorists who view the penchant of traditionalists to exaggerate the differences between men and women as an effort to "keep women in their place," that is, in the home and in the lower echelons of corporate management (Bem, 1993). Feminist theorists, on the other hand, are viewed as promoting the position that men and women are essentially the same in their capabilities for purposes of invalidating the stereotypes that have hampered women's opportunities for advancement in the professional marketplace. That there was a widely enjoyed scientific consensus in the 1970s that any sex differences that existed in the realm of cognitive, emotional, and psychomotoric attributes were negligible served a valued political objective – the objective of creating a level playing field for women (Matlin, 1993).

Stereotypes

Swim (1994) found evidence that rank-and-file judgment of sex differences accurately corresponded with meta-analytic effect sizes reported in the scientific literature. Eagly adds, "considerable evidence suggests that laypeople, once maligned as misguided holders of gender stereotypes, are fairly accurate observers of female and male behavior" (1995, p. 154). That there exist important, albeit small, sex differences has reignited the debate about gender stereotyping. Usually when personality theorists ask if gender stereotypes diverge from what men or women are *really* like, the "*really*" makes reference to biology, the "stereotype" to social constructions. The sub-text in these comments is that men and women are biologically constituted in one way, but are both *gendered* and popularly construed in warped, divergent ways.

Consistent with this latter view is that of Susan Fiske (1993), who stated that stereotyping "operates in the service of control" (p. 623). This is a view that would be endorsed by J-J. Rousseau who initiated the eighteenth-century view that society does violence to nature to promote the vested interests of select segments of the political economy. That the stereotypes are consistent with the prevailing needs and organization of society was more recently affirmed by Williams and Best (1982):

> If females are to have principal responsibility for the care of the young, it is reassuring to believe that they are – or can become – affectionate, gentle, patient, sympathetic, and so on. If males are to serve as hunters and warriors, it is comforting to believe that they are – or can become – adventurous, aggressive, courageous, energetic, independent, self-confident, and the like. (p. 237)

Men and women both *tend* to describe themselves in terms of the prevailing standards of socially appropriate traits. Whether one considers this to be an accurate or a *false gender consciousness* (Moghaddam, 1998, p. 424) depends, of course, on whether one considers the stereotypes to be accurate or distorted. If one assumes that stereotypes do not exaggerate the differences between men and women (*cf.*, Martin, 1987), that may, according to the Symbolic Interactionist School, simply reflect the influence of social expectations in inducing compliant behavior in men and women – especially in such traits as assertiveness and nurturance, and in nonverbal and agentic behaviors. If lay people are rather good assessors of the attributes of various demographic groups, the reason may in part be that they in effect shape that behavior.

The two-cultures thesis redux

Although patterns of socialization are imposed from the first day of life, it seems that sex-typical patterns of behavior spontaneously appear in the earliest days of life. These patterns are so pronounced and distinct, and evidence themselves in such divergent areas as dominance vis-à-vis cooperation, roughness of play, competition, turn-taking in social activities, play themes, discourse in groups, and affiliative and distancing behaviors relative to adults, that Maccoby (1998, pp. 32–58) characterized the divergent lifestyles of the two genders as two cultures. Maccoby's analysis led her to conclude that "the socialization account has not proved adequate to the task of explaining gender differentiation" (p. 9). She therefore argued, as has Brizendine more recently, that biology must explain a large proportion of these differences. This line of argument is plausible, as the presence of various hormones in varying proportions in the two sexes, pre- and post-partum, influences profoundly and differentially not only the behavior of boys and girls but the morphology of their central nervous system as well.

In child care centers children as young as 18 to 20 months of age have been observed to make positive, playful approaches toward other children, and as they grow older, these approaches are increasingly made niche-wise to same-sex partners (Fagot, 1991, cited in Maccoby, 1998; Jacklin and Maccoby, 1978; LaFrenière, Strayer, and Gauthier, 1984; Lloyd and Smith, 1986). This pattern is more pronounced among girls than among boys. Pitcher and Schultz (1983) demonstrated in their study of large numbers of children that by the age of 3, girls initiated playful encounters with other girls much more often than they did with boys. Schofield (1981) discovered that in unstructured classrooms where young children can place themselves where they wish, they segregate themselves by sex. Similarly, in the school playground children appear to drift spontaneously into sex-segregated games. "Boys usually claim the larger, more central spaces, while girls have more peripheral spaces for jump-rope or hopscotch, or sometimes claim their own jungle gym" (Maccoby, 1998, p. 25). Benenson (1993) further demonstrated that patterns of socialization differed among pre-school boys and girls. The former invested in larger male groups whereas the latter invested principally in reciprocated same-sex friendships.

The above studies suggest that the sex-specific interactional patterns that are found in adults have precursors in the earliest years of life. A developmental course is set in motion that ensures the acquisition in adolescence of various and distinct social scripts for boys and girls. The implications of this during a stage of life when career and education options are being considered, and in some cases crystallized, are evident and significant. These studies suggest that the gender roles that emerge from a society's division of labor by sex are not simply socio-cultural constructions. That the requirements of a society's productive activity spur the streaming of men and women into careers that are complementary to their habitual personality and socialization patterns is consistent with social role theory. But it allots a heavier role to biological priming of these patterns than is politically acceptable in Euro-American societies.

Spontaneous *gender segregation* and varied between-sex *interactional styles* appear to occur in the most diverse cultures across the globe. That segregation occurs at a more heightened level in Muslim than, say, in most Protestant communities does not call into question the existence of this nomothetic trait – only its variability. On the other hand, that sub-cultures evolve in all-girls groups that are different than those of all-boys groups cannot mask the fact that there is wide divergence *among* girls and *among* boys in the expression of these differences. This divergence is especially evident among adolescent boys. Some boys become "jocks," others "nerds." Some are shy and loners, others "party animals," hypergregarious, and boisterous. Risk-taking is behavior that is displayed (like all others) on a continuum, and although as a group boys are more risk-taking than girls, a large proportion avoid the bravado of inter-gang rumbles and the risky behavior of motorized games of "chicken." To conclude from this within-group variability that we can disregard between-group variability is an argument that has been amply refuted by the statistical disciplines and by common sense.

Descriptive research enlightens us as to the sex-divergent social behaviors of a Western population – and, in some respects, of a divergent array of multicultured populations. It does not, however, explain it. As there are (in a rational world) no effects without causes, and as there is little hope of altering effects unless we can act on the causes that maintain and perpetuate them, personality and developmental scientists continue to devote their professional lives to unraveling the complex and interactive causal web that may account for gendered behavior. Evolutionary psychology, social role theory, Maccoby's two-cultures theory, and several feminist approaches to personality development are only some among the more notable attempts to explain, as well as describe, the determinants of gendered personality. On a larger screen, we should, of course, key in other variables, as gendered personality is only a part of the picture. Genically determined temperament, parent–offspring conflict, birth order, rampant alcoholism or sexual abuse, catastrophic changes during children's formative years (for example, the death of a primary caregiver or the dissolution of the home), broad social attitudes, belief systems, religious values, and socio-economic imperatives – all have an impact on the development of personality.

Conclusion

We live in a world of mirrors. We come to evaluate ourselves the way others perceive us (see Chapter 6, The puzzle of the self). The house of mirrors at the carnival is interesting and amusing because it sends us distorted and unflattering images of ourselves that we can safely ignore. But when we enter into an institutional house of mirrors the consequences are different. Not only is such a looking-glass self a product of how others view us, whether in a flattering or unflattering way, it can have serious consequences for our success, given the institutional taboos and the criteria for advancement within that house. Applicants for an academic position or a job hope that there will be no prejudicial judgments about them that will compromise their ability to equitably compete with the other applicants. The fact is that some women and, increasingly, some men feel that when they enter a corporate structure they have entered a house of mirrors; their images are distorted not only in their own eyes, but also in the eyes of the decision makers.

Below the reader can find some social correctives that may strike some as having a moralistic tone but are not intended to be such. They have been widely disseminated and may be of some use.

* * *

Afterword: some social correctives

Halpern (1997) drafted a number of research-based suggestions bearing on constructive attitudes that educators might adopt in treating sensitive issues of gender psychology. Here paraphrased, adapted, and elaborated are several.

1. Where differences are found between women and men, taken collectively, do not characterize those differences in terms of better or worse. "There are intellectual areas in which females excel and others in which males excel; these data do not support the notion of a smarter sex, nor does it mean that the differences are immutable" (Halpern, 1997, p. 1098). Carl Gustav Jung proposed that men and women differ significantly. There is a complementarity in these traits that enhances the development of both sexes. Using research to advance biological politics and "spinning" doctrine to satisfy canons of political correctness demean the professional ends we wish to achieve.

2. Although sex-based group averages indicate a difference between males and females on a number of traits, intellectual and simply personological, there is never a justifiable presumption that any individual is inferior or superior to others by virtue of belonging to one group or another. Although perceived or demonstrated group differences can have negative political consequences (which need to be guarded against) for members of these groups, individuals must be judged on the basis of

their performance and promise. "We cannot afford to write off anyone or allow group membership to limit talent development" (p. 1098).

3. "Recent research has shown that beliefs about group differences exert powerful effects on thoughts and behaviors that occur without conscious awareness" (p. 1098). Stereotypes are ubiquitous furnishings of the human mind (Dumont and Fitzpatrick, 2001), as people cannot live without fashioning expectations and personal theories of how others will behave. For this reason, we must develop an ethos in which we readily sanitize the scripts that subtly prepare us to anticipate unpleasant actions by others. Negatively prejudicial stereotypes need to be unmasked as destructive rather than constructive.

4. Research on human cognition will eventually reveal the neuro-hormonal basis for several of the threats to mental health that differentially affect males and females. This is particularly important in the psychological effects of aging where the decline in production of "male" and "female" hormones can have wide-ranging effects, not least in delaying debilitation of cognitive functions.

5. Male infants at birth, on average, are neurologically 3 to 5 weeks less mature than female neonates. Boys labor with this handicap well into adolescence. If this fact is not well understood, unrealistic expectations can be harbored by parents, unrealistic demands made by teachers who use inflexible and uniform criteria for boys and girls, and intimations of inadequacy and failure communicated by members of the community at large to sensitive and vulnerable children who are struggling to fashion a personal identity.

6. Female performance is underpredicted and male performance over-predicted by most of the standardized tests that have been designed to stream applicants into college, graduate, and professional schools. Scholarships, bursaries, fellowships, and other opportunities for academic advancement have not been equitably awarded by relying on traditional psychometric instruments. "Multiple measures of ability and achievement must be used in these decisions, especially course grades and tests of writing, which tend to favor females" (p. 1098). Other educational initiatives need to be examined, such as sex-segregated education, which provides other, valorizing experiences for girls as well as boys (Halpern, 1997; Kleinfeld and Yerian, 1995).

7. "Empirical research does not create stereotyping, as its critics imply; the systematic study of sex differences using scientific rules of evidence is the only way … to understand legitimate differences" (Halpern, 1994, p. 524). Conclusions about a topic as complex as cognitive sex differences will rest on the cumulative results of many studies. Although each person has socio-political preferences for particular types

of explanations, it is also necessary to strive to maintain an open-minded fairness when assessing a variety of theories and empirical findings ... Explain the many pitfalls and pratfalls of this sort of research whenever simple statements like "It's all in the hormones" or "It's all in the mother's attitude" are made (p. 1098).

8. Much can, and does, go wrong between conception and maturity, but the human organism is highly plastic and self-healing. Nurturant and supportive environments promote growth and remediation. A clean, healthy, and nourishing environment allows the human organism to do the rest. Rather than restricting opportunities and imposing choices on the young we need to allow them to make free and informed choices. Given the malleability of humans, whatever sex differences exist today are subject to significant modification. "The workforce of the future will need to be better educated than any past generation. Our best hope for the future is a well-educated citizenry where individuals can develop their intellectual potential to its fullest." The welfare of both men and women will be optimized by allowing them to engage in those activities that they genuinely feel inclined to perform. Socially constructed, gendered straitjackets can be discarded.

9 Emotions and reasoning: a definition of the Human

> Consciousness is the key to a life examined, for better and for worse, our beginner's permit into knowing all about the hunger, the thirst, the sex, the tears, the laughter ... the flow of images we call thought, the feelings, the words, the stories, the beliefs, the music and the poetry, the happiness and the ecstasy. At its simplest and most basic level, consciousness lets us recognize an irresistible urge to stay alive and develop a concern for the self. [And] at its most complex and elaborate, consciousness helps us develop a concern for other selves and improve the art of life.
>
> Antonio Damasio (1999, p. 5)

Setting the tone: some prolegomena

Emotions have traditionally been thought of as distinct from reason, just as thinking was considered to be distinct from feeling (LeDoux, 1996, pp. 15, 24).[1] This hallowed philosophical view goes back to the Hellenism of more than two millennia past. Plato, in *Phaedo*, affirmed that emotions made clear thinking difficult. On the other hand, he gave short shrift to the view that emotions were useful to improve thinking, especially the thinking that leads to practical judgments. Today most psychologists are convinced that our thinking about serious social problems will be significantly distorted if we try to exclude consideration of the emotions associated with those problems. Impairment or neglect of either one diminishes them both as, indeed, they seem to constitute a single faculty. This chapter is dedicated to exploring these propositions.

Personality psychology has largely focused on expressive behaviors, especially facial behaviors that are evident to observers. As noted in

[1] Various models of emotion will be treated as we progress through the chapter, and their definitions will take shape in that context. The term *affect* in the psychology literature is often used interchangeably with emotion as well as with feeling and mood. Unless context calls for affect, we will usually stick with the less ambiguous term, emotion. On the other hand, we will normally use *affect* and *affective* states to refer to mood, feeling, and tone.

Chapter 1, the very origin of the term personality (derived from *persona*) betrays this bias in our science.[2] We rarely refer to individuals' personality in terms of, say, their neuro-endocrine production – and other infrastructural components of their visibly expressive behavior. Instead, personality analysts study posture, greeting styles, handgrip, choice of dress, handwriting, phatic communication,[3] and, at a certain remove from interpersonal behaviors, penciled reactions to questionnaire items. Non-obtrusive measures are valued because they do not trigger the self-conscious defenses and subterfuge that individuals use, even unknown to themselves, to gild the self-image they are projecting. But even such unconscious defensiveness is an ever present personological variable – a guardian of privacy that, like an overzealous valet, controls everything it was meant to serve. The mind is thus occasionally controlled by faculties that have exceeded their "mandate." (That is where psychopathology enters the picture.)

In the past two centuries theories of personality were largely fashioned by the great pioneers of psychotherapy. Their theories emerged from an interaction of formative childhood experiences that gave shape to their values and belief systems, their professional education, the experiences that became woven into the fabric of their clinical practice, and especially the particularities of their innate temperament that influenced all of these situational variables. Some of these clinicians, like Freud and Jung, had troubled emotional lives of their own, euphemistically characterized by historians as *creative illnesses*. Just as important, their theories of personality were formulated during many years of working with seriously ill people. Two biases were at work in this context: the first was the traditional view of the human as *animal rationalis*, whose essential quality was the primacy of reason – a quality that was exalted over emotion. The second bias arose from the fact that these theorists of personality subscribed to the principle that humans could heal themselves by clear thinking (and by extension by *insight* into the causes of their disorders). In any event they neglected the exploration of all but the negative emotions – most notably guilt, depression, and anxiety.[4] Deficient models of personality resulted.

[2] In a contemporary setting: Kahneman (cited by Feltham, 2007) asked the "subjects of his study if they presented a smiling face to the world but were privately unhappy. He stopped asking this question because it caused many people to burst into tears" (p. 178).

[3] Phatic communication consists of greetings and other verbal forms that are intended primarily to convey and elicit feeling and attitude (for example, Have a nice day!, How're you doing?, Yours truly, etc.), rather than to convey information. Misunderstanding a greeting as an invitation to engage in conversation usually leads to disappointment.

[4] Unlike classical psychoanalysis the psychotherapies of Paul-Charles Dubois, Pierre Janet, Alfred Adler, Carl Jung, and others variously integrated home tasks, good diet, civic involvement, readings, and exercise into a cognitive treatment program.

Do we learn emotional reactions?

In twenty-first-century psychotherapy one can view the swing to evoking, analyzing, and processing the emotional components of experience, especially those experiences that are at the root of personal distress. Therapists have made important contributions to the research, theory, and understanding of emotion (for example, Greenberg, 1993; Greenberg and Safran, 1989; Mahrer, 2000). Although there is considerable diversity in the way they construe emotion and emotional schemas, there is a growing convergence of views on the importance of processing emotion in order to help troubled individuals. Some therapists distinguish emotional feelings from emotion schemas. Safran and Greenberg (1991, p. 8) explained: "From birth, the individual begins developing memory codings for specific events and the emotional responses that are evoked by them." Over time people abstract "generic representations of specific types of events and specific expressive–motor behaviors, and patterns of autonomic arousal that have been evoked by these events … [Such representations] influence all future experiences." These emotion schemas enable complex emotional responses (visceral, neural, and hormonal), as well as action tendencies and their consequent behaviors and feelings. One is always experiencing them at an unconscious level and, of course, from time to time when they surge into awareness. It is the unconscious character of these schemas that accounts for the automaticity of our responses to events that trigger intense feelings.

Researchers have determined that we do not *learn* how to be fearful, or angry, or joyful (Pervin, 1996, p. 312). Emotional behavioral systems are innate endowments that we share with a multitude of other animals. Joseph LeDoux (1996) stated that all animals have neural systems that help them obtain the resources they need to survive and procreate. "Within the animal groups that have a backbone and a brain (fish, amphibians, reptiles, birds, and mammals, including humans) it seems that the neural organization of particular emotional behavioral systems … is pretty similar across species" (p. 17). What we *do* learn is what should (and does) cause us fear, what excites (and justifies) anger, and what, given certain achieved or anticipated benefits, will give us pleasure and – its emotional spin-off – happiness.

The expression of emotions is by its nature a physical process that plays out in all parts of the organism. Fear resulting in the awareness of a physical threat, say, the sound of strange footsteps in the middle of the night, can result in accelerating adrenaline production and heart beat, freezing, and a raft of visceral reactions including heightened blood pressure, the urge to void bladder and bowels, goose pimples (that is, *pilo-erection*), and sweating (see LeDoux, 1996, pp. 138–78). Consider the impact on a student who had just received a failing grade in an important subject. She reported:

> My mouth just dropped open. I stared at [my grade] like, "Oh, No!" I felt like a bomb just went off in me; just devastated ... that drop in my stomach, that big drop; and kind of like a numb feeling ... something that you get with shock. (Turner, Husman, and Schallert, 2002, p. 83)

Scholars who are more neuropsychologically grounded will search for expressions of emotion throughout the neuro-endocrine and visceral systems of the organism. Social psychologists, on the other hand, will look for emotional expressions that are apparent in social discourse – the *epiphenomena* of upheavals taking place within the organism. The most important of these are facial expressions. Beside facial expression, there are other superficial but important perceptible correlates of emotion such as voice tone, paravocal cues, postural rigidity or flaccidity, trembling, pupil dilation, and sweat glistening on the forehead and upper lip. The former characterize the domain of investigation of those who are primarily interested in the biology of emotion; the latter of those most interested in the social psychology of emotional life. Needless to emphasize, all such *signs* are part of a *single, holistic organismic response* to a situation that portends possible negative consequences for one's future.

Behavior, although grounded in biology, is shaped by culture and dominant social *mores*. The most vivid example of this occurs in gut-wrenching human dramas – for example, the funeral of a treasured member of one's family, especially a child or young adult who was accidentally killed while in excellent health. Although the death of loved family members is wrenching for humans of every culture, the *display rules memorializing this* vary widely. Some cultures prescribe uninhibited, florid displays of emotions, while others require stoic, self-composed endurance. However, all recognize the intensity of the suffering and emotion expressed by the bereaved.

Your call may be recorded for quality control purposes

The qualities alluded to in this familiar commercial refrain are revealed in the vocalizations of customers seeking information and help (as well as of call-centre employees). Emotion and feelings seep into mentation and are expressed involuntarily in our voices and other behaviors. How we speak – the timbre, cadence, pitch, intensity, and tone of voice – reveals our mood, emotion, level of concern, and the urgency of our message. Customers usually call because they are experiencing a difficulty with a product or service. The calls are monitored so that managers can better grasp the generality of needs and satisfaction of their client base. *The Economist*

(March 8, 2008) in its *Technology Quarterly* noted the sophisticated software that has been designed and used by firms to tease out relevant information from client phone calls. Much of the analysis bears on the choice of words, the metaphors used, and the changing emotional tone of the caller as she walks through the problem with the call-centre respondent. *The Economist* states "Cisco's voice analysis system monitors parameters including volume, cadence, tone, pitch and inflection, and then sorts callers into six personality types" (p. 14). Cisco, a telecom firm, refers to this software as "the bleeding edge" of caller profiling – an evocative metaphor in its own right.

Music and emotion

Music can evoke powerful emotions, as any concert-goer can experience. This has been known from the time of Homer, the early Vedic chants, and the healing rhapsodies of David's harp. Indeed, Pythagoras (580–500 BCE) is reputed to have discovered the relation of music to emotion, mathematics, and physics and tied these disciplines to the "music of the spheres" and the harmonies of the human spirit – a work of consilience that even Edward O. Wilson (1999, pp. 229–59) would applaud. These principles have not been lost on call-center managers, for as one is placed on "hold" one is exposed to various genres of music according to one's presumed level of culture. We may further suppose that one's call-recorded-for-quality-control purposes will be analyzed for traces of exasperation if not anger – and to determine one's level of emotional lability and psychoticism (see Chapter 5 relative to these two factors). In this context one can appreciate how critical is the choice of appropriate background music.

To listen to the vocalizations of primates in their dangerous ecological niches around the undeveloped regions of the world is to suspect that *variously inflected* hoots, yells, screams, and shrieks are the primordial language of our biological Order. The wide range of tones associated with food discovery, mating, alarm, aggressive competition, and nurturing suggests the presence of protean elements of music and speech. Even on the surface of it, these variously pitched howls would seem to serve an adaptive survival device – and to human ears, be laden with the emotion associated with mating and danger. To characterize music as "auditory cheesecake" or as an evolutionary spandrel (*contra* Steven Pinker and Stephen Jay Gould, respectively) appears unfounded if not flippant to non-experts (*cf.* an authentic expert, W. Tecumseh Fitch, 2005).[5]

[5] Some birds have a range of "musical" notes spanning more than two octaves. The reader who wishes can access them in the sound archives of the British Library at: www.bl.uk/collections/sound- archive/listentonature/specialinterestlang/langofbirds19b.html.

Music has long been known to inflame as well as calm the emotions, which is why Plato and later Augustine and the other elders of the early Church monitored its use carefully. Indeed, the medieval Catholic Church authorized only plainchant in its rituals and forbade the blending of male and female voices and other uses of complex polyphony. As Rosenfeld (2008) has noted, ecclesiastics eventually appreciated the power of music and rhythmic beats to enhance persuasion and belief. They later "began to sponsor music that was ornamental and expressive ... like the oratorio, which used soloists, chorus, and instruments" (p. 91). There are individual differences, of course, in humans' emotional responsiveness to music of various genres. Indeed, some are born with one form or other of *amusia* – the reader will have noted this among acquaintances who are tone-deaf. The personality ramifications of greater and lesser aptitude for the generation or even the appreciation of music are significant. The healing powers of music have been noted in the shamanistic rituals whose origins (we infer) are pre-historic. And as a contemporary composer, Michael Torke, has observed "Why waste money on psychotherapy when you can listen to [Bach's] B Minor Mass?" (Rosenfeld, 2008, p. 91; also consult Sacks, 2007, pp. 295–301).

Basic emotions

What are the basic emotions?

There are two principles that need to be stated at the outset. First, it is not possible to study personality only from a perspective of cognitive schemas, algorithmic programs, and conceptual models. Emotion not only confers a human patina to everything that we do and think, from acts of generosity and self-sacrifice to celebrations of marriage, birth, and human achievement, it informs all our behavior. Any valid model of personality needs to include emotion as well as cognition, components that motivate and drive every significant human endeavor. Motivation, conceptually distinct from emotion, also needs to be included in any valid model of personality. Second, as there is little cognitive activity that does not have an inchoate emotional substrate to it, the *motivation* that interacts with and directs cognition is itself powered by a constellation of emotions – some of them vivid; others faint and pallid, often unconscious. Finally, the science of emotion needs to integrate principles of *cognition*, for emotions are enmeshed in all intellection. Any science of the mind that begins by rigidly distinguishing feelings from reason, emotion from cognition, is tidy but unworkable. It obscures the essential interpenetration of the functions underpinning these constructs in a mental life that is never without both.

"Contrary to traditional scientific opinion, feelings are just as cognitive as other percepts" (Damasio, 1994, p. xv).

As soon as we begin to sketch the essential outlines of emotions, we are working in a fuzzy problem space (*cf*. Plutchik, 1994). Whenever one uses a colloquial language to categorize a body of experience in science, the discourse becomes laden with connotations, values, and surplus meanings that are irrelevant to the scientist. This is the case with classifying subjective human experiences, and especially when working with the construct of emotion. This problem has led theoreticians to formulate the concept of prototypical emotions, characterized by specific features (Gross, 1999). First, an emotion erupts when a new stimulus field portends something important in one's life. Importance can arise from many sources. It can be genetic, such as fear of falling from a height, or cultural, such as embarrassment over a social *faux pas*; it can be conscious, as in achieving a studiously pursued objective, or unconscious, such as the vague discomfort we experience in locations that remind us of painful episodes from our childhood.

Second, emotions are whole-body experiences. They engage all the *viscera*, especially the hormonal agents, the autonomic nervous system (ANS), and the most extensive viscus of all, the skin of the organism. The brain is also a prominent player in the cascade of emotions, and determines the subjective, conscious experience that is the end stage of an emotional event. These cognitive experiences are usually called affective responses, and they involve both the cortex and primitive strata of the brain.

Third, although there is a great variety of emotions and hundreds of words in the English language to denote them (just as there is a great variety of personality traits and thousands of words that denote them), emotions are categorized into a small number of basic higher order types. These are called *basic emotions*. Like all mental constructs such categories of emotion have a quality of artificiality. They are gross bins for holding subtly and widely varying emotions, such as envy, jealousy, resentment, covetousness, romantic love, and so forth. They cease to be simply constructs when they can be scientifically linked to the distinct neural systems that potentiate them (LeDoux, 1996). Further, each of these emotions can be assessed as to their intensity, motivational power, painfulness (or pleasurableness), and internal controllability. Each emotion has a number of such properties that can, in principle, be measured. In other words, each emotion is a continuous variable assessed in terms of the units in which it is measured. Although theoreticians differ in the classification systems they use, we cannot regard any one of them as right or wrong. They have been developed to respond to personal preferences shaped by the research questions proponents wish to ask and answer, as well as by the aesthetic and theoretical elegance scientists find in them.

Fourth, emotions are not primarily logical, ratiocinative, higher cortical processes. Although cortical cognitive processes are involved in our emotional life – and barely distinguishable from it – emotions are to a great extent sub-cortical and unconscious in nature. The old (reptilian) complex that we share with other vertebrates allows us to make rapid responses to environmental threats long before we have conscious access to them (LeDoux, 1996). The old brain allows us to use a "rather safe than sorry strategy" that results in many *false positives*, but in the long run prolongs our lives. For example, we sharply and reflexively react to stepping on something in deep grass that feels like a snake, although a further (and conscious) examination reveals it to be only a vine.

The basic emotions controversy

The perspective described above has been challenged by those who wish to give greater importance to higher cognitive processes in organizing our emotional responses to stimuli. Ortony and Turner (1990), for example, questioned whether there are *primary*, *fundamental*, or *basic* emotions. At first glance this appears to be a trivial taxonomical problem: that is, can we categorize the numerous emotions for which ordinary language has labels as either basic or nonbasic? The basic emotions camp marshals several arguments. For example, if there are basic emotions, then they are universal and can be found in every group and culture. If they are universal, they are biologically determined, with the panoply of physiological and behavioral mechanisms that enable them. This obliges us to search for the central nervous system substrate in humans that exists in evolutionary continuity with other organisms on the phylogenetic scale; in fact, these cerebral structures have been identified (for example, Panksepp, 1998).

Ortony and Turner have not been swayed by the pragmatic advantages associated with this perspective. Assuming that there are genetically determined mechanisms, we are faced with the scientific problem of determining how these irreducible affective states can serve as building blocks for complex and culturally influenced *displays* of emotion, both adaptive and maladaptive. But their case rests principally on two arguments: (1) the basic-emotions theorists have not proven their case; and (2) there is a plausible alternative to their theory, namely, that an emotion such as fear is a psychologically constructed set of responses to a unique (and threatening) stimulus field. Ortony and Turner maintained there are as many distinct emotions as there are distinct appraisals of ever-changing environments.

Enumerating basic emotions

Sylvan Tomkins (1962) proposed that there are eight basic emotions: *joy*, *fear*, *surprise*, *anguish*, *interest*, *shame*, *disgust*, and *rage*. Paul Ekman (1984), one of his students, developed a typology with six core emotions: *anger*, *fear*, *disgust*, *sadness*, *surprise*, and *happiness*. Carroll Izard (1992), another of Tomkins' students, argued that these emotions are like primary colors which, when mixed in varying proportions, give us the rich array of *affective* states that find expression in the poetry, drama, and sagas of world literature – as well as in everyday life. For example, jealousy may be a mix of fear and anger, yet with its own distinct organizational and motivational properties. Similarly, anxiety may result from the convergence of fear, sadness, and anguish, although the source of our anxiety is often obscure or hidden from our awareness. Their expression in varying intensity helps us define an individual's personality.

Theorists have inferred basic emotions from their association with distinctive facial expressions. This association may result from the need to communicate feeling states quickly and accurately. This is important for the prosperity and survival of humans (and other primates) both within their community and in hostile environments. The adaptational challenge of understanding one's social environment and being understood by others favors the transparency that facial expressions provide. Adaptation is maximized by externalizing such markers of the most important emotions.

Basic emotions and facial expressions

In 1972, Ekman, Friesen, and Ellsworth wrote:

> The human face – in repose and in movement, at the moment of death as in life, in silence and in speech, when seen or sensed from within, in actuality or as represented in art or recorded by the camera – is a commanding, complicated, and at times confusing source of information. (p. 1)

This window on human emotion has never lost its fascination for psychologists. Darwin ([1872] 1965) wrote a seminal work on the expression of emotions that anticipated much of the research that has since defined our approach to emotional behavior. However, this pioneering effort (there were many that preceded it) did not truly link Darwin's evolutionary principles of natural selection and mate selection to the facial expression of emotion (Ekman, 1973). It is only recently that measurable progress has been made in understanding the links between the facial behavior that reveals (and may with effort conceal) emotion and the psychophysiological "engine" that generates it.

Facial musculature is exquisitely tuned to emotional events in primates, especially humans. Indeed, it forms one of the constituents of some definitions of emotion. Even before we are fully aware of our feeling state, friends may ask us, for example, "what's so funny?" or "what's wrong?" There are innumerable shadings and variants of the fundamental emotions. Whether there are just as many different facial expressions as there are variants of such emotions is an empirical question still being investigated. Social interactions among humans are constantly modulated by our read-out of *others'* emotional responses that appear on their faces (Gleitman, 1996, p. 348), and people look for evidence of how others are reacting to their statements or behavior by intently studying the expression on their faces. Facial expressions become social markers of emotion, a window that we peer through in an attempt to get at the "real" feelings of others. We assess personality to a great extent by the facial expressions others habitually "wear" across a wide assortment of situations, both pleasurable and difficult.

Facial expression has become a significant part of the communicative repertoire of the higher primates, and it is especially important for humans. The ability to "read faces" is important in ordinary social intercourse, and it is a critical skill for psychotherapists. The ability to infer underlying affect from facial expressions offers a significant advantage in those social spheres where sensitivity to emotional cues is critical. There are also many human activities in which people attempt to alter facial cues in order to place competitors at a disadvantage. Examples include poker players, certain used-car salesmen, jade dealers, politicians, and "con" men. Given this window on the mind, people understand that the blinds can never be fully lowered.

The linkage between basic emotions and facial expressions is buttressed by cross-cultural studies using language-independent techniques (Izard, 1992, 1994). These studies support the hypothesis that these linkages are *innate* and *universal*, despite the fact that emotion-display rules vary from culture to culture and in many cases these displays can be simulated for purposes of deception. Although this issue is rife with controversy (see, for example, the exchanges between Paul Ekman and James Russell that appeared in the 1990s in *Psychological Bulletin*), support for the universal, innate position can be found in research in comparative psychology, *ethology*, and *primatology*. The importance of facial expression for communicating affect reached its pinnacle in the human species. As we descend the phylogenetic scale, "facial" cues become less important, and other physiological expressions of emotion – which can be picked up by sensory receptors other than eyes – become increasingly critical. The innate character of facial expression of emotion suggests that there are many other physiological emotion-specific changes that occur.

Secondary emotions

Damasio (1994) has proposed a second tier of emotions that implicates different cognitive responses to emotional stimuli. For example, people are often flooded with emotion when they recall the memory of a much-loved member of their family or circle of friends who has died. This recollection may bring to mind joyous memories. On the other hand, the memory may simultaneously cause them to feel the pain of experiences that they still deeply regret – lost opportunities to show affection; unkind or abusive response to expressions of dependency or love; or simply a profound sense of loss.[6] Damasio (1994, pp. 134–39) explained the complex neurological basis for these memories that are organized by mental evaluative processes, both nonconscious and conscious. They are structured differently from cerebral processes that the primary emotions engage, involving as they do the prefrontal cortex and the various sensory cortices that organize the smells, sights, and sounds of the past. Secondary emotions may blend into background feelings as one revisits the neighborhood in which one grew up, or walks through the ball parks and picnic sites that were scenes of a past life that has totally disappeared. One may feel bathed in feelings of nostalgia – the feelings evoked by these memories may be positive or negative, but they are never absent (see McAdams, 1994, for a poignant example, pp. 717–19).

Arousal and value: a two-dimensional paradigm

Some researchers have found that they could plot subjective reports of emotional experiences neatly into a two-dimensional space (Barrett and Russell, 1998; Russell, 1980; Russell and Barrett, 1999). One dimension plots the intensity of arousal; the other plots the valence, that is, the hedonic value, of the experience (consult schematics in Gross, 1999). For example, when one is highly aroused but experiencing little positive or negative valence, one is tense and alert. On the other hand, when the

[6] "In every man's memories there are such things as he will reveal not to everyone but perhaps only to friends. There are other such as he will reveal not even to friends, but only to himself, and that in secret. Then, finally, there are such as a man is afraid to reveal even to himself, and every decent man will have accumulated quite a few things of this sort" (Fyodor Dostoevsky, [1864] 1993, p. 39. A nod to LeDoux for this quote).

intensity of arousal is medium (neither high nor low), but one is experiencing highly negative feelings, one is upset or sad. Thus, all emotions, if properly calibrated on these dimensions, can be uniquely situated in this conceptual space.

This *circumplex model* posits two handy descriptors as the hallmarks of all emotions; however, it may not be sufficient to help us understand the true nature of emotions. The descriptive properties of an entity may not define its essence, and, indeed, descriptions can be irrelevant, no matter how interesting they are or how elegantly they are configured. For example, one can plot all species of birds in two-dimensional space: (a) vividness of color of bill and plumage; and (b) wingspan. This will yield interesting configurations of *speciation* in birds, but the exercise will not help us to understand their distinctive natures. Relative to emotions clear distinction needs to be made between identifying them and defining them. The more interesting question is, what precisely makes grief different from depression? Do different mixes of arousal and valence account for the difference? Beyond self-reports there needs to be research into the contributions of the endocrine and autonomic nervous systems, perceptual, neo-cortical, and thalamic systems, as well as personal history and other socio-cultural variables (*cf.*, for example, Damasio, 1994, 2000; LeDoux, 1996).

Phase-locking

Christiaan Huygens[7] discovered that two pendulum clocks with close but discrepant frequencies would gradually assume the same frequency, and shift into a common phase (Mayne and Ramsey, 2001; Strogatz, 2004). It is important to note that both clocks must be attached to a common board, thus forming a single system. The physical principle that any objects in systemic contact with each other will synchronize their actions appears in a number of natural systems. For example, "phase locking" appears among spouses, dorm partners, crickets, and fireflies. The nocturnal mating displays of fireflies appear as flashes and in time become synchronized. The workday schedules, menstrual rhythms, and various diurnal activities of humans who have lived together over a period of time also tend to become harmonized. Two

[7] Christiaan Huygens was an eminent seventeenth-century Dutch mathematician, mechanist, physicist, and astronomer, who associated with the great European scientists of his era, principally in Paris, London, and his native land. He developed the wave theory of light, advanced the science of optics, discovered the true character of the rings of Saturn, invented and patented his pendulum clock, defined pendulum motion, and made other scientific advances.

individuals – call them "coupled oscillators" – synchronize their movements in such a way that a harmony of energies supersedes initial chaos. Importantly, there is a convergence of emotional states among two or more persons working in the same space.[8] Gaia, a metaphor for planet Earth, is always in this autopoietic mode – self-organizing and revealing itself to the inquiring intelligence as a self-equilibrating system. Unless one is interested in measuring emotion at a point in time, to wit, "the weighting and combining of multiple variables: tension of specific facial muscles, blood pressure, serum epinephrine, amygdala activation …" (Mayne and Ramsey, 2001, p. 9), understanding what events precipitated it and how it is evolving is what is useful. In this perspective a dynamical systems theory (such as chaos theory) would seem a promising approach for developing an emotion metric.

Moods: a sub-type of emotion

Moods are emotional states that have less specificity than the "big eight" universal emotions, differing from them in a number of respects, not least in that they have little explicit intentionality associated with them. Often arising from hormonal and other biochemical changes of which we have little awareness or over which we have little control, they make us a mystery to ourselves. Moods, *per se*, do not motivate us to do *something*, as does, say, fear, anger, or love. They do, however, make us more or less sensitive to evocative stimuli (say, rudeness or sexual overtures). The individual who feels irritable will react more quickly to a personal slight than she would if she felt on top of the world. Mood can thus serve as an anchor to slow reactions or as a catalyst to potentiate emotional responses.

Moods are often characterized as good or bad, and like episodic emotions they fluctuate in intensity (Caprara and Cervone, 2000, see pp. 287–90). Like muted, slurred emotions, they are sustained over time. We habituate to them as we attend to the routine demands of life and lose our clear, immediate awareness of them. But people routinely

[8] This phenomenon has given birth to the principle of *reciprocal affect* (see, for example, Hatfield, Caccioppo, and Rapson, 1994), which is the tendency of individuals to respond to others with the same affect with which they have been approached. This is arguably a process analogous to the phase-locking of pendulum clocks and room mates. If approached by friendly individuals, one tends to respond in a friendly way; if approached with hostility and coldness, one tends to respond negatively. This tendency to act like those around you reduces conflict and disorganization (desirable, for example, when there is a merger of corporations with divergent cultures) and can have important survival value. On the other hand, this dynamic may reduce the cultural diversity that is valued in some societies. The political injunction that immigrants should melt into their host society, avoiding cultural isolation, encourages this kind of phase-locking.

take notice that they feel "upbeat" or "funky," irritable or euphoric, but with little appreciation of the cause of, or even curiosity about, this affective state of mind. Because it is often difficult to pinpoint the event or the cognition that has brought on a mood, people spend little time trying to cope with moods directly, at least until they become seriously disruptive. Trusting that a "bad" mood will soon dissipate, individuals cope with their discomfort as they cope with bad weather, and go about their daily lives as best they can.

In popular parlance, we use the term *moody* to describe individuals who are prone to prolonged periods of depressed or somber feelings. Only in clinical circles are individuals who are consistently exuberant, manic, upbeat, or cheerful likely to be called moody. Damasio (1999) defines *background feelings* as those of which we are not fully conscious. He enumerates some of them: "fatigue; energy; excitement; wellness; sickness; tension; relaxation; surging; dragging; stability; instability; balance; imbalance; harmony; discord" (p. 286). These emotional feelings are mostly subliminal as they reside in a cerebral register that monitors what is happening in the body at large. The body, which is the theater of the emotions, holds the brain, with its mind, as a captive listener. As long as the neural and hormonal channels are healthy and functional the flow of information from body to brain is constant and inexorable. Moment to moment, background feelings, whether we attend to them or not, are an accurate barometer of the state of the body as well as its response to the emotional stimuli impinging on the organism as a whole.

Normal baselines for background feelings are those that allow people to perform the routine tasks of life in a moderately functional manner. Chronic moods, however, that cause distress to victims and those who live and work with them are labeled disorders. These are grist for the clinician's mill – and increasingly for the physician's prescription pad.

Emotions and instincts

Fixed action patterns

Fixed action patterns (FAPs) are relatively constant behavioral programs that are elicited by specific environmental stimuli. The concept and the term were developed by *ethologists* who wished to escape the fuzzy, broad, and emotion-laden term *instinct*. FAPs are specific, behavioral, and stereotyped responses triggered by a specific external stimulus or pattern of stimuli. Unlike the broad constellations of behavior covered by terms like "maternal instinct" or "migratory instinct," FAPs are units of behavior – discrete, relatively brief, highly operationalized, and spontaneous (for example, Nash, 1978, pp. 66–68). Although the classic examples of FAPs are behavioral sequences, such as the response of a hen to the distress call of her chicks, the courting behavior of the male stickleback fish, and the fledgling jackdaw's pursuit of its parent when it takes flight at a specific angle, the construct has been extended to human behavior.

Ethologists believe FAPs are (a) unlearned, (b) spontaneously triggered, (c) once triggered, independent of external control as in consummatory sexual behaviors, and (d) genetically stereotyped. However, they acknowledge that as one ascends the phylogenetic scale, learning and flexibility modify such instinctoid behaviors. The neat conceptual boxes of ethologists must give way to the blended and continuous characteristics of humans who can exercise voluntary control over even the most powerful of their impulses. The more evolved and complex the central nervous system, the more flexibility one can expect to find in what heretofore were called instincts.

These response patterns may be more appropriately called *innate response tendencies* or *modal action patterns* (Barlow, 1977); they can be more easily modified and shaped by education, culture, and training than can instincts (Lazarus, 1991). The *rooting* behavior of the infant at the breast is an example of the more primitive, instinctive expression of such behavior;[9] the *agonistic* gestures in which strangers engage when in hostile environments is an example of modified FAPs, that is, behavioral response tendencies. The concept of the *innate release mechanism* (IRM), a presumed neurological device that triggers approach or avoidance behaviors, is related to the FAP. There are specific sign stimuli, such as the "kewpie doll" features of newborns, which promote instinctivist parental responses. Eye gaze by those trying to establish eye contact, coupled with a smile, can be a powerful affiliative sign stimulus that others may find either attractive or aversive. And, of course, buttocks, thighs, calves, breasts, and well-formed torsos are well-recognized triggers for sexual emotions.

Instinctoid human behavior allows for considerable individual variability. However, Dewsbury (1978) shunned the term FAP because of the rigidity of this "concept." This apparently semantic quibble is analogous to the controversy that has fermented about the use of the term *stage* in developmental psychology. Because there is (a) considerable variation in

[9] Klaus, Kennell, and Klaus (1996) found that by using time-lapse photography infants could be seen several minutes after birth engaging in a process called *entrainment*. As the mother held her infant *en face*, about 8 to 10 inches from her face, gazed into its eyes and began to speak and move, the child moved in synchronous rhythm. The interaction became a dance of communicational gestures (see also Myers, 1984).

maturational patterns in humans, and as there is (b) evidence, if one looks closely enough, of gradual, incremental change, one can conclude that there are no stages in human development. Whether one finds such phenomena as *FAPs* or *stages* depends on how microscopic is one's examination. As Piaget would say, both are correct depending on the scale one uses. Damasio speaks of the detail that we find in certain phenomena when we look up close, say waves crashing on the rocks, while the sight of a body of water seen from a height of 35,000 feet gives the impression of a neatly configured and placid surface.

Psychologists have traditionally used the term *instinct* to describe stereotyped behavior that is consistently triggered by the same stimulus when it occurs in lower orders of vertebrates; they have preferred the term *emotion* when these responses occur in higher orders of animals. Neural structures have been demonstrated to mediate emotions in humans that exist in homologous forms in, say, reptilian vertebrates, which leads us to assume that there are similar processes at work. The nature of the processes that we call instincts in animals and those that we characterize as emotions in humans differs largely in flexibility and volitional control. We assume, however, that the processes differ in terms of the *subjective feelings* that follow on the whole-body response to harmful or pleasurable events or prospects. Whether these differences are simply dimensional or differences in kind is still a controverted matter.

James–Lange theory of emotions

William James drew on the work of a contemporary Danish physiologist, Carl Lange, to develop a theory which has become known as the *James–Lange theory of emotions* (Rubin and McNeil, 1987). In essence, it is captured in the following words of James ([1890] 1990):

> Common-sense says, we lose our fortune, are sorry and weep; we meet a bear, are frightened and run; we are insulted by a rival, are angry and strike. The hypothesis here to be defended says that this order of sequence is incorrect, that the one mental state is not immediately induced by the other, that the bodily manifestations must first be interposed between [the perception and the feeling] ... Without the bodily states following on the perception, the latter would be purely cognitive in form, pale, colorless, destitute of emotional warmth. We might then see the bear and judge it best to run, receive the insult and deem it right to strike, but we should not actually *feel* afraid or angry. (p. 743)

The emotions that James was referring to were the *emotional feelings* that we experience after corporal emotions have been aroused. When this is understood much of the confusion over his theory dissipates. This is an example of the mischief that the misuse of popular language causes in psychology. Unless it is clear what *to be afraid* means or, more fundamentally, what an emotion is, it will be difficult to discuss the scientific basis of these phenomena. When William James wrote, "we are afraid because we tremble," it is obvious that he understood fear in a different sense than that used by some other psychologists. He was, in fact, referring to the *felt* experience of the emotion – the "hot" cognition of fear. The controversy over the nature of emotion arose as much over the matter of what elements should be included in the construct, emotion, as in the character and temporal sequencing of those elements. What operations of the neurohormonal systems enabling emotions are implicated in generating a human emotional feeling rather than a robotic or sham reaction? For example, does the left hemispheric lateral cortex operate on the emotional stimulus before, with, or following feedback from the "visceral brain" and the rest of the body? Or is it not involved at all? If we refer to emotions exclusively in terms of the feelings that accompany an organismic response to an exciting stimulus field, we will endorse a different theory of "emotion" than someone who refers to the total organic response – cognitional and physiological – as an emotional response.

James' theory presupposes that each emotion has its own distinctive set of physiological responses, which can be recognized and then cause an appropriate feeling. James did not simply presuppose this; he attempted to demonstrate it, citing the tediously detailed descriptions of, say, grief, hatred, and fear: "rigidity of this muscle, relaxation of that, constriction of arteries here, dilatation there, breathing of this sort or that, pulse slowing or quickening, this gland secreting and that one dry, etc." [1890] 1990, p. 742). In a later passage, James spoke of the entire organism as a *sounding board*, where every change of consciousness, however slight, may reverberate. He wrote: "The various permutations and combinations of which these organic activities are susceptible make it abstractly possible that no shade of emotion, however slight, should be without a bodily reverberation as unique, when taken in its totality, as is the mental mood itself" (p. 743).

These emotion descriptions, however, apply only to the *average* person, as we all have idiosyncratic ways of expressing our emotions, whether they result from jokes or accounts of accidents or tales of altruism and heroism. Reflective of the unique personalities of individuals, the "internal shadings of emotional feelings … merge endlessly into each other" (p. 743). And the language of emotion only imperfectly reflects these subtle distinctions: "hatred, antipathy, animosity, dislike, aversion, malice, spite, vengefulness, abhorrence, etc." (p. 742). The emotional feelings we each experience flow from the capture of the system-wide, whole-body reverberations triggered by the perception of stimuli, whether internal or external to us. To the extent that we eliminate these bodily cues, to that extent do we diminish our ability to have appropriate feelings (Damasio, 1999, pp. 289–95).[10] In the preceding quotation from James' *Principles*, the operative words are "bodily manifestations must first be interposed." Interposed between what? Between the stimulus field and the meaning our life history has given to it, on the one hand, and the feeling that we experience, on the other hand, must come the physiological response. Among many other examples that James used is that of a mother imagining herself caressing her child, which releases a flood of responses culminating in a feeling of parental longing. (He obliquely refers to coarser, presumably erotic, examples that illustrate the same point.) It has remained for later neuropsychologists to explain the mechanisms linking perceptions to physiological responses.

Damasio (1999) has argued that the century-old attack on James' theory of emotion "is just not valid" (p. 288) because we now understand that the theater for the play of emotions is the entire body, including the brain. He concluded that William James' model of emotions is so plausible that one might be tempted to disregard his critics. But not to respond to them is a mistake as misconceptions embedded in such critiques have gained currency and escaped explicit refutation. The neurology that James had at his disposal caused his model to be incomplete rather than erroneous.

The Cannon–Bard theory of emotions

The first major assault on the James–Lange theory of emotions came with a publication of Walter B. Cannon (1927). Whereas James emphasized the role of physiological feedback in generating the sense of what the total organism is experiencing, Cannon and his laboratory associate, Philip Bard, initiated a powerful tradition that emphasized the cognitive variables that shaped emotion and its feeling dimension. Their experiments led them to believe that emotion originates and plays out entirely within the brain. Cannon's attack was initially based on the experiments of Charles Sherrington, an English physiologist who studied the behavioral sequelae of severing the spinal cord of dogs, and, in some instances, the *vagus nerve*

[10] In a segment entitled "Emotion and Feeling after Spinal Cord Transection," Damasio refers to studies done with patients with spinal cord damage. Because the spinal cord is a *partial* conduit for bodily feedback, emotional feeling is impaired when input to the brain is reduced. "The higher the placement of damage in the spinal cord, the more impaired feeling is" (p. 289).

that bypasses the spinal cord. Sherrington found that dogs could still make emotional responses following the transections of their spinal cord and vagus nerve. Cannon found similar results in his experiments with cats. The reason is quite simple: the displays of emotion that appear in the head of the animal, say, the growling and baring of teeth, result from nerves, afferent and efferent, that run directly between the brain stem and other parts of the head. The transection of the spinal column and the vagus nerve has no effect on certain facial muscles.

Cannon's position was also founded on the belief that the emotional responses that get played out in sub-cranial regions of the body are not differentiated according to the particular emotion that one feels. He only considered the physiological responses that are mediated by the autonomic nervous system (ANS). A visceral sub-system of the ANS, the sympathetic nervous system, generates such emotional responses, for example, as dry mouth, accelerated pulse, and sweating brow, armpits, feet, and palms. Cannon thought these to be a uniform, undifferentiated response pattern to a wide range of emotions. From this perspective, as LeDoux (1996) notes, one could not conclude that James was "right about why different emotions feel different, since all emotions, according to Cannon, have the same ANS signature" (p. 46). In addition, responses such as dry mouth and sweating are not instantaneous, and appear to trail feelings that surge on the cognitive register.

Rage displays, evoked by electrical stimulation to the appropriate brain site, seem like poorly "faked" demonstrations of rage, and are commonly referred to as *sham rage*. This conceptual problem encapsulates the challenge that every actor must overcome: how can I convey joy, sadness, grief, anger, and so forth, when the display of these emotions results from a conscious and deliberate decision to fake it for the cameras? More poignantly, it is a daily problem for legions of humans who realize that there are certain displays of emotion that social convention calls for, even when they are not deeply experienced and felt. The aspirants for an Emmy or an Oscar must display appropriate graciousness and smiles when the lone rival who gets the award parades triumphantly up on the stage. After all, everyone is on view, and the cameras and probing eyes are recording their personal reactions to disappointment. The rejected suitor for the partner that a friend has won over must salve his wounds in private and not be seen to harbor rancor, bitterness, jealousy, or pettiness. Regardless of one's feelings, society demands that one fake magnanimity.

Generating the feeling of an emotion we have not experienced is even more difficult. James J. Gross (1999) has reported the words of a college student who was asked to describe her effort to generate the feeling of being "in love":

> This past week I have tried to feel "romantically" about a friend of mine, who apparently is in love with me. I had previously only thought of this person as my friend and one of the nicest guys I have ever met, but never sexually. Well he confessed his true emotions to me, since I like him so much as a friend, I wanted to feel the same way about him; to make him happy. So I tried and tried and tried to think of him that way, but I just couldn't. I tried to think of how sweet he was and what a great relationship it would be, and how good he would be to me, but it just didn't 'turn me on.' I would try to view him in a sexual light and find things that would appeal to me sexually, but these things were just the components of my good friend, and not of a future partner. (p. 542)

Emotional feelings are not phenomena that one can will or directly control.

Cognitive labeling theory

Schachter and Singer's contribution

The evolution of this debate took a distinct turn with a remarkable (and perhaps no longer ethically or easily replicable) experiment of Schachter and Singer (1962). Like Cannon and Bard, these social psychologists assumed that the patterns of visceral responses to emotional stimuli were all similar. Like William James, they theorized that the physiological response to emotional stimuli precedes the subjective experience that we call feelings. But they gave a distinctly social psychological spin to the issue by postulating that the subjective feeling that one experienced resulted from the meaning that one attributed to the arousal. That meaning is a product of our socially constructed expectations arising from our personal history and the social context in which one finds oneself.

To test their theory Schachter and Singer devised an experiment in which no meaning could reasonably be inferred from the events that preceded the rush of physiological arousal. They arranged for the participants in the study to be injected with epinephrine[11] that induced a high level of arousal. Some of the students were informed that they were receiving this stimulant; the others were not. Each of those who

[11] The term *epinephrine* is a derivative of two Greek words, *epi*, which means upon or over and *nephron*, which means kidney. The adrenal glands, which produce epinephrine (otherwise known as adrenalin), rest on top of the kidneys. The student may recognize in the term adrenal glands the Latin-root term, *renal*, an alternative expression for kidney (as in renal failure).

were not so informed was unknowingly paired with a confederate who was "planted" by the researchers to create either (a) an angry, or (b) a "high-jinks" exhilarated emotional social context for the partner. The data purportedly supported the theory that it is our understanding of what is socially happening about us (and is expected of us) that leads us to interpret unthinkingly and consistently the *proprioceptive* cues that our body is feeding back to the brain. Those subjects who had been paired with an "angry" confederate felt angry; those with a "loopy" and wild confederate felt elated; and those who had been simply informed that they had received a stimulant, attributed their internal state to the only plausible cause, the injection.

Although the results of this study have not been replicated, it was a widely cited and influential study and propelled emotion into scientific visibility that had not been evident in previous decades. The new thrust was enhanced by a book that long influenced researchers in this field, *Emotion and Personality*, by Magda Arnold (1960). One of her core principles is now widely accepted in twenty-first-century emotion psychology: *before people can react to an emotional stimulus, they must appraise it as being possibly harmful or beneficial*. In the often used example of meeting a bear at one's camp site, before one takes flight from a bear there needs to be an initial appraisal of how this animal can affect one's welfare. If there were not an antecedent appraisal, why would one not embrace the bear rather than run from it? It is when a bear is perceived as a threat that a fight or flight[12] response occurs. This response may be a simple felt action tendency, a priming for action, so to speak, or a fully fleshed out visceral response. This appraisal and response is only *then* experienced as a subjective feeling of fear. Thus, in Arnold's view, distinctions have to be made between initial appraisal (at an unconscious level), the whole-body response, and the more subjective (cognition-involved) feelings that ensue.

Thinking can never be completely distinguished from emotion. They are two modalities of knowing, but each is thoroughly enmeshed with the other. To think is to excite a feeling, in addition to generating an idea. Every idea is clothed with a feeling. Moreover, to suppress the expression of an emotion is quite distinct from obliterating it. Emotions happen whether one suppresses their expression or not. Although men are socialized to suppress or conceal emotions that may reveal weakness, these emotions still may surge, even rage, within them.

[12] Walter Cannon coined the term *fight or flight response* to describe the optional bodily responses to a perceived threat (Carlson, 1998, p. 549).

The need to feel content with one's self-image

Confabulation

Humans are often not fully aware of the reasons for doing what they have done. They may simply fabricate what they consider to be plausible explanations for their affectively toned behavior. Not surprisingly, when their behavior draws criticism (or when they expect that it will), these explanations often have the character of excuses. In all cases it has a justificatory bias, even when it is masochistic. Ellenberger (1970) described a clinical demonstration of post-hypnotic suggestion done by the celebrated nineteenth-century neurologist, Jean-Martin Charcot. During one presentation Charcot hypnotized a "hysteric" patient. While she was in trance he instructed her to perform several post-hypnotic tasks. One task was to open a window in the hall; another was to seat herself at a piano in the hall and to play a tune. Following the trance she had no recall of anything that had occurred during the hypnotic episode. While Charcot continued with his talk, his patient, in a wakened state, spontaneously got up and went over to a window and opened it. Charcot turned to ask her why she had done that, and she explained that the room seemed stuffy and needed fresh air. She fabricated an equally plausible reason for playing a melody on the piano, which she did without prompting. We have no evidence, of course, that the reasons she gave for her behavior were not among her true motives, independent of the instructions she had been given by the physician. Perhaps the hall was stuffy. We can make the plausible inference, however, that she engaged in these aberrant behaviors in part because she had been instructed to, although she could no longer recall those instructions.

Telling more than we can know

More rigorous demonstrations of this human penchant for rationalization appeared in Nisbett and Wilson's (1978) article entitled, "Telling More Than We Can Know." They concluded from their experiments that the participants in their studies were unaware of the stimuli that actually motivated their behavior, though that did not prevent these subjects from actually proffering explanations. For example, in one illuminating experiment the researchers placed several pairs of identical nylon stockings on a table and asked female participants to select the pair that they liked best. They made their choice and were then asked to give the reasons for it. They spoke of the various properties of the stockings that motivated them. Although it may be argued that there are no absolutely identical artefacts in this world, if only at a microscopic level, it is unlikely that the participants could have detected such differences – which in any event did not prevent them from imagining some. After all, appraisal of sheen, transparency, color, texture, and sheerness of nylon stockings, or even imagined resistance to runs, always involve considerable shopper subjectivity.

What we have learned is that we have little access to the mentation that goes on "behind the scenes" of conscious awareness, and we are not able to identify and articulate all the motives that inspire our acts and thoughts, even though we speak as if we did. This makes it problematical for others to trust what we put forward as grounds for our actions, as we ourselves cannot fully trust what we have proposed – sincere as we may be. Nisbett and Wilson suggested that we concoct explanations based on "implicit causal theories" of why we would want to do what we did. To a greater or lesser extent, these explanations may be correct. But as the source of our appraisals of a stimulus field begins in the headwaters of the unconscious and provides the impulse for us to entertain reasoned (or impulsive) courses of action, there is little if anything in our lives that is not saturated or tinged with emotions – those positively or negatively valenced judgments that provoke a whole-body response. On the other hand, some of the events that precipitate a response from us may be so vivid and unmistakable that neither we nor witnesses, nor even an audience remote in time and place, would be inclined to doubt the essentials of our account – unless they suspected that our intent was to conceal, embellish, or deceive.

A further variable that needs to be introduced here is individuals' *personae*. People generally prefer to give a favorable account of themselves rather than a denigrating one. There is almost always a self-serving aspect in any description of our behavior that we present to others. The temptation to gild our image is typically present, as is the persona *called for* and *elicited by* the group we are addressing. For this reason we make excuses or rationalizations when recounting actions that we are ashamed of and embellish those accounts of behaviors in which we take some pride. Of course, these dynamics exist on a continuum. Those approaching a saintly humility may try to be fully transparent and even self-deprecating. Those who feel insecure and vulnerable may confabulate and distort the accounts of their behavior, even unbeknownst to themselves. For this reason, research founded on the introspective accounts we give of our thinking processes is largely unreliable (Wilson and Nisbett, 1978).

Autobiographies and personality

The implications of Nisbett's research for personality assessment and psychotherapy are far-reaching. If we cannot trust the accuracy of the stories that individuals tell about themselves, how can we make use of their autobiographical stories? The most plausible answer is that when clients plumb the recesses of their personal histories, say in a therapy session, their account is not expected to have the qualities of a news documentary. Clinicians understand that they are listening to verbal "portraits" of what their clients may fear is the truth, as well as what they wish were the truth. Partial life histories that emerge as personal narratives in a clinic, a laboratory, or a late-night tryst in a romantic piano-bar reveal the meanings that the speaker wishes to convey *at that moment*. Different audiences at different times elicit different versions of the same events, which vary in the details included or omitted, the color and tone of language, and the self-image the speaker wishes to convey (McAdams, 1994, pp. 719–36). Akira Kurosawa captures the complexity of such interpretation in his classic 1950 film, *Rashomon*.

Support for this view stems from several streams of research. In one set of studies it was shown that individuals who had been initially exposed to novel stimuli later expressed a preference for these objects rather than those they have never seen before (Zajonc, 1980). What is even more confounding for rationalists is that this "*mere exposure effect*" operates when the exposure has been *subliminal* (Bornstein, 1992). The most befuddling aspect of this is that we can have preferences for things of which we have no knowledge. In fact, people can even become emotional about objects they think they've never seen before (LeDoux, 1996). LeDoux wrote: "The subliminal mere exposure effect has been confirmed by many different laboratories and the idea that preferences can be formed for stimuli that do not enter consciousness seem rock solid" (p. 54). Indeed, he makes a strong case for the thesis that cognition and emotion are two sides of the same coin. This scientific challenge to cognitive science's traditional view of thinking as rationality in action brought respectability to those who gave primacy to the emotion-drenched subconscious.

Zajonc is characterized by LeDoux (1996) as the "psychologist who came in from the cold" (p. 53). Pursuing that metaphor, Gordon Bower can be considered the psychologist who turned up the heat (for example, Bower, 1981; Clark and Isen, 1982; Singer and Salovey, 1988). Their work on memory and learning that are mood dependent was paradigm busting. As I have previously noted (Dumont, 1993), "There is evidence of a 'state-dependent learning effect' such that, for example, when one is in a 'bad mood' one is more prone to hear and encode material that is consistent with that mood than material that inspires optimism (Bower, 1981; Isen, 1984)." Other findings indicate that mood "selectively biases the recall of affectively toned material" (Gilligan and Bower, 1984, p. 547). It appears well established then that emotion seriously affects one's ability to make accurate social judgments as well as to recall the evidentiary data that will influence decisions relative to one's own life and, just as important, the lives of others. A solid body of research has demonstrated that "feeling good or feeling bad can have a major influence on social judgments, such as the way we perceive others, because of the inherently complex and ambiguous character of most social stimuli, and the necessarily selective, inferential and constructive nature of social perception" (Forgas, 1990, p. 442).

Constructivist psychologists such as Mahoney (1991, 1995), and *historiographers* generally, have maintained that histories of nations, ethnic groups, clubs, fraternities, and universities change over time. Indeed, constructivist personality psychologists go so far as to assert that the autobiographical self of each person gets slightly altered from day to day. Every retelling of a personal story reveals a new wrinkle, an altered shade of meaning, a somewhat different moral. In this perspective, *humans continually recreate the truth. Narrative therapy* has gained wide currency in the past generation precisely because it exploits this human penchant to change one's personal history. What the narrative therapist does is to help clients re-author their lives so as to minimize the damaging aspects of their self-image and enhance the constructive aspects of their personality (for example, Richert, 2002; White and Epston, 1990).

Changing the past: our plastic memories

Personality is shaped in part by numerous experiences that in the distant past precipitated a cascade of emotional responses and feelings that have now faded into the mists of our personal history. Although they are lost to conscious memory, they remain in our implicit memory stores. The incidental learning that occurred at the time of those experiences – the melody perhaps that we were listening to when we received some tragic news, the musty smell of the hallway leading to the apartment of a much loved grandmother, the sound of the alarm preceding the attack of enemy aircraft, the soft sounds of the bells announcing vespers or a marriage that drift from the village church near which we spent our early and blissful childhood – all seem to be coded in our unconscious. Often these cues have the same power and intensity they had in the past. Marcel Proust (1913) provided a memorable description of the way an experience as seemingly innocuous as tasting a cookie dipped in tea can trigger a flood of cascading memories. These memories form the basis for Proust's classic work, *Remembrance of Things Past*.[13]

Gantt (1966) discovered that decades following veterans' wartime experiences, they still resonated viscerally to the sound of a klaxon announcing a "call to stations," even though they "knew" themselves to be in a secure and benevolent environment. This supported the Pavlovian principle of *schizokinesis*, positing that various sub-systems of our physiology may work in conflict with each other. Moreover fear can be *autokinetically* incubated in an individual such that fearful cues take on increasingly ominous meaning even though nothing in the environment has reinforced that meaning. The fear responses the veterans had originally experienced in combat had not faded, even though they had never been subsequently reinforced. LeDoux (1996) suggests that emotional memories appear never to die (pp. 200–4), nor do conditioned fear responses ever seem to lose their strength; they survive tenaciously even when subjected to systematic extinction procedures. The waking of these ancient memories can evoke the same painful feelings that accompanied

[13] And soon, mechanically, dispirited after a dreary day with the prospect of a depressing morrow, I raised to my lips a spoonful of the tea in which I had soaked a morsel of the cake. No sooner had the warm liquid mixed with the crumbs touched my palate than a shudder ran through me and I stopped, intent upon the extraordinary thing that was happening to me. An exquisite pleasure had invaded my senses, something isolated, detached, with no suggestion of its origin. And at once the vicissitudes of life had become indifferent to me, its disasters innocuous, its brevity illusory – this new sensation having had on me the effect which love has of filling me with a precious essence; or rather this essence was not in me it *was* me. I had ceased now to feel mediocre, contingent, mortal. Whence could it have come to me, this all-powerful joy? I sensed that it was connected with the taste of the tea and the cake ...

It is plain that the truth I am seeking lies not in the cup but in myself. The drink has called it into being, but does not know it, and can only repeat indefinitely, with a progressive diminution of strength, the same message which I cannot interpret, though I hope at least to be able to call it forth again and to find it there presently, intact and at my disposal, for my final enlightenment. I put down the cup and examine my own mind. It alone can discover the truth. But how: What an abyss of uncertainty, whenever the mind feels overtaken by itself; when it, the seeker, is at the same time the dark region through which it must go seeking and where all its equipment will avail it nothing. Seek? More than that: create. It is face to face with something which does not yet exist, to which it alone can give reality and substance, which it alone can bring into the light of day.

And I begin to ask myself what it could have been, this unremembered state which brought with it no logical proof, but the indisputable evidence, of its felicity, its reality, and in whose presence other states of consciousness melted and vanished. I decide to attempt to make it reappear ...

And suddenly the memory revealed itself. The taste was that of the little piece of madeleine which on Sunday mornings at Combray (because on those mornings I did not go out before mass), when I went to say good morning to her in her bedroom, my aunt Léonie used to give me, dipping it first in her own cup of tea ... (Proust, *Remembrance of Things Past*, 1913, pp. 48–49)

the original experiences and, indeed, may even exacerbate them. LeDoux expanded on this principle when he noted that people often feel uneasy, troubled, or tense, but do not know why. Every effect, of course, "needs" a cause. Although our explicit memory cannot reveal the trigger for those emotions, our implicit memory of some cue that has just reappeared (and that we perhaps have not noticed) must certainly account for it. From an evolutionary point of view, this legacy from our past retains obvious advantages. Though our conscious memory loses track of the details of traumatic events, our unconscious memory appears to retain indelible traces of many telltale aspects of such experiences. What we call our intuition is quite plausibly this "wisdom" garnered from our past that whispers "better safe than sorry." These "gut feelings" are nature's best guesses as to who or what can harm or help us.

Lie detection and the control of emotions

The expression of emotions cannot be totally controlled. Emotions arise at an unconscious level, and we only gradually become aware of them. Even once we are aware of them, our voluntary nervous system can only imperfectly modulate their expression, as it is the entire organism that is reacting to a stimulus field *which itself is imperfectly comprehended.*[14] The imperfect control of the expression of emotions is the assumption that underpins the use of the polygraph ("lie detector") technique and other psychophysiological detection methods. The reactivity of an organism's autonomic nervous system to the onset of affectively charged stimuli is sudden, involuntary, and visceral.[15] The limitations of this approach to lie detection and to emotion assessment are technical rather than theoretical in the view of those who utilize it. If our instrumentation were sufficiently sensitive to the visceral reactions that accompany and constitute an emotion, this approach to forensic investigation of the criminal personality would be justified and acceptable as evidence by our courts. Pavlidis, Eberhardt, and Levine (2002), in an article entitled "Seeing Through the Face of Deception," described a high-definition thermal-imaging technique for recording facial temperatures varying with fluctuations in

[14] Goleman (1995) observed that Erasmus lost confidence in the robustness of reason in the governance of human behavior. "How far can reason prevail against [emotion]? Reason does the only thing she can and shouts herself hoarse, repeating formulas of virtue ... until at last she's exhausted, gives up and surrenders" (p. 9).

[15] The most obvious way to sabotage the lie detection process is for individuals to preemptively mask their reactivity by subjectively, vividly, and randomly introducing incriminating, emotion-drenched images into their thoughts. The correlation between emotional reactions and any category of incriminating cues then becomes highly imperfect.

subjects' emotions, suggesting not only that reliability is improving but that emotion saturates our entire body from our epidermis to the covert interstices of our central and autonomic nervous system.

There are still other problems with technologies such as polygraph tests, no matter how accurate they may be. The most important is that emotional arousal can be triggered by many elements of one's environment other than the intention to deceive. Specifically, how can one theoretically link physiological arousal to the intent to deceive, when there are innumerable and irrelevant features of the stimulus statements and interrogation setting that could trigger a psychophysiological response? It is difficult to link the intensity and type of reaction to a specific emotion. Further, more than one emotion can be catalyzed by a statement independent of one's intention to tell the truth, deceive, or otherwise mask our real convictions.

Emotional and social intelligence

Intelligence and the unconscious

No one denies that emotions have a role in an organism's adaptive responses to life challenges. Whether they should be construed, however, as constituent components of a larger construct, intelligence,[16] is a matter of theoretical convenience rather than ontological reality (Caprara and Cervone, 2000, pp. 90–98). In the last decade, a flood of studies have purported to demonstrate the existence of emotional intelligence (for example, Mayer, Salovey, and Caruso, 2000; Salovey *et al.*, 2008) and developed measures for it (for example, Bar-On, 1997). Contrarians may affirm that intellect and affect are conceptually distinct, but they do not deny that they interact or form part of a larger integral system. Those who argue that we must integrate emotions into our understanding of adaptiveness are also assuredly correct. Whether one wants to integrate emotional adaptiveness (*cf.* Izard, 1994) into a larger construct, emotional intelligence (EI) as many now do is a matter of personal preference.[17] The

[16] *Intelligence* is etymologically related to *intellect* and not to *emotion*. The Latin terms from which the former are derived, *inter*, meaning between and among, and *legere*, meaning to read, to choose, or to gather, are clearly related to cognition rather than emotion.

[17] There are at least two ways to proceed in creating constructs. One can start with a popular term, like intelligence, which has a rich cultural history and an abundance of surplus meanings and shoehorn still other meanings into its definition; the other method, which is less contentious, is to formulate a scientifically grounded construct with the precisely operationalized meanings one wishes it to have and give it a new, neutral label. The former method results in struggles for rival camps to retain ownership of a label; the latter results

theory of emotional intelligence (Salovey and Mayer, 1990, 1994) which Salovey *et al.* (2008, pp. 535–37) analyze into four distinct functions (perceiving emotions, using emotions, understanding emotions, and managing emotions) is a mega-construct bearing primarily on people's ability to monitor and regulate their interpersonal relations.[18] The level of EI determines the adaptive capability of individuals to engage in social problem solving, deal with human environmental challenges, and modulate the emotional impact that one's behavior will have on others. These dynamics have a boomerang effect, and systematically affect the emotional tone that envelops all participants in any group setting.

The principal benefit of emotional intelligence is that it helps us to get along with people. To promote others' positive feelings about us (and themselves) is a major part of succeeding in life, dependent as we all are on the collaborative support of our fellows. Self-regulation is an inherent aspect of this capability. Emotionally intelligent people understand the principle of *reciprocal affect*, that is, that their affective responses potentiate cues that can evoke similar emotions in others. This is captured in Salovey and Mayer's (1994) definition of EI: "the ability to monitor one's own and others' feelings and emotions, to discriminate among them, and to use this information to guide one's thinking and actions" (p. 312). This capability clearly connotes practical intelligence more than academic intelligence (*cf.* Sternberg, 1985, 1988). It allows us to engage in moral volition[19] and make solid practical judgments, as distinguished from engaging in the abstract thinking involved in complex problem-solving.

The concept of EI can be traced back to Thorndike's (1920) theory of social intelligence, and more recently to Howard Gardner's (1993) theory of multiple intelligences. Gardner has delineated seven kinds of intelligence: linguistic; spatial; logico-mathematical; musical; bodily-kinesthetic; interpersonal; and intrapersonal. These latter two include all the elements of the later-postulated EI. Interpersonal intelligence includes acquired skills for adaptive social relations, and sensitivity to the feelings of others and to the multiple cues that, in any culture, signal people's reactions to the behaviors of others. It helps people (a) to tune in to how others are

in reluctance to accept a novel term. (This points to the interesting notion that symbols may be more important than the meanings they assume for different people. Adopting new labels is more difficult than adopting new meanings.)

18 To attribute intelligence to emotion would appear to mean that intuitive and viscerally conditioned responses to specific cues are adaptive, that is, they advance the safety and well being of the individual. Of course, one can thus aver that dogs and reptiles have emotional intelligence, perhaps superior to many humans.

19 The popular aphorism, "The right thing to do is usually the smart thing to do" exemplifies the role of emotional intelligence in moral conduct.

feeling about how they have been treated, and (b) to appreciate others' long-term affective needs, and how these needs can be harmonized with one's own. This intelligence encompasses a multitude of adaptive schemas for dealing with a large range of environmental challenges. Those who can get along well with very different kinds of people have high levels of this acquired intelligence, and it is most evident in those who succeed in such person-related professions as psychotherapy, politics, personnel management, pastoral counseling, and marketing. Intrapersonal intelligence, on the other hand, refers to sensitivity to what is happening within one's skin, that interior universe of mood, feeling, emotion, and thought that reflects whole-body experiencing. It further includes the utilization of that knowledge (and other resources) to promote one's personal and social goals.

Wisdom

Overlapping the constructs of social and emotional intelligence is that of *wisdom* (Sternberg, 1990), which involves an understanding of how emotions, attitudes, values, and human weakness influence a well-ordered social existence. It implies a "warm" as well as a "cold" cognition, the accumulation of schemas built on whole-body and social dimensions of daily life, through variegated and critical experiences, and the "savvy" counsel of experienced mentors (Baltes and Staudinger, 2000). Emotional and motivational dimensions of our pragmatic schemas for dealing with the vicissitudes and challenges of life are never absent. All of our plans and projects are saturated with emotion at some level of our interior life, and these emotions inform all of our pursuits.

The pragmatics of living the good life involves a coordination of knowledge, implicit causal schemas, moral convictions, and prudential judgments that involve, in the view of Baltes and Staudinger (2000), the following properties: "(a) strategies and goals involving the conduct and meaning of life; (b) limits of knowledge and uncertainties of the world; (c) excellence of judgment and advice; (d) knowledge with extraordinary scope, depth, and balance; (e) search for a perfect synergy of mind and character; and (f) balancing the good or well-being of oneself and that of others" (p. 132). The wisdom, however, of any one person is fragile if cut off from input from the rest of the community. That is one reason why professionals, faced with a moral dilemma, are well advised to consult with other respected professionals in their community. Courts routinely ask of defendants in professional tort cases, "what would the typical accredited professional in your larger community do if faced with the same circumstances in which you acted?" Taking counsel with respected and diverse colleagues when confronting a vexing and complex problem is always advisable, and all the more so if the decision will be heavy with consequences. The wisdom of the group provides the essential "distributed cognition" that cannot be found in any single individual. This is one of the reasons that appellate courts are constituted by more than one judge, and supreme courts by many.

Motivation and emotions

One cannot understand the drama of a chess match between two grand masters without first understanding their drive to win. The impact of a crushing move by an opponent has a far-reaching effect on either contestant's state of mind. The emotions that are generated flow from the thwarting or the achievement of a desire to win. When Big Blue, the IBM program that played Garry Kasparov for a "world championship" in 1997, overwhelmed its opponent in nineteen moves, it was unmoved, nor did it enjoy its victory. On the other hand, the searing emotional impact on Kasparov was evident. Although a computer can be made to simulate bare cognitional processes, the visceral, endocrinological, and other humoral components that are complexly enmeshed in logical processes cannot be replicated artificially, at least with the level of subtlety and fineness that eons of biological evolution have achieved. The rush of satisfaction in success and the exhilaration from the applause of millions was reserved, following that historic match, for those individuals who programmed Big Blue's victory. Without a thought to its future, or any self-indulgent reflections on its brilliance, it returned to a deep sleep in the hard drive that was its home. Big Blue was not a self-motivated player. This creature did precisely what its designers obliged it to do.[20]

One may wonder whether any organism (or any intelligent robot) can truly experience emotions without motivation. Each organism, from the unicellular to the human, drives itself to optimize those conditions that assure its well being. As Damasio (1999) reminds us, even the amoeba, "brainless and mindless" as it is, engages in activity that reveals "the form of an intention ... to keep the chemical profile of its internal milieu in balance while around it, in the environment external to it, all hell may be breaking loose" (p. 136). This dynamic can be as basic as maintaining a chemical balance in each cell of our own bodies or as complex as choosing

[20] The question of self-motivation raises the issue of free will. If one rejects Cartesian dualism, affirms that there is no effect without a cause, and postulates that every act of will is the effect of an organically constituted agent, one can be fairly asked whether the difference between a human and Big Blue is merely one of degree rather than of kind.

a marriage partner or getting a college education. The fundamental drive, of course, is to survive. "Emotions evolved as behavioral and physiological specializations" (LeDoux, 1996, p. 40) to serve that drive. On the other hand, "many theorists have agreed with [Charles] Darwin, [William] James, [Robert] Plutchik, and [Silvan] Tomkins that emotions have motivational functions that give them critical adaptive qualities" (Izard, 2001, p. 253). They not only support this instinct for self-preservation (and its corollary, self-replication), but as we move up the phylogenetic scale they can even enhance it in increasingly elaborate and refined ways.

Fear leads us singly and collectively to avoid or take (often complex) defensive measures against threats to our life and well being. Anger energizes us to counter malign and predatory individuals and groups that threaten us. Guilt motivates us to repair relationships that have been damaged by the violation of consensual codes of conduct. All these and other so-called negative emotions, though not intrinsically pleasurable, exercise highly positive, adaptive functions. By their very nature, they motivate. That they are more numerous than the pleasurable basic emotions, say joy (or happiness),[21] suggests that they have served an equally adaptive and positive evolutionary "purpose." In that light all of our emotions are positive. Only the failure to regulate and control them can render them *un*helpful and "negative" (Gross, 1999). Negative emotions and positive emotions are not apt expressions for experiences that can be difficult or easy, pleasurable or discomforting. Indeed, some people may find the experience, even the expression of so-called negative emotions like anger, a positive, therapeutic, and rewarding experience. The value system of the taxonomist is evident here (see Ekman and Davidson, 1994; Lazarus, 1991).

The existence of motivation and goals is essential to the generation of emotional states (this is presumably why many personality texts treat these two topics in the same chapter). If one has no interest in achieving a goal, one experiences little emotion when it is placed out of reach. Imagine the man who has lost all interest in living and seriously contemplates taking his life. When faced with an intruder who threatens to take it for him, this individual would experience less fear than would someone who is experiencing life joyfully and fully. Level of emotion is always proportional to the

[21] *Interest* and *surprise* or *startle* are often classified as "positive" emotions, but, of course, unpleasant surprises and interest in threatening events are not less common than pleasant surprises and hints of good fortune. Love is also an ambiguous term for an emotion whose valence, good or bad, is determined by the appropriateness of the love-object. (This evokes semantic issues, as many languages have distinct words to designate various kinds of love: filial love, erotic love, altruism, etc., each of which catalyzes different action tendencies and potentiates differential somatic involvement.)

value of the goals we are motivated to secure and maintain. If an environment projects no meaningful stimuli onto our internal "registers," we react with little emotion. When we say that someone is highly motivated, we are conflating two different variables. In the first instance, we are raising the issue of the value we place on the goal(s) he or she is pursuing. One normally places a higher value on preserving one's life than in securing a parking place. Second, how much higher the values placed on such idiosyncratic objectives are depends on the varied meanings they have for us.[22] In pursuing either of these goals, we can experience highly volatile emotions that propel us to extraordinary lengths to achieve them. This second source also varies for idiosyncratic reasons, some of which relate to our trait structure.

The history of human endeavors is a record of the goods and benefits that we have been motivated to preserve and secure. The anticipated joy of achieving these objectives has drawn us to dutiful, daily toil under harsh and boring conditions, as well as to extraordinary heights of creative labor, heroic endurance, and even martyrdom. Emotion appears to be part of the motivational package that energizes the pursuit of our goals, but it has not always been treated as such. Linnenbrink and Pintrich (2002) observed that "although historically affect has been an important aspect of motivational theories ... the role of affect has been largely ignored in current social cognitive theories of motivation" (p. 69). The history of arts and sciences, and of political, religious, and economic reformers, is replete with vivid portrayals blending motivation and emotion. Less celebrated, but in equally vivid hues, are the emotion and motivation that play in everyone's life and that shape their personalities.

Conclusion

Emotions, temperament, and personality, as Izard (2001) reminds us, are, at a biological level, interrelated systems that cannot be understood in isolation. Innate emotions, such as fear and anger that depend on limbic system circuitry are variously configured in an individual's temperament, each with a different threshold for activation. Personality traits, such as

[22] Personality traits orient individuals to niches and activities that evoke certain emotional responses they enjoy. The movie hall or theater is one such place. William James described the Russian noble lady who enters the theater to weep over the hardship of one of the characters in a play while her coachman sits outdoors in the bitter cold and suffers patiently until she leaves the theater to climb back into her carriage. Although we normally avoid situations that plunge us into painful experiences, we can also enjoy the experience of vicariously sharing someone else's pain. Aristotle addressed this conundrum in his *Poetics*.

optimism and pessimism, as Scheier and Carver (1993) affirm, influence the character and intensity of the emotions we experience, and the expression of emotion reflects personality. Clearly, the individual differences in the expression of emotions found in infra-human species, as well as in *Homo sapiens*, form a major aspect of personality. Indeed, emotional expression in toddlers predicts the personality traits that will clearly take shape in early school years (Abe and Izard, 1999). Emotion organizes the pathways of our thoughts and actions, influences our life projects in a large sense, and shapes our ambitions, fears, aspirations, hopes, and needs. It has to be central to an understanding of personality in general, and of each person in particular.

Emotion includes anything that excites a feeling, whether of surprise or disgust, curiosity or indifference, religious awe or disdain. Whether as background feelings or raw emotion, humans are continually experiencing their affective state. They respond in a whole-body mode to emotional stimuli. Each adaptive person is a tightly integrated system comprising many sub-systems. Motion in one reverberates throughout all the others. Although we make inferences about how others are feeling, principally from their facial expressions – their frowns and winces, eye rolls and laughter, groans and yawns – as well as subtly nuanced expressions of feeling, there is always much, invisible to the eye, that we can never know about one another. This is not surprising when one recognizes that there is much about *ourselves* that we will never know, buried as its sources are in the mists of our personal histories, much of it burnt off by the heat of our recent passions and struggles, and the normal decay of ancient memories.

Emotional life has long been denigrated in the West, largely because it does not have the same formal, abstractive, logical, aesthetic, and problem-solving qualities of rationality. Emotions have heretofore been regarded as visceral properties that humans share with animals, and that needed to be dominated by dint of strong volition. Reason was seen as an intellective faculty that relegates us more to the realm of pure spirits than to that of our inferior phyletic cousins. In any event, it was as rational spirits that we were traditionally thought, in the West, to enter the afterlife, not as emotional and embodied minds. This dualistic tradition reigned unchallenged from classical Greece to the Enlightenment – and in some quarters to our own times.

The science of emotion matured in the second half of the twentieth century and continues to be a flourishing area of investigation. Recent advances in neuropsychology have revealed that it is precisely the most highly evolved and abstract of our mental capabilities, our consciousness, which allows us to experience the full range of human emotional feelings that are the pride of our species. This legacy of our long human evolution is captured by Damasio (1994):

> Feelings let us catch a glimpse of the organism in full biological swing, a reflection of the mechanisms of life itself as they go about their business. Were it not for the possibility of sensing body states that are inherently ordained to be painful or pleasurable, there would be no suffering or bliss, no longing or mercy, no tragedy or glory in the human condition. (p. xv)

10 Taking the measure of the Human: benefits and inherent limitations of personality measures

Of all things the measure is man, of the things that are, how they are, and of things that are not, how they are not. Protagoras (490–420 BCE)

Antecedents of personality measurement

Humans have always needed to size each other up in view of selecting mates, protecting themselves, engaging in commerce, and furthering their collective and private interests. Such assessment activity never reached the scientifically formal levels that it did in the twentieth century. We know that the Chinese have used formal tests for over 2,000 years for selecting candidates for their civil service,[1] but there was something inherent to contemporary large-scale industrial societies, both Western and Eastern, that necessitated this explosion of interest in instrumentation for measuring humans. As modern society has become a *testing society*, it will be useful to examine this appetite for testing, which has spawned both a science and an industry.

Virtually no one in our post-industrial society will not have been measured countless times, in mind and body, in the course of his or her growth to maturity. Most of this assessment is done using formal tests. The pioneer of formal testing who is given the greatest credit for launching this social revolution in our modern era is Alfred Binet. He was commissioned in 1904 by the Ministry of Education in Paris to develop a methodology for *triaging* those children whose lack of success in the public schools of the time suggested that they might be aided by special education programs. In

[1] Although testing of public officials in China has been traced back to 2,200 BCE, little is known about it (Cohen *et al.*, 1988, p. 25). There are historical records of Chinese civil service exams developed and used in the Tang Dynasty (618–907 CE), available at: www.bcps.org/offices/lis/models/chinahist/dynasties.html, which continued until 1905 when they were abolished.

short, he was charged with developing means for measuring children's aptitude for learning independent of the quality of the education that they had heretofore received.[2] In collaboration with a team of researchers, most notably his student, Théodore Simon, he developed an instrument for measuring individuals' mental age, known today simply as the Binet Scale. Though he refused to consider his Standard Age Scale – later transformed by others into a deviation intelligence quotient, that is, IQ – as a measure of innate intelligence, he did recognize that it could be used for diagnostic purposes and for designing programs that remedied individuals' more tractable learning problems.[3]

This nascent technology, which has become widely known as psychometry,[4] meshed well with the needs of the United States, which in the early part of the twentieth century was faced with the task of integrating millions of immigrants into its work force, enlarging the scope of public education, slotting recruits to the armed forces into suitable roles, and providing the public and private sectors of the economy with a means for assessing the aptitude of candidates for various career paths. The popularity of the Binet–Simon instrument in America was largely the work of Lewis M. Terman, a professor at Stanford University, who transformed it from an instrument for selecting pupils who could benefit from special education programs into an all-purpose general intelligence test (Terman, 1916). In fact, the test purported to measure many attributes of intelligence (indeed, many intelligences; see Thurstone, 1938). Gardner (1993) observed that "in the United States especially, with its focus on quantitative markers and its cult of educational efficiency, there has been a virtual

[2] It is interesting to note that the US President's Commission on Excellence in Special Education (PCESE, 2002) has issued a policy statement discouraging the use of intelligence tests for diagnosing disabilities (and by implication for sorting children into special education programs). They quote Sharon Vaughn: "There is no compelling reason to continue to use IQ tests in the identification of learning disabilities. And, if we eliminated IQ tests from the identification of individuals with learning disabilities, we could shift our focus on to making sure that individuals are getting the services that they need and away from the energy that's going into eligibility determination." In this perspective it is unlikely Alfred Binet would have been asked to launch the revolution that this chapter describes. It is not clear from this report, moreover, how the use of *any* instrument that accurately identifies individuals with learning disabilities is inconsistent with ensuring that these individuals get the services they need.

[3] For a lucid but tendentious account of the origin of the Binet Scale, consult Stephen J. Gould (1981), pp. 146–58.

[4] *The Dictionary of Psychology* (Corsini, 1999) defines *psychometry* as "the science, fields, or process of measuring abilities and personality through psychological tests and statistics. Also known as psychometrics." As well, it comprises the science of *developing* scales and tests. The etymology of the term connotes the measurement of the psyche and its operations, but this science extends far beyond that narrower focus and encompasses all aspects of individual and group behavior.

mania for producing tests for every possible social purpose" (p. 163). The umbrella discipline for psychometry, assessment psychology, has become a highly sophisticated set of techniques, underpinned with theory that has grown and evolved with the professions it serves. As a result there exist today enormous databases of psychosocial measurement instruments[5] supporting the scientific, educational, and economic needs of post-industrial societies.

Relevance of testing to the social sciences

The prediction of behavior and the accurate symbolic representation, often numerical, of the constructs that have been thought to define human personality has motivated legions of personologists to create instruments for this purpose. Assessing personality has become an industry and has found a prominent place in such varied specializations as psychotherapy, forensic psychiatry, industrial/organizational psychology, and career psychology, among many others. Carson notes (2008) that "assessment for matching potential dating partners has emerged as one of the major growth areas for personality testing, with the advent of online testing."

Approaches to measuring humans are necessarily influenced by the test builders' conceptions of human nature. Advances in cognitive science, developmental psychology, neuroscience, emotion psychology, perceptual psychology, and psychopedagogy inevitably influence our assumptions about the constructs we purport to measure by our tests. Wide differences in one's conceptualizations of human nature will result in different principles of testing and the inferences we draw from the results of such testing. If *Man* is regarded as a complex machine, a mechanistic testing approach makes sense. If *Man* is regarded as a plastic, malleable, and soulful organism, testing assumes a clinical, artful, highly contextual, and individualized aspect. Interacting with the scientific paradigm in which tests are designed are the values of the dominant culture that permeate all aspects of test protocol and interpretation.

The measurement of people's behavior, aptitudes, and ideation is an intellectual minefield, as it engages issues that are intrinsically social and political. There is no psychological discipline whose products have more

[5] Among the most complete is the database *Health and Psychosocial Instruments* (maintained by Behavioral Measurement Database Services) that provides end-users with 104,000 distinct instruments, along with a panoply of ancillary services. The Buros Institute's Test Reviews can be found in *The Seventeenth Mental Measurements Yearbook*, widely regarded as the most authoritative source of English-language test reviews. Available at: www.unl.edu/buros.

profound consequences for the welfare of individuals than testing, used as it is to influence profoundly the course of their lives. Serious consequences are also evident for the social and ethnic groups to which those tested belong. For this reason, all deficiencies, both real and merely suspected, in the instruments that educators, psychologists, employers, and law enforcement agencies use generate allegations of socio-political abuse.

Practical and scientific objectives

Testing serves to further the economic, educational, artistic, and vocational guidance objectives of society. Thus, the design and use of testing instruments reflect the often rapidly changing needs and interests of the cultures in which they are used. Test results help to mesh the needs of society with the needs of numerous individuals. For this reason, population averages of scores that over time purportedly measure identical constructs change dramatically as succeeding cohorts of the same population are tested. New tests, besides measuring the achievements of examinees and specific skills they have differentially developed, reflect the character of the culture in which and for which they have been designed. Furthermore, tests serve as instruments of change. As legions of people prepare themselves to take tests that will in part determine their future, those very tests shape as well as reveal the nature of educational experiences that society programs for them.

When testing is intended to serve basic, scientific objectives rather than immediate pragmatic goals, assessment instruments are developed for administration to numerous grouped subjects in standardized environments and, optimally, in carefully timed trials. These tests purport to measure innate cognitive potential, temperament, fluid intelligence, and other human properties (Gustafsson and Undheim, 1996) that are less the product of culture than of nature.[6] They are qualities, argue evolutionary psychologists, that the human genome has disposed them to develop. They are dispositions writ into a *tabula* that is less *rasa*[7] than copiously inscribed by our evolutionary history. Scientists typed as environmentalists minimize the number of qualities and schemas that Nature determines in our collective *ontogeny*. Toward the other end of the spectrum are biological determinists who tend to maximize the predisposing and inchoate schemas and behavioral programs embedded in our nature from the

[6] For contemporary schemas of human cognitive abilities (beyond the purview of this chapter) and an introduction to the three-stratum model of John B. Carroll consult McGrew and Flanagan (1998).

[7] I have dealt with the concept of *tabula rasa* (the blank slate) in Chapter 1.

moment of conception. Though many disputants in these doctrinal conflicts state their position in strong language, most do so in moderate and relativist rather than absolutist terms.

Testing and values

Formal, mass testing reflects the values of nomothetically oriented scientists (Gardner, 1993) as well as the needs of large corporate and public programs. Of necessity the tests they design are intended for standardized administration. They are brief, nationally normed, largely machine-scored, and algorithmically interpreted. Gardner cited as vivid examples of this school of thought Arthur Jensen's ([1987] 2003) "reaction time" measures and Hans Eysenck's (1979) analysis of brain-wave patterns. On the other hand, individual clinicians, school psychologists, and psychotherapists, schooled as they may be in clinical neuropsychology (see Chapter 4 on biology and personality), relate to their clients very much as master artisans relate to their apprentices. Like artisans, these professionals establish a cordial but formal relationship with their "charges." They observe them carefully over a period of weeks if not months as they go about their tasks. They measure their baseline performance in various life skills, informally assess their deficiencies, lead them step by step in the acquisition of new skills, and encourage them to levels of self-confidence that accelerates their learning curve. Assessment is formative and continuous, idiographic and particularized, informal, nuanced, and (ideally) humanized.

Clinical observation

Henry Murray (as noted in Chapter 5 on traits) was one of the great pioneers of personality measurement as well as personality psychology (for which he coined the term, *personology*). His research bore more on the personalities of normal persons than of the personality disordered. He was interested in the consistency in people's behavior in natural settings and the manner in which they mobilized their resources to meet unusual challenges. He was focused on the applied aspects of this discipline and the ability of psychometricians to assess the current state of individuals' behavioral patterns and to predict their future behavior (Murray, 1958).

Murray's interests in personality theory brought him to the attention of the United States military during the Second World War. As a member of the Army Medical Corps he was charged with recruiting and screening candidates for military and paramilitary undercover missions that required a set of well-honed personality traits among which are emotional stability under stress, steely resolve, conscientiousness, and high levels of independence and personal resourcefulness (see Office of Strategic Services Assessment Staff, 1948, cited in Hall, Lindzey, and Campbell, 1998).

Personality assessment during the nineteenth and early twentieth centuries had fallen largely within the domain of dynamic psychiatry, and succumbed to its disciplinary penchant for modeling disordered and abnormal personalities. Murray, as well as Gordon Allport among many others, intended to change that. Hall, Lindzey, and Campbell (1998) observed that "[Murray] has been sharply critical of psychology for projecting a negative image of humans ... [He] stood firmly for a humanistic, optimistic psychology" (p. 226). His work gave an impulse to wellness models of the human that still serve us well.

Steeped by training and interest in psychoanalytic theory, Murray nevertheless emphasized the importance of studying "well" individuals as they went about their daily tasks *in the natural settings of their home culture*. His humanistic leanings, his interest in the arts and in the most creative and profound philosophical inventions of artists and thinkers[8] led him to shun not only the study of less evolved organisms (such as the Norwegian white rat) for purposes of extrapolating behavioral principles to human beings (we need to remember that he was a contemporary of B. F. Skinner whose *Behavior of Organisms* appeared in 1938) but also the study of individuals in laboratory settings. He preferred to rely on the sensitive and skilled clinical eye of the human observer. This, of course, put him at odds with Meehl's thesis that *actuarial approaches*[9] to measuring human traits are more reliable than any observational one (Meehl, 1954; Sines, 1970). He was convinced that the well-adjusted and trained observer is the most reliable instrument for appraising the personality of another person. *A fortiori*, when one has several capable observers, each assessing an individual from a different standpoint and in various sets of circumstances, the judgments made are all the more reliable. Consistent with this principle, Murray promoted the use of *diagnostic councils* in which the pooling of insights by highly qualified clinicians could be synthesized, and differences resolved. The case

[8] Henry Murray was one of the great commentators of his generation on the works of Herman Melville. His analysis of the themes and personalities in the novels *Moby Dick* and *Pierre* are of interest to this day to literary critics and psychologists alike.

[9] Actuarial methods rely on statistical data for making predictions of human behavior. Granting the relative reliability of such data Meehl (1954) postulated that diagnostic decisions predicated on this solid, quantified base are more reliable than decisions made by simple clinical observation (Meehl, 1973). This has implications for the usefulness of interviews for making judgments about individuals' suitability for a job or admission to a professional program.

conferencing that occurs in many clinical centers and hospitals is a contemporary expression of Murray's principles and the confidence he felt in the judgments of well-trained clinicians.

Despite Murray's penchant for studying "normal" humans, it is ironic that, among the many psychometric instruments he developed and inspired others to develop, the most frequently used is the *Thematic Apperception Test* (TAT) (Morgan and Murray, 1935, discussed below), a projective device that, for all its limitations, is extensively used by a large segment of the psychotherapeutic profession. Like his student, David McClelland, he proposed that we are often unaware of our true motives and that in order to gain access to them we need to approach this quarry in indirect ways. By definition, self-reports and self-conscious behavior are not a sound basis for making inferences about personality traits.

Gordon Allport, in contrast to Murray, was little influenced by the psychoanalytic principles that were regnant during the 1930s. His personal encounter with Freud, who seriously misread him during an interview by placing an illness template on his personal narrative about his tram ride to Freud's office, convinced him that "psychologists would do well to give full recognition to manifest motives before probing the unconscious" (Allport, 1967, p. 8). His deepest conviction was that, other things being equal, one received the most accurate representation of individuals' personality by asking them explicitly what they believed about themselves. Although all individuals may be tempted to dissemble and present themselves in a favorable light (especially in high-stakes interviews), their consistent responses to varied, real-life situations are a more reliable indicant of underlying personality structure than the clinician's or the researcher's construal of fantasies made in response to ambiguous stimuli presented in an artificial environment.

Allport theorized that by observing individuals in a variety of natural settings one could reliably detect the principal features of their personality. The expressive style in which individuals act and react to the exigencies of life was one such construct. Allport and Vernon (1933) wrote:

> It appears that a man's gesture and handwriting both reflect an essentially stable and constant individual style. His expressive activities seem not to be dissociated and unrelated to one another, but rather to be organized and well-patterned. Furthermore, the evidence indicates that there is a congruence between expressive movement and the attitudes, traits, values, and other dispositions of the "inner" personality. (p. 248)

Of course, it can be argued that the study of an individual's responses to unstructured and ambiguous life challenges is as much a projective device as the TAT. The psychologist reads the underlying dynamics of individuals' personality in their spontaneous, indeliberate responses to the *press of environment*. The intuition and clinical sense of psychologists are the grounds for judgments about the personality of those being observed. Such assessments and the use of projective devices generally, as attractive as they are theoretically, have not been substantiated to date by research.[10]

Test validity

Before examining the subject of personality measurement, it will be useful to briefly review the various principles that govern the validity of psychometric instruments. It is not possible in this chapter to detail the strategies for establishing validity, a very complex issue in itself, but I will try to alert the reader to the issues that need to be addressed. In 1999, a joint committee of three professional associations[11] published a revised edition of *Standards for Educational and Psychological Testing*. Their purpose has been to cut through the confusion that existed among professionals who used tests. Among their concerns were standards for establishing test validity. They concluded that the validity of an instrument is the *level of certainty* one can reasonably have about inferences derived from test results. Consistent with their findings, Kaplan and Saccuzzo (2005) stated there are several types of evidence to support such certainty, most importantly: (a) content-related; (b) criterion-related; and (c) construct-related. Although there are many sub-types of validity (say, *concurrent* and *predictive*, *empirical* and *statistical*, *convergent* and *discriminant*, *trait* and *factorial*), all can be subsumed under these three. Note that none of these validities is a dichotomous variable. They are all dimensional – which is not to deny that each can, at least theoretically, be decomposed into increasingly refined components.

Face validity

We often hear of the purported *face validity* of various instruments. To say that an instrument has face validity merely affirms that it "looks like" it measures what it is supposed to measure. The critical question is: to whom does such a test look valid? To the naive, it may just reflect the lore and village wisdom that infuses many a psychology program. Depending on our acculturation to various folkloric schemas a test item can look "self-evidently" relevant or absurd. But as the validity of an instrument depends

[10] Consult Gieser and Stein (1999) for a wide-ranging examination of the TAT and the strengths, deficits, and "future" of projective devices (*pace* Robyn Dawes).

[11] The three associations are: American Educational Research Association; American Psychological Association; and National Council on Measurement in Education.

on the quality of the *evidence* that is brought to support that claim, it is fair to say that face validity, superficial as it is, is not validity in any scientific sense. Nevertheless, when experts in the fields for which an instrument has relevance have judged its contents favorably, this contributes to the face validity that justly increases the confidence one can have in it.

Content-related validity

Test items, which form the content of many instruments, are developed to represent as accurately as possible the knowledge and those dispositions, schemas, skills, values, or attitudes that govern the behaviors and experiences of the person being tested. Although the ideal is for the test or questionnaire to generate results that are a mirror image of the psychological construct or constructs the test designer and (at least implicitly) the test users wish to tap, the stark reality is different. The marks that a person makes on an answer sheet, dependent as they are on that person's construal of the meaning of the words and sentences that are the stimulus items of a test, are many removes from the neural engrams in which conscious behavior originates. The transmission of data from one medium (neurological) to another (say, paper-and-pencil) admits *some* noise and distortion into the recorded signals. Moreover, the symbolic character of human speech is highly abstract and rarefied, even when its referents are highly concrete. Words can never fully capture the reality being measured, or the flawed representation of the reality, even when it has been clearly conceptualized by both test designer and test taker.[12]

Content analysis of an instrument is easiest when we are trying to measure an examinee's mastery of certain facts, a body of knowledge, a set of skills. The greater the clarity in the definition of the constructs we are trying to measure, the easier it is to establish the validity of an instrument. But when we are trying to determine, say, the level of depression in

[12] In my review of C. Robert Cloninger's book, *Feeling Good: The Science of Well-Being*, Dumont (2005) I wrote that: "The problem with paper-and-pencil tests and other lexical media that Cloninger uses is that there are several levels of abstraction that both precede and follow the causal pathway that Bertrand Russell [has] described. Thoughts must take shape in psychometricians' minds, themselves products of their personal history and psychobiological influences. Those thoughts must be formulated in terms of words of a certain language, symbols that are variously concatenated and construed by the test-maker into queries, and then accessed by those individuals more or less familiar with that language who will look at them and 'mentally' see them. The meanings that the test-taker projects into those complex sets of stimuli are only roughly similar to those that psychometricians had in mind as they sought apt ways to express them" (p. 5). The epistemological challenges presented by instruments that purport to measure ideation at such rarefied levels are staggering.

an individual who suffers from it, the clarity of our definition of this construct, about which there is still controversy relative to its constituents and etiology, determines the ease with which we can establish content validity of the instrument that measures it. What complicates matters is that each instrument measures different things depending on the various shades of meaning examinees attribute to words and phrases, as well as the conditions under which they took the test. Furthermore, many idiosyncratic variables (for example, motivation, sleep deprivation, anxiety, discouragement, and blood sugar levels) influence examinees as they engage in this complex task and obscure in part the meaning of final scores.

Criterion-related validity

The psychometrician's dream is to have a well-anchored criterion with which he or she can compare the ratings that a new instrument generates. Medical specialists, for example, aspire to produce tests that will predict the probability that one will (or will not) contract a particular disease. There are well-established criteria in the medical sciences that serve this purpose. The incidence of strokes, prostate cancer, or heart attacks is crisply defined conditions that have been clearly linked to certain risk factors. Tests to measure those risk factors, say, the levels of blood pressure, prostate specific antigen, or high-density lipoproteins, have been developed – and their validity established. In other words, the *predictive validity* of such tests can be tested (and corrective procedures undertaken).

We often develop tests to be "stand-ins" for other measures that we cannot take (Kaplan and Saccuzzo, 1997, p. 134). As we obviously cannot measure people's future performance until they have performed, the best approximation to that is to take measures right now that are highly (though never perfectly) correlated with future performance. Future performance is one *criterion* for establishing the validity of a test. The clearer and better established the criterion is to which test designers wish to relate their measures, the better grounds one has for establishing the level of validity of such measures. How well our instrument can predict that performance is the measure of its predictive validity. For example, *Scholastic Aptitude Tests* (SATs) are regarded as having predictive validity if they (usually in combination with other variables[13]) indicate with reasonable level of probability how a high school student will perform in college. Similarly, measures of personality variables such as

[13] Obviously, if anyone takes a day-long test having slept fitfully the night before and is suffering from a splitting headache, SAT results will be negatively affected. Innumerable other random events can also compromise the objectives of the testing.

"conscientiousness" have predictive validity if highly correlated to future job performance. We reason that the more conscientious people are, the more committed they will be to doing a good job, whether it be studying, educating children, or working on an assembly line. The strength of the relationship of a test result with performance occurring *now* rather than in the future is called, quite plausibly, *concurrent validity*. Obviously, such tests are not stand-ins, but, typically, instruments that help explain, and cast more light on, what is happening elsewhere.

Although we prize spontaneity and free will, we are comforted by the deterministic assumptions underpinning human behavior. After all, predictive validity is predicated on that assumption. The social machinery of a harmonious community depends on our ability to rely on people acting according to "type." Unpredictability is generally considered to be a negative personological feature. When we hear people noting of someone, "this is so unlike him" or "this is so out of character for her" we do not initially assume that something good has been said about that person. Unpredictability and spontaneity are valued in trivial and inconsequential activities, such as parties and other leisure events. People do not value erratic behavior on a high-speed highway or in the commercial transactions of an industrial society. Reliability in performance, stability of moods, consistency in behavior, and hewing to contractual commitments in commerce and rules of engagement in warfare are valued in all cultures. Carefree as well as deceptive deviation from them is punished. Although we praise "free spirits" and encourage spontaneity, this can often be sand in the gears of finely calibrated social mechanisms.

Construct validity

Personologists and professionals who work with people in correctional institutions, mental health centers, and rehabilitation clinics need to assess the personality traits of the individuals they treat. But as we have seen, personality traits are constructs, that is, products of creative imagination that have certain inherent plausibility. Soon after the Second World War, it became evident to psychometrists that they needed a procedure to determine to what extent the tests they were designing were valid – and no aspect of test validity was judged to be of greater importance than its construct validity. Indeed, some theorists came to regard all psychometric validation as a component of construct validation (for example, Cronbach, 1980; *cf.* Messick, 1989). A precursor to that principle was the design of a multitrait–multimethod matrix (MTMM) procedure by D. T. Campbell and D. W. Fiske (1959) – essentially the conflation of the convergent and discriminant validities of the same instrument. MTMM is now considered by some as a superannuated approach to determining the construct validity of an instrument (see, for example, Borsboom, Mellenbergh, and van Heerden, 2004).

Defining meaningful constructs and validating instruments assessing their behavioral referents is a difficult process, but it is of the greatest importance in personality psychology. As the constructs with which personality psychology is preoccupied (for example, anxiety, intelligence, creativeness, openness, spontaneity) are intrinsically fuzzy, scientists have had difficulty operationalizing their definitions and establishing unambiguous criteria for validating measures of them. The task has been further complicated in that many constructs are laden with meanings derived from incompatible personality theories within which they have been defined. Depending on the theoretical orientation of the designers of a test, the results of that test may have meaning only in terms of the theory that framed its construction. For example, researchers who are psychodynamically oriented can simultaneously define a construct, say, sexual repression or need-achievement, in a theory-specific way, and develop an instrument to measure it. Researchers in a self-actualization tradition will define these constructs differently and consequently develop a different instrument to measure them. Clearly, the interpretation of the results of such tests will reflect the underlying personality theory that inspired them. McAdams (1994) illustrated this principle with the example of developing a test of "friendliness" (as he understands that construct). He theorized that people who would score high on "friendliness" smile more easily than those who score low on that construct. He then suggested that a measure of friendliness can be validated by establishing that high scorers will smile more in a strange laboratory setting than low scorers (p. 255).

When a network of investigations demonstrates the coherence and internal consistency of a test, and experts in the domain agree that it does, indeed, reflect the construct that it was supposed to measure, one can begin to claim that the test has construct validity. Validity claims are purportedly buttressed when the test correlates to a greater or lesser degree with other tests that are purported to measure similar constructs – a process of cross-validation that Meehl (1978) subjected to a withering analysis. Still other validational methods are experimental. For example, Aiken (2000) highlighted the correspondence between findings on the *Taylor Manifest Anxiety Scale* (TMAS) (Taylor, 1953) and predictions arising from Hullian learning theory. Hullian principles affirm that anxiety is a drive and, as such, those people who are most anxious should be most easily conditioned when threatening stimuli impinge on them. Aiken makes the point that the acceptance of the construct validity of the TMAS was facilitated when those who scored high on the TMAS acquired a conditioned eye-blink response in an air-puff–eye-blink experiment more quickly than those who scored low (p. 96). The very nature of personality measures requires this construct approach to test construction.

Construct validity is less the property of a test than it is the character of the inferences we can make about how examinees will behave based on their test performance (Ozer, 1999; *cf.* Borsboom, Mellenbergh, and van Heerden, 2004). One alleged reason for this is that every bona fide test certainly measures *something*, even though we may not be sure precisely what it measures. (There are, in fact, no effects without causes – and a constellation of causes certainly exists that can explain the results of a test.) But given the nomological system in which the construct is embedded a test score may represent a latent personological variable which, in fact, is only a remote variant of what the psychometrician construes it to be. Be that as it may, construct validity is the *purported* relevance of the test or inventories psychometrists have developed to the personality traits or attributes being assessed. Further, it depends on the presumed correspondence of those measured traits with the future behavior of the examinees in a representative assortment of real-life situations. There is a fragile skein of inferences leading from one's performance on a personality test to conclusions about one's performance in a distant situation.

A divergent approach to establishing construct validity has been proposed by Borsboom, Mellenbergh, and van Heerden (2004). They contend that one must demonstrate that the presumed ontological variables (say, "g" or "conscientiousness") that purportedly *cause* individual test differences *really do exist* and adequately account for those differences. That is a tall order. Following the causal arrow forward (from the referent to the test score) can be more difficult than following it backward (from the test results to the personological attribute). We know that the latent causes we "measure" exist but cannot be certain how exactly to define or where to situate them. On the other hand, good psychological detective work involves studying the evidence, often experimentally produced, and reasoning back to the identity of the agents, and to approximations of the nature of the causes. We can do this as all effects participate in the nature of their causes.[14] Increasingly informative neuro-imaging technologies promise to render this argument moot as, say, test results, mental states, attitudes, and patterns of behavior become linked to synaptic networks, "cell assemblies," or other neurological referents.

[14] A similar procedure exists in the physical sciences. For example, Einstein and other astrophysicists were long mystified by the anomalous precession of the perihelion of planet Mercury. They speculated as to the "latent variable" that might be the cause of this "individual difference." Though the matter is still in dispute, Einstein's "armchair" research methodology has been found acceptable.

Test validity coefficients

A validity coefficient measures how two variables vary in relation to each other. The more closely a variable moves in tandem with another, the closer the coefficient approaches + or −1.0 (see footnote 10). For example, age is typically highly correlated in children with measures of reasoning skills; years of schooling or training programs with performance on achievement tests; physical height with making the cut for selection to NBA basketball teams. The closer a coefficient of correlation between two variables approaches + or −1.0 the more confidence we can have that knowing the value of one allows us to realistically estimate the value of the other, in the latter's absence. However, as there are innumerable variables that contribute to any human behavior, and no instrument can measure all those relevant variables, every measure is somewhat deficient for predictive purposes. For that reason, validity coefficients in the social sciences rarely exceed 0.60.

Scientists are generally pleased to find personality measures with validity coefficients in the 0.30 to 0.40 range. SAT scores, for example, are not highly correlated with later performance in one college or another, but they do give us a probabilistic assessment of future success that is greater than if we flipped a coin. Likewise how individuals score on the *Sensation Seeking Scale* (SSS) gives us *some* grounds to judge how willing they may be to ride a world-class roller coaster compared with the general population's baseline for taking scary rides. Thus, SAT scores or SSS scores need to be supplemented with other gauges of success. Gathering other indicants of future behavior can increase the accuracy of our judgments. What each additional instrument adds to the accuracy of our judgments is characterized as *incremental validity*. Anomalously, some personality measures are so poor that when their results are added to better information they actually reduce the validity of earlier judgments (for example, Kaplan and Saccuzzo, 1997; Lilienfeld, Wood, and Garb, 2000; Meehl, 1978).

Test reliability coefficients[15]

Test reliability "refers to the accuracy, dependability, consistency, or repeatability of test results" (Kaplan and Saccuzzo, 1997, p. 12). The

[15] Similar to the validity of a test its reliability is numerically computed and expressed as a coefficient by using a set of test scores for that test. Coefficients are always a positive decimal number that range from 0.00 to 1.00. Coefficients above 0.80 indicate that one can have relative confidence in the accuracy of the test. See Chapter 5 on traits for my treatment of this statistical measure.

two principal approaches to assessing reliability involve: (a) comparing individuals' scores on the same test taken at different times (the interval between tests should not be so long that one could reasonably fear that examinees might have significantly changed between tests, nor so short that they could well remember items from the first testing that could bias the second measure); and (b) comparing individuals' scores on two randomly divided halves of the same test. The first approach is called measuring *test–retest reliability*, namely consistency over time. The second is called measuring *split-half reliability*, namely internal consistency. These derive from classical test theory. *Item Response Theory* (IRT) and a Raschian variant of it employ different mathematical models to gain greater precision in assessing latent traits (for example, Fischer and Molenaar, 1995).

Homogeneity of test items improves the reliability of a test. Should a test be embedded with items measuring variables irrelevant to the construct at issue, the probability increases that the reliability of the test will be reduced (as well as its validity) – and the more so as the number of irrelevant items is increased. Poor construct reliability means that the test is tapping into ill-defined variables, which in turn can reduce the consistency of results. If an instrument is not reliable, one can only draw tentative conclusions about its validity. It is not possible to determine what a test measures, as intuitively certain as we may be, if it yields scattershot results when one uses it.

Response sets

Social desirability

The bane of all psychometrists is the penchant of test takers to respond to items on the basis of *response sets*, that is, ingrained tendencies to respond systematically to items in ways that have less to do with what they believe about themselves than with what they wish were true – and want others to believe about them. One such set is *social desirability*, which inclines test takers to "fake good." This is an overarching need that drapes a shroud of respectability over the actual needs the inventory is calculated to reveal. The results of personality inventories can thus become more a measure of test-takers' perceptions of prevalent societal values than of their own individual needs, values, attitudes, and schemas. This is particularly significant when the inventory is being used for *selection* purposes. Subjects are prone, at an unconscious if not conscious level, to display personality features that will gain them approval rather than rejection. Interestingly, Eysenck and Eysenck (1968) observed that when college students were instructed to "make a good impression" on the *Eysenck Personality Inventory* their scores on the extraversion scale were not significantly altered, but the measures on the neuroticism scale were significantly depressed. One can only conclude that college students have a greater investment in not appearing neurotic than in being judged extraverted, even when little is at stake.

Acquiescence and oppositional sets

An acquiescence set is the tendency to answer *yes* or *true* to items, independent of their content. Of course, like all personological traits, acquiescence is a continuous variable. One may have minimal or extremely high levels of it. The polar opposite of this set would be an *oppositional* set evidenced in high levels of "nay saying," possibly enmeshed with a penchant for nonconformist and defiant attitudes. One obvious, though partial, remedy for this, used in recent instruments, is to word the items in such a way that the number of items positively keyed false is equal to the number keyed true. Other response sets such as preferring "*cannot answer*" or "*?*" responses (that is, noncommittal responses) have been obviated in recent inventories by eliminating response options that have heretofore permitted the bias to be expressed. The noncommittal set is also seen in rating scales where respondents will disproportionately choose a number midway between two extremes (for example, 3 in a five-point scale or 4 in a seven-point scale).

An interesting aspect of response sets is that, like Matryoshka dolls – Russian dolls, as they are commonly known[16] – they represent personality characteristics that an inventory may not have been primarily constructed to measure. These traits nevertheless influence results and, to a greater or lesser extent, diminish the accuracy of the instrument and its usefulness (for example, Wiggins, 1962). The so-called "validity scales" embedded in inventories, whose purpose is to measure invalidating response sets, may also be subject to ever more deeply embedded personality styles – and response sets peculiar to them. The challenge of disabling test-takers' biases that distort measures of *other* traits that are of more immediate interest to the test administrator is one that psychometrists are still addressing (Piedmont *et al.*, 2000).[17]

[16] Matryoshka dolls (carved, hollow dolls hidden within a series of increasingly larger dolls) is a useful metaphor for phenomena that conceal ever more subtle, inconspicuous "realities." The biggest "doll" is the metaphysical construct, "being."

[17] Readers who wish to explore recent developments in this field can usefully study the VRIN and TRIN scales of the MMPI-2.

How to measure personality

Observational methods

There are several ways to measure personality, including structured interviews, systematic observation of individuals in public and semi-public areas, self-report questionnaires, projective techniques, and objective tests (for example, Aiken, 2000; Kline, 1995). The simplest way to measure personality, and the most prone to error, is for one person simply to observe another, preferably in a naturalistic situation, and keep a record of that person's behavior of interest. One can later code one's notes and analyze them. Passive and *unobtrusive* observers, especially those who themselves remain unobserved, reduce the self-conscious reactivity of those they are studying and assessing. Needless to say, recording the behavior of people who are unaware that they are being observed raises ethical issues. Even should researchers wish to gather data on individuals or groups as they pursue their personal interests in the street, the bazaar, or the mall, they need to have thought for those subjects' dignity, and their right to anonymity and to peaceful, undisturbed public concourse (see Kaplan and Saccuzzo, 1997, pp. 606–17).

Clinical and counseling psychologists, as well as school psychologists and psychiatrists, routinely engage in personality assessment as they work with their clients. Clinicians often engage their clients in unstructured conversation with a view to leaching out information about their character and problems – usually about disorders that need to be remedied. Client transference elicits idiosyncratic *countertransferential* responses from the therapist. On the other hand, therapists' *transference* of their own schemas and biases that are culture- and theory-based products of their own personal history affects the kinds of information the client will divulge. The net result of this complex, unstructured data gathering, based not only on the words spoken, but on the gait, dress, vocal intonations, posture, grooming and cleanliness, facial expressions, and client eye contact, is subject to clinical distortion. This occurs also among school psychologists as they go about their work interacting freely as *participant observers* in classrooms and other school activities. Their very presence in the child's milieu initially elicits atypical behavior. This is no less true of industrial and organizational psychologists who work in corporate domains studying, say, the critical incidents that reveal, and have an impact on, employee morale, efficiency, managerial professionalism, and team cohesion.

Interviewing is a complex interpersonal skill often used for purposes of selecting individuals into, say, academic programs or job positions. There are obvious validity and reliability problems with this method of assessment as interviewees' presentation of self is generally self-conscious, mannered, rehearsed, and atypical of their more casual behavior. Further, there are great individual differences in the way interviewers proceed in accessing information, accentuated by the often informal and non-standardized way that they record and process the information. One can only conclude that interviewing is among the *least* reliable of the methods used for making judgments of people's personality.

Personality inventories

There are serious validity issues arising in inventory construction. Self-report questionnaires are among the most popular means for assessing personality. They consist of a list of items to which one reacts one at a time, indicating to what extent each item corresponds to one's values, feelings, or typical behavior. Tens of thousands of these inventories have been constructed, all succeeding to a lesser or greater extent in measuring what the test constructor had in mind to measure. These instruments are designed to assess traits and attitudes that are relatively stable over time and across situations. When such a personality test allows us to draw useful inferences about test-takers' behavior in a variety of real-life situations, we consider the test to have construct validity. Ideally, high construct validity should have high predictive validity, for that is precisely what the test has been constructed to do – to presage non-test performance. Of course, the better it does this, the greater validity it has.

A number of problems, however, compromise the usefulness of construct-valid instruments. In the first place, we can never be certain what a specific hypothetical construct means, as it is always in evolution. The construct *intelligence* is a good example of this. Not only is there ongoing controversy about the constituent elements of intelligence, but every test of intelligence can be criticized for *not* measuring the elements that many notable scholars consider essential to its definition. Whenever a society has difficulty describing what is signified by the semantic label for a construct (say, love, or curiosity, or honesty) developing a measure for it is challenging. Kaplan and Saccuzzo (1997) state that establishing the construct validity of a trait requires that "a researcher simultaneously define some construct and develop the instrumentation to measure it. This process is required when 'no criterion or universe of content is accepted as entirely adequate to define the quality to be measured' (Cronbach and Meehl, 1955, p. 282)" (p. 144). In the absence of consensual definitions of the many psychological variables that personologists wish to assess, establishing criteria by which to assess the validity of a test for each is not feasible. This means, as we noted above, that we have been condemned to the limitations inherent in a *construct approach* to test construction. These limitations have been brilliantly demonstrated by Meehl (1978).

Personality inventories are invaluable to researchers in trait psychology who typically work with large groups of individuals. Using instruments with high construct validity to conduct large-population research of an epidemiological, basic-science character can lead to many useful findings. However, just because an instrument has high construct validity does not mean it has high predictive validity. This is of particular concern to clinicians and others who work with individuals and use personality instruments for diagnostic purposes. To validate an instrument, researchers often *begin with a known group* of, say, children with ADHD and contrast their performance on their test with another group known to be free of the disorder. Such an instrument can serve large-scale studies where one wishes to determine the incidence of ADHD or some other disorder in a population where baseline occurrence is small. But when individual clinicians use this test to screen for low baseline phenomena among their clientele the probabilities are high that they will come up with an inordinate number of *false positives*. In the limiting case in which none of their clients has a particular disorder, clinicians will still "detect" a considerable number who do, even when using instruments that have high construct validity (for example, Dawes, 1993). The fallibility of most diagnostic instruments is the reason why more than one device should be used to determine the disorder from which a client is suffering – a principle as useful in medicine as in psychotherapy. The fundamental question, as Michael Wertheimer (2008) reminds us, is whether the validity of any test (or tests) is high enough to warrant their use in individual diagnosis.

Trait–situation interaction

Mischel's thesis. We have long known that individuals behave differently (and moreover adaptively) in a variety of different situations.[18] In a funeral parlor extraverted individuals tend to behave in a sedate, subdued, and constrained way; aggressive and impulsive individuals typically behave in a meek and docile way when stopped by a police patrol car or when appearing in court; and as convicts walk into a maximum security penitentiary, they not only don new apparel but a different persona and style of interaction. This chameleon-like quality is not only a property of the fraudster, but a characteristic of every adaptive and well-adjusted

[18] We discussed this at greater length in Chapter 6 on the self.

person. This "situationalist" position was presented by William James in the nineteenth century, but it was Walter Mischel (1968, 1973) who gave it renewed visibility. He marshaled an abundance of evidence, both historical and contemporary, to demonstrate the malleability of personality traits according to the demand characteristics of the various situations in which individuals find themselves.

Mischel (1984) acknowledged that some individuals, particularly those who are immature, less competent, or maladaptive, will demonstrate greater consistency across situations that call for flexible response. But consistency of response would also seem to be a function of the consequentiality of a course of action. When one perceives momentous consequences for an action that one must take on short notice, habitual values and schemas tend to come into play. Taking a systemic view of human behavior, one must assume that in any situation there are a multitude of (moderator) variables that impinge on the organism and its decision making. As in the nature–nurture debate, the analogous trait–situation debate has been reduced to varieties of interactionism. This does not cast doubt on the usefulness of personality measures, but it does place constraints on the breadth of the inferences we draw from their results, such that caution and care are always advised.

Mischel's thesis was most challenging to those who upheld the use of personality measures to measure large populations and then presumed to predict on the basis of the results how they would behave in the future. Trait psychologists, largely of a "dispositionalist" orientation, pushed back. A persuasive counterargument was launched by Bem and Allen (1974) in an empirical study reported in an article entitled, "On Predicting Some of the People Some of the Time." They demonstrated that various traits assume greater importance for some people than for others. This idiographic perspective leads us to conclude that some people view honesty as a very important aspect of their self-image and that they will act with scrupulous honesty in all situations. But there are others who are more relativist in their approach to a variegated world and will adjust their application of principles from situation to situation. In this paradigm everyone can then theoretically be placed on a continuum of "interest" or investment in the trait at issue. The more relativist people are, the greater cross-situational variation will they exhibit; the more absolutist they are, the more rigorously and single-mindedly will they hew to lapidary principles of conduct.

Over the ensuing decades the tension between the "dispositionalists" and the "situationalists" was attenuated in a series of studies by Mischel and others (for example, Mischel and Shoda, 1999). In a fine-grained schema they classified the types of cognitive-affective units differentially accessed and evoked in various situations that influence the type of responses an individual may make – characterized by them as a processing dynamic (Mischel and Shoda, 1995). In their 1999 chapter, they ask in a Jamesian spirit (and then answer in the affirmative) "Can the two approaches to personality – dispositional-trait and processing dynamic – be reconciled and integrated within a unifying framework?" (p. 213). Within their cognitive-affective personality schema (coded CAPS), they integrated underpinning genetic–biochemical, pre-dispositional variables. However, they did not express hope that this integrative framework would succeed in reconciling the dispositionalists and the situationalists, given their divergent "underlying philosophies implied by different conceptions of personality ..." (p. 213).

Attributionism. Nisbett and Ross (1980) demonstrated that when people in North America try to explain the behavior of others they are inclined to attribute that behavior to their disposition (that is, a personality trait), not to the situation that possibly elicited it. This is a *metatheory* of human behavior that explains the way lay people (and many psychologists) make sense of the world. Based on their observation of people functioning in a limited number of roles, they admit their puzzlement when they get inconsistent information. (Something must have "snapped" to turn this gentle, courteous neighbor into an axe murderer.) People have an abiding faith in the consistency of the behavior of others in a variety of situations. Having viewed them in a few roles, they assume they behave similarly in all roles. Citing the novels of Horatio Alger, Jr. and Henry Fielding, Nisbett and Ross stated that this view of personality is widespread in the literature of the West. The theory that one's character will govern one's behavior and in the long term be the mirror of one's success or failure is imbedded in our culture. These are "beliefs which Max Weber (1904) long ago identified as a precondition for the rise of capitalism, and it is consistent with the many philosophical positions that have assigned central roles to the concept of personal responsibility and free will" (p. 31). Anomalously, however, people have a tendency to excuse their own behavior by alluding to extenuating circumstances (that is, situational variables) rather than to a flaw in their character.

Structured personality tests

Varieties of self-report inventories

The human penchant for classifying the myriad data that flood our senses or that characterize our behaviors runs into the same problems when organizing assessment instruments as when categorizing other products of human ingenuity. The problem in principle is simple, the solution is bedevilingly imperfect: whereas all properties of things are distributed in nature as continua, their analysis often requires that they be digitized – broken into units, and then labeled. Personality assessment instruments of which there are many thousands and numerous varieties are no exception.

Deductive vis-à-vis empirical approaches. Many strategies have been devised for constructing personality tests, but there are controversies not only about the relative value of these strategies but also about how they should be understood and labeled (for example, Ben-Porath and Butcher, 1991). At a basic level we can divide strategies into two categories: rationalist, deductive approaches and data-based approaches. The former approach is often grounded in a formal theory of personality. The latter approach is more empirical, utilizing the technique of factor analysis[19] and so-called *criterion keying* in which test items are devised that relate to the behavior of special groups of individuals. The brief discussion that follows will be restricted to several examples of some widely used inventories.

Reason – specifically deductive logic – is at the basis for the construction of the earliest and some of the best personality tests. The procedure used in constructing such tests makes sense to rank-and-file individuals, a fact that contributes to their strength and validity. The items tap into the cultural schemas that are psychologically sedimented in the psyches of a particular society's members. We normally make the assumption that a personality test measures a trait or set of traits that approximates a given society's common-sense definitions of such traits. We say "common sense" as it is precisely the vernacular understanding of terms and their referents that test designers have in mind when designing an inventory to be taken by rank-and-file individuals. Given the common cultural heritage of those who are judged to be suitable subjects for one test or another, the concepts and the language that accurately reflect their day to day experiences are those that form the warp and weft of the instrument.

The items of any inventory, of course, are only a minuscule sample of the broad spectrum of behaviors, attitudes, interests, and beliefs presumed to be variously shared by the test-takers. They patently seek information that test designers intuitively think is related to the construct they are investigating. For example, to pick up diagnostic signs of *eating disorders*, they might ask questions about the use of laxatives or other purgatives. Or if

[19] See Chapter 5 in which this technique is used for reducing a large number of specific traits to a small number of general personality types.

gregariousness is a personality trait considered to be of interest, and important to assess, then a test designer might intuitively judge such items as *I enjoy going to parties* and *I prefer small intimate gatherings* as suitable. An item such as *I like informal attire* would seem to have less relevance to that construct, although an empirical investigation of, say, party lovers' life-style might reveal a substantial relation between type of attire and gregariousness. If we assume the willingness of test takers to give frank and candid responses to all of these items, we can assume the test measures a trait that approximates our common-sense understanding of that trait. We emphasize "common sense" as it is precisely that understanding of the language and its referents, as we mentioned above, that psychometrists have in mind when designing a test for rank-and-file individuals.

Some instruments are designed to detect traits or measure levels of trait expression that are normal in a specific culture, whereas others are intended to detect levels of dysfunctionality. The first personality test devised, *The Woodworth Personal Data Sheet* (Woodworth, 1920), was of the latter type. During the First World War psychometrist*s* developed a paper-and-pencil instrument intended to give a global measure of functionality of personnel highly important to the war effort. If a score revealed an unusual number of suspect behaviors, the examinee was asked to present for an interview. Other personality tests were soon developed that targeted a well-functioning clientele, less concerned with discerning patterns of poor functioning than normal and healthful patterns, measuring, for example, population variances in levels of shyness.

But the problem with measures of dysfunctionality and well-functioning is that the subjects may have incentives to fake socially desirable behaviors and interests. Obviously, the motivation of individuals in taking a test is important for interpreting the results of the test. Those who are seeking a better understanding of themselves and of where they fit in the broad spectrum of a specific population will be less concerned about making a fine appearance and "faking good results," at least consciously, than if they were engaged in a competitive process to gain, say, admission into an officer training program or to be hired into a prestigious corporate managerial program.

The more subtle the test items are, the more the moderator variable intelligence comes into play and skews results in favor of the shrewdly intuitive "player" – and away from a veridical representation of other personality variables, such as emotional instability or uncooperativeness. The temptation for enuretics to conceal from test interpreters that they occasionally urinate in their sleep may be overwhelming in view of the career implications of an honest answer. "Psych-ing out" the intent of the test designer in the formulation of any item compromises the validity of results to the extent that it warps the response of test takers. The examinee who "manages the results" is the implicit adversary of the psychometrist. But for the past half-century at least, it has been noted that dissembling on personality tests does not necessarily reflect an attempt to deceive (Anastasi and Urbina, 1997, p. 375). Researchers (for example, Crowne and Marlowe, 1964; Frederiksen, 1965) "have presented evidence to suggest that the strength of the social desirability response set is related to the individual's more general need for self-protection, avoidance of criticism, social conformity, and social approval" (Anastasi and Urbina, 1997, p. 375).

On average, individuals describe themselves with attributes that correlate highly with the traits that the population in which they consider themselves to be embedded considers desirable. This, of course, is what culture signifies in part and what a dominant culture prescribes. To the extent that individuals describe themselves differently, to that extent they are, and consider themselves, culturally marginal. These limitations have been of such concern to psychometrists that they have largely, but not entirely, abandoned the simple, logico-deductive approach to test construction (Kaplan and Saccuzzo, 1997). Multi-modal approaches to test construction have been developed such that the former dichotomization of tests into empirically vis-à-vis logico-deductively developed tests is less useful than formerly.

Edwards Personality Inventory. Elements of the deductive, theory-based approach *and* the empirical method can be found in the *Edwards Personality Inventory: R* (EPI:R) (Edwards, 1990). It is a good example of a content-validated instrument, which means that the items were initially formulated and selected because they were intuitively thought to tap the attributes the developer wished to measure. Edwards wrote item-statements that seemed to represent a given personality trait, as well as others drawn from published biographies and interviews with selected individuals. For example, an item signifying *sensitivity* is: "Am sensitive and easily hurt by others"; one signifying *avoidance* is "Try to avoid trouble of any kind"; and another signifying *helpful* is "Do many things to help others" (p. 1). However, like the great body of contemporary psychometrists who have trended toward an empirical basis in test development, he used factor analytic techniques as well as criterion-keyed methods in the development of his latest instrument. He formulated 2,824 items, which were then sorted into 216 categories presumed to relate to various personality traits. Through a series of factor analyses these 216 scales were reduced to 18 (Edwards, 1990, p. 19) that enable the assessment of examinees' interpersonal-relations skills, personality traits, interests, and values.[20]

Myers-Briggs Type Indicator and the Eysenck Personality Inventory. The *Myers–Briggs Type Indicator* (MBTI) (Myers *et al.*, 1998; Myers and Myers, 1995) is a widely used personality test whose conceptualization was driven by Carl Gustav Jung's theory of personality types. The MBTI measures responses along four axes comprising the following bipolar scales: (a) introversion–extraversion; (b) sensing–intuition; (c) thinking–feeling; and (d) judging–perceiving. Eysenck's Personality Inventory (EPI), on the other hand, measures responses along two axes comprising the following bipolar scales: (a) introversion–extraversion; and (b) emotional stability–instability (neuroticism). It is interesting to note that although theory-derived, the MBTI taps constructs that Hans Eysenck derived using factor analytic techniques and which he measured with the EPI (Eysenck and Eysenck, 1968). Eysenck concluded that his analyses confirmed the existence of temperamental traits that jibed not only with the constructs of Jung, but with the ancient schemas of Hippocrates and Galen. Consistent with this finding, he insisted that it is essential to go beyond the merely statistical approach and to link up personality dimensions with the main body of theoretical and experimental psychology (Eysenck and Eysenck, 1968, p. 13). Crossing introversion–extraversion with neuroticism–emotional stability yields the four temperaments posited by Hippocrates and Galen: (a) sanguine; (b) choleric; (c) melancholic; and (d) phlegmatic.

Criterion-group strategy. An effective approach to developing a personality test is to discover the characteristic behaviors and attitudes of distinct populations and then to formulate items that tap those characteristics and attitudes. The strategy often used to achieve this is the contrasted-groups method in which a sample of individuals who it has been clinically established are, say, risk- and sensation-seekers are compared with a random sample of *hoi polloi.* Test items that distinguish the risk-takers from the generality of people can be compiled into a sensation-seeking scale or inventory (Kaplan and Saccuzzo, 1997, pp. 402–3). Even

[20] Following are paraphrasings of the terms the EPI:R Manual uses to describe the constructs that each of the 18 scales measures: (1) approval seeking; (2) social anxiety; (3) sensitivity to criticism; (4) sympathy-seeking; (5) conflict-avoidance; (6) docility–assertiveness; (7) cooperativeness and team loyalty; (8) competitiveness; (9) verbal fluency; (10) responsibility; (11) friendliness; (12) negative self-image; (13) irritability/prickliness; (14) helpfulness/altruism; (15) tactfulness; (16) dependency; (17) shyness; and (18) kindness/forbearance.

the earliest of personality inventories made use of this strategy. For example, the Woodworth Personal Data Sheet retained only those items from the original pool of items that clearly discriminated between a sample of "normals" and those who were seriously troubled. If 25 percent or more of a sample of normals found a behavior, idea, or value acceptable, it was not considered a good discriminant of "abnormals."

A widely used personality test constructed by criterion-keying was the *Minnesota Multiphasic Personality Inventory* (MMPI), which has been upgraded in two recent versions. Originally developed in the 1930s to diagnose mental disorders of varying degrees of severity, it has over the decades expanded its purview to assess normal individuals for vocational guidance, military assignments, marriage counseling, and so forth. The object of the inventory is not to get an accurate fix on what is happening in subjects' lives but on *what they believe is happening* in their lives. Thus, they are asked to respond *true* or *false* to statements such as, "I believe I am being plotted against," or "I am sure I am being talked about."[21] Whether the statements are true or not is of less interest than the assessees' perceptions of the world, internal and external, that they inhabit. What is significant is the pattern of responses that have been empirically established to correlate with certain criterion behaviors – behaviors that convention has it are more or less adaptive (or maladaptive) in a particular society at a particular time.

As the MMPI aged it became evident that certain items reflected a vanishing lifestyle and the earlier norms of a population in rapid change. An upgraded inventory, the MMPI-2, included items probing political and social attitudes as well as health and medical issues, career and educational interests, sexual and marital issues, and family dynamics. The 576 items in this newer inventory enable examinees to be scored on ten clinical scales[22] and three "validity scales." The latter scales estimate an inferred level of lying, "faking good," and simply botching the results for reasons that may not be clear to either the test taker or the test giver. The usefulness of the clinical scales is compromised, of course, to the extent that the validity scales indicate significant levels of deception, deliberate or otherwise. Although the sample used for normative purposes in this later version of the MMPI was more representative of the present

[21] No reader can accurately respond "false" to this statement, as the preponderance of social conversations, on a daily basis, concern our interactions with others. Thus, the content of the item is of less importance than its ability to statistically discriminate between a criterion group and a control comparison group.

[22] These scales are labeled: depression; hypochondriasis; hysteria; mania; masculinity–femininity; paranoia; schizophrenia; psychasthenia; psychopathic deviate; and social introversion.

North American population than was the earlier one, it has been criticized nevertheless for under-representing Hispanics and Asian-Americans.[23] Despite the flaws inherent in this instrument, when used with caution, it can serve to indicate generally adaptive as well as maladaptive patterns of mental functioning.[24]

Millon Clinical Multiaxial Inventory (MCMI-III). Like the MMPI whose development preceded it the *MCMI-III* is a notable example of a personality instrument that was constructed using a range of methods transcending the theoretical–empirical divide. Criterion-keying for the selection of items was one method. That is, items were formulated that characterize the behaviors of selected populations, independent of our ability to establish the causal and intuitively meaningful pathways linking them to a syndrome, or to the trait expressed by the population in question.

Millon, among others, was influential in transforming the *Diagnostic and Statistical Manual* (DSM), once grounded in select psychodynamic theories of personality, into the significantly less theory-biased DSM-III (American Psychiatric Association, 1980). Nevertheless, his own *bio-psychosocial* theoretical orientation in personality assessment (Millon and Davis, 1996) strongly informed not only the DSM-III and its successor editions, but especially the Axis II (personality) disorders as defined in the DSM-IV-R. In addition to the serious clinical syndromes, such as delusional disorders, the MCMI-III enables scoring on the eleven personality disorders reviewed in Chapter 12. Using the MCMI for purposes other than clinical assessment is not advised, but Millon did develop other instruments that are more forgiving of non-clinical use. One in particular is the *Millon Index of Personality Styles* (MIPS) (1994), which was developed with a normal population in mind and can be used for general counseling purposes.

Important to note is the fact that the Millon scales can no more be said to have been constructed using a single method than can other recent psychometric products. As Anastasi and Urbina (1997) note, traditions of research and test development that have coexisted for several decades "have coalesced in recent years" (p. 363). The MCMI is a good example of an instrument that resulted from logico-deductive, theoretical, factor

23 Numerous translations and adaptations of the MMPI-2 have been prepared for Hispanic populations including those who live in North America (Butcher, 1996).

24 The scope of this chapter has not permitted more than a brief description of the MMPI, but Anastasi (1988) provided an excellent abbreviated description of this heavily researched instrument in the 6th edition of her textbook.

analytic, and criterion-keyed/contrasted groups approaches to item formulation, selection, and scaling.

Projective techniques

Projective tests are "procedures in which a person reveals characteristic traits, feelings, and behavior patterns through responses to relatively unstructured stimuli such as ambiguous pictures, inkblots, or play materials" (Corsini, 1999, p. 767). The assumption underpinning the value of this procedure is that persons being tested will reveal in their responses, schemas, and values they unconsciously harbor but which are not apparent to them. This *projective hypothesis* assumes that test data are a veridical representation of those schemas and that they will not be unconsciously screened or bowdlerized by the examinee as they are accessed. A further assumption is that those who are doing the testing will administer the procedures in the scrupulously standardized mode in which they were constructed and validated. There is an intrinsic principle of unreliability embedded in projective techniques such as the TAT where a high degree of clinical subjectivity is involved in the interpretation of the results. Even veteran clinicians cannot fully escape the influence of their own cultural, theoretical, and personal schemas when interpreting test results, no matter how long they have been trained to recognize them.[25] The best that the profession can hope for is that test interpreters have a deep understanding of their own cultural biases, which they can *bracket* and isolate as they interpret the fuzzy and ambiguous data that the examinee produces.

Unstructured, natural settings have many of the features of projective devices. Natural settings in which we are confronted with challenging problems, say, in a city of a culture different from our own, usually comprise uncertainty, missing information, ambiguity, and pressures to make inferential leaps about the meaning of changing situations. In a manner of speaking we can say that life is rife with "projective tests," for we rarely have a comprehensive grasp of the meaning of stimuli to which we are obliged to react. From that standpoint it is fair to say that all psychometric instruments can be placed on a continuum ranging from the most ambiguous and intrinsically meaningless to the most explicit, univocal, and unmistakable. Inkblots would be near the former pole; a clearly written trigonometry exam near the latter.

[25] Indeed, an intrinsic hazard of becoming experienced is that one becomes more proficient in applying one's possibly biased heuristics or flawed schemas (Dumont, 1991).

Scientific status of projective devices

Over the past half-century, there has been a growing avalanche of evidence that projective techniques have neither the validity nor reliability to justify their use in the mental health professions. Donald Peterson (1995), in a widely cited article, related his reflections on his struggle with this issue when he was at the University of Illinois. In an article of pertinence to those who train professional psychologists, he recorded an internal dialogue, in which he engaged while reading Dawes' (1994) *House of Cards*:

> Critic [of projectives]: Dawes says that all or most of the projective tests that psychologists use, like the Rorschach, are worthless. Do you agree?
>
> Educator [Peterson]: I'm still not absolutely sure about that, but in general yes, I agree. I have said so in print on several occasions ... When I was directing the clinical PhD program at the University of Illinois, we were acutely aware of the accumulating negative evidence on the Rorschach and most other projective tests. Some of the sentence completion devices might have been exceptions, but the negative evidence on the major projectives, like the Rorschach, finally got so strong that we decided to stop teaching it. (p. 981)

The passage in Dawes' (1994) book that catalyzed this reaction can be found on pages 146–54.

Proponents of projective techniques question the value of the studies that purport to demonstrate their lack of validity and reliability (for example, Wade and Baker, 1977). Typically they allege that the studies are flawed and inaccurate. In reaction, Dawes (1994) commented: "This last rationalization should be all too familiar by now; the many studies whose conclusions are negative are dismissed as inadequate for this or that reason, and then a positive result is invented from the resulting vacuum – in the absence of any actual studies showing positive results" (p. 150). Although Dawes' commentary has a polemical tone, it reflects the views of later, more subdued assessments (for example, Lilienfeld, Wood, and Garb, 2000) of the Rorschach and other such devices: "It doesn't work as it is supposed to. Nor do the other projective techniques" (Dawes, 1994, p. 150).

Although the research evidence bearing on this matter when taken together appears compelling, the question arises as to why internship supervisors, in many sites where our novice therapists gain their predoctoral experience, insist that interns receive training in the Rorschach and other projectives. This disconnect between academia and field settings, and research and practice, is one that has still not been effectively addressed. Peterson (1995) asked, "if the main problem in the cognition of professional psychologists is that they separate practice from research, how can we teach them to do otherwise?" (p. 981).[26] The answer resides in part in the inertial momentum of those who have been trained in their youth to use projectives, which overrides the paradigm-busting of succeeding generations of researchers and educators. As the enthusiasm for projectives remains high among rank-and-file practitioners, we can only conclude that these instruments satisfy the professional needs and intuition of these individuals, and that they support the clinical purposes that still have not been well delineated, let alone assessed.

We know that therapeutic outcome is in large measure determined by the level of confidence that patients and clients have that the treatment plan will work. The use of an instrument that is not fully understood by the therapist is assuredly less well understood by the client. Confidence in esoteric psychometric measures one does not understand, belief in the higher science of the therapist, trust in the public credentials authorizing the use of those instruments that are in many therapists' armory inspire the hope that heals. As we expect more from our scientific disciplines than large doses of common sense, a methodology that uses cryptic instruments to tell us about ourselves has a potency of its own. We expect effective treatment reflecting the technological expertise that twenty-first-century science can afford us. Pedestrian techniques dull the patina of high-science with which our procedures are imbued. The science reflected in the panoply of dials and lights in the cockpit of a jumbo jet increases the expectation that there is a commensurate expertise in the practitioner.

Rorschach inkblot test

Among the projective tests, none has captured the romantic imagination more than has the Rorschach.[27] Making inkblots is child's play, though making sense of people's reactions to them is not. One only needs to place several drops of ink on a blank sheet of paper and then fold the sheet in half. This results in symmetrical designs which can suggest a limitless

[26] The split between science and practice is not specific to psychotherapy. Beutler *et al.* (1995) present evidence that it afflicts other professions. They cite articles indicating that dental clinicians fail to integrate scientifically derived advances into their practice in treating, for example, temporomandibular disorders, or nurses who neglect empirical findings that impact on the quality of their practice.

[27] The origins of making inkblots and fantasizing about the images they suggest are obscure. The German philosopher and experimentalist, Justinus Kerner (1786–1862), engaged in the pastime of blotting ink on paper and folding it into quarters to generate weird figures. He wrote verses and narratives about the monsters he imagined them to represent (Ellenberger, 1970, p. 81).

number of meaningful (and meaningless) images to an observer. The blot designs are as close to fully ambiguous images as can be fabricated (computer-generated random designs are superior in this respect), although the symmetry of the inkblots places an initial imaginal constraint on the observer. Moreover, as the developers of inkblot tests have made a choice of certain inkblots among the many that they generated, one can only suspect that they chose those which appeared suggestive *to them* of images that would be diagnostically useful. Of course, their personal history as well as the particular school of psychotherapy to which they belonged would have influenced these choices.

Freudians who use the Rorschach look for image-responses that correspond, say, to their constructs of anality, orality, or genitality. Jungians, on the contrary, would look for religious or other archetypal symbols. What would look like a church steeple to a Jungian patient would presumably remind Freudian patients of a phallus. Other schools, typically psychodynamic, would look for expressions of a variety of unconscious fears, hopes, beliefs, schemas, and concerns they have theorized to be meaningful. The interpretation of inkblot test responses can be no more valid than the theory that underpins their interpretation. Masling (1986), a proponent of the use of Freudian constructs in inkblot tests, avers that theories serve to generate self-fulfilling prophecies. His own system seems to exemplify this (1986; Bornstein and Masling, 1985). The "oral" psychosexual stage in the psychoanalytic system focuses on behaviors that have radiated out from the "mouth-that-seeks-the-breast" to all food-seeking, talking, kissing, passivity, begging and praying, wishing, and nurturing. Baby-talk responses that include "bunny rabbit" and "pussy cat" suggest a level of "orality" to some psychoanalysts.

Much skepticism exists among psychologists relative to the validity of projective techniques. Lilienfeld, Wood, and Garb (2000) state:

> Jensen's (1965) famous quotation, although 35 years old, still captures the sentiments of many contemporary scientists toward the Rorschach Inkblot Test and numerous other projective techniques: "… the rate of scientific progress in clinical psychology might well be measured by the speed and thoroughness with which it gets over the Rorschach." (p. 67)

This is a strong position to take relative to this instrument. Moderate positions are those in which the value of projective techniques is maintained for serving nonmetric purposes. For example, many clinicians use the Rorschach, not because they have an objective scoring method for evaluating the trait structure of the examinee or have trust in its psychometric properties, but because it is useful as a foil against which the examinee can joust, in the process of which covert motives and penchants come to light. Talking points emerge as the examinee ruminates about the various forms that are "seen." The test becomes an arena in which clinician and client interact and wrestle over issues. It is less a metric than a device for generating data that serve as grist for the clinical mill. In this context the clinician as artist supersedes the clinician as technologist. Lilienfeld and colleagues (2000) state that projectives can be moderately valid indicators of artistic ability or intelligence. If so, one must argue that there is an intrinsic link between the character of the projective responses as expressions of covert schemas, fears, and beliefs and the intelligence of the examinee. Using a projective device as a measure of artistic ability arrests its function *qua* projective test just as using a stapler as a doorstop arrests its function as a stapler.

The effort of examinees to assign a meaning to an ambiguous stimulus never fully disambiguates it. Sane individuals always recognize that they are fantasizing meanings when they look at an inkblot and say they can see, for example, "two bearded gurus." Ambiguous stimuli normally remain ambivalent stimuli to examinees. On the other hand, the less meaningfulness built into the stimulus, the more freedom are examinees given to pull from their own hidden and creative resources the stories and images that are meaningful – often unknown – to them. Clearly, there is a continuum of ambiguity on which any test can be situated (Meehl, 1945). Although Exner's "Comprehensive System" of scoring responses to the Rorschach has diminished the level of subjectivity used in more theory-based interpretations (Exner, 2002) and is used by a small but respected cohort of aficionados of this instrument, the generality of diagnosticians in North America still regard it as contaminated by uncontrolled variables, not least respondents' creativity, imaginativeness, intelligence, and playfulness. Asking people to interpret dreams (their own or others') is a comparable projective technique. Dreams are intrinsically ambiguous, some more than others. As clients talk about them some clinicians aver that they "get a sense" of the dominant motifs of their lives, and the concerns that are currently engaging them.

Clients in a treatment situation are a projective test for clinicians; they present the latter with an ambiguous set of stimuli. Clinicians have the difficult task of construing the significance of innumerable, confused, and polyvalent stimuli presented by the client. Inevitably this elicits *transferential*, sometimes *countertransferential*, reactions from clinicians, whose own unconscious is tapped for the multitudinous schemas that have accumulated through their personal history. On the most generalized level, life itself is a panoramic projective, never fully understood, always with elements of ambiguity, constantly eliciting responses shaped by experiences that have faded out of conscious memory.

Thematic Apperception Test (TAT)

This projective test was the brain child of Henry Murray and his student, Christiana Morgan (1935). As with other projective tests, it is based on the premise that examinees will *project* their values, needs, and schemas, of which they may or may not be aware, in response (in this case) to images of an ambiguous social situation printed on cards. Because they are asked to fabricate stories whose principal players are shown to them on a succession of thirty cards (a thirty-first card is a blank), it is assumed (a) that they will covertly identify with a principal figure in the picture, and (b) that the events they recount, fictitious as they may be, will reveal the hidden aspirations, latent fears, repressed impulses, and urgent needs that are unrecognized fixtures of their psyche.

Unlike the Rorschach, the images in the TAT are not without intrinsic meaning. Card 1 portrays a boy sitting at a table on which there rests a violin. There is no doubt that he is well groomed, that he is looking toward an instrument, that the instrument is on an unadorned table, and that it is a violin. If, instead of a Caucasian male, the card showed a neatly dressed African-American youth observing a basketball on the floor of an apartment balcony, respondents' narratives would vary dramatically depending on the socio-economic segment of the population of which they were a part. The obvious signal properties of the TAT that are *not* ambiguous create the culturally biased aspects that compromise its validity for diverse populations, as distinguished from diverse individuals. Card 12F shows an elderly woman, her hair covered by a kerchief, standing behind a younger woman, and peering over the latter's shoulder.[28] Their gaze is directed toward divergent visual fields; their coifs and clothing are drab. The picture has the potential to elicit a variety of schemas in the mind of the observer: generational differences in presentation of self, attitudes to marriage, feminism and independence, feelings about aging, quality of relationship between mother and daughter or counselor and protégée.

Because projective tests are premised on the hypothesis that unconscious schemas will spontaneously emerge as respondents tell a story about the figures depicted on the card, the better that examinees are informed about the purposes of the test, the more self-conscious they become as they spin their narrative. Self-consciousness provokes screening of material one is ashamed of and highlights material that seems socially admirable. The more intelligent the respondent is, and insightful about the covert motives of the examiner using assessment measures, the

[28] Matching stimulus cards to the sex of the examinee does not appear to enhance the discriminative power of the TAT (Katz, Russ, and Overholser, 1993).

more that knowledge interferes with the effort to present a spontaneous narrative. Likewise the more competitive, say, or paranoid respondents are, the more they tend selectively to conceal material that places them at a (self-perceived) disadvantage with the clinician or a reference group. The great disparity that exists among many covert and irrelevant moderator variables that affect examinees' responses presents an interpretive problem for the examiner.

Reliable interpretation of test results depends on how carefully the professional giving the test hews to the directives governing its use. This has been the Achilles' heel of the TAT, for many examiners take liberties in the administration of this test. Ryan (1985) asserted that "in common usage, the interpretation of the TAT is based on strategies of unknown and untested reliability and validity, a potentially dangerous outcome" (p. 812). For example, many practitioners provide a personalized description of the TAT cards along with the image stimuli on the cards. Others typically present only ten or twelve of the cards (see, for example, Kaplan and Saccuzzo, 1997, pp. 451–57; Garb, 1998). As any variation in the structure and administration of a test in essence *creates a new test*, the norms for interpreting the scores of the standardized tests are no longer applicable.

The TAT was expressly designed for use with well-functioning populations rather than for diagnosing personality disorders, and though Morgan and Murray intended their device to be used in non-clinical settings, it was clearly seen to have uses that clinicians would find attractive. The TAT is theory-based, in that the interpretive guidelines that the inventors provided emerge from Murray's psychodynamic approach to therapy, his own taxonomy of human needs (which number about twenty-eight), his theory of environmental *press*, and the construct *achievement motivation* (alternatively designated in the literature as *need achievement*).[29] In spite of the limited uses to which the authors of the test destined it, some clinical and counseling psychologists use it for psychodiagnostic purposes, although these practitioners "have largely abandoned a psychometric approach to interpretation [of the TAT] as well as any attempt to use an objective scoring system" (Rossini and Moretti, 1997, p. 394).

The Draw-A-Person Test

There are several human figure drawing techniques, the purpose of which is to make reliable personological inferences about those who have made

[29] David McClelland built a career on his investigations of this construct (for example, 1958, 1961, 1965, 1985; McClelland *et al.*, 1953), frequently using the TAT as a research instrument.

the drawings. Machover's (1949) Draw-A-Person Test (DAP) is representative of this genre. Originally developed to assess children's intelligence, it was later used to include the assessment of personality and psychopathology. Machover (1949) summarized the rationale for using the DAP for such purposes:

> The human figure drawn by an individual who is directed to "draw a person" relates intimately to the impulses, anxieties, conflicts, and compensations characteristic of the individual. In some sense, the figure drawn is the person, and the paper corresponds to the environment. (p. 35)

No compelling evidence, however, exists that the figures drawn do in fact represent the persons who have drawn them, nor that the paper they were given to draw on is a meaningful representation of their environment. No one can deny that DAPs *may* be intimately related to the drawer's "impulses, anxieties, conflicts, and compensations"; however, no established correlations between features of drawings and personality have been scientifically established. Among the many confounding variables that contaminate interpretations of the DAP is that individuals differ in their ability to draw. Individuals' level of artistic ability can account for distortions in their self-portraits. These may have no relation to their impulses, anxieties, or conflicts. Further, five decades of research have failed to demonstrate, for example, that large eyes or ears signify paranoia or that other properties of the drawing reflect personality and emotional disturbance. Other presumed indicants, such as thickness of the lines in the drawing, the size of the figure relative to the size of the paper surface, the number of buttons in the apparel, the size of belt buckles, or the number of erasures, have not been correlated with features of subjects' personality (Smith and Dumont, 1995). Kahill (1984), in her review of 15 years of research on human figure drawings, concluded that "the evidence regarding both the content and the structural and formal aspects of drawings fails to support the majority of Machover's hypotheses or is contradictory. Only two of the 30 indices [reviewed] were supported" (p. 288).

The argument is often made that there is not a convincing body of evidence demonstrating that human figure drawings or other projective tests are invalid (for example, Waehler, 1997; *cf.* Lilienfeld, Wood, and Garb, 2000, p. 48). But historians and philosophers (see, for example, Lakatos, 1970) have demonstrated that it is virtually impossible to prove beyond cavil that any test or any theory is without value (*cf.* Dumont and Smith, 1996). On the other hand, whether or not this is true, logic and the philosophy of science affirm that the burden of proof rests with the affirmative. Those who affirm the value of an instrument – and use it – are the ones responsible for assuring its validity. One cannot develop a test, speculate on its usefulness, and proceed to use it until such time as *others* have demonstrated that it has little or no value.[30]

As with the Rorschach, DAPs are rarely used in isolation to make a diagnosis. They constitute one in a battery of interview procedures and assessment instruments, but are seldom the first instrument to be used in such assessments. An *anchoring effect* complicates matters further as the traits that clinicians inferentially attempt to leach out of a drawing are those suggested in the first moments of an interview (*cf.* Meehl, 1960;[31] for example, Spengler and Strohmer, 1994; Strohmer and Shivy 1994; Strohmer, Shivy, and Chiodo, 1990) – which they then "discover" in the drawing. These issues aside, the principal concern is the lack of cogent evidence of a correlation between structural aspects of the drawing (like size of figure relative to, or its location on, the paper), or the content of the drawing (like length or stubbiness of fingers, or size of belt buckles), on the one hand, and personality traits of the drawer, on the other hand (see Lilienfeld, Wood, and Garb, 2000).

Symbols on paper, schemas in the mind

Levels of abstraction

A formal, printed test that rests on a desk is a blunt device intended to bridge the mind of the consumer with the mind of the test designer, whose schema-driven objectives are symbolically represented in that test. The test presents stimulus symbols to patients or clients asked to respond to them. It is important that there be a close correspondence between what the symbols signify to the examinee and what they signify to the psychometrician. To the extent that there is a divergence in these two sets of meanings – and it is inevitable that they do diverge *somewhat* – to that extent is the examinee answering questions or responding to stimuli that the designer of the test had no intention of posing. Conclusions based on the test results are accordingly flawed.

Scope for distortion is much greater when probing for personality attributes than when probing for knowledge of, say, physical or chemical

[30] Analogously, pharmaceutical firms that develop medications cannot proceed to market them until such time as others demonstrate they are ineffective. Prior to promotion and use of a medication, the burden of proof demonstrating that it achieves what it is purported to achieve rests on the seller.

[31] Meehl demonstrated that the diagnostic hypothesis clinicians formulate in the first few minutes of an intake interview is typically the same as the diagnosis retained after many sessions (and on which treatment has been predicated). This process, of course, can eventuate in treatment for an imaginary problem.

principles. The reason for this is complex: in measuring personality, there is considerable fuzziness around the edges of the constructs at issue, not to mention controversy relative to the constituent elements of the entity measured. Further, the emotional connotations of the stimulus items influence their responses, not least the unconscious motivation to "gussy up" their self-image. Needless to emphasize, these unconscious factors complicate the inferential process of stating conclusions in qualitative as well as quantitative terms. On the other hand, the psycho-cultural differences that plague psychometricians working in the domain of social psychology or the other social sciences are less important when working in the domains of the hard sciences. Testing individuals on the laws of physics, say, does not easily admit of these distortions as they are stated in the univocal, international language of mathematics. Wilson (1998) writes, "The laws of physics are in fact so accurate as to transcend cultural differences," and we would add, personological differences. "They boil down to mathematical formulae that cannot be given Chinese or Ethiopian or Mayan nuances. Nor do they cut any slack for masculinist or feminist variations" (p. 53).

All individual characteristics lie along a continuum ranging from the concretely measurable (such as one's appetite for sweets and fats) to the highly elusive and intangible (such as the level of trust one has in total strangers). The very words used in the inventory serve as discriminative cues to childhood schemas that took on meanings in episodes that are lost to memory. The words nevertheless still trigger psycho-affective responses that alter the meaning of the phrases that form the literal content of the personality measure. Although one can exaggerate the difficulties of getting at the truth, critics like Gould (1981), though no psychometrist, and Gardner (1993) need to be heeded. Epistemological problems remain that have spawned a legion of post-modernist skeptics. Wilson (1998) has noted that the *constructivists* say to the rest of us: "Maybe, just maybe you are wrong." He continued, "Their ideas are like sparks from firework explosions that travel away in all directions … soon to wink out in the dimensionless dark. Yet a few will endure long enough to cast light on unexpected subjects" (p. 47).

Personality and intelligence

Strangely, *intelligence* and *personality*, two megaconstructs that have been variously defined and poorly integrated, continue to be treated as independent and only remotely related domains of investigation. Although psychologists have, over at least the past half-century, investigated the links between intelligence and various personality traits (Most and

Zeidner, 1995), they have treated them as two distinct sets of constructs that interface rather than become enmeshed in each other. Anastasi and Urbina (1997) aver that "in interpreting test scores, personality and aptitudes cannot be kept apart" (p. 300), yet they make a clear distinction between personality traits and abilities (p. 380).[32] Personality and intellectual development are thoroughly enmeshed and interactive, yielding substantial intercorrelations (p. 302). Intelligence, in fact, plays an integral part in personality in some theories (*cf.* Cattell, 1971; Eysenck and Eysenck, 1985; Guilford, 1959), although its measurement and influence have evolved separately because it became the object of *scientific* investigation earlier than did personality, was the object of more scrupulous and public-policy driven definitions, developed more cogent theoretical underpinnings, and appeared more amenable to reliable quantification. These attributes seem to have paved the way for its establishment as a separate (cognitive) discipline. Personality, as a congeries of mental constructs that appeared more socio-philosophical than psychological, was separated out from intelligence.

This separation is superficial (Anastasi and Urbina, 1997) and, like all constructs, artificial. Intelligence and its multiple expressions are seamlessly woven into the fabric of personality, both as a science and as an existential reality (Saklofske and Zeidner, 1995). Whether one wishes to give an individual differences (that is, idiographic) slant to their exploration or a human developmental (nomothetic) slant, each necessarily implicates the other. The interaction of intelligence with the other dimensions of personality can be studied on a cross-sectional basis measuring individual performances at a moment in time, or longitudinally, measuring their development along a number of dimensions over one or more life stages. Nevertheless, as Davies, Stankov, and Roberts (1998) argue, there appear to be cogent grounds for distinguishing intelligence and personality traits. They point out that most self-report personality scales fail to correlate with intelligence, and there is little basis for the view that any conflation of personality traits into a megaconstruct would do so either.

The relatively recent appearance in the literature of emotional and social intelligence (Davies, Stankov, and Roberts 1998) has provided an impetus for exploring their validity. If one wishes to refer to personality traits as behavioral preferences rather than as abilities (Salovey *et al.*, 2008) then one is rebuilding the distinction, whether artificial or not, between these two domains of inquiry. If, in fact, the typical mode of assessing personality is through self-report devices, the proponents of a

[32] For a divergent view consult Ackerman and Beier (2003); also explore the work of Andrew Carson, a vocational psychologist and psychometrist of note.

divide between personality and intelligence aver that they must show that any form of intelligence is distinguishable from personality traits as measured, for example, by the Eysenck Personality Questionnaire – Revised, or the NEO-Personality Inventory (Costa and McCrae, 1992).

The boundaries between the constructs of personality and intelligence are fuzzy, so much so that it is difficult to pigeonhole some personological features as belonging to either the former or the latter. The rule of thumb: "if an ability then intelligence, if a preference then personality trait" fails in some important cases such that it may cast the validity of the rule into doubt. *Creativity* is an example (Saklofske and Zeidner, 1995; Wallach and Wallach, 1983). Is it a personality trait or is it an intelligence factor? The creative person is one whose schemas are flexible, who *can* – and *does* – think innovatively, who transfers schemas from one domain to another, "thinks outside the box," or better still builds novel, more adaptive boxes. On the other hand, the creative person is also one who is *inclined* to innovate, build divergent models, and enjoys using adaptational abilities. If one theorizes that personality and intelligence are simply fluid aspects of personhood and that all the constructs we have devised to define personhood are on an ability–preference continuum, creativity appears to be near the center of this axis. Curiosity, somewhat like creativity, is another trait that would seem to load on both the intelligence factor and the personality factor. The curious person is one who is inclined to explore, to search out new meanings, niches, possibilities, a person who expresses strong motivation to be cognitively stimulated and to learn. This has the appearance of an endowed ability as much as an innate and acquired preference. In a humanistic perspective, it satisfies the drive to actualize one's potential, meeting the criteria for potential *and* for motivation. In a dimensional perspective curiosity would seem to lie near the center of the axis in which ability is at one pole and preference or inclination is at the other.

Professional, social, and ethical issues

Use of assessments

The American Psychological Association (APA) as well as other professional associations (for example, American Counseling Association) have developed formal codes of ethics governing professional practice, instructional services, and research of all their members. The APA's most recent version, "Ethical Principles of Psychologists and Code of Conduct" (2002), which took effect on June 1, 2003, contains directives for the proper use of assessment techniques, and the standards that must be respected to assure the protection of those who are evaluated and assessed. In particular, Section 9 of the Code addresses specific concerns bearing on the values and the scientific rationale for administering these devices. Their appropriateness depends on the scientific evidence supporting their usefulness for particular purposes. That means that psychologists are restricted to the use of "instruments whose validity and reliability have been established for use with members of the population tested" (Article 9.02, b).

Tester competence. Persons who are not qualified by education, training, and experience to administer tests are not permitted to do so except for training purposes and under proper supervision (Article 9.07). The interpretation of personality tests requires maturity of judgment as well as intensive background not only in the use of specific instruments but in the psychological domains – developmental, emotional, social, cognitive, and behavioral, among others – that support the clinical practice in which the tests are used. The implication of this is that even otherwise competent test administrators and psychologists may not use an instrument in which they have not received requisite training, nor should they use personality inventories whose properties are not familiar to them. A strong position is articulated by Weiner (1989) in this regard: "Competence is prerequisite for ethicality, and psychologists who practice or teach psychodiagnosis without being fully informed concerning what tests can and cannot do are behaving unethically" (p. 829).

Requisite test properties. Only those instruments should be used that research has demonstrated have the requisite validity and reliability and other psychometric properties appropriate to the individual or populations being tested. This evokes two issues. First, to the degree that outdated tests are not representative of current populations or were normed on population samples that diverge greatly from the population at issue, caution must be exercised. Second, psychologists must take into account "various test-taking abilities, and other characteristics of the person being assessed, such as situational, personal, linguistic, and cultural differences, that might affect psychologists' judgments or reduce the accuracy of their interpretations" (APA, 2002, p. 1072). In other words, in terms of the purpose of the test, the purposes of the assessment, and the character of the examinee, only such instruments can be used as are appropriate for the individuals being assessed.

Informed consent. Gathering data about the behavior of individuals unbeknown to them can be considered an invasion of their privacy even if it is gathered in a public place (see Kaplan and Saccuzzo, 1997, pp. 606–17). With few exceptions psychologists must obtain informed consent from those they presume to evaluate, assess, or diagnose (Article 9.03). This process "includes an explanation of the nature and purpose of the assessment, fees, involvement of third parties, and limits of confidentiality and sufficient opportunity for the client/patient to ask questions and receive answers" (APA, 2002, p. 1072). The obligation to obtain informed consent from clients *before* beginning to gather personal information from them, and about them, is at the core of a professional relationship. The operative term here is *informed*. The language one uses in explaining the scope of an inquiry into their personal lives must be clear, easily understood, inclusive of essentials, and clearly circumscribed. Seeking information that is *irrelevant* to the objectives of the relationship is unwarranted and possibly invasive. An important implication of this is that the use of broad-ranging instruments whose validity is predicated on constructs that are not pertinent to the needs of the client should be shunned.

To conclude this segment, I refer the reader to brief, nicely summarized treatments of this complex subject by Anastasi and Urbina (1997, pp. 533–50), Ozer (1999, pp. 671–86), and Pervin (1999, pp. 689–704). It needs to be emphasized here that tests and other psychometric devices, despite their well-researched limitations, have proven to be useful for personality assessment and for advancing laudable societal objectives. Findings that flow from test data should always be placed, however, in the context of information gathered by other modalities, including clinical interviews, third-party case histories, and personal client narratives. The bias and subjectivity implicit in them, as well as in letters of reference and unstructured interviews, can be balanced by the triangulations provided by a well-chosen battery of tests.[33]

Conclusion

Our lives, and the tasks life presents to us, can be thought of as a vast, unstructured projective test that continually elicits our responses to stimuli that are never *fully* understood. Every communication we receive and every event that impinges on our consciousness oblige us to make

[33] For a thorough discussion of informed consent and such issues as audio- and video-recording, access to one's clients' test data by third parties, and familiarity with legislation on privacy that is in force in the jurisdiction in which one is doing assessments, consult Kenneth Pope and Melba Vasquez's (2007) *Ethics in Psychotherapy and Counseling: A Practical Guide*.

inferences about their meaning and significance. Most of our inferences are solidly based on experience, but many are not. And some false presumptions about the loyalty and honesty of others we have trusted can have serious repercussions in our personal lives. The projective hypothesis is one that has applicability to non-test situations. We are constantly disambiguating the streams of information that play on us in our daily lives. Much of this is done unconsciously, conditioned as it is by long experience and the expertise we have developed in judging people (and the stock market, fashion trends, and other impersonal forces). Clinicians faced with clients they only imperfectly understand inevitably utilize schemas from their own past experience and personal history to make sense of their clients' needs. Those readers who have participated in psychiatric grand rounds and case conferences will have witnessed the divergence in clinical assessment that professionals from various cultural backgrounds manifest (see Meehl, 1973).[34] Clinical assessment presents therapists with a projective test of their own.

For many professionals, assessing others has become a vocational responsibility. Personnel managers, organizational/industrial psychologists, guidance counselors, and others who function as psychology-based professionals do extensive personality assessment. The rest of us must test the social environment on a daily basis to assure ourselves that we are minimizing the sources of threat to, and optimizing opportunities for, our well being. (Even crossing a busy street requires that we make a judgment about the sanity and alertness of others driving vehicles that have lethal potential.) Just as many tests have "lie scales" to detect inconsistencies that can suggest whether or not the examinee is trying to mislead the tester, most of us have developed deception detectors to protect ourselves from people who may be trying to gain an unfair advantage over us. Some are less adept at this than others. In any event, personality assessment is an activity in which everyone must engage. Whether structured or unstructured, formal or informal, clinical or actuarial, lay or professional, whether done by diplomats, politicians, corporate employers, teachers, stockbrokers, auditors, law enforcement agents, on-line dating site research directors, or casino croupiers, personality assessment is an increasingly challenging and important fixture of twenty-first-century life.

[34] Paul Meehl (1973, pp. 225–302), in his ch. 13 entitled "Why I do not attend case conferences," has pungently detailed thirteen arguments for his position, and multiple sub-arguments, grounding his skepticism about clinical assessment. Clinicians must address them, for each has more than a ring of truth.

11 Can personality change? The possibilities of psychotherapeutics

> Your patient is one person today, quite another person tomorrow, and still another person next week, next month, next year. Five years from now, ten and twenty years from now, he is yet another person. We all have a certain general background, that is true, but we are different persons each day that we live.
>
> Milton H. Erickson

> It is easier to act yourself into a new way of thinking than to think yourself into a new way of acting.
>
> O. Hobart Mowrer

Defining the question

Before people address the subject of altering and ameliorating personality, they need to establish that personality change is even possible. Scientists, students, and lay people have repeatedly asked the question: barring serious brain lesions and disease, can an individual's personality change in more than a superficial sense? Answers have varied depending largely on (a) how broad a definition of personality one uses and (b) the scientific lens through which personality is examined. The answers most often heard start off in the affirmative. Some developmental psychologists will answer, "Of course personality changes. After all, personality is the product of numerous interactions between a changing body and a changing environment. As they both change so does one's psyche." Others would answer, "If you accept that no one is *born* with a personality, you must conclude that humans either acquire it spontaneously and fully effloresced at some moment in time (hardly likely) or that its development is a gradual and largely contingency-based process."

Still others proffer the thesis that a personological script is encoded in the genome, which places constraints on radical personality change. In addition, the human brain and the larger nervous system of which it is a part are fashioned by myriad pre- and post-natal experiences – all made possible by our individual as well as species-typical endowments.

The synaptic structures that result from these experiences (LeDoux, 2003) constitute one's "self." This self is like a palimpsest, which maintains its identity and yet is continuously overwritten by later life events. The earlier events do not get fully erased, but decay, fade, and are largely overshadowed by organic development and environmental impact. The proto-personality, however, remains indelible beneath the persona and governs throughout one's life. Lastly, a dwindling number of preformationists (see Ausubel, Sullivan, and Ives, 1980, pp. 14–34, for the theological antecedents of this theory) render a negative verdict, which, indeed, has a certain resemblance to the positions of some behavioral and personality molecular geneticists.

These various responses have reflected popular "doctrines" in Western psychology. This chapter will examine these responses in the light of research done by *hereditarians*, on the one hand, and staunch *environmentalists*, on the other hand, both of whom recognize the evolving character of personality in its broad sense.[1]

There are implicit, often hidden, assumptions underpinning the position one takes on this issue. Certain scientists have defined personality strictly in terms of biological traits. Such a definition allows them to conclude that for all practical purposes humans are in fact born with personalities – personalities that unfold like a fern as the individual matures. Their position differs from those who affirm that personality changes across the life span result from the impact of socially prescribed beliefs, values, mannerisms, and behavioral scripts. All these change as individuals move into various environments and progress from one life stage to another. A fuller explanation of this principle will result from studying the characteristic ways people change through the interaction of social demands with internal neurohormonal processes, and gene-expressive events.

An applied psychology perspective

Mental health professionals, such as clinical or counseling psychologists, work on the assumption that personality can be modified, if not entirely restructured. They are in the profession of helping people to achieve balance in their cognitive and affective life, find peace of mind, and

[1] *The Dictionary of Psychology* (Corsini, 1999) defines an hereditarian as "a person [who assumes] that genetic inheritance is the major influence on behavior." Care must be taken not to lump all hereditarians with predeterminists, who have removed all autopoietic, self-shaping willfulness from the ability of humans to freely plan their future and become the kind of person they wish to be.

develop functional patterns of interacting within their social environment. Their professional goals range from changing a single dysfunctional habit to altering a systems-wide pattern of values, attitudes, social behaviors, personal routines, and career goals. Corrections and prison psychologists have jobs that are predicated on the assumption that ingrained attitudes and patterns of behavior can be radically altered, and the rehabilitation of convicts and youthful delinquents implies that personality can change. Consistent with this view is one enshrined in the jurisprudence of the West, predicated on the assumption that felons, for example, can change into law-abiding citizens. If we assumed that the "felonious personality" cannot change, we would be obliged to incarcerate all convicted felons for life, for a sane society does not release people from prison if it is *certain* that they will heinously re-offend.

Forces driving change

Mental health professionals, many of them educators and school psychologists, routinely assume that, in collaboration with home and community, they can shape malleable individuals into citizens who can assume the responsibilities of civic and domestic life. The well-known manifesto of John B. Watson ([1924] 1930) begins, "give me a dozen healthy infants, well-formed, and my own specific world to bring them up in" (p. 104). He claimed that he had the ability to take any of them at random, like so many Oliver Twists, and independent of their penchants or endowments, shape them into cabinet makers, doctors, merchant-chiefs, or even thieves and beggars. The argument put forward by Watson is the most radical of the environmentalist approaches to this issue. His excessive claim was predicated on the behavioral assumption of the day that heredity has a negligible role in shaping personality.[2] Stretching what seems obvious out of recognition, he maintained that thought patterns, such as belief systems, philosophies, and complex skill sets are generated solely by the environment in which one has been born and educated. In the United States a mitigated form of this view is referred to frequently as the *Standard Social Science Model* (SSSM). This model has been dismantled, notably by Tooby and Cosmides (1990), by their able admirer, Steven Pinker (1997, pp. 44–52; 2002, pp. 67–69), and a host of hereditarians and evolutionary psychologists. Except in the most conservative educational and social work circles, the SSSM has been largely discarded. There are, of

[2] This is not without support by Nietzsche who averred that cognition, worldviews, and value schemas are enabled and constrained by one's first language, surely one of Watson's conditioning devices.

course, others – moderates who concede the limits to what school, family, marketplace, and reformatory can accomplish, yet are sensitive to the political and social repercussions of touting a politically sensitive approach to education. The twenty-first century has seen the pendulum start to swing back from genetic predeterminism. This staunch thesis of molecular geneticists is being modified not only by social cognitive neuroscientists but also by epigeneticist biopsychologists themselves, as pointed out below.

In the first half of the twentieth century a respected school of developmentalists presumed that inertial, biological forces are at work. This contrary point of view was endorsed by formerly influential theories of personality, namely *preformationism* and *predeterminism*, which were related to the biosocial theories of Herbert Spencer and the later social Darwinists. Though both of these schools have emphasized the importance of innate, biological determinants of personality development, the preformationists essentially rejected the possibility for significant change in the personality development of humans. They postulated that all aspects of personality are preformed and undergo no qualitative change through life. Traits and response potentials may appear to change, but in reality they are just expressed at various stages in the *ontogeny* of individuals as their bodies mature (Ausubel, Sullivan, and Ives, 1980, pp. 14–27). This *homuncular*[3] view has largely disappeared with the advance of a scientific developmental psychology. Distinguished from preformationists, predeterminists like Arnold Gesell (1954) argue that qualitative changes occur during the life course, in perception, cognition, and emotions, as well as in the values and motives that build on innate traits. Although they do not believe that all these components of human personality are preformed in the human *zygote*, they do affirm that they are largely genetically determined – hence predetermined. Unlike preformationism, predeterminism thrives in the work of some *Three-* and *Five-Factor Model* theorists and researchers.

Many developmentalists aver that there *are* major aspects of personality that are relatively constant through life and are in large part genetically determined (for example, Costa and McCrae, 1988; Thomas and Chess, 1980; also see Chapter 3, above). This is true of personality as formulated

[3] *Homunculus* means "little man." The homuncular view is that the infant is a preformed human in whom one can find all the reaction potentials of the adult. Ausubel, Sullivan, and Ives (1980) observed that the "homuncular theory of human reproduction" is "obviously related to the theological conception of instantaneous creation and to the widespread belief in the innateness of the individual's personality" (p. 15). An example from a contemporary theory is the psychoanalytic belief "that infantile and adult sexuality are qualitatively equivalent" (p. 15).

in the molar perspective of the Five-Factor theorists. Basic traits involving adaptability to novel situations, psychobiological regularities, distractibility, intensity of reaction to stimuli, and other *higher-order traits* such as emotional stability and introversion remain relatively constant throughout life. For example, although the extraversion or shyness of 5-year-olds is expressed in increasingly mature ways as they approach adulthood, Five-Factor theorists assert they always maintain their extraverted or shy character.

Genotypes are ontogenetic givens and presumably stable; *phenotypes*, which are related to the world of appearances, and which express the genomic structure of an organism as it is translated and variously expressed within the hurly-burly of daily life, can and inevitably do change – through the entire course of life. When we expand the definition of personality to include numerous molecular, lower-order components such as generosity, gaming proclivities, occupational interests, sense of humor, recreational and sports interests, religious practice, playfulness, and gullibility or cynicism, it appears that lifestyle flexibility is the norm, for experience can powerfully shape such traits, behavior, and even gene expression. But this flexibility is never so extensive, assert the hereditarians, as to drown out the undertones of temperament that are determined by genotype.

There are other inertial forces, such as the constraints of social pressure and the force of habit. Habits can be stubborn. With each repetition, behaviors become more indelibly programmed in "the nerve-cells and fibers" of the organism, to use William James' expression. In consequence, to carry one through the innumerable mundane and occasionally dangerous decision points of one's personal history it becomes easier to rely without thought on such neurological programming than to engage in a renewed set of deliberations. Habit is a friend just as it may be an enemy.

Social cognitive theory of personality

Albert Bandura is renowned not only for his landmark book, *Principles of Behavior Modification* (1969), but for his influential social cognitive theory, a theory that views individuals as innately *agentic* – capable of taking charge of their lives. In 1999 he wrote "In the agentic sociocognitive view, people are self-organizing, proactive, self-reflecting, and self-regulating, not just reactive organisms shaped and shepherded by external events. To be agentic is to be an intentional doer selecting, constructing, and regulating one's own activity to realize certain outcomes" (p. 154). That personality can be changed is a principle that suffuses his theory. He avers like LeDoux that agentic behavior can shape brain development, foster brain cell growth, promote cognitive development, improve memory, and enhance other "faculties" that subserve the intellective functions of the person. He acknowledges, however, the role of chance ("fortuity" in his lexicon) and explores the myriad paths that even trivial events that one had not foreseen can open to the individual who is alert and aware.

Inveighing against the determinism evident in the work of the Five-Factor Model proponents, he proceeded further by assailing the simple bidirectionality of those theories that propose that behavior is a function of the interaction of the person and the environment. His triadic schema affirms: (a) the dynamic interaction of personal determinants (the biological, affective, and cognitive endowments of the individual); (b) the responsiveness of individuals to perceived opportunities and ambient challenges; and (c) the involvement of a reactive environment to those behaviors. One needs in his view to consider the behavior of the individual who precipitates the rewards and punishments of an environment that has been activated. These triadic factors in dynamic, complex interplay must be considered to explain adequately the course of a person's life and the development of his or her personality.

Sorting varieties of change

Pervin's schema

Lawrence A. Pervin (1996, pp. 172–73) has noted that there are several ways of conceptualizing change. First, there is phenotypic as distinguished from genotypic change (alluded to above). Aside from laboratory-induced alterations of an organism's genome, there is little evidence that genomes, *per se*, change over the course of one's life. On the other hand, phenotype, the expression of genotype, is in a continuous process of evolution through a continuous barrage of patterned stimuli, both internal and external, and of social interactions and situational stressors. Genotype, moreover, accounts for the systematic, "programmed" changes that are species-typical, as well as some that are idiographic. A second kind of change is conceptualized as *absolute change* and *relative change*. For example, children, initially dependent on their parents or guardians for survival, normally adopt more independent habits as they mature, venturing further afield from the protective confines of their homes. But shy and withdrawn children, for example, who become more risk-taking and bolder as they mature, may not change *relative* to other more extraverted children who become proportionately more daring as they follow their own developmental trajectory. One can say that such children have either changed or not changed, depending on one's perspective.

A third kind of change can be conceptualized in terms of *quantitative* and *qualitative*[4] change. Young adolescents in, say, the "bright–normal" range of performers on well-standardized intelligence measures think quite differently than they did in elementary school. Their minds have developed the capacity for abstract, hypothetical, multidimensional, and theoretical ideas as well as for simple, concrete, sensory-based thought. The change is less pronounced for those in the "dull–normal" range of school-age populations. Although all children grow on quantitative measures of accumulated knowledge,[5] there is a specially notable change in both the quantitative as well as the qualitative character of those who are among the brighter and more gifted of students.

A fourth distinction has been made between *continuous change* and *punctuated change*.[6] Punctuated change is evident when a large swath of an individual's life course is examined at a glance. Various life-course stages appear to be initiated by abrupt leaps in cognitive abilities, bodily structure and size (each entailing psychological sequelae), attitudinal patterns, socio-recreational preferences, and interpersonal skills. These changes seem well defined and abrupt when we stand back and look at the big picture. But those who examine the canvas of a person's life from close up tend to see changes not as punctuated but as gradual and continuous. Even stage-to-stage transitional periods, where change is most rapid, appear to evidence continuous change when looked at under a powerful enough microscope. The more fine-grained and precise our measurements, the less conceptual clarity we appear to gain. Too much information obfuscates.[7]

[4] The *quantitative–qualitative* distinction is a construct of convenience. In principle, all qualitative variables are presumed to be scientifically measurable – or so Pythagoreans would argue – otherwise differences could not appear in the first place. If the instrumentation with which we establish differences should become sensitive and subtle enough, a unit of measure and a grading method could theoretically emerge. Other equally apt descriptors for *quantitative–qualitative* are: "more easily quantified–less easily quantified."

[5] *Crystallized intelligence* (Cattell, 1943) is the term applied to accumulated knowledge and skills, which increase with education and random experiences. *Fluid intelligence*, on the other hand, is genetically determined and can be defined as our "on-the-spot reasoning ability, a skill not basically dependent on our experience" (Belsky, 1990, p. 125). The genetics of human aging determines that fluid intelligence fluctuates through the life course as a function of programmed changes in the central nervous system (CNS). As CNS efficiency declines over time, there is a corresponding decline in fluid intelligence.

[6] Compare this analysis with Stephen Jay Gould's evolutionary theory of *punctuated equilibria*, which is examined in Chapter 4 on biology and personality.

[7] For this reason Piaget called this conceptual problem a problem of scale.

Kinds of continuity

The other side of the developmental coin is the one that stresses continuity, as discussed in Chapter 5 on trait psychology. The focus is on personological consistency over time and the measurement of departures from one's own previous personal and group norms. Just as there is room for the notion of continuity in a change-oriented focus in personality, there is room for an appreciation for evolutionary change in a constancy-oriented theory of personality. Researchers in this tradition (see Caspi and Roberts, 1999; *cf.* Morizot and LeBlanc, 2003) make several distinctions in kinds of continuity.

Differential continuity (also referred to as *rank-order* continuity) refers to the consistency with which individuals in any group that one is studying maintain their relative standing in the group, that is, relative to each other. The initial scores on, say, impulse control may change as they proceed toward mature adulthood, but if all members of the group improve in a comparably proportionate way so that their rank-order remains unchanged, differential continuity exists. Intuitively it seems that the greater the heritability of a trait the more resistant it will be to change as a result of socio-educational factors. In this perspective, rank-order consistency in level of intellectual performance is high, even though there are dramatic changes in quality of performance as children move through adolescence into adulthood. It is likely that there is more stability in intellectual performance than in values, interests, or vocational orientation.

Absolute continuity[8] refers to the measured levels of a trait in any individual. In the words of Caspi and Roberts (1999) it refers "to constancy or amount of an attribute over time" (p. 305). As individuals mature from late adolescence into mature adulthood, they typically manifest greater impulse control and emotional stability, less recklessness, and greater responsibility and foresight (Carmichael and McGue, 1994; McGue, Bacon, and Lykken, 1993). But Costa and McCrae (1988) among others report that there are "few consistent age-related personality changes in adulthood" (Caspi and Roberts, 1999, p. 305). Between early adulthood and middle age, stability appears to be in evidence.

Structural continuity is another conceptual model of personality change and stability that is suitably studied in advanced courses in personality. It

[8] Also referred to as *mean-level* continuity, as research on this variable is usually done with group samples. Needless to emphasize, individuals' perceptions of personality change in *themselves* do not exactly correspond with objective measures of change. Given the imperfection of all psychometric instruments and the error that enters into all objective as well as subjective computing, this is not surprising. This discrepancy between objective and subjective assessments of change is probably diminished by people's tendency to behave in accordance with their self-perceptions.

refers to the relationship that exists *among* personality traits. In this model one studies the extent to which the traits under consideration maintain the same measured relationship to each other at succeeding stages of human ontogeny. One may even ask if the nature of a personality trait is altered as individuals work through their life stages. Finally, there is *ipsative*[9] *continuity*, which refers to the stability or change experienced by any individual independent of what is occurring in the reference groups to which she or he might be compared. The subjects of such ipsative assessments are their own reference point; they are not compared with one or another reference group to which they may be related. One might characterize this as a person-centered or idiographic approach to personality assessment.

Factors in personality stability

Much controversy in the social sciences springs from confusion in choice of labels for the constructs that are being studied. This problem has bedeviled personality theory, for the same label is often used for rather different entities. There is virtually no dissension among chemists and physicists on the definition of a hydrogen atom or a water molecule, labeled H and H_2O, respectively. But there are many variously shaded definitions of personality. One example may be helpful. McAdams (1994, pp. 717–45) has a particular view of the nature of personality (as revealed in life stories), which has not been widely adopted. He rejects the notion that personality should be restricted to core dispositions such as certain trait psychologists propose (for example, Five-Factor theorists McCrae and Costa, 1990) and suggests that *two* other levels be included in a definition of personality. Beyond such notions as extraversion, emotional stability, and openness to experience, a *second* level comprises, in his view, consideration of individuals' life projects, their social relationships with family, friends, and fellow workers, avocations and recreational interests, and many other personal, contextualized factors.

McAdams adds a *third* level that is concerned with "the making of the self" (1994, p. 306) and the generation of an evolving identity. People's identity, as a component of personality, is best captured in this model by their life narratives. McAdams found that individuals, beginning in late adolescence and progressing through life, are prone to "narrativize" their lives to make sense of who they are and who they are becoming. In short, humans need to fashion their identity in terms of the developmental passage they are making through the various crises and decision points

[9] *Ipsative* is a term derived from the Latin term, *ipse*, which means "himself" or "he, himself."

that life presents. Our life-narrative is fashioned and re-edited as we embark on new adventures and enter new life stages. Each new edition of our life story, and there are many, marks an evolutionary change in our personality profile. Choosing the most positive revisions, none, of course, fully veridical, would seem to augur a sunnier future. Describing the manner in which he selected the subject for his Man Booker prize-winning novel, *Life of Pi*, Yann Martel (2002) said that "reality is a story and we can choose our story and so why not pick 'the better story' ... ?" (p. b1–1). Indeed, this is the affirmation of a generation of narrative therapists who have helped their clients and patients "re-story" their lives. This quotation underscores the power inherent in all of us, as we confront the existential challenges in our lives, to define our selves, shape our lives, and "author our stories."

Physicians as far back as Hippocrates have appreciated the complex relationships between personality and disease. However, a number of prominent physicians have recently recognized that we can also conceptualize disease and injury in terms of narration, and to truly understand their patients, physicians must learn to hear the stories patients tell. An understanding of narrative and the ways in which patients use narratives to make sense out of the dramatic life changes that frequently accompany illness or injury can significantly enhance the practice of medicine (Brody, 1987; Greenhalgh and Hurwitz, 1998; Kleinman, 1988; Nicholas and Gillett, 1997; Sorum, 1994).

Some trait psychologists have rejected a narrow framework for their conceptualization of personality (for example, Helson and Stewart, 1994). Beyond the basic traits of the Five-Factor Model, they postulate motives, value systems, and various situation-appropriate scripts. Examples are personal concerns, coping strategies, vocational interests, and other lifestyle factors that become transformed by education, fortuitous experiences, and physical aging. They reject the reductionist approach of the Five-Factor Model that obscures the rich array of changes that personality otherwise defined would reveal.

Stability as artefact

As intimated above, the stability that trait psychologists find in human personality after age 30 may be an artefact of the methodology used for assessing personality. As McAdams has pointed out, questionnaires typically ask subjects to rate themselves on a dimension that ranges from low to high. When their responses to a questionnaire indicate, for example, that they are very uncomfortable mixing at large parties or, on the other hand, that they are moderate risk-takers in managing their finances or riding new-age roller coasters, they are really comparing themselves with their peer groups. Their *relative* standings in these implicit comparison groups tend to remain the same. The apparent stability of their traits may in large part be the stability of *perceptions* by the subjects and by those intimates who observe them over time. These perceptions bear not only on the subjects of the research but on the particular reference groups with which they are compared. As the subjects respond to a questionnaire they may be making implicit reference to out-dated and ill-conceived reference groups – like a fraternity or sorority that they left 10 or 20 years earlier.

Life narratives

The storytelling in which all individuals engage can reveal a good deal about the story teller. Astute psychotherapists learn much about their clients from the narratives that unfold in session. But as an instrument, storytelling has all the limitations of other instruments with the added drawback that it is fluid, unstandardized, and unreliable. It is not surprising that there is a relative dearth of empirical evidence to support the usefulness of this method for assessing personality change. Traditional *quantitative* approaches are less useful for analyzing narratives than are *qualitative* approaches. But most qualitative approaches are dependent on the idiosyncratic construals of meaning by researchers who admittedly let their own life stories interact with that of the subject. Depending on the period in researchers' lives when they are listening to or rereading subjects' narratives, many different, perhaps inconsistent, meanings can be teased out of the transcripts.[10]

Life narratives and identity

The life narrative is more than a research instrument for getting at the traits and lower-order dimensions of personality. It is used by clinicians to get at the identities that constitute the core of personality. Indeed, the story, if well fashioned, *is* that individual's identity. McAdams (1994) wrote, "For a personologist who believes that the stories are not simply interesting avenues for getting at traits and strivings but that the stories are indeed the identities themselves ... identity development in these terms represents real and exciting change in personality" (p. 308). The principal limitation in this view is that people reconstruct their life stories to reveal

[10] We have long known that whatever communication is received must pass through the filters of the receiver. This is one reason why researchers normally try to make their research instruments as standardized, objective, and unambiguous as possible.

an approximation of their ideal self rather than their real self. How often do people selectively attend to the aspects of themselves in telling their stories that they take no pride in or that they are ashamed of? Though they have no intention of deceiving the listener, their stories reveal, rather, the persons they would like to be rather than the persons they really are. But it cannot be denied that that better self any individual would like to be reveals an aspect of the character being described – and a perspective on his or her values, goals, and ambitions.[11]

Environment and heredity

The most mathematically gifted persons will never achieve their potential unless supported by their environment. For example, Erwin Schrödinger, a Nobel laureate in mathematics, whose work created the foundation for twentieth-century science and technology (Bernstein, 1993),[12] would have been a relative innumerate had he been born into a Stone Age society in western Borneo. Those individuals who are most comfortable in a sedentary setting such as a classroom or a laboratory would have found it difficult to cope with the adaptational challenges of a pre-information-age society. On the hypersedentary–hyperactive continuum, those who are closer to the hyperactive pole would have had an easier task evolving an adaptive strategy for survival.

DNA does not contain "programs" for particular activities. Skills are not, strictly speaking, heritable. When Buss (1999) speaks of "heritable skills" (p. 45), he is suggesting, at most, that humans inherit the *potential* to acquire certain skills. Individuals' genetic endowment pre-existed the patterns of behavior that the society and culture into which they were born enabled them to acquire. Speed typists have developed a skill that their ancestors of 10,000 years ago could not develop. But even hunting and gathering skills are not inherited. Genes are inert chemical substances that program the synthesis of proteins. But there are no genomic templates for ping-pong, accounting, space walking, or violin playing. Of course,

[11] The impression that there is little scientific rigor in qualitative research needs in any event to be dispelled. As Kazdin (2008) affirms, "Qualitative methods meet the desiderata of science; the methods are systematic, replicable, and cumulative" (p. 153). He noted for the benefit of those who are quantitatively oriented that "qualitative research even has software for quantitative data analysis" (p. 154). An excellent set of examples of such research in psychotherapy – even anthropology – can be found in the *Book of Abstracts* of the 38th International Meeting of the Society for Psychotherapy Research (2007). Of particular interest are the investigations in one of the most challenging areas: core conflictual relationships themes (CCRT) and its reformulations.

[12] Bernstein (1993) wrote that Schrödinger's work "has had more to do with the evolution of twentieth-century science and technology than any other discovery in physics" (p. 54).

genetic determinants of human neuro-endocrinology and temperament incline individuals to develop the skills that these endowments make possible. Such physiological qualities can enable them to become draftsmen rather than cabaret singers, or racing-car drivers rather than psycholinguists. What genes provide is the substrate that can initiate and support a skill-development process to which only environment and chance, historical circumstance, and available technology can give shape, character, and proficiency.

Change and the life cycle

Gradualism in personality change

Nature does not change by leaps. Even when there is a cataclysmic event, such as a volcanic eruption or an earthquake, the causes have long been building up *the potential for change.* When alcoholics or tobacco addicts stop their self-abusive consumption "cold turkey," we cannot attribute this to spontaneous remission. In the background there have been the rumblings of the social machinery that militates against self-wounding and socially damaging behavior. Change occurred because there was a crescendo of long-term medical, financial, career, and social costs that had reached a level that was no longer bearable. Passage through the different life stages reveals the same gradual evolution of more idiosyncratic aspects of human personality: people's strivings, self-concept, career aspirations, values, coping strategies, and interpersonal scripts. In their meta-analytic studies of (rank-order) change through the life cycle, Roberts and DelVecchio (2000) found that personality does change over the lifetime, the peak of change occurring at age 50. Many personality psychologists assert that the view taken in certain psychodynamic theories that personality gels by the age of 3 or 5 or even by early adulthood is without empirical foundation.

As humans age, the character of their interaction with their environment undergoes change. There are three reasons: first, their bodies change; the psychobiological sequelae of simply maturing necessarily affect the way people think of themselves, moderate their approach to a shifting and frequently unforgiving environment, and reduce or enhance their ability to perform cognitively, sexually, vocationally, recreationally, and socially. Second, one's environment, whether physical or sociocorporate, is always changing – perhaps glacially, perhaps rapidly – but the Heraclitean aphorism about not being able to dip one's foot twice into the same stream has never seemed truer than today. Third, these sets of changes necessitate new patterns of person–environment interaction.

The necessity of this principle is illustrated at its simplest by sleep-depriving those who must run an obstacle course, while gradually increasing the difficulty of that course. In a stage-theory perspective imagine the recklessness of a teenage male maturing over time into the cautious, subdued, and less flexible lifestyle of a middle-aged father of teenage sons. Sixteen-year-old and 45-year-old males have very different ways of assessing dangers in the physical environment, their responsibilities to others, and the challenges posed by their social and career aspirations. Life experiences have typically taught the older male not to take physical or social chances with his own and other people's lives.

Masculinism and feminism

David Gutmann (1987) proposed that there are strong aging effects in personality development. In later middle age he found that men tended to "go with the flow," and to adopt a less aggressive, more passive interactional style in dealing with social stressors. Men can be said to capture a feminine side of themselves (assuming a popular Western stereotype of women), which was previously suppressed while they pursued the socially prescribed goals of economic provider. Women, on the other hand, as they move into later middle age, and become more independent, grapple aggressively with financial challenges, face the adaptational challenges of solitary living, adopt a managerial style in their social and domestic life, become, in brief, more "masculine." Gutmann concludes by stating:

> "Masculine" and "feminine" qualities are distributed not only by sex but by life period. Men are not forever "masculine"; rather they can be defined as the sex that shows "masculine" traits before the "feminine" pattern. The reverse is true for women. (p. 312)

Social-clock norms

Age-status conventions have been important in influencing the style in which individuals carry themselves and behave socially. It is epitomized in the admonition to "act your age." When a 50-year-old at a party behaves like an adolescent, onlookers regard it as a transient aberration brought on by a "high-jinks" environment or by too much to drink. But when the same person behaves like this across situations and settings, it causes concern about judgment, emotional balance, mental health, and vocational future (see Helson, Mitchell, and Moane, 1984; Helson and Moane, 1987).

The social-clock paradigm for understanding stage-based norms for proper behavior has been greatly eroded over the past 30 years. The social mores revolution of the 1960s, captured in the cult film, *Harold and Maude*, has worked its way into our collective psyche. Neugarten (1979), an influential developmental psychologist, argued that our "social clock" that governed age-appropriate behaviors was getting reset if not disregarded entirely, and she demonstrated that age-related norms were becoming irrelevant. Fluidity had been introduced into the life cycle of contemporary North Americans, and in the timing of educational, career retirement, and even parenting initiatives. The social clock is being reset by large segments of the population who are individualizing their life projects. The fixed order of the past has been banished as people sense that they are increasingly free to engage in developmental tasks that are not tradition compliant. We may expect this pendulum to swing back some day.

The influence of social convention is weakening. Watching the dress codes shift toward the informal pole of a "formality" continuum at professional conferences, university seminars, anniversary dinners, and other previously rigidly controlled social events provides anecdotal evidence that role-based norms are being loosened. The practice of "dressing down on Fridays," even in law firms, is growing (whether this continues in difficult economic times is an empirical question yet to be answered). Nevertheless, the social rewards and punishments for compliance and noncompliance with age-based norms continue to exercise great force in our lives. Brent Roberts (1997) tested the "plaster theory" of William James (that is, that personality does not change after the age of 30) as opposed to the "plasticity theory" that personality does change as a function of the demand characteristics of the jobs to which one aspires. He studied eighty-one women between the ages of 27 and 43 who were at various stages of their career development and found that there was a strong correlation between their work experiences and personality change. Of particular interest is his finding that these women became more agentic and "masculine" as they grew older and more embroiled in the hurly-burly of the business world.

Simply aging does not effect changes. It is the interaction of a maturing body with educational, occupational, and socio-political imperatives that brings about alterations in the way humans address adaptational challenges. Bernice Neugarten, the estimable developmentalist, stated that "age itself is an empty variable, for it is not merely the passage of time, but the various biological and social events that occur with the passage of time that have relevance for personality change" (1977, p. 633). The most adaptive individuals make effective changes without unduly sacrificing what is most personal and distinctive about themselves. Although there is a commonality about the way in which individuals face major milestones in their lives, say, the death of their parents or their graduation from college, each copes and adjusts in ways that prompt one to say, "yes, I recognize her in these coping responses."

Constraints on change

The *behavioral genetics* of personality apply brake and gas pedals to personality change. As was noted in Chapter 4 on biology and personality, there are many genetically programmed patterns in the dispositions and traits that evolve slowly over the life span. Preprogrammed alterations are evident in biologically based life transitions, for example, the onset of *menarche* or menopause, or in the physical decline of the aging human. Other changes can be the result of a simple willfulness on the part of the individual, compulsory exposure to various value systems such as are found in a prison or reformatory, a traumatic experience such as a near-death accident or diagnosis of a life-threatening cancer, or in a transformative psychotherapy. But evidence exists (for example, Costa and McCrea, 1988; McCrae and Costa, 1990) that the broad outlines of character involving conscientiousness, risk-taking, emotionality, and other *basic traits* change, if at all, at a glacial rate.

Evolutionary psychology

Heatherton and Nichols (1994) stated that "adaptations and traits that have developed over the course of human evolution should be relatively ingrained and unchanging" (p. 4). If species-typical, these traits form an aspect of our human nature. But can we characterize human traits that are universal as personological? Grieving at the death of a loved one, for example, is considered to be a universal human response to such an event; all human societies have funeral rites that aid the bereaved to adjust to their loss. Similarly, everyone experiences shame when humiliated. If one were asked to describe a friend's personality, one would not say, "well, he's the sort of person who would grieve at his mother's death," or "he feels deeply embarrassed when he's humiliated." "Ingrained and unchanging" traits are not personological *per se*. They become useful only when they express idiographic departures from human universals and individual differences in a clearly defined referent group. Magnanimity and vindictiveness, generosity and miserliness are not universal traits. Hence, they become useful personality descriptors, as when we characterize one individual as vindictive and another as magnanimous, one person as miserly and another as generous. *Individual differences* in modes of response to adaptational problems and other social, cultural, political, and economic demands are what allow us to define someone's personality easily. Those human activities in which there is little evident difference from individual to individual and from culture to culture cannot easily be brought under the umbrella of personality.[13]

Evolutionary psychology's structural mechanisms are classic hypothetical constructs. They are inferred physiological mechanisms that presumably exist in the organism. We have never examined, or even seen, any of these mechanisms.[14] Buss (1999) and other evolutionists have argued they are necessary, leaving themselves open to the criticism (for example, Eagly, 1995) that their model of human behavior is less empirical than speculative. They have built this edifice on Darwinian premises, reasoning that because people exhibit patterns of behavior that vary consistently in a variety of situations, these dispositions must have evolved from age-long responses to persistent adaptational problems.

David Buss (1999) suggests that a useful way to bridge the theoretical domain of universal human psychological mechanisms, on the one hand, and individual differences, on the other hand, is through an analysis of sex differences. This is particularly effective, as mate selection helped to determine the evolution of human emotion; it also reveals salient and easily verified individual differences. In this perspective the phenomenon of jealousy, for example, is triggered by a "mechanism" that can be plausibly hypothesized to result from the intrasexual rivalry that was part of the mate selection and retention challenges of humans through eons of evolutionary adaptation. Moreover, it gives rise to several adaptive strategies for beating out the competition, including displays of physical prowess, demonstrations of fidelity and long-term commitment, subversive gossip about one's rivals, and evidencing ability to be an exceptional provider. The choice of one or more strategies and the range of skills that one can deploy in any one of them gives wide scope for classifying personality. But other adaptational challenges arrive as one moves from life stage to life stage, social sector to social sector. The characteristic ways in which individuals cope with them is what Tooby and Cosmides (1990) label "reactive heritability" (p. 59). One chooses those domains in which one has inherited special gifts.

[13] Distortions of nomothetic human traits are dealt with in Chapter 12 on disordered personality.

[14] A distinction has long been made between intervening variables (IVs) and hypothetical constructs (HCs). For a sophisticated analysis of the types of constructs used in psychology, see Kenneth Spence (1963).

Habits

Stability in personality is in part the result of inherited temperament.[15] Although temperament is inherited by humans, it can still be radically modified by education and a variety of social and institutional pressures. The favored method for altering a habit is repeated practice of an alternative incompatible behavior. One does not become an expert at parallel parking by doing it once. Although visualization is an important part of the repertoire of sports psychologists, one does not develop the skill of serving aces in tennis by thinking about, and then executing the maneuver a few times. Complex skills such as concert-standard piano playing require years of practice. The same can be said of interpersonal skills, regardless of one's temperamental endowment, for impetuous people can be taught to keep a grip on themselves; lethargic people can be taught to be task-oriented. This raises the matter of habitual behavior.

Acquisition of skills

The great advantage of an acquired habit is that it allows us to perform certain acts with ease. Of course, the more complex and difficult the act, the more practice is required to perform it expertly. Learning to play Chopin, golf below par, or deliver a motivational speech are examples of skillful habits. Moreover, if it were not for our ability to form habits, life would become a ceaseless and tiring cycle of clumsy, deliberate efforts to walk, dress, feed ourselves, and do the countless other chores that we do with hardly a conscious thought. Toddlers devote years to learning how to walk with agility and talk with ease. By later childhood, they can carry on conversations hardly giving a thought to where or how they place their feet. But there is a negative aspect to this. Once we pick up a maladaptive habit, it is difficult to unlearn it as anyone who has learned to incorrectly bow a violin or swing a golf club understands.

Learning to speak a foreign language as an adult is extremely difficult, and learning to speak it with the accent of native speakers is even more challenging. There are sensitive, indeed critical, periods in one's childhood for the learning of language and the ability to perceive language contrasts. Researchers have demonstrated that it is by the second half-year of life that infants need to be exposed to the phonemes of their first language in order to develop the ability to discriminate speech contrasts

[15] Dog breeders may say "He has the temperament of a Lab," meaning that the dog is friendly, gentle, playful, perhaps boisterous, and highly sociable. However, even strains of dogs that have been bred for their attack abilities can be taught to be gentle and friendly; we label as habits these relatively permanent changes in behavior.

easily (Nelson, 2006, p. 70). By the age of 12 months infants have largely lost the facility to make such discriminations in languages to which they have not been exposed. Charles A. Nelson (2006) notes that "English-speaking adults who have not been exposed to such languages as Swedish, Thai, or Japanese experience great difficulty in discriminating speech contrasts from these languages" (p. 70).

Henry Kissinger has lived in the United States for over 60 years yet still speaks with a thick German accent (Nugent, 2002). If he has put personal effort into acquiring a Boston or Washington accent he has clearly failed – and conceivably he prefers not to. The fact that he arrived in the United States after the age of 12 partially explains this fact. Elderly Italian and Swedish immigrants who have spent their entire adult lives in New Jersey or Minnesota still speak with discernible Italian or Swedish accents. But what is true of such psychomotor skills as speaking a non-native language is no less true of depth perception, face perception, and pattern perception (see Nelson, 2006, pp. 67–72 for references), not to mention the many personal mannerisms, thought patterns, and belief and attitudinal systems that constitute one's personality. Merleau-Ponty wrote that both personality and culture become habitual and "sedimented" in one's body. Personality, indeed, partially reflects the culture in which one has been educated.

William James ([1890] 1990) spoke of this phenomenon in his luminous chapter entitled, "Habit." He quoted a mid-nineteenth-century psychologist, L. Dumont: "Everyone knows how a garment, after having been worn a certain time clings to the shape of the body better than when it was new ... [and how] it costs less trouble to fold a paper when it has been folded already" (p. 69). Dumont goes on to speak of the habituation of the nervous system to certain environmental stressors such that even after a lapse of time, it responds in the way to which it had once become accustomed. Behaviors, whether adaptive or not, once begun, seem to perseverate by simple inertia.[16]

The plasticity of the brain and the rest of the CNS allows it to adapt to new information, to establish new reflexes, to form new dendritic

[16] William James ([1890] 1990, p. 77) wrote about the accomplishment of magicians in simultaneously performing complex cognitive and psychomotor activities. "[Robert] Houdin early practised the art of juggling with balls in the air; and having, after a month's practice, become thorough master of the art of keeping up four balls at once, he placed a book before him, and while the balls were in the air, accustomed himself to read without hesitation. 'This,' he says, 'will probably seem to my readers very extraordinary; but I will surprise them still more when I say that I have just amused myself with repeating this curious experiment. Though thirty years have elapsed since the time I was writing, and though I have scarcely once touched the balls during that period, I can still manage to read with ease while keeping three balls up'" (p. 77).

branchings from neuron to neuron, and to initiate new patterns of behavior. For every sensory neuron that feeds fresh data to the brain and spinal column there are about ten motor neurons that essentially drive the rest of the organism (Mahoney, 1991, p. 101).[17] But what is truly remarkable is that for every motor neuron there are about 10,000 interneurons that form a complex maze of interconnections that bind the organism to itself and its own experiences. Mahoney (1991) comments on the sheer *ipsativity* of this network: "If we accept the traditional notion that one's sensory receptors constitute one's contact with the outside world, we are forced to conclude that one is much more extensively connected with oneself than with the external environment (at a ratio of 100,000 to 1)" (p. 102). This relative insularity ensures stability in the face of the flood of ideas and numerous social pressures that would otherwise have us vacillating like a weathervane in the wind.[18]

Although the amount of variance in personality traits within any population that can be accounted for by heredity is larger than we thought in the past, it is a mistake to underestimate the power of education and acculturation in shaping behavior, particularly in the early formative years. And although individuals often self-select themselves into those professions that are most congenial to their temperament, the power of one or another workplace to homogenize the behaviors of those working there is self-evident. The mannerisms of the longshoreman are not those of the jazz musician doing nightclub gigs, nor those of office managers in a financial center. James ([1890] 1990) observed:

> Already at the age of twenty-five you see the professional mannerism settling down on the young commercial traveler, on the young doctor, on the young minister, on the young counselor-at-law. You see the little lines of cleavage running through the character, the tricks of thought, the prejudices, the ways of the "shop" in a word,

[17] The sensory neurons that carry data toward the CNS are also termed afferent neurons, and the motor neurons that carry signals within and away from the CNS are called efferent neurons. A mnemonic to help students remember the meaning of these terms is to understand that the Latin prefixes *ad* and *ex*, meaning toward and away, respectively, have been combined with the verb form *ferrens* meaning carrying – which give us *afferent* and *efferent*.

[18] Williams James devoted a chapter to the subject of "habit" in his nineteenth-century classic, *The Principles of Psychology*. He concluded, "Habit is thus the enormous fly-wheel of society, its most precious conservative agent. It alone is what keeps us all within the bounds of ordinance ... and prevents the hardest and most repulsive walks of life from being deserted by those brought up to tread therein. It keeps the fisherman and the deck-hand at sea through the winter; it holds the miner in his darkness, and nails the country-man to his log cabin and his lonely farm through all the months of snow ... It dooms us all to fight out the battle of life upon the lines of our nurture or our early choice, and to make the best of the pursuit that disagrees, because there is no other for which we are fitted, and it is too late to begin again" (p. 79).

from which the man can by-and-by no more escape than his coat-sleeve can suddenly fall into a new set of folds. On the whole, it is best that he cannot escape. It is well for the world that in most of us, by the age of thirty, the character has set like plaster, and will never soften again. (p. 79)

Personality change through psychotherapy

Since the middle of the nineteenth century, psychotherapists have led the effort to formulate viable theories of personality, personality change, and human development. Because therapists help people to reshape not only their lives but, in many instances, the personalities which serve their evolving needs, they find few scientific questions that are as salient as those addressed in this chapter. The issues of what human nature is, what constitutes personality, what is normality (and abnormality), how personality is shaped, and, once shaped, how it is changed, confront therapists on a daily basis. The approaches they have developed to effect change in their clients have always presumed a certain vision of the Human, and psychotherapy has always been enmeshed with personology. Epigenetics, alluded to at the end of this chapter, provides a window for therapists to treat clients who need a partial personality makeover.

Group strategies to change

Milieu therapy is predicated on the principle that the most effective way to modify the structure of personality is by radically changing people's social environment. The milieu therapy movement emerged following the Second World War and was influenced by several causal streams. Social and community psychiatry began to flower when the notion that people are shaped and misshapen by the social environment in which they live – the clubs, gangs, families, schools, neighborhoods (poverty-stricken as well as affluent) – gained wide acceptance. However, as early as 1924, Trigant Burrow had maintained that it was society that pathologizes its members; people internalize the images of the fashion setters and come to think of these as accurate representations of normality. As it is through social groups that people are damaged, it is likewise through groups that they must be restored to health. Therapists who have followed in his tradition promote the integration of those who have been misshapen by society into healthier, validating milieus.

Milieu therapy assumed a number of forms (Jones, 1969, 1984) following the Second World War. One of its early manifestations was the *therapeutic community* (TC), which in its most evolved forms required clients to take up residence in an enclosed, often secluded, setting. A medical version of this was the hospital ward in which the residents were given some authority to govern themselves within an established organizational and therapeutic framework. Its origins can be traced to a group of nurses working in Mill Hill Emergency Hospital, north of London, with shell-shocked, war-battered veterans during the Second World War. These independent, non-authoritarian women discovered the value of bringing patients together to discuss the nature and course of their illnesses. Their work caught the attention of a young Scottish medical resident, Maxwell Jones, who realized that these group discussions were transforming the entire social structure of the ward. As the practice spread to other wards of the hospital, evidence began to accumulate that there were numerous therapeutic benefits that flowed from this social approach to patient care. The communal nature of the procedures evolved into the prototype of the "therapeutic community" and soon began to spread to other hospitals under mandate from the British Ministry of Health. It replaced pure custodialism, on the one hand, and psychoanalysis, the dominant form of psychotherapy at the time, on the other hand (Shorter, 1997, pp. 232–38).

Individuals who enter a contemporary TC for treatment are usually invited to immerse themselves in a community or social group that is very different from the one in which they have habitually malfunctioned, to escape from what is familiar, and to geographically displace themselves: in brief to commit to a total environment that requires a change in the daily routines of hygiene, work, and recreation. The rationale for this is that to effect behavioral change it is helpful to eliminate the situational cues that trigger and support a dysfunctional lifestyle and to engage in a comprehensive program in which different tasks and different rewards are the order of the day – every day. This treatment modality typically schedules work sessions, enforces standards of personal hygiene and grooming, requires participation in group therapy of several kinds, programs various kinds of social events, exacts public personal admissions of delinquencies during community meetings, enlists members in educational and vocational programs, and generally involves residents in a comprehensive socialization program that reaches even into the details of kitchen service and bath and toilet maintenance.

TCs typically regard their residents as forming a family whose members are responsible for maintaining the harmonious and efficacious functioning of their household and community. The changes this indoctrination has been demonstrated to achieve result from bringing one's self-attributions and identity into line with one's institutional behaviors.[19] The insights that

[19] The *cognitive dissonance* (Festinger, 1957) created by the *socio-therapeutic* pressures exerted by the TC may possibly account for some of the behavioral changes that are observed in almost all abstinent addicts, at least as long as they remain in residence.

flow from the community experience and the feedback that is provided to individuals on a daily basis purportedly effect a makeover of personality that it is hoped will last long after the TC experience is over.

An early, but typical, example of studies that have tested the effectiveness of this treatment modality was conducted by Skolnick and Zuckerman (1979). They compared the progress made by fifty-nine drug addicts who had been treated in a TC with thirty-seven drug addicts who had been confined in prison, untreated, for the same length of time. Using the Minnesota Multiphasic Personality Inventory (MMPI) and the Eysenck Personality Inventory (EPI) as their measures, they found that the fifty-nine treated addicts showed marked reductions as a group in measures of general psychopathology, whereas the thirty-seven untreated prisoners remained, not surprisingly, unchanged. Further, the TC group evidenced a decrease in sensation-seeking and corresponding increases in extraversion and tendencies to "self-actualization." They did not, however, show improvement on the psychopathic deviate and hypomania scales and on measures of impulse control.

Recidivism among graduates of these programs nevertheless remains high. The deeply-rooted habits of an earlier life are often revived in the re-entry process, as the old environments renew their pathogenic grip on them. The old companions, the shooting galleries, the smells and other sensations of earlier experiences can pull reformed addicts back into the social coils that cause them to spiral into the dysfunctional lifestyles of old. The more the re-entry scene approximates the setting of a former addictive lifestyle, the greater the danger that reformed persons will relapse. For this reason it is always advisable for addicts to re-enter mainstream society in settings that differ as much as possible from those in which the crippling dysfunctions took birth. The roots of this problem lie in the fact that it is impossible to *unlearn* a bad habit. As Baumeister (1994) says in a different context, "once the self has been defined in a certain way, this identity cannot be completely undone" (p. 293). Chronic and skilled bank robbers never unlearn in prison how to rob a bank, any more than they can unlearn how to ride a bicycle. The treatment of choice is to *train and incentivize them to do something else*, to engage in productive and socially rewarding occupations that give them more satisfaction and fewer serious disincentives.

Personality change through individual therapy

Models of personality change

Mahoney (1991), in a book entitled *Human Change Processes*, contrasted various models of human change (pp. 255–57). He took the therapeutic vantage point, like Alfred Adler before him, of helping change patients' lifestyle. Psychotherapy is predicated on the assumption that people *can* change, in the surface attributes of values, interpersonal style and manners, vocational interests, and recreational preferences – as well as in the deep recesses of the psyche in which unspoken, often tacit, trait structures are embedded.

To facilitate their work, psychotherapists have elaborated ambitious models of human change that implicitly endorse divergent conceptual models of human nature and how it fits into the larger scheme of life and cosmos. The notion of the self is absent or considered a fictive construct in some models (for example, behavioral), a real entity in others (for example, Rogerian). Psychologists take different stances on the plasticity of the human psyche and the various possibilities for change in different life stages. Certain psychodynamic models, such as the psychoanalytic, regard personality as relatively fixed and resistant to change after the age of toddlerhood or, at the latest, adolescence. Humanistic models, on the other hand, regard people as malleable and developing throughout life. Psychotherapies envisage different possibilities and sources of change, some emphasizing physiological variables, others environmental forces. These sources are considered and in some cases demonstrated to contribute in varying proportions to different categories of personality descriptors. They model the mechanisms of change in divergent ways: that is, by manipulating environmental contingencies and "working with" internal change agents, such as reason, neuro-endocrinological "levers," emotions, and other genetically influenced variables.

Various *clusters* of psychotherapeutic systems, whether cognitive, behavioral, humanistic, or psychodynamic, take different positions on all of these issues. Within these clusters, various schools of thought (say, rationalist and constructivist sectors within the cognitive cluster) have developed divergent positions (see Mahoney, 1991, pp. 255–57), although there is a powerful integrative movement in this century that mitigates these divergences. Clearly, the resolution of these divergences will depend, but only in part, on the success or failure of each to validate its claims empirically. Additionally, however, there has to be a convergence of philosophical principles. The differences are so critical in some instances that one can reasonably expect that these will never be bridged. Humanists who take a materialist position on the critical issues are a world apart from those who take a spiritualist position, and both are radically divergent from those who take a predominantly biomechanical and strictly naturalist view of human nature.

In the following sections we will address some broader issues in psychotherapy and restrict the discussion, for illustrative purposes, to two perspectives on change.

Search for meaning and personality change

Sweeping changes in one's environment can alter one's attitudes and values as well as one's view of society. Likewise, life-altering, near-death experiences can affect the pattern of one's friendships and the familial and occupational relationships that had formerly given a specific meaning and coherence to one's life. A classic example for such an epiphany is the mystical experience reported by Paul of Tarsus (Acts 26:12–20) as he traveled the road to Damascus. The result of this subjective experience was a change in his cognitive frame that set him on a course that had a fateful impact on Hellenistic (and later, Roman) civilization. A comparable modern example is Victor Frankl's (1959) experiences in a Nazi death camp, which he survived, unlike many of his companions. Frankl chronicled those events and the meanings in which he took shelter. These meanings developed into a philosophy of life and a system of psychotherapy that he eventually labeled *logotherapy*.[20]

Logotherapy

What is instructive in the experiences that Frankl described is that the impact of the sufferings that individuals endured in the death camps caused far-reaching personality changes – fundamental changes of attitude, values, and social interactional patterns. These changes were for the better as well as for the worse. The *persona* of respectable and honorable people, ordinary by any standards, were transformed to a level of nobility, even grandeur, of which one would never have thought them capable, as they faced implacable tormenters in a drama that they knew would terminate in their death (Frankl, 1965, pp. 84–93). With varying levels of generosity and heroism they refused to betray fellow prisoners to save themselves or to gain transient relief from hunger and pain. Other prisoners, originally indistinguishable from them, self-selected themselves to become *capos*, allied with their captors and "prepared to use every means, honest and otherwise, even brutal force, theft, and betrayal of their friends, in order to save themselves" (Frankl, 1959, p. 7). The transforming power of the environment to mediate the most deep-seated and lofty inspirations of our common humanity – for survival and for self-sacrifice – is documented in the great literature of all nations (for example, Tolstoy's *The Death of Ivan Ilyich*, [1886] 1987).

[20] Frankl initially called his system, largely anthropological in nature, *Existenzanalyse* (Tweedie, 1961, as cited in Patterson and Watkins, Jr., 1996), but he later gave it the title logotherapy for two reasons: to emphasize its clinical purpose and value and to distinguish it from other existential systems.

Frankl's system of psychotherapy attributes health-giving properties to the meanings we develop for our lives. These meanings emerge in our lives at critical points, called, in existential psychology, a *kairos* (Corsini, 1999, p. 521; Ellenberger, 1958). Then one experiences one's humanity at its most fragile; with crystal clarity one views the most treasured prizes slipping from one's grasp, and one is confronted with some ultimate issues: life's transience, one's self-definition as a person enmeshed in a rich web of social relationships, the need for renewed commitment to ends that transcend oneself, making meaning of one's passage through this world, valuing the present moment, and assuring that one does not sleep-walk down a corridor of years that permits no turning back.

Spiritualist perspectives

Reminiscent of Carl Gustav Jung's principle that people who seek therapy eventually bring their spiritual and philosophical questions to the fore, Frankl affirmed the need to deal with such questions. Every crisis, every fork in the road, is a test. These tests cannot be resolved by others, but only by the individual who is living through them in the here-and-now. This goes far beyond the pleasure principle extolled by the Utilitarians and Freudians. Resolving them brings a healing dynamic into play. Finding meaning in the crisis often results in a drastic re-ordering of one's life. Such meaning is the most powerful remedy for the temptation to despair that afflicts people in their worst crises. Rather than query the universe with "What is the meaning of life?" Frankl (1959) stated that it is we who are questioned: "Each man is questioned by life; and he can only answer to life by *answering for* his own life; to life he can only respond by being responsible" (p. 172). The meaning that people find in life is transformed when: (a) they achieve or perform something of significance; (b) they experience love for another human being; (c) they appreciate the value with which a work of nature is imbued; and (d) they find meaning in the losses and suffering that inevitably overtake every human (pp. 176–87). These events are enduring. The only things that are transitory are opportunities neglected and potentialities unrealized.

In *War and Peace* Tolstoy ([1869] 1990) observed that Count Bezúkhov, a prisoner of the French, had come to realize that:

> a superfluity of the comforts of life destroys all joy in satisfying one's needs, while great freedom in the choice of occupation – such freedom as his wealth, his education, and his social position had given him in his own life – is just what makes the choice of occupation insolubly difficult … (p. 577)

The experience of being transported into a new, often terrible, environment can bring us to the realization that there are more important values in life than those that previously enthralled us. Further, Frankl found that altering the meanings we attribute to tragic life events can have a profound impact on personality. His experiences in the death camps provided insights that he extended to his psychotherapeutic practice and formed the basis for the development of logotherapy. These findings have been supported by the research of social psychologists such as Shelley Taylor (for example, 1989) and her colleagues.

Positive illusions

The pursuit of meaning, Shelley Taylor (1983) found, takes some strange twists. She and her colleagues conducted interviews with women who had suffered from a life-threatening cancer and subsequent surgery, who had been raped, and with still others who had been devastated by a bitterly contested divorce. Finding a new and transforming meaning in life became the foundation for their recovery. The researchers had expected to find that the victims would slowly rehabilitate themselves and regain the balance, level of interpersonal functioning, and readjustment of personal goals that characterized their former lives. To their surprise, among the strategies employed by the victims was the fabrication of a number of illusions. They found that many of the victims aggressively addressed the adaptational problems that resulted from their personal tragedies.

Taylor presented her research findings on the role of positive illusions in the lives of well-adjusted people in an influential article that she later elaborated into an often-cited book, *Positive Illusions: Creative Self-deception and the Healthy Mind* (1989; see also Taylor and Brown, 1988, 1994). Like Frankl, Taylor focused her investigations on the manner in which highly stressed individuals dealt with calamity in their lives. Three strategies emerged. The first strategy, and potentially the most important in terms of personality change, is searching for an answer to the question: "What does my life mean now?" (1983, p. 1161). The answer to that question involved understanding the *significance of the event* in their lives. They asked themselves: "What caused these bad experiences?"; "Why did this shit happen to me?"; "And where should I go from here?"

The second strategy involved themes that were conditioned by this search for meaning. Gaining a sense of mastery over the disease, the sexual assault, or the destructive divorce proceedings (and those consequences that turned their lives upside down) led them in an intense search for remedies. Getting a grip on the causes of, say, a cancer, and control over the environment that was a presumed condition for it to occur, became a passionate quest for many of them. The quest led them to underestimate the influence of variables over which they had no control and to ignore the fact that there are always conditions that shape our behavior of which we cannot possibly be aware. The complexity of the biological reality that *is* our body, with the innumerable billions of organisms that have colonized it and that, indeed, sustain it in life, allows us only limited control of what happens within it. Additionally, many social and geophysical variables, with all their subtle but compelling influences, operate outside our personal awareness.

Humans choose to focus on those variables that they can control and (of course) of which they are aware. Unfortunately, evidence that those variables have significant relevance to their condition is often missing. In a process of reasoning that one can only characterize as grasping at straws, humans who have faced life-threatening diseases (without remedy) seek out methods of treatment that as yet have little or no foundation in science. They find individuals who, like themselves, have embarked on an altered lifestyle and undertaken a radical biopsychological program. Striving to gain control over their cancer, victims try a variety of psychological techniques, such as self-hypnosis, imaging, meditation, and positive thinking. As one woman stated, "I'm realigning the cells."

In a desperate effort to gain a *mastery* over a threat to their very life, others seek out physiological approaches that have little scientific basis. Many fall in the category of treatments generally known as complementary and alternative medicine (CAM). Some may sell their houses and move to a neighborhood that is less "toxic." Others will make drastic dietary changes. Although any of these changes may be beneficial for some, they are pursued with a zeal and commitment that bespeak a new belief system. What characterizes them all is that they mark a dramatic discontinuity with a lifestyle that preceded the cancer diagnosis and surgery, and they provide a new cognitive structure in which to fashion a world over which they have some control. These radical new measures provide an illusion of control. Those who engage in this process reveal an optimism that may appear highly unrealistic.

The third theme Taylor identified was the need to *regain self-esteem*. Research had demonstrated that there is an initial disabling loss of self-esteem following a divorce (Thabes, 1997), the death of one's child (Chodoff, Friedman, and Hamburg, 1964), or going on welfare (Vosler and Page-Adams, 1996). Within the timeframe of Taylor and her colleagues' study in which they interviewed seventy-eight women with breast cancer, they found that many of them engaged in esteem-enhancing cognitions. Surprisingly, some not only saw themselves as well-adjusted humans, they regarded themselves as better adjusted *than before they became ill*.

For some there was a drastic re-ordering of priorities in their lives. Personal relationships were increasingly prized and fostered. Their material possessions became less important to them. Fashion and social status were placed in a different, down-graded perspective. Many adopted a "count-your-blessings" attitude to meet their need for peace of mind in the face of the diminished resources and opportunities that life could afford them. Freed from former cravings, they sought pleasures in the here-and-now, with a heightened awareness of the brevity of life.[21] Most interestingly, many engaged in downward social comparisons. Cancer victims compared themselves with those women who had had more serious operations at more compromising stages of their life cycle. On the one hand, some saw themselves as more stoic and psychologically stronger. On the other hand, those who had had lumpectomies sympathized, perhaps self-servingly, with those who had had an entire breast removed (a simple mastectomy). The elderly woman who had had a mastectomy felt blessed compared with those who had been mastectomized at a young age. Those who were younger and married felt privileged that they were not single and faced with the task of explaining their condition to a suitor. There was always someone whom fate had treated less kindly than oneself.

Meaning and self-schemas

The meaning that we give to certain events in our lives derives from our *self-schemas*. Self-schemas give us a coherent, unified sense of who we are; they are belief systems that bear on such personality attributes as we know we have. In brief, a self-schema defines our identity. When an event is traumatic, if not catastrophic, it precipitates an abrupt change in the routines of our daily life that cascade in various directions. The more profound the changes, the more people sense the need to reshape the self-schemas that define them. The person who has just passed through the ordeal of prolonged and bitter divorce proceedings, resulting in a new civil status, a change in financial health, material well-being, and separation from loved family members, necessarily makes corresponding adjustments in her self-perceptions. Life suddenly becomes a more serious project, entailing a new set of values, new objectives, new ways of relating to institutions, friends, family, and strangers. People who have endured a life-shattering ordeal selectively attend to events in their environment to which, theretofore, they had paid scant attention. Those who fare best *tend*

[21] Consider the parallels between these cognitive shifts and the principles of some Asian contemplative disciplines (see, for example, Walsh, 1999).

to place their relationships in a new light, lose their fascination with the latest gossip, the latest fashions, and the glittering but transient social scenes in which they had previously participated, if only vicariously.

There is an abundance of research demonstrating that self-enhancing biases ward off depression and that those who are chronically negative about themselves are prone to depressive episodes (for example, Abramson and Alloy, 1981). Just as individuals can adopt a *helplessness* schema in the face of challenge, they can, contrariwise, adopt a helpfulness schema. Both are personological features of individuals, which Seligman (*cf.* 1975; see Seligman and Csikszentmihalyi, 2000) argues can be learned. Those who experience traumas and attribute them to stable, pervasive, negative internal qualities in themselves incline toward depressive states. Conversely, those who in similar situations attribute self-enhancing and positive qualities to themselves arrive at more adaptive, healthy levels of daily functioning. Putting a positive spin on their circumstances is characteristic of optimistic individuals, which provides a salient prognostic sign for those recovering from calamities, whether medical, financial, or interpersonal. The conclusion drawn from the research in positive psychology is that positive illusions, those subtle biases that allow us to take an optimistic view of our future even in the face of terrible adversity, foster recovery and mental health (Taylor, 1989, pp. 199–226; Seligman and Csikszentmihalyi, 2000). This calls for a revision of Carl Rogers' famous dictum, *the facts are always friendly*. The age-old principle that accurate and unflinching perception of reality supports recovery of mental health needs to be modified. The implications for personality psychology are significant.

The crystallization of discontent

When individuals experience a large number of unpleasant events in connection with a current life situation, they may regard them as distinct and unrelated. These unsatisfactory, perhaps distressing, episodes may coalesce over time into a well of resentment, which at some point overheats and boils over. Roy Baumeister (1994) developed this commonplace sequence of events into a construct that he called the *crystallization of discontent*. This construct is important for it is operationalized as the shifting of personal attitudes that effect major, often sudden, changes in lifestyle. Like Shelley Taylor and Seligman, Baumeister regards the interpretive activities in which people engage as important in bringing about changes in their social, occupational, and personal lives. These subjective processes mediate the emergence of new meanings in life, noticeable personality change, and often drastically altered life projects.

Although crystallization of discontent can be considered a special subset of conditions that motivate individuals to improve their psychosocial condition, it has wide ramifications for explaining any change in one's life that one initiates. The reason is that every change that is made is necessarily related to some valued goals – and the effort one makes to achieve them. The motivation to improve one's situation comes from assessing the history of interactions with an employer, a spouse, one's neighborhood, a social club, a professional association, or any other social entity with which one has had an ongoing relationship. Discontent can arise simply from viewing the gulf that separates the benefits afforded by a current situation and the significant improvement that could result from an alternative one. But change always entails some costs, and in some cases these costs will outweigh the benefits.

In Baumeister's perspective, many individual episodes that remain subjectively unrelated can have little impact on one's self-perception and one's very identity. However, when they coalesce into a Gestalt and are clearly seen as parts of a bigger picture, creative alternatives snap into place. The individual begins to seriously consider terminating a relationship, relocating, changing his or her occupation, or adopting a major new role in life. Baumeister (1994, p. 282) presents the example of a married couple who have been drifting toward mutual alienation and divorce, although neither may have consciously considered it. Suppose a long series of disagreements has troubled them, but these fights have been regarded as isolated and unrelated events. The couple may have had squabbles about who should have attended to paying a bill, doing the laundry, or bringing the youngster to the hockey rink at 6 a.m. A pervasive undercurrent of dissatisfaction with the sharing of chores may periodically erupt into a fight. These may all be ascribed to the other's having a bad day or some indisposition. But at some point the wife, say, may reflect to herself, "Hey, this guy is chronically insensitive to my needs. Are we just incompatible?" This crystallization of discontent can entail a rethinking of one's life course, the adoption of a new persona, and some serious consequences for major facets of personality.

Another example that Baumeister presents bears on the disillusionment that some individuals experience with a particular political or religious group to which they had formerly adhered. Such members had been willing to overlook the failures within these groups by attributing them to the delinquency of individuals rather than to the institution, say, a religion or political party or fraternal order. When they became subjectively convinced that the problems were systemic, their disenchantment with the institution gelled into a decision to separate (see Bromley, 1988). Baumeister insists "there does not have to be any change at all in the objective circumstances" at the root of the disillusionment, "what matters is the sweeping change in the way the person perceives and interprets them" (p. 286). The personological impact of such a far-reaching life-decision on those who abandon closely-knit groups is often heightened by their being ostracized by those who close ranks within the wounded institution.

Gradualism in personality change

There is a reasonable presumption that dramatic changes in individuals' personality do not happen spontaneously. There are multiple and lengthy chains of causality that converge to generate the action-specific potential that enables an individual to engage in a tragically crazed behavior or to reach a radical, far-reaching yet well-reasoned decision. To crash a plane into an office tower, assassinate university colleagues, or kill several classmates while they are at lunch or in a prayer service can only issue from a fountainhead of rage that was many months if not years in building (*cf.* Lipsitt, 2002a). To sue for divorce or change careers in mid-life, decisions that millions of mentally healthy people have made, must in this perspective have been long in gestation. In 2001 Andrea Yates drowned her five helpless children. Mental health professionals, as well as the general public, have been at a loss to give more than speculative explanations for her tragically aberrant behavior. But Lipsitt argues that it is not plausible to think "yesterday she was okay; today's she's not." He continues, "the human psyche is not like that. Minds don't snap and nerves don't break down all of a sudden."[22] Consistent with Baumeister's thesis, he argues that the lines of causality stretch deep into the past.

There are many reports, on the other hand, of dramatic changes in individuals' patterns of behavior that appear to have been precipitated by a unique and momentary constellation of events. Reports of radical changes in personality structure as a result of a single therapeutic experience are numerous. Raymond Corsini (2000) tells of the experience he had with a convict in Auburn Prison in New York. The convict was about to leave on parole and wanted to thank Corsini for what he had done for him. The latter had no recollection of ever having spoken to the man. The convict reported that following an interview with Corsini he had immediately

[22] Lew Lipsitt (2002b) reminds us that dichotomies are always suspect and that he does not hold to any polar position. "Crucial 'defining moments'" can precipitate profound life-style changes. "Sudden change in some alcoholics' behavior (like drunk today, and sober tomorrow and forever)" is evidence, on the one hand, that change can be sudden. On the other hand, the change requires a prepared mind, and the incubation of the personality change could have been lengthy.

sought a job in the prison machine shop, began hanging out with a different group of men, pursued a high school equivalency diploma, started writing to his family and attending church services, and took a correspondence course in drafting. He even had a job lined up with a drafting company (which, in fact, he took as soon as he was discharged). He continued, "my family is grateful and remembers you in their prayers." Stunned, Corsini, said, "Are you sure you have the right man?" "It was you all right, and I will never forget what you said to me. It changed my life." "What was that?" asked Corsini. The convict replied: "You told me I had a high IQ" (pp. 4–5). This comment, based on the result of an aptitude test that Corsini had administered to him, tipped him into a cascade of initiatives that resulted in a radical alteration of his self-concept, lifestyle, and personality. Corsini relates other life-altering changes that have occurred after 15-minute therapist–client interviews (1997, 2002).

Quantum personality change

Sudden, profound, and enduring character transformation is often reported in the psychotherapy literature, not because it is a common occurrence – on the contrary it is relatively rare – but because it has significant *heuristic*[23] value. Miller and C'deBaca (1994) report a number of these conversions of lifestyle, from the pathological to the adaptive and healthy or vice versa. Some involve a sudden degradation of personality, as when a 15-year-girl awoke late and found her parents' bodies – the victims of violence – strewn in different rooms. She was placed in an uncaring and depersonalizing foster home. She reported, "Every aspect of my life, from getting up and brushing my teeth to going to church to knowing who I was, was over. I remember thinking I shouldn't even have a name, because I'm not a person anymore" (p. 270). She became mute and did not utter a syllable for a year.

The renowned psychotherapist, Milton H. Erickson, suffered as an adolescent from advanced poliomyelitis. One evening he overheard his mother speaking outside his bedroom with three physicians. One of the doctors said to the mother, "the boy will be dead by morning." After the physicians left, he asked his mother to rearrange the furniture in his room in such a way that he could watch his "last" sunset. This cameo experience served as a model for much of his later therapy, a guiding principle of

[23] A heuristic is a discovery-oriented device the purpose of which is to facilitate the investigation of problems. Heuristics are tentative models that have an exploratory character; they often lead to more rigorous scientific hypothesis testing.

which was "Always have a realizable goal in the near future" (Rosen, 1982, pp. 52–53). Erickson survived polio, although he always suffered from it, especially following a relapse during the last decades of his life. But the "sudden" therapy approach that he used in his successful career can be traced back to this event. "Realizable goals in the near future" were the hallmark of his procedures. Often therapy with Erickson was accomplished in a few brief (occasionally single) sessions, and clients left with only the instruction to engage in a single activity (say, an afternoon visit to a botanical garden). Reflections on the visit precipitated profound lifestyle and personality changes.

Mesmerian crises and abreaction

The history of clinical psychology and psychiatry is replete with accounts of profound personality change following powerful, shattering emotional upheavals. As it is widely recognized that traumatic and environmental upheavals in our personal life can bring on personality disorders, such as post-traumatic stress disorder, it is equally evident (for example, Greenberg and Safran, 1987) that emotional experiences can effect changes in the personality of the "worried-well" and healthy. Personality change does, in fact, arise from therapy as well as from near-death experiences. These changes can be for better or for worse (for example, Lieberman, Yalom, and Miles, 1973). More recently Yalom (1995) stated that his successful clients often reported that the "defining moment" in their psychiatric recovery was an emotional episode in which they exploded in anger against the group in which they were "working." Continued acceptance by their therapy group and their sense of personal validation despite having temporarily "lost it" *convinced them* of their intrinsic human worth. These critical moments hold the promise of reintegration into the normal society in which they felt they were aliens.

These life-changing experiences almost always occur in the context of human relationships. Life-altering events require the presence of caring others who can share vicariously and empathically in the pain, anxiety, and quandaries of the sufferer. The simple experience of upheaval, of course, is not sufficient to effect improvement; exploring the meaning of the experience through reflection, analysis, and discussion is also necessary. In McAdams' terms, one must "re-story" (1994, pp. 719–36) one's life. Numerous experiential psychotherapies (for example, Mahrer, 1989, 1996) are founded on this approach to helping individuals reconnect with their authentic selves.

Continuity versus instability, redux

This chapter has focused on the possibilities for (and benefits of) change rather than on constancy and continuity. The student of research methodology knows that the way in which we frame the issues that we wish to research subtly influences the outcomes that we get. Should we include in personality structure such variable elements as idiosyncratic scripts, interests, belief systems, defense mechanisms, and cultural affinities, our conclusions will tilt toward the possibilities for change. Science has traditionally sought regularity, lawfulness, and predictability in the study of nature. It is understandable that developmental psychologists are inclined to find the regularities in human personality rather than the random, multifarious, and unpredictable determinants of a contingency-based personology. Psychotherapists, who daily study the effects of *fortuity* in shaping people's lives and attitudes, are drawn to understand how hurtful chance events can *alter* personality and how systematic treatment can restore it to higher levels of functionality. As noted above fortuitous events can be intrinsically mind and lifestyle altering (Bandura, 1999, pp. 159–60).

Epigenetic change

The plasticity of the human organism is self-evident. What is not self-evident is that there are many alterations of the CNS and other gene-influenced sensory, endocrinological, and visceral systems that are the result of environmental factors (for example, Champagne and Curley, 2005). Epigenetics accounts for some of this plasticity and is of particular interest to the psychologist and mental health service provider. The term refers to a process whereby events within the organism and in its environment turn genes on or off, and at varying operative levels. This explains why one twin of a pair of identical twins (who can be considered clones of each other) may develop a disease, such as diabetes or schizophrenia, and the other may not (for example, LeDoux, 2003, pp. 260–300). The process of change often involves signal environmental events, such as the pleasurable experience of receiving a compliment or a gift. As Onur Güntürkün (2006) has written, "Perceiving and attending to some environmental stimuli (for example, a novel smell) is accompanied by neuronal interactions that can induce the activation of immediate early genes … that show a rapid and transient expression immediately after resting cells are stimulated" (p. 387) by, say, hormones and neurotransmitters. Introducing a novel idea to someone (indeed, to oneself) may or may not cause a novel expression of a latent gene, but it certainly entails synaptic changes – what LeDoux refers to as a rewiring of the brain

(p. 299). Klaus Grawe (2007) noted in *Neuropsychotherapy: How the Neurosciences Inform Effective Psychotherapy*, "Psychotherapy, as far as it leads to substantial behavior change, appears to achieve its effect through changes in gene expression at the neuronal level" (p. 3, quoting Kandel, 1996, p. 711). Clinical interventions by a skilled psychotherapist are part of the environmental press on a patient that brings about neurohumoral and personological changes. The domains in which this scientific discovery has immediate application are numerous: international diplomacy; personnel management; the mental health service professions, and others.

Conclusion

Trial lawyers have long taught us that the way they formulate questions that they address to witnesses can bias the answers given. "Can personality change?" is one such question. Pervin (1996) asked "would discussion follow a different course if we asked, Can personality be stable?" (p. 315). The mainstream position of personality psychologists today is that there are some aspects of personality that are relatively stable, and there are other aspects that are relatively unstable and evolving. The controversies boil up over the elements to be included in the construct, personality, and the degree of stability and change longitudinally found among these elements *within* individuals and *between* groups of individuals.

Likewise Caspi and Roberts (1999) have asked, "When is personality fully developed?" Although they are fully cognizant of the fact that there may be no answer to this question, does the question imply that any change after "full development" requires some degradation of one's existing personality? Or at best only marginal and insignificant changes? In any event, the limits on growth and improvement are more constraining than the limits on decline and degradation, and whereas the latter may generate apathy, lethargy, self-indulgence, and hopelessness, the former requires discipline, stamina, hard work, and hope.[24]

[24] That it is far easier to destroy than to build, to degrade than to construct, to prevent an accident than to raise from the dead is captured in the following aphorism. Rosten (1968) informs us that *mazel* is a Hebrew term that means luck. *Schlimazl* is a Yiddish term denoting a chronic loser. In that light he cites the observation about schlimazls: "From mazel to schlimazl is but a tiny step; but from schlimazel to mazel – oy, is that far!" (p. 352).

12 The disordered personality: evolution of nosological systems

> Disorders of personality are not medical entities; nor should they be seen as human perversities either. Viewed from an ecological and evolutionary perspective, we conceive them as problematic styles of human adaptation. They represent ... individuals whose constitutional make-up and early life experiences have not only misdirected their development, but have also constructed an unsatisfying sense of self, a problematic way of expressing thoughts and feelings, as well as a troublesome manner of behaving and relating to others. Theodore Millon (1996)

Universal norms of behavior

A book that deals with *personality* as its core construct cannot avoid dealing with disordered personality as well, especially as the latter is to a certain extent socially and variously defined. *Personality disorders* were defined in the third edition of the *Diagnostic and Statistical Manual* (American Psychiatric Association, 1980) as "*personality traits* [that] are inflexible and maladaptive and cause either significant impairment in social and occupational functioning or subjective distress" (p. 305). Norms for civil behavior are constantly evolving. From era to era, each society imposes shifting normative behavioral schemas on its members as they engage in routine social and occupational activities. If we accept this principle, the behavioral patterns defining various personality disorders evolve historically but also vary cross-culturally. Drastic, collective divergence from those norms gives rise to new, ultimately respectable cultural models of personality disorder – as well as to "scandal" and moral outrage. In the light of recent advances in molecular genetic and epigenetic research, these principles have become suspect.

Universal norms of human behavior *do* exist (Moghaddam, 1998; Pinker, 2002b, pp. 435–39), and individuals who consistently violate them in a disruptive way can be justifiably labeled personality disordered, no matter to which culture they belong. Turn-taking in conversation among peers is an example. Individuals who persistently deprive others of the opportunity to respond to them and monopolize conversations in a self-absorbed way fit cross-cultural models of personality disorder. Another example of a norm that exists in all cultures is truth telling. Chronic liars are shunned in all societies and are regarded in Western nosologies as personality disordered, all the more so when their falsehoods have serious, socially disruptive consequences. Of course, levels of tolerance for this behavior vary from one society to another. We must not, however, exaggerate the cultural relativity of norms. By definition, universal norms for acceptable behavior transcend cultures, at the level of the family and the society. Such norms are grounded in the physiology that defines the nature of the organism and that can be integrated into a global psychology. That aspect of this subject will now be addressed.

Divergent criteria for similar disorders

Pre-scientific schemas of madness

A vast, relevant topic of exploration that must be passed over in this chapter is how the moral virtues of medieval Europe and its theological notions of the virtuous life shaded gradually into the bourgeois notions of proper living styles. The dark side of those virtues was the disorders of character and mentation that were evident in the nineteenth century not only at the margins of society but among the working and middle classes of Western societies. These disorders were endued with a secular and medical aura as trained professional cohorts emerged to treat them, using such psychiatric devices – often of dubious value – as were available to them.

In Renaissance Europe the distinction between fools and miscreants was not always clearly drawn. Folly became a vice in its own right, indeed, foremost among "the joyous throng of all human weaknesses" (Foucault, [1965] 1988a, p. 24). Mental illness was a moral infraction of right living and needed to be treated as such. For a period some of the more seriously demented were consigned to boats which plied "the rivers with its thousand arms and the sea with its thousand roads, to that great uncertainty external to everything" (p. 11). The *Ship of Fools* furnished passage to nowhere except perhaps to the "vasty deep."[1]

[1] A brief, excellent synopsis of this phenomenon is given by Michel Foucault ([1965] 1988a) in his enchanting chapter, *Stultifera Navis* (pp. 3–37) (unfortunately the book lacks a good index).

Others insisted that the best treatment for the entire spectrum of dullards, social misfits, thieves, vagrants, dements, and screamers was a house of confinement such as the Hôpital Générale in Paris where inmates were obliged to work in a disciplined way. The obligation for humans to work was a Biblical injunction that issued principally from the second epistle of the Apostle Paul to the Thessalonians (3:10): "If anyone will not work, neither let him eat." This notion that thoroughly infused the ethos of the early Church can be traced back to the book of Genesis – and forward to our workfare programs of this day. But a biblical warrant was not necessary in Foucault's worldview. Jean-Baptiste Colbert, Comptroller-General at the court of Louis XIV, and a burgeoning mercantile class saw that putting the idle poor to work not only contributed to the welfare of the state, especially those already privileged, it staunched the drain on the public treasury – while ensuring a more tranquil social ambience for all.[2] Foucault's animus against the privileged classes of the Renaissance (and their successors) may have inspired his conviction that the royal houses of Europe, their aristocracies, the landed gentry, and a sedentary but comfortable clergy conspired to conscript the idle poor into the ranks of the working poor.[3] It is evident today that these idle poor comprised the personality disordered "riff-raff" of the day, like the homeless legions of the twenty-first century who have largely been evicted from their contemporary asylums and now people our streets.

The family has always been the first, but not always a compassionate, refuge of the mentally disordered. Supplemental to families were custodial asylums for the troublesome and demented. Edward Shorter (1997) has demonstrated that these asylums have existed since the Middle Ages (see his chapter, "The Birth of Psychiatry," pp. 1–32). On the other hand, there is an abundance of examples of compassionate men and women, and their foundations, such as the eponymous Saint Vincent de Paul Society, the Society of Friends, Knights Hospitaller of Jerusalem, and other religiously motivated folk, lay and clerical, who to this day provide food and shelter for wanderers and beggars – independent of their mental health. The analysis of the asylum movement of the late eighteenth and early nineteenth centuries, and the transition from institutional care to a biological psychiatry that coexisted with it has been cogently treated in Shorter's *A History of Psychiatry*.

2 The Duc de Sully (1560–1641), minister to Henri IV, famously observed that the two breasts of France were *labourage et pastourage* (tillage and pasturage). He neglected to mention hard work, which, of course, tillage still is.

3 See the estimable book by Edward Shorter, 1997, on Foucault's ideologization of history in this regard. A complementary view of Foucault is presented by Philip Cushman (1992, pp. 26–27) bearing on the evolution of "the self."

Changing lenses to study personality disorders

Societies around the world have formulated highly divergent criteria for diagnosing specific personality disorders – or for establishing that certain personality traits even constitute a disorder. That personality disorders are in great measure socially constructed has found wide agreement in transcultural psychiatry and among other social sciences. Controversy consequently exists as to whether there are any universally valid types of such disorders (Marsella and Yamada, 2000). On the other hand, no doubt exists that personality disorders described in the major diagnostic manuals of the world are grounded in reality, and that their severity is, in principle, measurable. Within any particular society, there is usually a degree of consensus relative to the traits and behavioral patterns that constitute disorders, although as more research is conducted on them, their definitions continue to be more crisply defined. These definitions evolve as we learn more about them, and as the culture from which they take their character changes.

Part of the definitional problem revolves around how to conceptualize disorders. Are they discrete entities like a medical illness or a socially defined class of behaviors that, given prevailing mores, are excessive or deficient expressions of normal behavior? In the latter perspective behaviors are distributed along a (generally) bell-shaped continuum, the tails of which by definition lie outside the normal range. For example, individuals who are extremely *gullible* or *paranoid* are those who maladaptively move outside a normal range of prudent risk assessment. If one opts for the dimensional view, the question arises as to how extreme the trait must become before it can be regarded as a component of a personality disorder. The answer depends on how much the behavioral trait at issue compromises one's ability to live amicably and productively among one's neighbors and co-workers. This issue is discussed in more detail below.

As a consequence of the relativity of these constructs, some scholars prefer to consider personality disorders more as fluid and convenient social schemas than as *syndromes*[4] comprising scientifically verifiable "symptoms." The non-dimensional view simply affirms that there is a radical discontinuity between populations that have a serious psychiatric illness and those who have an attenuated form of it labeled a personality disorder. This distinction is treated in some detail below.

[4] A *syndrome* is a constellation of symptoms that when taken together define a recognized illness or disorder. The adjective is *syndromal.* The root term, *drome*, comes from Greek meaning *to run*. Symptoms that *run* together constitute a syndrome. *Prodromal* refers to events that precede the onset of the disorder.

Western perspectives on personality disorder

A physiological perspective

There are few experiments that have been so extensively discussed and had such momentous implications for personality psychology as the catastrophic accident suffered in 1848 by *Phineas Gage*, a rail worker. Though oft recounted, the story bears repeating here. Gage was in charge of a train gang that was laying track in Vermont for the Rutland & Burlington Railroad. Preparing to detonate an explosive charge in a hole that had been drilled in a rock escarpment, he was using a tamping iron to pack sand on top of explosive powder and a fuse. He accidentally created a spark that set off an explosion, driving the three-and-a-half-foot long iron rod through his cheek, up through the roof of his mouth and into his brain. It exited through the top of his skull, landing more than a hundred feet away. This 13-lb rod was 1.25 inches in diameter and did catastrophic damage to Gage's frontal lobes.[5] The profound changes that this caused to his personality provide insights into the neurological basis of human character and cognition that no laboratory experiment or autopsy could have revealed. Although the trauma left much of his cognitive abilities unimpaired there were profound consequences for the regulation of his emotions and social relationships.

Twenty years after the accident, Gage's personal physician, John Harlow, published his record and recollections of its psychological as well as medical consequences to the victim. He wrote that the balance between Phineas Gage's intellective and "animal propensities" had been seriously untracked and he observed that his patient was:

> fitful, irreverent, indulging at times in the grossest profanity which was not previously his custom, manifesting but little deference for his fellows, impatient of restraint or advice when it conflicts with his desires, at times pertinaciously obstinate, yet capricious and vacillating, devising many plans of future operation, which are no sooner arranged than they are abandoned ... A child in his intellectual capacity and manifestations, he has the animal passions of a strong man. (quoted by Damasio, 1994, p. 8)

Gage's former employer, the Rutland & Burlington Railroad, refused to rehire him, not because he had lost his technical skills and physical vigor – he obviously had not – but because of the dramatic change in his

[5] A computer simulation of Phineas Gage's skull (which physically resides in the Warren Medical Museum, located on the Longwood campus of the Harvard Medical School) reveals that the likely trajectory of the iron bar that pierced his brain was in the ventromedial prefrontal region, which is critical for normal decision making. The bar skirted the region that is implicated in language and motor functions (Damasio *et al.*, 1994).

character. The new Phineas was not the old Phineas, and a new spirit now animated the restored body that the rail company neither recognized nor respected. The lovable, respectful, socially skilled gang leader of several months previous had disappeared into the thickets sown in a careless moment by an aberrant steel rod. He never emerged from those thickets, living as a vagabond brawler, stagecoach driver, circus performer, and day laborer, and dying of a seizure in San Francisco at the age of 38.

Minutes after Gage's dreadful accident, he spoke coherently with his co-workers, as well as with the physician and local folk who had come to his aid. He sat quite calmly as he was treated by a physician on the porch of the local hotel to which he had been brought, and he answered all the questions that were directed to him rationally and coherently. His attention to detail, his perception, memory, sequencing of causes, grasp of complex physical and social relationships, and facility with language were all unimpaired. What *was* impaired, history reveals, was his ability to organize them in the service of larger social goals.

This "experiment in nature" strongly suggested what had long been denied and contested by scientists. There are systems in the brain that enable the development of personality, that mediate the characterological types that trait psychologists had long postulated, and that serve to translate social learning into appropriate behavior in the hurly-burly of our daily interpersonal, financial, and occupational lives. The gripping fact is that when these brain structures are destroyed, whether by accident or by disease (admittedly accidental events in themselves), one is not left without a personality, one is left with *another* personality, which is disordered and alienating. For better or worse, all humans have a distinctive style for accommodating the needs of others, planning and executing short- and long-term projects, respecting prevalent social conventions, expressing personal values and communal ethics, using profane and polite language, and conforming to the other *mores* that govern social interactions.

The bedeviling question that this raises is, if the irresponsibility that Gage indulged in can result from a catastrophic lesion in a prefrontal lobe, are there comparable physiological explanations for the wanderlust, laziness, malingering, fanciful projects, daydreaming, rudeness, violent behavior, feckless decision making, and other incompetent behaviors that resulted in his fall from social grace (and that we may witness in an attenuated form in our friends and colleagues, possibly even ourselves)? As Gage's brain damage resulted in patent characterological deformities,[6] what of the

[6] Place this in the perspective that Edelman and Tononi (2000) present: "Some people have had almost half their brains removed (in a so-called hemispherectomy) because of tumors or intractable epilepsy, yet their cognitive abilities are only marginally affected" (p. 53).

myriad lesser, and covert, lesions that all humans are heir to as they ply the highways and byways of life? Do such minor "hits" take a proportionate and cumulative toll not only on our perceptual, memorial, attentional, and reasoning capabilities, but on our social competence as well? Further, there are significant individual differences in people's brains just as there are in people's faces and physiques. But just as people normally do not take responsibility for the flaws in their faces and the social impact these entail, should they be obliged to take responsibility for an accumulating assortment of anomalies in their brains and the deviant behavior that that entails? Damasio (1994) writes that there are "Gage's" all around us. "Some have had no overt neurological disease and they still behave like Gage, for reasons having to do with their brains or with the society into which they were born" (p. 19). What, in fact, are the grounds for imputing responsibility to individuals for behavior that is largely physiologically determined? And what possible grounds exist for our belief in *free will*? A renascent Cartesian dualism (Lipton, 2004)?

Genetics and personality disorders

Research has now conclusively demonstrated that there are genetic determinants of many mental diseases (for example, Baltes, Reuter-Lorenz, and Rösler, 2006; Plomin, 1991). Genetic psychiatry has tentatively situated the genes causing schizophrenia on chromosome 6 and those implicated in manic-depressive disorder on chromosomes 18 and 21. The struggle to establish the *physiogenic* principle in psychodiagnostics was not easy and has been troubling to psychodynamically oriented therapists whose methods are predicated on the belief that psychological illnesses have primarily psychosocial causes. The myth of the *schizophrenogenic* mother is emblematic of the philosophical principle (sorely in need of revision) that as all humans are conceived with an integral human nature, structural deformities in their psychic life must be derived from the uterine environment or from the influence of a pathogenic family upbringing.[7] The reality is that every person carries many sub-lethal genes that may not have been expressed but that carry the potential for future disorders – if not for the carriers then certainly for their descendants (Cosmides and Tooby, 1999).

More unsettling to traditionalists is that these findings are reminiscent of *degenerationism*, a school of thought that flourished in nineteenth-century

[7] The earlier view that the body was material and the intellect spiritual – often referred to as Cartesian dualism – was supportive of the principle that mental illness is of psychic, even daemonistic origin.

Europe. The proponents of this position, Emil Kraepelin among them (Hoff, 1998), were firmly grounded in neuroscience and recognized what is now commonly accepted: disposition to many serious mental diseases is genetically transmitted. The school became discredited, however, by the proposal of some of its adherents that not only is mental illness transmitted from forebears to offspring, but that the illnesses become increasingly severe in succeeding generations. This idea was further besmirched when it was co-opted by politicians who assimilated it into racist ideologies – and ultimately used it to "justify" the state murder of "mental defectives" and ethnic populations they hatefully presumed to have high levels of psychopathology.

Like personalities that are socially approved, disordered personalities result from the interaction of genetic and environmental variables. This interaction "is evident from the very consistent finding of huge individual differences in people's responses to all manner of stresses and adversities, however severe" (Rutter and Silberg, 2002). We can measure the main effects of these genetic variables, most of which fall in the healthy range. But even where there is significant abnormality, as in obsessive-compulsive disorder (OCD), the expression of the disorder will be influenced by other traits that interact with it, singly as well as in concert. Normally extraverted individuals with moderate OCD manifest their compulsiveness differently from those who are highly introverted. Even where environment seems to play an important role in the expression of a trait, behavior geneticists recognize that individuals may simply have selected and shaped the very environment that facilitates the expression of their trait. That line of reasoning leads us to conclude that causality is bidirectional; it can proceed from genetic variables to environmental correlates as well as from the environmental to the genetic correlates (Plomin and Rutter, 1998).

Subsequent to the work of Richard von Krafft-Ebing,[8] Emil Kraepelin, and the Scottish asylum master, Thomas Clouston, scientific studies emerged to demonstrate biogenetic determinants of major mental illness. By the 1920s a number of studies were initiated to determine what the contrasting concordance rates of major mental illness are among monozygotic (MZ) twins and dizygotic (DZ) twins who had been reared apart.

[8] Richard von Krafft-Ebing (1840–1902) was a brilliant forensic psychiatrist and author of the authoritative *Psychopathia Sexualis* ([1886] 1937), which was initially written in Latin and went through numerous revised editions. Though tainted with degenerationist notions and denigrated by less noteworthy contemporaries, such as Moritz Benedikt (Shorter, 1997, pp. 95–96; *cf.* Cushman, 1992, pp. 21–64 for the American perspective on these "Victorian" issues), this influential and highly esteemed book *preceded* Breuer and Freud's *Studies in Hysteria* ([1895] 1960). It followed, however, Schopenhauer's (1819) *The World as Will and Representation*, which provided psychoanalysis with its template for a theory of human development.

Hans Luxenburger in Germany and Aaron Rosanoff in the United States carried on methodologically cogent twin studies in the inter-war period that supported the genetic thesis (Shorter, 1997, p. 142). Rosanoff, Handy, Plesset, and Brush's (1934) study is particularly compelling. Using a data base of 1,014 twin pairs in which 142 pairs revealed a history of schizophrenia, they discovered that among the MZs, both twins had schizophrenia in 68 percent of the cases, whereas among the DZs, only 15 percent did. Rosanoff published a comparable study of manic-depressive disorders in which this disorder appeared in both twins among 70 per cent of MZs and among 16 per cent of DZs (Rosanoff, 1935).

The work of Franz Kallmann was no less instructive in this respect. Thoroughly familiar with the work of Rosanoff and his predecessors, he began his own twin study drawing on a population of 73,000 patients in the asylum system of New York State. Among those patients, he was able to identify 691 schizophrenics who had a traceable co-twin who could be assessed. His results were even more compelling than those of Rosanoff. The concordance rate for the identical twins was 85.8 percent; for the fraternal twins it was 14.7 percent (Kallmann, 1946). He presented his results several years later at the First World Congress of Psychiatry in Paris in which he drew conclusions that are today widely accepted: there are genetic determinants to this disorder.

The furor that erupted over this presentation is a testament to the entrenchment of traditional explanations of the origin of mental disease that reigned at that time. Since then, the mental health professions have recognized that genes account for a significant proportion of the individual differences in the incidence not only of schizophrenia, but of obsessive-compulsive disorder, hypochondriasis, depression, and other "family-transmitted" disorders. As with every other behavioral characteristic, nature *and* nurture determine the type and severity of the mental illnesses that we formerly thought had an exclusively psychosocial cause. Once molecular geneticists have identified the various genes that influence the onset of an illness, a labor-intensive and lengthy process in itself, such markers help social cognitive neuroscientists focus on the developmental patterns for these disorders, as well as identify the types of environmental risk mechanisms that maximize and minimize their eruption.

Studies of the *epigenetic* basis of disorders (noted in Chapter 4 on biology and personality), induced by environmental events that deregulate the expression of genes, cast a broad shadow over the simple presumption that a broad array of mental disorders is coded in the genome. Such discourse about the *epigenome* bears nevertheless on *genes*, the expression of which is deregulated by environmental events, such as pre-natal rubella exposure, traumatic pre-natal stress, and post-natal events to which individuals have been exposed involuntarily. Of course, we have long known that severely stressed gravid females generate an abundance of adrenocorticosteroids that pass into the embryo and stunt its development – especially the particular system such as the heart or the dental arch that is in formation at the time. We were not aware decades ago of the epigenetic mechanisms that vectored these disruptions. A recent wave of research (consult, for example, the doctoral thesis of Ramses F. J. Kemperman [affiliated with Kema *et al.*, 2007, pp. 36–54] into the diagnosis and causes of disorders resulting from aberrant gene expression rather than gene mutation) attests to the scientific interest in further explorations in behavioral genetics. This has effected a drastic shift in the "nature and nurture" schemas explaining human development.

Evolutionary psychology and mental disorders

Another hybrid science that sheds light on the biological underpinnings of psychosocial behavior is evolutionary psychology. Modern Darwinian biology has begun to alter the foundations of psychology and transform the approaches to treating mental disorders (Cosmides and Tooby, 1999). The frame of reference that we need to build requires that we acknowledge that many reactions to illness that our *hominid*[9] ancestors evolved may appear to us as abnormal or dysfunctional but, in fact, are highly useful and effective. An illustrative medical example (such examples are usually simpler than psychological ones) is that of a fever, which we all know is an effective defense for weakening harmful bacteria and viruses. If we allow the fever to continue (keeping it within safe bounds of course) we will recover more quickly than if we reduce it drastically with antipyretics such as aspirin. A different example that Cosmides and Tooby use is that of the bioavailability of iron in the human system. There are strains of bacteria that flourish in an iron-rich environment. A natural defense is to temporarily reduce iron availability in the body. Treating patients for anemia as they combat iron-dependent bacteria is counterproductive.

Generalizing these ideas to the domain of mental health and to normal social behavior can lead to productive theory and research. Cosmides and Tooby (1999, p. 454) prime this notion by stating:

> One of the central insights now available to psychologists, psychiatrists, and neuroscientists from this integration [of evolutionary biology into mental health

[9] Corsini's (1999) *The Dictionary of Psychology* defines *hominids* as "Bipedal primate mammals of the family Hominidae, including *Homo sapiens* (humans)" (p. 449).

disciplines] is that the human psychological architecture consists of a constellation of adaptations or devices, each of which evolved among our foraging ancestors to perform specific computational functions necessary or useful in such ancestral conditions.

Examples given are:

The cost-effective avoidance of bites from venomous snakes is the function of the computational circuits underlying snake phobias (Marks, 1987); disinvestment of effort and hedonic attachment to unprofitable activities or relationships is the hypothesized function of the computational systems underlying depression (Nesse & Williams, 1994; Tooby & Cosmides, 1990b); a countervailing distaste for sexual contact with close genetic relatives who are otherwise available and attractive potential sex partners is the function of the Westermarck incest-avoidance mechanism (Wolf and Huang, 1980).

Nineteenth-century psychiatry evolved theories, most notably psychoanalysis, founded in large part on Nietzschean notions of repression, sublimation, projection, denial, and other defense mechanisms, to explain how people manage the emotional disorders from which they chronically suffer, as well as respond to daily stressors over which they have little mastery. Stripping patients rudely and abruptly of such defenses could easily precipitate a crisis and bring on a worsening of their condition.[10] To regard dysfunctional behavioral patterns, then, as evolutionary mechanisms that kick in when usual adaptive strategies are not working may be a better therapeutic approach than to think of them as neurotic coping strategies. It may also be a better personological explanation of troublesome behavior.

Harmful dysfunction analysis (HDA)

The concept of mental disorder is critical to the disciplines whose *raison d'être* is to diagnose and treat it. Not least of the reasons for fashioning a definition of it is the need to obviate suspicions of cultural, racial, gender, and sexual-orientation bias that have been alleged to exist in the *Diagnostic and Statistical Manual* (DSM), the principal North American taxonomy of mental disorders. Anomalously, until recently there has been little agreement on how to define them, although definitions have abounded (Wakefield, 1992). These definitions have ranged from the most superficial (for example, a mental disorder is any condition that a mental health professional treats) to the most comprehensive and finely reasoned, such

[10] *Implosive therapy* and *flooding* are two procedures that force individuals to abruptly confront their fears, drop their defenses, and adopt "conventional" attitudes to the negative stimuli in their lives. The former therapy integrates psychoanalytic and behavioral elements in the treatment; the latter is entirely behavioral.

as Jerome Wakefield (1992, 1999) has proposed. Wakefield has argued that the definition must embrace two domains: the natural world of the human organism as it has evolved; and the constructed social world and culture in which clients live. These social worlds variously construe a condition to be normal or significantly "disordered" and accordingly assess them as being harmful to victims or indifferently benign. Thus, in Wakefield's view a disorder involves two components: (a) the functional failure of an evolutionarily selected mechanism or capability that normally enables adaptive behavior in the hurly-burly of expectable life events; and (b) the perceived harmfulness in any society of the consequences of such failure.

That scientists cannot accept a pure value concept for disorder (that is, a disorder is whatever a culture judges to be a disorder) is evident from the fact that these values are constantly changing in kind and intensity. Scientific principles do not change normally as political winds shift and social tensions rise and fall. In the memory of many readers, *premenstrual syndrome* and *homosexuality* have been judged to be mental disorders. In an earlier period of Western history, *drapetomania* was a disorder attributed to slaves who had become so unhinged that they sought to escape their bondage. *Hyperactivity* is a condition of individuals whose spontaneous activity and restlessness disrupts the social unit in which they function. They are uncomfortable spending 5 or 6 hours each weekday sitting quietly behind a desk, focused on mental tasks. That a majority of teachers disapprove of restless behavior in the classroom is not sufficient grounds for alleging that their pupils have a disorder. Uncongenial environments can prime uncongenial social behaviors. Clearly, a pure value concept of disordered behavior is not a solid base on which to build a therapeutic profession dedicated to changing such behavior.

Although it is not necessary that we have a clear understanding of the biological system that is malfunctioning, we should be able to make reasonable inferences from non-normative behavior that function failure has occurred. This requires some clinical acumen. Wakefield states that if we see a man talking aloud to himself, a number of clinical hypotheses arise as to whether this is or is not evidence of a disorder. This man may be trying to commit to memory an agenda or a shopping list; he may be wired to communicate with an accomplice who is out of sight; he may be engaged in a charade or other game whose circumstances are not apparent to some onlookers; or he may be hallucinating the presence, voices, or sight of distant acquaintances. Wakefield (1992) proposed a finely reasoned argument to support the notion that disorders are failures of biological mechanisms (as well as dysfunctional social judgment), to respond to environmental demands in the adaptive ways that *evolution has equipped them to respond*. He joins many personality psychologists who take a bioevolutionary approach to understanding disorders and "dysfunctional" defenses. They assume that there are selective evolutionary factors that have programmed certain default responses to environmental challenges. In that light one must assume that there are objective standards of adaptive behavior corresponding to Darwinian selection criteria that have been built into our human nature. Harmful dysfunction analysis (HDA) requires that disorders not only meet objective scientific criteria for judging that an organismic failure has occurred, but a subjective value judgment must be made that individuals have thereby suffered harm.

Darwinian principles are, of course, nomothetic. However, value judgments bearing on perceived excess or deficit in behavior patterns suggested as normative by evolutionary psychologists shift from culture to culture. If extraordinary levels of emotional lability, for example, are not considered harmful in one society or another, we cannot conclude, in a Wakefield-type schema, that a disorder exists there. In this perspective, departures from "normal" behavior may be better understood by *first* studying functional impairments of the regulatory mechanisms encoded in our central nervous system (CNS) before studying variation in, say, stress responses judged by cultural or local community standards. Establishing a "species-typical" Darwinian standard may be a sounder initial heuristic for understanding aberrant human behavior than using data that are derived from an unrepresentative cultural sampling of the human race – the latter approach resulting in "many ethnic minority patients being inaccurately diagnosed and inappropriately treated" (Marsella and Yamada, 2000, p. 6; Pedersen and Marsella, 1982), on the one hand, and human nature inaccurately defined, on the other hand.

A countervailing approach to HDA is the eponymous Roschian prototypal schema that derives from the work of Eleanor Rosch (1973). From data gathered among a Stone Age people, Rosch concluded that there are natural semantic categories that reflect "natural prototypes," for when she attempted to teach these people, say, geometric forms that violated norms for traditional cross-cultural prototypes, her subjects had much more difficulty learning them than when she taught them forms that respected them.[11] Using this Roschian schema for categorizing mental disorders,

[11] Contemporary evolutionary psychologists also propose this view. For example Steven Pinker (2002b) insists that there are prototypical CNS-based schemas that determine standards of beauty, among other cognitive dimensions. He states that we have lost sight of this, to our great detriment: "In the 20th century, modernism and post-modernism took over, and their practitioners disdained beauty as bourgeois, saccharine, and lightweight. Art was deliberately made incomprehensible or ugly or shocking – again, on the assumption that people's tastes for attractive faces, landscapes, colors, and so on were reversible social constructions" (Pinker, 2002a).

Lilienfeld and Marino (1995) proposed that we should not define disorders but, rather, construe them as prototypes, each comprising a set of attributes. Membership in any category of disorder would accordingly depend on the perceived similarity of a syndrome to its prototype. The opposing view that disorders are of a continuous nature, that their incidence manifests the characteristics of a normal curve, introduces a high level of subjectivity into clinical judgment, particularly as one departs increasingly from the classic features of the Roschian prototype. The boundaries between normal and abnormal are fuzzy and ill-defined. This position entails an obvious difficulty: what is ill-defined must be, in many specific cases, *arbitrarily* defined.

Emil Kraepelin and his legacy

Kraepelin is considered to be the founder and central figure of modern psychiatry in the West. The third edition of his influential textbook (1893), which went through many further editions, steered German (and eventually world) psychiatry away from its biological moorings into a predominantly psychologic–clinical stream. Kraepelin was born in northern Germany, but attended medical school in the charming Franconian city of Würzburg. As a young man he was an avid reader of works on psychology and had a great admiration for the experimental psychologist, Wilhelm Wundt, with whom he later chose to work in Leipzig. In his early work with the brain anatomists of Munich, where he did his residency, serious eye problems limited his ability to work with microscopes. Further, he developed a disdain for the strongly medicalized and symptom-based approach to psychiatric diagnosis practiced in Munich by brain biologists. Acutely aware that a half-century of intense research into the neuropathology of mental illness had yielded little beyond a clinical picture of neurosyphilis (Shorter, 1997), he became convinced that the soundest approach to mental illness was studying the *course* of various illnesses as well as their *outcomes*. Shorter (1997) writes that "Kraepelin's attraction to psychology, and his repulsion at anatomically oriented psychiatry, were strengthened by … his work with Paul Flechsig, whose only interest lay in the microscope and dissecting table" (p. 101). Flechsig was notorious for his cold, technical bedside manner; Kraepelin was reputed to be a caring physician.

Kraepelin: diagnostician

The data-gathering method that Kraepelin favored consisted of keeping records on "diagnostic card" files of his patients' disease course in order to clinically observe the progress of their disease. Periodically (often on his vacations) he would take them out, sort through his clinical observations with a view to finding increasingly consistent patterns of behavior, mentation, and mood, and then formulating more precise disease definitions. It was a longitudinal rather than a cross-sectional approach to defining disorders. He respected the need to discover the causes of diseases, biological as well as psychosocial, but the science of neuropathology at that time was of little use other than for pigeonholing diagnoses. Its value for helping patients and their families was virtually nil. The psychiatric treatments available at the time, medical as well as psychological, were poorly founded and of very limited usefulness, but at least the work of Kraepelin enabled patients and their caregivers to get a glimpse of what they could expect – in the distant as well as near future. What is particularly notable about the sixth edition of his psychiatric manual (1899) is that it provided the basis for a classification of mental illnesses later adopted in the American Psychiatric Association's *Diagnostic and Statistical Manual of Mental Disorders* (1980), now used extensively by clinical psychologists as well as by medical practitioners.

The investigations of mental disorders in Europe in the early 1800s were split between those who believed that they were caused by brain pathology (this group is referred to as the *Somatiker*), and those who believed that they had emotional and social causes (referred to as the *Psychiker*). The most renowned psychiatrist preceding Kraepelin was Wilhelm Griesinger (1817–68) who, although he acknowledged the emotional etiology of many mental problems, nevertheless came down firmly on the side of the organicists who claimed that "mental diseases are brain diseases."

Kraepelin was a proponent of a balanced approach to the etiology of mental disorders. He believed that manic-depressive "insanity" and lesser affective disorders had *exogenous* causes: that is, the personal history of the patient would reveal emotional traumas that resulted from more or less severe and abusive events of a socio-familial character. He contended that such disorders are amenable to treatment. Premature dementia, on the other hand (then known as *dementia praecox* or *d.p.* – later as schizophrenia)[12] had *endogenous* causes. No treatment existed at the time for such a disorder. Dissatisfied with the scientific basis for the many alleged causes,

[12] The term schizophrenia was coined by the eminent Zurich psychiatrist, Eugen Bleuler (1857–1939), who mentored Carl Gustav Jung at the Burghölzli, a psychiatric facility in Zurich. The term is not an apt one, for etymologically it signifies a "split brain," a concept used by a very small number of mental health professionals for "dissociative" disorders, and which finds little place in the abnormal psychology literature.

both neurologic and psychologic, of the prevalent disorders of the time, he simply abstained from postulating such causes until firm proof was provided that they existed.[13] He divided all psychotic illnesses that had no specifically established organic cause into illnesses involving an affective (mood) component and those devoid of it. The former was called manic-depressive psychosis; the latter, premature dementia.

Although Kraepelin is berated by modern psychiatry for being simply a nosologist[14] (an expert in the discipline of disease nomenclature), a scientist who was possessed by a passion for creating accurate diagnostic labels (Ellenberger, 1970, p. 285), he was more than that. He set psychiatry on a different course that has endured in part to this day. His principle that mental illnesses are distinct entities analogous to measles, pneumonia, and syphilis is not as widely accepted today as it was in the first half of the twentieth century. Psychiatry in that era rejected the, as yet unfounded, linkage of psychiatric syndromes with brain disorders. Kraepelin prematurely stamped "paid" to the biological psychiatry of the time. It resurfaced only late in the twentieth century with the arrival of the benzodiazepines and a sophisticated neuroscience (of which Kraepelin would have approved), abetted by brain imaging technologies.[15] On the other hand, he and his numerous disciples rejected the *Psychiker*'s complex theory building and their years-long probing of the content of a patient's disordered (and normal) mentation. He attacked wild speculation about the psychosocial causes of mental illness, insisting that practitioners suspend judgment until there was *proof* of the linkage of specific personal historical events with a psychoneurosis. The rampant speculation that came into vogue during that period arrested the movement toward the *biopsychosocial approach* that was attracting waves of psychotherapists looking for systemic explanations for the mental problems they were asked to resolve. The Kraepelinian school

[13] Romantic psychiatry is the label historically given to the school that focused analysis on the psychosocial and personal history of the patient. Passion, emotional trauma, and socio-physical abuse were considered by this group to be the root causes of neurotic *and* psychotic disorders. The principal figures in this early movement were Johann Christian Heinroth and Jean-Etienne Esquirol (1772–1840). The movement was largely eclipsed by what Shorter calls "the first biological psychiatry," the experimental and laboratory-based German neuroscience that swept the Western world in the mid-nineteenth century, only to be revived by Forel, Bleuler, Janet, and, above all, Kraepelin toward the turn of the century.

[14] Nosology, deriving from the Greek term for disease, *nosos*, is the medical science for classifying disease.

[15] Shorter (1997) wrote of this resurgence in the 1970s: "The neurobiological paradigm came roaring back from the grave in which Kraepelin had interred it – with medications that truly worked and evidence that psychiatric illness represented a biological phenomenon far deeper than troubled human relations or a schizophrenogenic mother" (p. 238).

rejected, of course, the sexual reductionism of the burgeoning psychoanalytic movement.

Kraepelin's contributions are substantial: an insistence on studying the course and outcome of mental illnesses rather than simply noting their initial clinical manifestation – this eventuated in a new way of classifying mental disease; masterly descriptions of such illnesses; an insistence on sanctioning only an empirically founded psychology; the rejection of a depersonalized anatomical psychiatry; an attention to making fact-based definitions of disease entities – carving nature at the joints, to use Plato's expression; and placing a high value on prognosis. Kraepelin believed that one inherits a disposition to violent and criminal behavior. He considered chronic offenders who proved to be resistant to educational interventions as having a mental illness. (Because they had a greatly diminished responsibility for their actions, he vigorously opposed the death penalty for capital crimes [Hoff, 1998].) These ideas have shaped the mental health professions of the twentieth and twenty-first centuries and underpinned that globally utilized diagnostic instrument (that is not without its enemies): the *Diagnostic and Statistical Manual.*

Kraepelin: cross-culturalist

According to Jilek (1995), Kraepelin was well aware that ethno-specific disorders exist and that socio-cultural norms profoundly affect the incidence and course of the disorders one studies. Jilek asserted that Kraepelin's international research gave an impetus to comparative *ethnopsychology* by his recognition that the epidemiology as well as characteristic symptom presentation of many mental illnesses are greatly influenced by the social conditions surrounding the patient. Kraepelin's psychology-mindedness led him to understand that endogenously caused mental disorders would further be shaped by the character of the family and society in which the patient had been raised. This view is corroborated by Marsella and Yamada (2000, p. 6) who acknowledge the importance of this scientist to a nascent cultural psychology.

> In the early years of the twentieth century, Emil Kraepelin (1904), the father of modern Western psychiatry, journeyed from his home in Germany to Asia and North America as part of a worldwide lecture tour. During the course of his travels, Kraepelin experienced difficulties diagnosing some patients. He noted that the patients in these lands failed to express their illness with the prototypical symptoms characteristic of his patients in Germany and Northern Europe. Puzzled by this situation, Kraepelin suggested a new specialty within psychiatry be created – *Vergleichende Psychiatrie* or Comparative Psychiatry – to study cultural differences in psychopathology.

That syndromes that were formulated by culturally encapsulated Western system builders are seriously limited has become a commonplace principle in today's culture-sensitive academic arena, as noted.

The diagnostic and statistical manual (DSM)

The architects of the DSM have crafted a manual that is fundamentally atheoretical and essentially Kraepelinian in its orientation. In other words, the descriptions of illnesses and disorders are framed in operational terms that do not presume a particular theoretical framework for understanding their etiology, treatment, or conceptualization.[16] The early versions of the DSM (DSM-I and DSM-II) were drafted by psychiatrists who were psychodynamically oriented (largely Freudian or neo-Freudian) rather than biomedically so. As the creation of a nosology implicates the philosophy and theoretical orientation of those who are drafting it, it was no surprise that these versions abounded in terms like "conversion reaction" and "dissociative reaction," reflecting Freudian constructs and terminology.

The official nomenclature of an influential professional elite draws the scrutiny of everyone who has a personal or political stake in it. Insurance companies were no exceptions, nor were Vietnam veterans suffering from what was to become known as post-traumatic stress disorder (PTSD), cultural psychologists, and, of course, gay and feminist organizations. They exerted a powerful influence on the evolution of this manual. Even more significant was the influence of a group of young psychiatrists who were determined to turn their profession away from the psychoanalytic movement they felt had hijacked it in the early part of the century.

As the builders of the DSM-III began their work in the early 1970s they recruited like-minded collaborators. Melvin Sabshin, a Young Turk (appointed medical director of the American Psychiatric Association in 1973) asked Robert Spitzer of Columbia University and Theodore Millon of the University of Illinois to undertake the revision of the DSM. That was a momentous decision for they (with Eli Robins, George Winokur, and Samuel Guze, a group of researchers from Washington University in

[16] The builders of the *International Classification of Diseases* (ICD) and the DSM have worked closely with each other over the past three decades and in most respects their manuals complement each other. Robert Spitzer and Janet Williams (1987) wrote in their introduction to the DSM-III-R: "Although representatives of the American Psychiatric Association had worked closely with the World Health Organization on the development of the ICD-9, there was concern the ICD-9 classification and glossary would not be suitable for use in the United States. Most importantly, many specific areas of the classification did not seem sufficiently detailed for clinical and research use" (pp. xviii–xix).

St. Louis) initiated a powerful current of thought that resulted in the ground-breaking DSM-III. Bayer and Spitzer (1985) wrote: "With its intellectual roots in St. Louis [Missouri] instead of Vienna, and with its intellectual inspiration derived from Kraepelin, not Freud, [this] task force was viewed from the outset as unsympathetic to the interests of those whose theory and practice derived from the psychoanalytic tradition" (p. 188).[17] Edward Shorter observed that: "The appearance of the DSM-III was thus an event of capital importance not just for American but for world psychiatry, a turning of the page on psychodynamics, a redirection of the discipline toward a scientific course, a reembrace of the positivistic principles of the nineteenth century, a denial of ... the myth of psychiatric illness" (1997, p. 302).

Personality disorders and the DSM

As personality theories have bloomed across the academic landscape, research findings have multiplied the number (and enlarged the scope) of personality disorders. Although the history of personality psychology reveals a persistent *clinical* focus on how personality is shaped by the powerful contingencies of family, community, and socio-religious events, there has been no dominant conceptual framework to explain how these same factors have shaped personality disorders. Those who have presented the most cogent explanations are the numerous developmental psychologists who have studied the effects of various levels and kinds of infantile deprivation (for example, Margaret Ribble, René Spitz – following the Second World War, and John Bowlby). On the other hand, the maturing discipline of personality psychology enables us to understand personality development gone awry. When we examine serious pathology in clients, a knowledge of their *pre-morbid* personality traits provides not only a context for explaining a disorder, but also a framework within which to remedy it.

Individuals with the same mental disorder can present with heterogeneous arrays of symptoms. For example, those with clinical depression manifest it differently depending on whether they were extraverted or introverted in their pre-depressive life. Widiger, Verheul, and van den Brink (1999) wrote that: "depression within a dependent

[17] Marsella and Yamada (2000) wrote that: "it is no secret that for the past few decades, psychiatry has sought to extricate itself from its Freudian heritage and to re-establish itself as a medical specialty. To accomplish this, it would be necessary to support a medical model of psychopathology, which sought etiological causes within reductionistic levels of explanation" (p. 6). Paradoxically, as psychiatry has trended in a rigorously biomedical direction, the American Psychiatric Association's flagship diagnostic manual, the DSM, has, since 1980, moved in a perceptibly countermedical direction.

(sociotropic)[18] person will be characterized in particular by feelings of deprivation, loss, loneliness, and unlikeability; depression within a narcissistic (autonomous, self-critical) person will be characterized in particular by feelings of defeat, failure, withdrawal, and self blame" (p. 348). When buffeted by high winds trees tend to fall in a direction determined by their root systems – in the direction of least resistance (*minoris resistantiae* to use William James' expression). Comparably, human disorders are always *preconditioned* by the strengths and weaknesses with which victims were endowed when the pathogens that precipitated their illness began to assault them.

Idiographics and nomothetics redux. The architects of several editions of the DSM have recognized the tension that exists between an idiographic and a nomothetic approach to personality disorders. This tension has transmuted into confrontations between multiculturalists who emphasize the impact of contextual variables as well as diversity of worldviews shaping personality, and the scientists of a Wundtian tradition (that is, *nomologists*), on the other hand, who search for universal laws of the human mind. This latter tradition is focused on the regularities that one finds across a class of persons. Such science looks for and explains *invariance* and *underlying order* among the dazzling complexities revealed by individuals. Multiculturalists, however, stretch the envelope in the opposite direction. These scholars are less inclined to investigate how people are alike or what laws govern species-typical development. They take an emic perspective on cultures. They prefer to understand how this community or that ethnic group feels and behaves in various situations, enforces its norms, negotiates conflicts, and promotes cherished communal attitudes and values.

Cultural psychologists are as interested in developmental processes as are nomologists, but in the limiting case, their qualitative studies are as much biography or personal history as psychology. Indeed, Millon and Davis (1996) warn that "in its extreme form, the idiographic approach militates against all taxonomy as an illusion that maintains the pretense of science, but is not in itself scientific" (p. 6). Peering in the opposite direction one views, in its extreme form, the nomothetic approach that results in paradigms that are *essentialist*, abstract, counter*existential*: that is, removed from the poignant, everyday particularities of life. In the polemics of these controversies, the *nomotheticists* are characterized as socially and politically irrelevant; the *idiographicists* as writing literature

[18] Social psychologists also use the terms *socio-tropic* and *socio-fugal* to refer to the tendency of animals including humans to (a) seek the company of their conspecifics or (b) keep them at a distance. *Socio-petal* is also used to designate those who seek the presence of others.

more than researching psychology. In order to skirt this double danger, the architects of the DSM introduced and enlarged its *multiaxial*[19] approach to their classificatory system.

The multiaxial system. The DSM continues to evolve as research findings shed light on the syndromal complexity of various disorders. The DSM first described *personality disorders* as such in its 1952 edition, distinguishing this category of pathology from *psychosis*. With the multiaxial schema that was founded in the third edition (1980) personality disorders were gathered in a category of disorders labeled *Axis II*, distinct from the psychopathologies of *Axis I*. But beginning with DSM-III (1980), the classification of pathologies[20] that were listed in Axis I was supplemented by other variables that helped diagnosticians to gain an understanding of the patient's overall functioning and the psychosocial and medical variables that might be contributing to an illness. Axis II encompasses personality disorders and other developmental disorders. The architects of this monumental taxonomical system clearly wished to distinguish even serious personality disorders from the illnesses listed in Axis I. However, they intended these two axes to comprise *all* mental disorders, recognizing that distinguishing a personality disorder from an Axis I illness often places diagnosticians in a fuzzy problem space, all the more so if they endorse the view that human mental functioning could be situated on a *single* continuum that ranges from the fully healthy to the abjectly demented.

Axis III and *Axis IV* form two other sets of variables, that is, physical illnesses and psychosocial stressors, respectively, which may be present with, and could significantly interact with, the disorders in Axis I and Axis II. Physiopathology, such as a stroke, can result in mental disorders by virtue of the fact that it may not only impair the neural regulatory mechanisms for emotional expression but generate disabling anxiety; this process is the reverse of psychosomatic – somatopsychic so to speak. On the other hand, there is evidence that certain personality types, whether they satisfy criteria for a DSM disorder or not, harbor risk factors for hypertension and cardiovascular problems. What is significant about Axis IV is that it formally recognizes the importance of *current* as well as historical psychosocial events in a patient's life that can precipitate, maintain, or

[19] The DSM comprises five dimensions, called axes, each of which represents a variable or set of variables describing the actual condition of a patient. Collectively assessing a patient on all five dimensions in a multiaxial procedure provides a holistic and integrative view of the patient.

[20] Although the DSM is a descriptive manual of *disorders*, Axis I disorders are commonly referred to as psychopathologies.

worsen a mental disorder. That axis provides scales for assessing the severity of the stressors that can bring on mental illness. This is an implicit correction of some psychodynamic therapies that had historically focused *only* on childhood sources of adult anxiety and other "psychoneurotic" disorders. Besides that, it is a counterweight to the increasing biological focus of medical psychiatry and its psychopharmacological remedies, valid and useful as those can be.

Axis V presents an array of symptoms, ranging from mild anxiety that one might experience before an exam (at the healthy, well-functioning end of the spectrum – calibrated with a maximum score of 90) to homicidal or suicidal gestures (at the most dysfunctional end – calibrated with a minimal score of 1). The therapist/diagnostician is requested to evaluate the level at which patients have functioned in the year prior to their evaluation as well as in the present. Psychological, social, and occupational competencies are assessed, using this 90-point scale (see table X-1 of the manual for the codes assignable to the experiential/behavioral criterial markers in Axis V). These measures serve as baselines not only to judge future progress but as indicants of the progress that patients can be expected to make. An individual who has functioned at level 80 in the recent past has a better prognosis than one who functioned at level 10.

Abnormal psychology in Western cultures

"Nature–nurture" and personality

Enthusiasm for an environmental explanation of human personality to the exclusion of the consideration of the essential, universal, and biobasic properties of, say, a characterological disorder such as paranoid personality or avoidant personality is no more acceptable than the contrary. Dispositional and neurological factors arguably play a salient role in the genesis of personality disorders. To deny the universality of such factors is to call our common human nature into question (Pinker, 2002b), a nature shared by all people from the most diverse cultures. But more importantly, "pop" psychology has considered environmental variables and genetic factors to be two independent causative domains shaping personality.

As noted in Chapter 4 on biology and personality, *nurture* is in many respects the child of *nature*. Social environment is a creature of human needs and the adaptive mechanisms that have evolved to ensure the survival of our forebears. Maternal behavior is a potent environmental shaper of infants' health and psychosocial well being. But the mother's impulse to tenderly nurture her baby, environmental though this is for the baby, is genetically determined. And from the first day of their extra-uterine life, babies are shaping the environment that many have erroneously believed was only shaping them. For example, the baby's social smile elicits a maternal smile. Both are environmental for the other, but both smiles correspond with ancestral programs for ensuring the psychobiological well being and survival of the other.

As hunting and foraging species hominids evolved in environments that varied considerably from those which exist today. Many of the genetic features with which humans are endowed enhanced human fitness in ancestral environments but do so poorly in modern environments. "Our modern world," state Cosmides and Tooby (1999), "diverges enormously in thousands of key respects from the world of our foraging ancestors, and so modern humans can be expected to be riddled with scores of dysfunctions caused by these environmental dysfunctions" (p. 459). Many dysfunctional personalities may be perfectly healthy, endowed as they are with the pristine mechanisms that allowed their forebears to survive and thrive. But if their natural endowments place them at odds in the twenty-first century with the prevailing values of their culture, they may (erroneously) be considered personality disordered by their fellows. Take typically boisterous, physically active, playfully aggressive early-adolescent males who have been placed in a sedentary environment such as a typical Western secondary school. If they act normally they will often be considered disorderly and hyperactive. Jealousy mechanisms are another example. The genetically determined neurological inputs that quite adaptively enabled this mechanism in ancestral environments are now considered maladaptive in many cultures. Psychotherapists *treat* jealous spouses who cause suffering to themselves as well as others. To call this evolutionarily healthy response a *dysfunction* unnecessarily generates another nosological taxon in the view of many (for example, Cummings, 2005).

Obversely, individuals who have a candor, trust, and self-disclosing demeanor that would have placed them at risk in the early Pleistocene era are not considered aberrant, disordered, and unhealthy in these times, not even if they place themselves at risk. Those who either collectively or singly embrace pacifism, refusing to defend themselves from lethal assault, have not in most instances improved their individual chances of survival, which may explain why so few societies have collectively endorsed this ideology. Legions of well-educated professional women, and no doubt some men as well, have forgone the option of raising a family and perpetuating their personal genic legacy. Valuing their career over their reproductive "destiny" would have been regarded as incomprehensible if not simply foolish in most instances, even in the Stone Age societies in which human nature assumed its most recent shape.

Pathoplastic relationships[21]

The etiology of any human pathology can be situated conceptually on a continuum that ranges from the most environmental to the most physiological. Psychobehavioral pathologies normally have many causes, and the distribution of those causes on such a continuum is problematic. That all pathology has genetic, organic causes as well as contextual, cultural causes is a commonplace principle in the sciences bearing on human behavior and illness. Even such disorders as diabetes or Tourette syndrome are triggered and expressed in ways that reflect the cuisine, mores, language, and lifestyle of a society, as well as our human physiology. Multicultural psychologists argue that the expression and evolution of personality disorders are in many respects less nomothetic than cultural artefacts. Axis III, as noted above, bridges in some measure the reality of mental disorders with the CNS-based science of contemporary psychiatry – especially as it involves cognitive neuroscience. Axis IV presents the psychosocial factors that a comprehensive multiaxial approach that began in the 1980s affords.[22]

The success of Western science is correlated with its penchant for (Platonically) essentializing, where possible, its scientific concepts. Scientists have fashioned studies with a view to arriving, in a lawful universe, at replicable findings. Their success is founded on establishing universal rather than culturally relative principles. As this principle has proved to be so beneficial in the "natural sciences," such as physics, chemistry, and biology, it is not surprising that medical science, and the psychiatry that is patterned on it, have favored diagnostics and treatment modalities based on universal behavioral patterns. Transcultural psychiatrists have argued that this traditional psychiatric approach to understanding mental illness has resulted in a process that decontextualizes the disorders it attempts to define, and it hampers our understanding of the personal experience of disease, as well as the concrete social and geographic factors that profoundly affect its course.

Pathoplasticity and culture

Personality disorders are catalyzed and shaped by social conditions that are specific to one culture or another. Stressors that precipitate and

[21] Pathoplasty is a compound term whose root elements are of Greek origin. *Patho-* refers to disease; *-plasty* means formation, shaping, or molding (as in dermoplasty, that is, cosmetic surgery). This term is usually used to characterize the influence that personality has on the expression of psychopathology and vice versa. It applies with equal aptness to the relationship of culture and psychopathology.

[22] Consult the formidable exploration of personality disorders by Millon and Davis, 1996.

conditions that mitigate disorders vary in character and severity from culture to culture, as do the familial and other institutional support systems that exist. Whether or not patterns of behavior even constitute a disorder depends on culturally divergent notions of selfhood and personality, and the standards for normal as well as acceptable behavior. The rapid social changes that have occurred since the Second World War in large sectors of the American rural southeast or the urban Rust Belt of the industrial northeast have created stressors that have generated psychological disorders specific to those regions. The pathoplastic character of the traditional responses to overwhelming stress have often been magnified in industrial societies by the relative paucity of familial and community support systems (Marsella and Yamada, 2000).

David Stannard (1992) has vividly described the pressures on indigenous Americans as they grapple with the complexities of a technologically complex society that envelops them. The rapid erosion of their own way of life, the evanescence of their most treasured values, and the spectacle of their children being socialized in the street, the playground, and the televised and alien world of the mass media have baneful effects on their mental health. The complex interaction of the psychological defenses that immigrants bring from their native soil with the particular challenges presented by the regional sub-culture to which they move give rise to idiosyncratic disorders. The fault lines as well as strengths of their traditional cultural armor become exploited and give rise to patterns that only members of their ethnic group can understand. Placing a Kraepelinian template on their "symptoms" results in diagnostic distortions.[23]

[23] This segment from *The New York Times* (Sarah Kershaw, January 18, 2003) illustrates the culture-saturated character of the mental problems that Asian immigrants evidence in New York City:

Therapy for Immigrants: Disorders From the East Emerge Here
The patients may suffer from classic mental ailments: depression, anxiety, schizophrenia.

But as they make their way to a sprawling mental hospital in northeast Queens, they also complain of problems that the average New York City psychologist has rarely encountered: pa-feng, a phobic fear of wind and cold that occurs in Chinese patients; hwa-byung, a suppressed anger syndrome suffered by Koreans; and Latah, a Malaysian and Indonesian psychosis that leads to uncontrollable mimicking of other people.

They are the kinds of illnesses that psychologists refer to as culture-bound syndromes. Experts say that while they are fairly common among New York's exploding immigrant population, they are often undiagnosed, or are confused with other conditions. But a growing number of mental health professionals are now focusing on patients' ethnicity and country of origin to treat their mental illnesses.

Westerners have their own culture-bound syndromes. Anorexia nervosa, for example, most often afflicts young women exposed to movie and television images of an idealized skinny female body.

The more host cultures and immigrant groups diverge in values and social patterns the greater will be the stresses on both populations. An example of this is the recent school-based turmoil in Montreal communities over the insistence of Sikh immigrant families that their children be allowed to bring ceremonial daggers to school with them. The controversy raged until a judicial judgment came down allowing the children to wear these daggers provided they were securely sewn into their clothes and rendered harmless. Stressful residues of this controversy persist.

Although the multicultural literature focuses principally on the stress that the enmeshment of alien cultures creates for immigrant groups there is a collateral price to be paid by the host society as it senses the loss of its own traditional lifestyle. This is understandable as immigrant groups experience a vulnerability and powerlessness that dwarfs that experienced by the host society. On the other hand, the anomie and alienation experienced by some European communities as they view their centuries-long rootedness in a treasured social–cultural and religious heritage threatened by large influx of as yet unassimilated immigrants from developing nations is well documented. This is analogous to the discomfort that members of a household feel when a large group of visitors overstays its welcome or even expresses the intention of moving in. Though this difficulty needs to be acknowledged it does not, of course, justify unjust treatment of the new arrivals.

Personality disorders are conditions that significantly reduce one's functionality in society. They reduce one's ability to negotiate the business tangles, social chicanes, and sexual charades of contemporary life. Dominant Western cultures prescribe respect for personal space, in varying degrees, and decibel limits of voice and laughter in public housing and spaces, such as theaters, restaurants, court rooms, and funeral salons. However, violations of these social codes do not connote personality disorder any more than does inappropriate behavior of naive travelers in foreign lands. At most such individuals suffer from ignorance and problems of adjustment.

The issue as to whether "culture-specific disorders reported from non-Western societies could be regarded as pathoplastic variants of disorders commonly observed by Western psychiatrists" (Cheng, 2001, p. 1) is still debated. The emic perspective affirms that personality disorders are often culture-specific precisely because they relate to one's ability to function in one's own society, which, indeed, influences the expression and course of psychological disorders. They are also era-specific in the sense that a society evolves different standards for mental illness over long periods of time. The mental disorders that were prevalent, even fashionable, in previous eras (say, *vapors*, *hysteria*, and *neurasthenia*, and a variety of fashionable factitious disorders in nineteenth-century Europe) have been differently conceptualized or relabeled in the West than have other post-industrial stress-related disorders.

The pathoplasticity one observes in "personality," whether normal or disordered, interacts with Axis I disorders and is analogous to the causal impact of culture on the course of mental illness. These interactions present vexing problems not only for the diagnostician but also for the transcultural nosologist.

Taxonomies of personality disorders

Theodore Millon and personality disorders

Millon is an important figure in contemporary clinical psychology. His influential book, *Disorders of Personality* (Millon and Davis, 1996), posits a view of personality disorder that is psychological rather than medical, and reflects the Kraepelinian philosophy that has inspired the recent architects of the newer editions of the *Diagnostic and Statistical Manual* published by the American Psychiatric Association. He dismisses medical approaches to personality disorders that presume either some external pathogen analogous to a virus that attacks the organism or, on the other hand, some *idiopathic* process that spontaneously erupts in individuals and debilitates their social competence. He states categorically that personality disorders are *not* disease entities, but "heuristic constructs," the imaginative framing of processes that are admittedly grounded in reality, but whose purpose is to guide further investigation into psychological phenomena.

This principle does not mean that personality disorders are fictions. The reality is that *something* has interacted with personality disordered individuals to influence their development into more or less dysfunctional people in the communities in which they live. These taxonomists recognize that there are no uncaused effects. They posit, however, that the causes of disorders are endogenous *and* exogenous, biological *and* environmental, and the products of their interaction – and that there is little more consensus about the etiology of many disorders than there was in Kraepelin's era. Among the five specialized (axial) lists of diagnostic factors in the DSM-IV, Axis IV provides the diagnostician and student of personality with the social–environmental variables that can be brought into play to explain the highly individual character of the pathology or personality dysfunction one finds in clients and neighbors.

Social environment is the background against which personality development plays out and needs to be assessed. Personalities are composites of heritable trait dispositions as well as of the habits, cognitive schemas, and values taught and learned in a specific social milieu. Wide discrepancies between an individual's behavior and social expectations of appropriate behavior engender amusement if not disapproval in the case of eccentric behavior, opprobrium and legal sanctions for destructive behavior. Both are grist for the psychotherapist's mill as either can cause intense distress for such people and their close ones.

What persons learn over a lifetime is complexly related to the experiences that their genomically determined dispositions incline them to seek out. Character becomes differentially configured as we seek out, traitwise, experiences and varied social milieus that are most congenial to us. Although it is widely recognized that many pathologies are in part heritable, madness is not, today, a condition with which one is thought to be born. The serious mental disorders that some people develop, and that cause them to be eventually hospitalized or incarcerated, are shaped by the society in which they have lived. They are also shaped by their *premorbid* personality traits. Extraverts and introverts typically are at risk for different disorders. For example, whether persons pre-occupied with their body image become anorexics or bulimics is undoubtedly related to the premorbid character traits that governed their relationship with the ambient society and their attitudes toward themselves.

Trait dimensionality

The *dimensionality* of character traits – the property of a bipolar variable that admits of graduated levels of intensity – is a contentious issue. Some personality psychologists claim that personality traits can be placed on a continuum (as explained in Chapter 5) such that those who show moderate levels of a trait are considered normal, those with excessive or deficient levels are considered disordered. For example, countless normal people are disposed to shyness and introversion; of these some display frequent moodiness and occasional depressive episodes. Should (and as) those traits become more prominent they shift along a continuum in the direction of a "disorder." Whether they are ever clinically diagnosable as a personality disorder depends on the social criteria that prevail in one society or another (see Widiger *et al.*, 1994). In this theoretical perspective, the highly introverted person and the person with avoidant personality disorder are both shy. The latter may simply be an extreme variant of a shy person. If one views personality disorders as variable social constructs, then there is obviously no absolute value – were it possible to calculate it with any certitude and accuracy – at which a pattern of traits is *ipso facto* transmuted into an illness syndrome, independent of whether or not one is ever so socially diagnosed. In this radical constructivist position no one is mentally ill unless he or she has been clinically certified to be such.

Other personologists assert that those clinically diagnosable as personality disordered truly have a condition that *differs in kind* from those with comparable traits judged as normal (Meehl, 1995; Siever and Davis, 1991). Their view is that the traits that are common to both normal and abnormal conditions are discontinuous variables. For example, obsessive-compulsive personalities usually have the premorbid trait of conscientiousness, a wholly admirable trait in some cultures. As that trait becomes heightened, shifting into the "red zone" of obsessive concern with order and perfection, it is transmuted into an essentially different *kind* of thing – a disease entity. The dutiful, well-ordered, and organized approach to life becomes transubstantiated into OCD. This school asserts a radical discontinuity between the conscientiousness, say, of the principled housekeeper and the obsessiveness of a world-class detective. Clearly, the discontinuity assumed here is not radical. Most, if not all, personality disorders have their roots in early premorbid dispositions that give direction to later illnesses that may arise. The tree falls in the direction in which it has always leaned. Both schools see personality disorders as precursors to clinical psychopathologies (enumerated in Axis I), just as normal traits can be precursors to personality disorders (Maher and Maher, 1994). This cascade of effects is always enmeshed with socio-cultural, as well as gender and genetic, variables.

This issue, referred to by some as *syndromal continuity* versus *discontinuity* (for example, Millon and Davis, 1996), hinges on whether or not we consider psychopathologies to be distinct disease entities analogous to medical diseases. The resolution of the controversy will depend on the extent to which less-than-optimal psychological adaptation to environmental demands and stressors is medicalized. Some influential psychology theorists in this area have refused to concede a sickness category as distinct from a wellness category for human personality, but if the evidence continues to accumulate demonstrating the increasing importance of biogenicity as distinct from the psychogenicity in the origin of personality disorders, a medical perspective may come back into vogue.

The implications of this for cultural psychology are clear. Medically defined diseases are assumed to be cross-cultural in their definitions. Just as meningitis, hypoglycemia, and breast cancer are the same diagnostic entities whether they occur in Moscow, Tokyo, or Kalimantan Barat, it may be argued that "psychological" disorders defined in Axis I of the DSM-III and its successor editions are truly the same disorders in all cultures, granting that the criteria are met – and independent of the varying psychosocial and personological factors that precipitated and conditioned their expression. As strenuously as Cartesian dualism is repudiated in the community of psychology scholars, there are yet overtones of this dualism in the firewall some wish to build between medical taxonomies and psychological taxonomies. (To categorize personality types along illness or wellness lines, and to create illness taxa and wellness taxa is considered a concession to medical science.) Ideological overtones have flowed into this domain, as psychology wrestled with the legitimate and pressing concerns of diverse perspectives on health and behavior. A clash of cultures is evident here. This school of thought has an affinity with an essentializing, medical tradition in the West and is not congenial to the relativistic, culture-dependent approach to nosology that is the politically more acceptable approach in professional psychology.

Some prevalent personality disorders in the West

Before describing the most commonly referenced personality disorders diagnosed in the West, I note some of the properties that they all have in common. Coleman, Butcher, and Carson (1984) have presented an array of these properties which enjoys wide consensus.

First, individuals with personality disorders are difficult to work and live with, or even to play with. They leave "a trail of personal relationships marked by difficulties they have caused others" (p. 236). It is precisely because people with such personalities consistently create discomfort and problems for others that they are thought to have a personality disorder. Although personality-disordered people usually regard themselves as unhappy, as do their families and associates, it is difficult to know to what degree the unhappiness comes from the condition itself and how much from the interpersonal conflict and alienation that it precipitates. Herein lies the nub of the disorder: the personality disordered attribute their social problems to the attitudes and behaviors of others.

Second, personality disorders are chronic conditions that have taken many years to develop. Unlike episodic pathologies such as PTSD, which arise from sudden personal catastrophes or from acute, unrelieved, and severe socio-psychological distress, personality disorders represent long-standing character traits that are seriously maladaptive. These disorders have incubated and become entrenched over the years and are evident in consistent, self-defeating patterns of behaviors and attitudes.

Third, the particular trait that lends its label to a personality disorder (an example is paranoid personality) pervades all aspects of an individual's life: his or her social interaction patterns; organization of finances; structuring of work space; interpretation of greetings; preparations for travel;

and so forth. Although life circumstances can change individuals' fears and suspicions dramatically, their hypervigilance taints all their interactions with their environment. There are large individual differences among individuals characterized as having the same disorder; however, the cardinal dysfunctional trait of their personality causes them to "burn through" one relationship after another. What is particularly unfortunate about personality disorders is that those suffering from them rarely learn from the painful experiences that their interactional style causes.

Fourth, a corollary to the above is that disordered individuals typically engage in denial about their condition. They recognize that their lives are larded with interpersonal conflict, occupational failures, and domestic strife, yet assume little responsibility for their failures and disappointments. Typically they allege that others do not understand (or listen to) them, that they have been misrepresented, or that they have had the misfortune to fall among an odd assortment of miscreants, opportunists, narcissists, and incompetents. When encouraged to get counseling or psychotherapy by members of their family or workplace, they may do so as a stratagem for gaining a bargaining advantage in future altercations.

Fifth, prognosis for change in persons with personality disorders is dismal, as they often have little subjective motivation to gain insight into their own dysfunctional mentation and belief systems. In the *psychodynamicists'* universe of discourse, personality disorder is a taxon that is still often referred to as *neurosis*. Much of the psychotherapy done in that tradition entails "personality restructuring." Many such practitioners envision therapy for such clients lasting for years. This is not surprising as the personality disordered, given their penchant for denial, are highly resistant to change – one reason that client resistance receives so much attention in the psychotherapeutic literature.

Typing personality disorders

The taskforce established in the mid-1970s to develop the DSM-III embarked on a program of wide-ranging reforms in the taxonomical system that was later published by the American Psychiatric Association. The signal accomplishment of this taskforce was the development of the *multiaxial schema*, which shifted the traditional classification of psychiatric disorders away from the *medical disease model*. In the classical medical model, therapists routinely pare away those variables, symptoms, and tangential data that do not seem to light the way to the discovery of a discrete underlying disease entity. This flows from a centuries-old view that assumes that there is a single cause for each disease. The early Freudian system, for example, assumed that a single sexual trauma, whose origins in early childhood could not be recollected, accounted for the psychoneuroses that plagued his adult patients. The architects of the DSM-III, on the other hand, recognized that the causes of mental disorders are multidimensional, can arise in discrete forms over long periods of time, and involve many sectors: social; biological; and familial. They consequently argued that a holistic rather than a narrowly focused assessment of individuals was necessary. Such an assessment involved consideration of the course of a disorder, personality traits (exaggerated as they were), social stressors, and aggravating medical conditions.

Of critical importance was the taskforce's decision to hive off personality disorders from the clinical pathologies that constituted Axis I and place them on a distinct axis of their own, Axis II, characterized by Millon (1996) as profiles of "the individual's lifelong and pervasive style of functioning" (p. 77). The taskforce grouped personality disorders into three clusters. One cluster encompasses disorders characterized by odd, eccentric, weird, flagrantly aberrant, or stunningly unconventional behavior. This cluster comprises *schizoid*, *schizotypal*, and *paranoid* personality disorders. A second cluster includes disorders characterized by behaviors that are theatrical, self-absorbed, flamboyant, wildly emotional, erratic, socially destructive, and unpredictable. This cluster comprises the *histrionic*, *narcissistic*, *antisocial*, and *borderline personality* disorders. The third cluster consists of disorders characterized by fundamentally fearful, anxious behaviors. In this cluster fall *avoidant*, *dependent*, *passive-aggressive*, and *compulsive* disorders.

That these influential theoreticians chose to group personality syndromes according to these symptom clusters appears to have been a matter of taste and intuition. There seems to be a fittingness to these groupings that satisfies the human need to impose an order of one kind or another on a scattered array of items. Though grouping helps us to remember the items in a list, the purpose here does not seem to be simply mnemonic. Mastery of our environment, even when mastery is illusory, presumes that we understand that environment. And few devices connote understanding better than creating taxa. Indeed, simply naming something or placing it in a category suggests that *we know what it is*. When pioneering trait psychologists began in the 1930s to reduce 18,000 personality descriptors to a handful of personality factors, it provided us with the comfort of "understanding" personality by way of this universe of discourse. These factors are, of course, mental constructs – creatures of human imagination. And though they are notoriously "fuzzy at the edges" they serve as powerful heuristic devices.

This grouping of personality disorders into three clusters based on salient characteristics that each group has in common resembles an informal, intuitively-based factor analysis. (Humans are prone from toddlerhood to do "eye-ball" factor analyses. For example, young children group organisms that are adult, male, and two-legged into the "factor" *daddy*; furry, four-legged animals of a modest size are *doggie*; all motorized vehicles are *cars*.) The intrinsic properties of the disorders are sorted into three distinct groups by virtue of their apparent intercorrelations. This makes more sense than listing all of them by alphabetical order, or even by their prevalence, or their subjective level of severity.

Synopses of eleven personality disorders

Cluster 1

Schizoid personality disorder. Kretschmer's[24] (1925) distinction between two variants of schizoid personality endures to this day. He characterized the schizoid condition, which he called *anaesthetic-schizoid*, as one in which individuals' emotions are flat – usually imperceptible. And behind the sterile façade they present to the world, one can glimpse psycho-emotional lives that are equally sterile, boring, and anemic. On peering into their interior world one sees "nothing but broken pieces, black rubbish heaps, yawning emotional emptiness, or the cold breath of an arctic soullessness" (p. 150). Their persona *qua* mask is a veridical representation of what it conceals. Their placid, numbed exterior is evident in their lethargy and aloofness, in a social style bereft of humor, light, and movement. They are reclusives, often labeled by their peers as *loners*.

Avoidant personality disorder. The second variant of the schizoid personality defined by Kretschmer is the *hyperaesthetic-schizoid* personality now defined as avoidant personality. The "autism" of such individuals springs less from an inability to feel the excitement of human interaction and the simple quotidian dramas that fill our lives than from a social clumsiness. They are inept in enacting the conventional schemas that govern relations between peers and elders in the various social situations that require nuanced behaviors: restaurants; parties; funerals; work sites; sports events; casual encounters; romantic dates; and introductions to a crowd of strangers. These *avoidant* persons sorely desire to "fit in" but lack the skills to do so. They often have fine sensibilities and nicely honed

[24] Ernst Kretschmer (1888–1964), a Tübingen-based neurologist and psychiatrist, is regarded historically as a constitutionalist: that is, he proposed a theory relating one's constitution or physical build to character. Although his theory is now dated, his perceptive insights into a variety of characterological disorders remain useful and reward study.

tastes and culture (hence Kretschmer's label, *hyperaesthetic*). On the other hand, they have a fear of rejection, scorn, and ridicule that leads them to engage in defensive behaviors that alienate others, and promote the very ostracism that they fear. They suffer from their social isolation, oblivious to the impact of their defensive, and repellent, social style.

Schizotypal personality disorder. Individuals with this disorder evidence traits that appear to be an attenuated or latent schizophrenia. They suffer from this disorder like those with schizoid personality in that they are socially avoidant and suspicious of others. They differ from the schizoid in that they reveal by their speech odd and eccentric thought patterns, often verging on weirdness. This disorder is among the most serious of the personality disorders and distinguishes itself by an unwelcome barrage of delusional, even homicidal, ideation, as well as bizarre behavior and frightening, violent impulses; it is a first cousin to schizophrenia.

Individuals with this disorder are socially ostracized and in consequence are deprived of meaningful feedback about their behavior. How meaningful or acceptable such feedback would be even if it were carefully proffered by either close associates or clinicians cannot be accurately gauged, given that they are inherently suspicious, misconstrue intentions and social signals, and have deeply entrenched schemas that regulate their response to social stimuli.

Schizotypal and schizoid personality disorders appear to be two variants of the same clinical disorder, the latter somewhat less serious than the former. The Greek stem, *schizo-*, that is used in the label for each is the keynote to a personality structure that causes victims to encapsulate themselves, autistically, in a private world. The evolution of these taxonomic constructs continues (Millon, 1996, pp. 617–23); needless to say, their demarcation and current formulation in the DSM is not definitive.

Paranoid personality disorder. In the interior world of the paranoid individual, meanings are imposed on the behavior of others and motives imputed to them that are totally groundless and, in many cases, fantastic. These "others" are both known and unknown, real and imagined. Paranoid individuals feel insecure because mentally they *are* insecure. They have no doubts about their precarious status in a global society. Their Kafkaesque world is rampant with faceless enemies, many in distant lands, whose mission is to "take them down."[25] What is

[25] I treated an individual in the 1970s who was convinced that the KGB *and* the CIA were in collusion to blow up his university dormitory.

stunning for those who live, work, or study among paranoid individuals is the ease with which they take insignificant events and invest them with a menacing character. Although some may respond passively to delusional threats, many can be self-assertive, take "appropriate countermeasures," engaging in "pre-emptive" attacks, and plan elaborate stratagems to outwit their enemies.

That paranoid persons lack a basic trust in others is to state the obvious; it also isolates the guiding principle for interpreting their lifestyle, the protective wall they build around themselves, the distance they keep from everyone whom they cannot control, their hypervigilance, lack of joy, tenseness, and, when the condition becomes pathological, grandiosity. (This grandiosity is evident in their belief that elaborate schemes are concocted at the highest levels of government.) Even in the presentation of modest paranoid symptoms, the stubborn and unreasonable character of their illusions is evident. As with other cardinal traits, paranoia can surface, at times vividly, in many other personality disorders: the passive aggressive; the narcissistic; the compulsive; and the sociopathic.

Cluster 2

Histrionic personality disorder (HPD). Individuals with HPD like to showcase themselves. To ensure that they are noticed, they stage minidramas in the presence of others. Although most people engage in this behavior on a moderate scale, it is done with a finer sense of proportion and a sense of the impact it will have on others. Self-dramatization springs from the human penchant for storytelling. Individuals love not only to tell but to hear the narratives that reveal the exciting and curious events that characterize their lives. But those with HPD do it to an extent that draws social disapproval – even opprobrium. Those who seem to be normal but are approaching the cusp of this disorder are watched on talk shows, newscasts, in the theater itself, as well as in other arenas where histrionics is approved.

Individuals with HPD seek approval in ways that exceed the norm. They make grand entrances, preferably late, to social events. These cameo appearances are scheduled by such individuals for moments when others are not likely to upstage them. They enjoy storytelling, excel in telling jokes, and get a "rush" from the laughter and applause they receive, even from an audience of one. Typically, they crave excitement, are dependent beyond the norm on the approval of others, and appear to be emotionally unstable and immature for their age. Not surprisingly clinicians characterize them as vain and self-centered, but their wit, affability, and social energy compensate, in small doses, for their social deficits. In the longer run, they do not maintain stable intimate relationships, except with a rare long-suffering, subservient, responsive, and nurturant companion. On the other hand, state Millon and Davis (1996), "Histrionic personalities often demonstrate albeit in caricature and mild pathological form, what our society tends to foster and admire in its members – to be well liked, successful, popular, extroverted, attractive, and sociable" (p. 366).

This syndrome was inserted into the DSM for the first time in 1980 (see DSM-III, American Psychiatric Association, 1980) and replaced earlier formulations that were labeled in the psychiatric literature as *hysteria*. Hysteria was one of the rampant disorders diagnosed in nineteenth-century Europe, thought by many in that era (but not by all) to be a female disorder. HPD has proven to be an equal-opportunity disorder, afflicting men as well as women. In detailing the development of the criteria that emerged in the DSM-IV (1994), Millon and Davis made the interesting observation that one criterion of HPD – the manipulative use of suicidal threats, gestures, or attempts – "was deleted in order to better differentiate the histrionic from the borderline personality" (1996, p. 365).[26]

Narcissistic personality disorder (NPD). Narcissistic individuals share some qualities with those with HPD. They feel they are entitled to benign attention from everyone, and a respect that they have not, in fact, merited. This preoccupation with their image makes it difficult for them to enter into other people's frame of reference, to understand the feelings of others, and to measure the impact they are having on those with whom they interact. Their self-centeredness and stunning self-conceit sour their relationships with their peers, and turn them into social pariahs in the playground and in the workplace, a rejection that further potentiates their self-absorption.

There are such large individual differences among those who fall into this category of disorder that various diagnostic models of NPD proposed in textbooks lack clarity and crispness. Because narcissists, as they are conceptualized, have salient character traits in common with other

[26] This is evidence that taxonomies are socially constructed devices whose components are structured for disciplinary purposes, and that those purposes shape the conceptualization of disorders. *To differentiate* is a mentalistic procedure. To alter definitions so that we can "make" categories as disjoint and crisply delineated as possible can be regarded as acceptable or not depending on how absolutist and Platonic one wishes to be in portraying reality. To put a finer point on this, moving criteria about to create neatly packaged *taxa* and to diminish the overlap between them can arguably be said to reverse the process of moving from the data to models – as if we can change nature by thinking about it differently.

personality disorders, it is more difficult to distinguish them taxonomically. This may be the reason that the *International Classification of Diseases* (ICD-10) has declined to include NPD as a distinct taxon.

Millon and Davis (1996) have distinguished four types of narcissistic personality: the *elitist*; the *compensatory*; the *amorous*; and the *unprincipled*. *Elitist narcissists* are, in fact, normatively competent individuals who are, however, arrogant, full of themselves, and convinced of their eminent worth and superiority. They compete aggressively with others, engage in one-upmanship, and strive with single-mindedness to achieve celebrity. These are the purest variant of the narcissistic persona, according to these authors. *Compensatory narcissists* differ from other narcissists in that their self-absorption springs from profound insecurity. They are *poseurs* who have adopted a façade of superiority to conceal the weakness that drives their search for approval. *Amorous narcissists* seek to affirm themselves through their sexual adventures. Exploiting a succession of sexual partners, Casanova-like, dooms any hope of normal social adjustment or a stable relationship with a domestic partner. Finally, *unprincipled narcissists* flout prevalent moral conventions and flagrantly exploit others for their own ends. Such narcissists are scheming, devious, and infest the underbelly of white-collar professions as well as the street-junkie culture that is easily unmasked in the therapeutic communities that have been established to rehabilitate them.

The grounds for this fourfold division of a construct that is already ill-defined by virtue of its inconsistent and floating criteria appear unstable and Procrustean. The criteria for amorous narcissists rest on the narrow *social expression of certain behaviors*, namely sexual "conquests"; those for compensatory narcissists reside in the *etiology* of the disorder; those for elitist narcissists are founded on a *mental trait* (the conviction of their actual superiority); and those for unprincipled narcissists are based on *moral turpitude*. There is no guiding principle for this fourfold division. The boundaries between these four types are porous, overlapping, and inconsistently related. Beyond that, however, they cannot be distinguished from one another, for the amorous narcissist is truly unprincipled; nor can compensatory narcissists be distinguished from amorous narcissists. Still less can elitist narcissists be distinguished from the other types, for their self-conceit leads them into the same unprincipled, sexual, and compensatory behaviors as their co-narcissists. The absence of a unifying principle among these four types weakens the usefulness of the taxonomy.

Antisocial personality disorder. Individuals with antisocial personality disorder (often referred to as sociopaths or psychopaths) create distress among those with whom they come into contact. In their most virulent form they are dangerous, and tear at the social fabric in which most people take their comfort. What is most remarkable about them is their inability to feel remorse or guilt for the suffering they have caused others or compassion or sorrow for the misfortunes of the innocent and most vulnerable. Their self-seeking and heartless exploitation of others, independent of the close bonds of family and community, is their salient personality trait. They are quintessentially unprincipled narcissists. The classic models of this genre of personality disorder depict them as unable to form friendships, as disloyal, and as betrayers of the trust that family and associates have placed in them.

Bowlby (for example, 1946, 1973) and other classical attachment theorists (see Scott, 1963) affirmed a neurobiological basis for this disorder. When infants are deprived of the care and affection of caregivers during primary socialization (that is, the period from the second to the seventh month of age, inaugurated by the social smile) they fail to imprint not only to their family but to the human species – analogously to other vertebrates.[27] This accounts for their inability to form friendships in later life – and to relate to others as if they were members of the same species. Their failure as infants to form an affectional bond with anyone cripples their ability as adults to form affectional bonds with mates. The effects of such deprivation are permanent, and, in this school of thought, account for the principal traits of the sociopathic personality. It even makes them, like the amorous narcissist, unable to engage in stable sexual relationships – based, as such relationships are, on mutual respect and love.

Sociopaths are found to be incapable of loyalty, though they can convincingly feign friendship. Their "acting" skills make them excellent "con men," and like the elitist narcissists, their charm, intelligence, sense of invulnerability, and overweening ambition allow them to pull off "capers" that the average person would consider foolhardy, if not virtually impossible. Their cunning allows them in many instances to remain on the safe side of the law, engaging in business and financial schemes with the ruthlessness and scalpel-precision that harvests hatred and enemies as well as large fortunes. Drug rehabilitation programs (like Daytop Village and Phoenix House) harbor many such individuals. Those who are most successful in treating them are former street-junkies who are themselves past masters of the confidence schemes that they easily detect in others. It often takes a thief to catch a thief, a psychopath to spot a psychopath

[27] The seminal works of ethologists Karl von Frisch, Konrad Lorenz, and Nikolaas Tinbergen (for example, 1953), are well worth study with regard to this matter.

(or a therapist who has spent a professional lifetime studying them and learning the behavioral cues that peek around the edges of their consummately staged acts).

The issue of the dimensional versus the categorical nature of personality disorders is important here as it has implications for treatment as well as social tolerance for deviancy. If psychopathy is dimensional, and is simply normal aggressiveness and game-playing pushed beyond the bounds of social acceptance, then we can entertain more hope for rehabilitating the social predators who masquerade as respectable physicians, lawyers, and businessmen (in the mode, say, of Frank Abagnale, Jr., whose social predations and rehabilitation were captured in the film, *Catch Me If You Can*). On the other hand, if sociopaths form a distinct population rather than an outlying segment of a normal population, whose moral sensibilities and other *human* qualities were amputated in infancy with brutally ruptured caregiving bonds, then prognosis is bleak. It is less a question of nudging the person with an anti-social personality back toward normality than restructuring a truncated personality.

Those who take a dimensional view of disorders try to conflate the criminal and inhumane behaviors of some antisocial and self-aggrandizing personalities with the socially acceptable and rewarded behaviors of those who manage to remain legitimate, yet are equally ruthless and cunning. It can be argued that if we wish to safeguard the meaningfulness of *personality disorder*, it seems useful to clearly distinguish reprobated behavior that is punished by society from behavior that is socially approved and rewarded. The critical point is not that psychopaths are antisocial, which they are, but that the moral issue is intrinsically irrelevant to them. They will typically choose the lawful path as they have the intelligence to understand that this presents the fewest obstacles to the achievement of their narcissistic goals. In brief, they often make the non-criminal choice, not because it is right but because it is smart.

Borderline personality disorder. Individuals with this disorder have both affective and mentational dysfunction that might formerly have been characterized as severe (Axis I) pathology. They acquire the designation *borderline* through twofold entitlement: (a) their disorder straddles the boundary between Axis I pathologies and Axis II personality disorders; and (b) their disorder encompasses both severe mood and emotional swings, on the one hand, and a tenuous grip on reality and their social identity, on the other hand. The label *borderline* was not a felicitous choice by the architects of the DSM as it suggests nothing of the intrinsic traits of the disorder, as do the terms narcissist and paranoid – only a suggestion as to its fit into a taxonomic schema.

Borderline patients experience erratic mood swings, explosions of anger, periods of irritability interspersed with euphoric spells. They seem to incarnate a potpourri of various disorders – paranoid at some times, histrionic and narcissistic at others, mixed with confusion about their certifiable resources and abilities, career objectives, and interpersonal relationships. There is little cohesive organization or integrity to their personalities. Their personalities have never crystallized or assumed a stable form. They lead chaotic lives, governed by random events, having no stable behavioral heuristics to guide them. They are a puzzle to themselves as well as to their acquaintances and colleagues as, for no proportionate reason, they are equally apt to dissolve in tears or erupt into a fury.

Persons who defy categorization and clear taxonic features become *ipso facto* "borderline" to their clinicians. The absence of a meaningful pattern in their syndrome is what makes it so difficult to define – and the reason why clinicians who cannot fit a patient into a crisply delineated syndrome drop him or her into this catch-all bin. It is this absence of lawfulness, regularity, or predictability that challenges the diagnostician. There is hardly a dysfunctional trait that they may not be expected to express, albeit in shades ranging from faint pastels to brilliant Halloween orange. Borderlines are ambulatory compendia of everything that can go wrong in human development. Their principal feature is that they are survivors. As one might expect, taxonomists have formulated many sub-types of this disorder.

Cluster 3

Compulsive personality. Individuals with this disorder engage in repetitive, stereotyped, and inflexible routines. Of course, many normal people do so as well. The distinguishing factor is that this preoccupation interferes with the timely completion of common tasks. The mentation that accompanies and precedes these ritualized routines is characterized as excessively cautious, perfectionist, hesitant, and scrupulous. The nettlesome question that always remains in these dimensional issues is: when does conscientiousness and carefulness become excessive?

Compulsive individuals are noted for their attention to rigorous schedules. They draft and redraft agendas,[28] return repeatedly to stoves they have turned off, windows and doors they have shut and secured, and sterilize instruments and hands far beyond conventionally reasonable

[28] *Agenda* is a plural Latin form meaning things that must be accomplished. The singular is *agendum*. Although the convention has taken hold to further pluralize this plural form, it is syntactically similar to pluralizing, say, "datas," "medias," and "phenomenas."

norms. They tend to be addicted to work and miserly with their leisure activities. They even ritualize the stages of the family's vacation. Personal relationships, family life, and avocations always yield before the need to finish the task and be "productive." Rules are rigorously observed, punctuality is regarded as a cardinal virtue, and reliability is next to godliness.

Obviously, individuals with these qualities are invaluable in work involving the control of heavy air traffic, the management of nuclear plants, the implementing of operating room protocols. Yet even here, excessive ritualization and inflexibility can interfere with problem-solving in situations requiring new solutions to random and novel problems. The compulsive person, in short, has trouble thinking "outside the box."

Dependent personality disorder. A certain dependency and orientation toward one's conspecifics is normal in all gregarious species. Humans live in groups, socialize in groups, and protect and nurture themselves in groups – those who are leaders as well as those who are led. Clearly a dependency disorder is only such where dependency on others has gone to excess and where those needs that individuals can fulfill for themselves are left to others. The *reasons* that individuals leave their needs to be filled by others lie at the crux of this disorder.

We allow ourselves to be served at restaurants, to have our meals cooked by chefs, and to have our cars parked by valets. This kind of dependency is replicated in many domains of human activity. The very complexity of human society with its divisions of labor necessitates profound degrees of interdependency. There are innumerable tasks that we are virtually incapable of performing for ourselves. But the dependent personality does not do for himself or herself what other members of society, according to prevailing mores, consider appropriate. The notion of dependency is, more than any other personality trait, a culturally conditioned attitude. Each society expects certain groups of people within society to do certain things for themselves. However, the frequency with which we exempt ourselves from responsibilities appropriate to our stage of human development must fall within certain social parameters, and our willingness to provide a *quid pro quo* is expected and demanded unless there are medical or developmental reasons why we cannot. Dependent persons who are physically incapable of doing for themselves what others routinely accomplish are not *per se* personality disordered.

This disorder clearly has less to do with what individuals allow or solicit others to do for them than with reasons for their behavior. Dependent personalities cater to the whims of others to secure their approval and affection, which then helps them to secure the services and the help they believe they need. They allow, even ask, others to make critical life decisions for them, thus letting themselves be infantilized. To feel and act helpless leads them to engage in clinging and cloying behaviors that alienate rather than endear them to others.

The dependent personality is usually viewed as immature, incompetent, and weak. This is a mirror reflection of their own self-image. Such individuals, for obvious reasons, seek professional help more easily than do other personality disordered people. The challenge for their clinicians and friends is to nurture the seeds of *autonomy*, which they have themselves neglected or which controlling family members have suppressed.

Passive-aggressive personality disorder. Individuals with this disorder express their hostility and negativism, as the label suggests, by failing to complete tasks that would benefit others (as well as themselves). Rather than actively attacking those they resent they harm them by undermining their projects, securing their failure by surreptitiously sapping the foundations of their projects rather than by "assaulting their ramparts." They "forget" to do what was critical to success. They find numerous reasons why one measure or another could not be implemented. Their calculated inefficiencies, under the guise of external constraints or hypothetical difficulties, cause deadlines to be missed, letters to remain unwritten, phone calls unplaced and unanswered, documents to be lost, invitations never extended. There is cunning in this, for passive-aggressive persons can only rarely be accused with certainty of being malicious – thus, they do not invite aggression; they are simply shunned as incompetent. That passivity allied to aggression is not a frequently diagnosed condition is unsurprising. It requires the *interpretive* attribution of motives to clever individuals, which is exceedingly difficult to prove.

Conclusion

Born to be happy

Though this chapter is dedicated to a consideration of disordered personality, such subject matter needs to be placed in a larger context of the nature of the Human and that aspect of it that we call *personality*. Scientists have spent a lifetime studying defective personalities – a worthy scientific endeavor – but their tracts on this subject display a complex structure of deformities suggesting that there is a lawfulness, an order, in the domain of "human nature gone awry," just as there is a lawfulness in the ontogeny of any well-functioning human being. The dark side of this preoccupation with malformations and fractures, diseases and age-related decline is that we risk clouding our understanding of what humans are and can become.

The renewed interest in a positive psychology, discussed in an earlier chapter, is expression of a yearning to focus on what is good in humans rather than excessively on what is flawed (for example, Seligman and Csikszentmihalyi, 2000). In this vein Carl Rogers, the eminent psychotherapist *cum* personality theorist, wrote in 1961: "One of the most revolutionary concepts to grow out of our clinical experience is the growing recognition that the innermost core of man's nature, the deepest layers of his personality, the base of his 'animal nature' is positive in nature – is basically socialized, forward-moving, rational and realistic" (pp. 90–91).

The noblest goal of life, philosophers have argued, lies in the pursuit of competence through the exercise of our natural endowments. And our happiness resides in engaging in this pursuit in a society that affords us the scope to achieve our goals. Needless to say, a certain longevity, the peaceful company of our conspecifics, and minimally good health are necessary for this purpose. But there are those, it seems, who have been gifted with a genetic predisposition to be happier than the norm. Their *joie de vivre* is less attributable to good fortune than to good genes. Their habitual mental state is one of optimism, patience, and happiness. Crises in their family relationships, serious illness, or financial distress cause suffering, of course, but they return to an *elevated baseline*, and more rapidly than do most.[29]

Given that all theories reflect the personal history and the temperament of their inventors, we can rejoice in the psychological diversity of those who have proposed personality theories. That positive psychologies have been relatively few in number and have only recently achieved high visibility may say something about the character of the psychologists who have been attracted to this discipline and fostered their idiosyncratic views of the Human. The attraction to a dark Kierkegaardian rather than a lightsome Rogerian conception of the world and of human nature may not be something over which we have complete control. In any event, scientists need to give more attention to the positive than to dysthymic aspects of human personality, as well as to prevenient approaches to maintaining health and forestalling pathology.

[29] This condition can appropriately be called *euthymia* – good mood. *Hyperthymia* is a nineteenth-century psychiatric label for people who are often excessively cheerful, highly active if not agitated, bubbly, frenetic in their work habits, and, in the extreme case, manic. It is contrasted with *dysthymia*, another personality disorder, which entails listlessness, lethargy, sadness, anhedonia, or the simple absence of joy. K. L. Kahlbaum developed the concept of hyperthymia in 1882, along with *dysthymia* and *cyclothymia* (Brieger and Marneros, 1997).

Art and psychology

Art frequently anticipates science, and visionaries light the way for engineers and architects. Movies, plays, and novels can provide us with a futuristic psychology and prototypes of personality that science may later "discover." The happy salesman in the film *Thirteen Conversations About One Thing* shapes the circumstances and personalities of those around him. He is an example of individuals who keep their eye on a future rich with possibilities for self-betterment while slogging through a present that is larded with meanness and tough luck. The public resonate to this theme and in fact demand it in their entertainment. They yearn to see virtue rewarded and hard work crowned with success – the stuff of personal satisfaction. Steven Spielberg said, when reflecting on his early cinematic career: "I didn't know at the time that a happy ending could make you a lot of money. I really had no idea that if you kill off your leads, you kill off your audience" (quoted by Onstad, *National Post*, p. b2, January 3, 2003). The negative psychologist may demean the "escapism" provided by this type of Dickensian narrative. The positive psychologist will see it as an identification with the Darwinian exemplars who, like daemons in the wings of the theater, inspire a cast, on the one hand, to perform at the height of their powers, and their audience, on the other hand, to resonate with what is most human in the spectacle they are internalizing (see Wedding and Boyd, 1998).

The human mind is a fertile ground for growing new and imaginative approaches to dealing with human problems. Our perceptions of human personality will change with the temper of the times, as do our perceptions of varieties of personality disorder. As psychology refashions itself in a more positive image and as the improved procedures for helping individuals with personality disorders reform their social personae, personality psychology will help restructure their interests, attitudes, values, career paths, and self-perception. New templates for conceptualizing both healthy and less healthy personalities will surely emerge. The mental health professions and their client base will benefit.

13 Eight appendices: at the margins of personality psychology

> The topics in psychology that most interest laypeople – love, hate, work, play, food, sex, status, dominance, jealousy, friendship, religion, art – are almost completely absent from psychology textbooks.
>
> (Pinker, 2002b, p. 21)

Gathering eight appendices into a single chapter requires some indulgence from the reader. Each appendix has important personological features but is not otherwise closely related to the others. As these topics did not fit seamlessly into the preceding chapters I decided to aggregate them in a final chapter. Some personality traits are widely recognized as important in our daily lives, and though well treated by scholars in handbooks, reference works, and stand-alone volumes, they are often given insufficient attention in books on personality psychology, or indeed in psychologists' research agendas, as Cacioppo *et al.* (2005) have observed. Sternberg and Lubart (*cf.* 1991, p. 678) lament that although significant research *is* being done on specific traits, the scientific community of personality psychologists, and particularly those who write textbooks, do not give them sufficient prominence in their own work.

In the segments that follow I will cursorily examine some of the personality traits that have been traditionally subsumed under more generic topics, such as the Big Five Model or Cattell's 16PF (see Chapter 5, above). Relatively little information is available on a number of them. For example, the preparation and enjoyment of food takes up a large proportion of our waking lives. Individual differences exist in the value people place on exploring and savoring various cuisines or the care they give to preparing their own meals, yet the psychology of food preparation and dining is sparsely treated in textbooks. This chapter begins with the psychology of food preparation and follows up with segments on playfulness, creativity, curiosity, altruism, sexuality, spirituality, and mourning. I chose these not because they are most important or because there have not been notable scholars who have done excellent research on them, but because they seem insufficiently treated in the generality of texts on personality psychology.

Food preparation and dining

The pleasure that comes from eating food, the enmeshment of that activity in the most solemn, significant, celebratory, and joyful social occasions; the satisfaction of preparing meals for one's loved ones, and seeing them eat with gusto; the artistic creations of highly touted chefs, not to mention the management of their cuisines; and the simple pleasure of sharing an evening meal with one's family and friends are so central to the good life that it is difficult to imagine that they would be overlooked in a personality psychology text. Nevertheless, they generally are.[1] Is there any other (noncareer) activity to which people *daily* give as much thought, study, preparation, and joyful anticipation (usually as hunger grows) as they do to their next dinner? Except for those in the throes of an amorous passion, a domain of experience to which they can rapidly habituate, there is no other dimension of daily life to which so much time and energy are devoted by so many. The aphorism "kissin' don't last, cookin' do" encapsulates this reality.

There are wide individual differences, nevertheless, in the interest and motivation that people devote to the preparation and consumption of food. Profound physiological reasons explain this, but to a great extent, these differences are accounted for by an organism's ability to smell and taste. A genetic predisposition to enjoy certain kinds of taste accounts for some of the choices of food and drink evidenced in various populations. One of the more intensively studied tastes is that of *phenylthiocarbamide* (PTC), studied precisely because a wide diversity exists in individuals' ability even to taste it and then agree on what it tastes like (see Logue, 1986, pp. 55–61; also Chapter 4, above, on biology and personality). Many individuals cannot taste PTC at all (Crandall and Spence, 1974), and most of the others find it so revolting they gag on it. Still others pick up a taste in PTC that can be bitter, salty, or, in less than 2 percent of a Caucasian population, sweet. These latter tastes, if sufficiently diluted, are not at all unpleasant. There is evidence that the tastes of many common foods, and the acuity of the senses for perceiving such tastes,

[1] The work of Paul Rozin is an exception to this generalized neglect. The American Psychological Association (2007) recently recognized him "for his groundbreaking work on the psychology of food selection and for showing how crucial the psychology of food is for understanding so much else about human nature" (p. 751).

differ widely within any given population.[2] That some of these differences are genetic is indubitable, and in some cases they are attributable to enzymatic deficits to their digestion. The bad experiences individuals have had after eating or drinking them will shape their diet as well as their cuisine. On the other hand, one can certainly learn to dislike foods as well as learn to enjoy them.

The physiology of hunger and thirst has been extensively studied, as well as the disordered indulgence in food and drink that results from genetic and (consequent) central nervous system (CNS) anomalies. An example is Prader-Willi Syndrome (PWS), which results from the deletion of certain genes on an arm of chromosome 15 – usually a random error in gene replication. There is a complex of symptoms resulting from this genetic disorder, one of which is the experience of constant hunger and a degree of obsession with, and uncontrolled consumption of, food that can result in obesity. Indeed, PWS is the most commonly recognized genetic causal pathway to obesity (see the work of Daniel Driscoll, for example, Cassidy and Driscoll, 2009; Glenn *et al.*, 1997; Goldstone, 2004; Miller *et al.*, 2006).

In a systems perspective, dietary and culinary preferences, as well as the sheer quantity of food consumed (when it is available), are determined in part by personality traits. A system by definition is a complex of components that are interactive, such that a malfunction in one compromises the system as a whole; in organisms the sensitivity, health, and aptness of any component contribute to overall well being. Personality is such a system comprising individual preferences that pervade all behavior and not least one's attitude toward food, and the context in which it is prepared and consumed. In all-you-can-eat buffet-style restaurants one sees people eating more, we think, than they normally would – often because of affordability. That socio-economic class (SES), at least the attitudes prevailing within it, determines the quality and quantity of food consumed in any society would seem self-evident as well as historically demonstrable (Bourdieu, 1984). The upper strata of societies will employ creative, expert chefs to prepare their meals. S. Irene Virbila (December 28, 2005) described the culinary virtuosity that only the rich can afford, in the cuisine of Joël Robuchon. The acclaimed film, *Babette's Feast*, based on the short story by Isak Dinesen, celebrates

[2] World-class chefs are typically convinced that no one should alter the taste of their confections by adding, say, salt or pepper – this is in apparent denial of the fact that our ability to taste certain ingredients is genetically, and differentially, determined, and, indeed, changes with age. One can argue that this violates the rights of individuals with sensory disabilities to address this challenge in a public restaurant.

the genius, artistry, and supreme sensitivity that produced aesthetic dining experiences that few could appreciate.[3] Prior to the French Revolution, for example, the nobility were able to employ expert chefs to prepare their varied and creatively seasoned dinners, while the proletariat and peasantry had to be satisfied with bread, cheeses, quiche, and stews. Today, Burger King and Denny's Restaurants draw a different clientele than do Michelin 3-star restaurants.

The film industry has appreciated the appetite of all classes of society for artfully prepared food, even if financially out of reach. (Magazines like *Elle*, which are published in various languages and cultures, appeal to those who have upper-class sensibilities, but only lower-class budgets.) Movie moguls are as aware as psychologists that great cuisine, like great art, springs from the impassioned personalities of those who are in touch with the meaning as well as the sensory allure of good food. The success of food films like *Eat Drink Man Woman*, *Big Night*, *Like Water for Chocolate*, or even *Tom Jones* attests to the fact that millions are fascinated not only by the consumption of savory food, but also by the emotion that is invested in its preparation. Niche cuisines and the popularity of recipe books are powerful countervailing influences against fast-food outlets (*cf.* Smith, 2002; Warde, 1997). Recipe books are a form of self-help literature that, no doubt, have a beneficial effect on the well being of those who use them. The impact of near-universal literacy in the West on the aspirations of the working classes who try to emulate the dishes and lifestyle of "the rich and famous" can be easily underestimated by academics.

An argument can be made that, within any class and ethnic group, the choice of an environment is also largely determined by temperament and personality. Individuals can either shun or choose to go to all-you-can-eat restaurants, cocktail or tailgate parties, picnics, intimate home-cooked

[3] General Loewenhielm (speaking of the first time he had *cailles en sarcophage*):

> One day in Paris, after I had won a riding competition, my French fellow officers invited me out to dine at one of the finest restaurants, the Café Anglais. The chef, surprisingly enough, was a woman. We were served Cailles en Sarcophage, a dish of her own creation. General Galliffet, who was our host for the evening, explained that this woman, the head chef, had the ability to transform a dinner into a kind of love affair, a love affair that made no distinction between bodily appetite and spiritual appetite. General Galliffet said that in the past he had fought a duel for the love of a beautiful woman. But now there was no woman in Paris for whom he would shed his blood – except this chef. She was considered the greatest culinary genius. What we are now eating is nothing less than Cailles en Sarcophage. (lines from *Babette's Feast*, transcribed by David Schimpf, 2005)

Caille is Cornish hen or quail. *Sarcophage* (that is, sarcophagus) is resonant with the spiritual dimension of a dinner prepared by Babette – the complete chef. Consult also a segment from Priscilla Parkhurst Ferguson's *Accounting for Taste*, available at: www.press.uchicago.edu/Misc/Chicago/243230.html#copyright, accessed July 29, 2008.

meals, smorgasbords, formal hall dinners, and festive wakes. These situations are not normally thrust upon them; there are personological, socio-economic, religious, fraternal, and culture-specific reasons why they would do this. The public versus private character of dining is one of the environmental variables that influence the behavioral patterns of eating. Although cocktail parties are rule-governed and peer-modeled events that evoke a generally consistent eating pattern among participants, how much they nibble and drink depends on their decision to attend the party in the first place and their personal preferences and self-discipline in the second.

Multi-course meals are common in most global cuisines. Main courses are usually preceded by an "antipasto" or "starter," which is preceded by pre-prandial drinks and nibbles. A second, main course can be followed by a salad, and then fruits, cheeses, or desserts, which are then topped off by a coffee and a final *digestif.* Hemingway notes in one of his letters that he often took meals in Montparnasse, a Parisian literary haunt of *littérateurs* of the 1920s, where he and his buddies would routinely finish dinner at one restaurant and take coffee and a cognac at a different one. That there are class-based norms for these patterns of dining-out is evident. As for weekday dinners eaten at home, many typical middle-class American families do not prolong this meal, after which some members repair immediately to a front row seat in the TV room to watch a sitcom, quiz show, or football game.

Eating and dysthymia. Folk psychology proposes that individuals eat more palatable "comfort foods" than they would ordinarily when anxious or depressed and, *ceteris paribus,* more than non-depressed individuals would. But folk psychology also paradoxically proposes that fat people are both jolly and more prone to overeating than are skinny people. Clearly such psychology does not provide clarity in these matters and needs to be informed by empirical studies of the relation of mood, personality, and physical constitution to food consumption. Plutchik (1976) found, a generation ago, that overweight as distinguished from thin individuals tended to indulge in eating when anxious or depressed. Comparably, Slochower and Kaplan (1980), having induced anxiety in both obese and normal weight individuals, discovered that the obese *increased* whereas the latter *decreased* food consumption. These findings partially corroborate folk psychology, but this is not surprising as popular lore proposes all manner of contradictory principles, some of which are bound to be correct. Rat studies also corroborate these findings. Given a succession of mild tail pinches rats thus stressed over-ate to the point of obesity, many of them going on eating binges (Antelman and Caggiula, 1977).

Current research suggests that personality factors correlate with, and influence, eating patterns. Individuals eat differently; some with gusto and relish, others slowly and sensually; some rapidly and distractedly, others with intense focus – autistically so to speak. But it is choice of the food that has the most serious implications for long-term health, and personality determinants of diet have begun to be studied. An illustrative example is a cogent study (Kelloniemi, Ek, and Laitinen, 2005) demonstrating that dispositional optimists eat a healthier diet than pessimists. These researchers assessed the diets (and alcohol and tobacco use) of 8,690 men and women, age 31, born in northern Finland in 1966. They found that those scoring in the highest quartile for optimism ate a healthier diet, that is, more foods rich in fibers and low in animal fat, than did those in the lowest quartile. This diet consisted of an abundance of berries, fruits, salads, legumes, and low-fat milk products. Furthermore, optimists were less prone to use tobacco and to indulge in heavy (alcohol) drinking. Contrariwise, individuals who had heightened expectations of bad outcomes in their work and social activities appeared to overindulge in "comfort foods" and to assuage their negative moods with heavy drinking. These are recognized lifestyle risk factors for cardiovascular disease and obesity. The causal pathways linking all these variables are not clear and in any event are not linear.[4]

Aside from the basic need that all organisms have to ingest nutrients to support physical existence, eating is an activity that humans configure in many complex group rituals to celebrate transitions through life stages. They are both secular and sacred. The wedding banquet, the religious

[4] Food, memory, and personality. A serendipitous finding relating personality to food preferences emerged from a study by Loftus (2003), who investigated make-believe memories and their influence on behavior. Alan Alda attended a conference at the University of California, Irvine, to explore fields of inquiry for his own journalistic contributions to Scientific American Frontiers. Loftus and colleagues, concealing the true objectives of their research, suggested to Alda that they were interested in the relationship of personality and eating patterns. She stated:

> Once in the lab, we tried to convince him that when he was a child he had gotten sick eating too many hard-boiled eggs. We explained that a sophisticated computer program had analyzed [the personal data he had submitted a week earlier] and discovered … the gotten-sick fact. An hour or so later, he had a picnic in the park with students, post-docs, and other members of my lab. There were many foods from which to choose: sandwiches, pickles shrimp cocktail, hard-boiled eggs, deviled eggs and more. He refused to eat the eggs. (p. 867)

Although Loftus cautions us that there are, of course. other plausible reasons for Alda's refusal to eat eggs on that particular day, what is *suggested* is that not only personal memories but our collective history may be partly responsible for our attitudes to foods.

supper, the birthday party, the tribal festival – all are *per se* dining events; birthdays, anniversaries, family reunions, and post-funeral receptions are organized around a communal meal invested with special meanings. Even on a weekly basis families and other tightly knit social groups gather for dinner. Although the rationale for this is not always made explicit, family dinners are a celebration of the family's togetherness, even when frayed by familial squabbles. Nowhere is that more evident than on occasions that memorialize the family's deceased members.

In a debate that unfolded at Princeton University in the fall of 2005, animal rights activist, Peter Singer, and philosopher, Roger Scruton (2006), debated the moral dimensions of sacrificing animals for food. Scruton wrote: "Rational beings are nourished on conversation, taste, manners, and hospitality, and to divorce food from these practices is to deprive it of its true significance" (p. 23). The larger issue here is that humans have invested meals, as social events, with moral, often spiritual, significance. Our respect for each other and for the lives of those that were sacrificed to nourish us imbues us with a sense of our rootedness in the earth. It may also account for the enmeshment of meals in that meaningful skein of rituals that evoke intimations of mortality and the possibilities of transcending it[5] – and this even among the most secular of humans.

Playfulness

In his treatment of *psychodrama*, Blatner (2001) referred to "one of the most fundamental (and often neglected) dimensions of human existence: playfulness" (p. 545). Philosophers and historians have noted that we have a primal need to play, that it precedes culture and informs human expressions of art, science, commerce, law, poetry, and philosophy. The historian Johan Huizinga ([1938] 1971) pushed this notion further in his classic work, *Homo Ludens*, and stated that all culture arises in the "primaeval soil of play" (p. 5). Ethologists, not to mention legions of pet owners, have observed that puppies and kittens joyfully gambol in their gardens and living rooms, engaging in mock fights, chasing each other about, yet never mauling or biting each other too hard. Humans also indulge in play simply to amuse themselves, usually voluntarily – and even though it brings no

[5] The Fitzwilliam Museum (FM) in Cambridge, England had an exhibit in 2005–6 on Chinese eating and drinking and its relation to immortality. Thousands of years ago, it is evident, meals played an important part in the family ceremonies that involved not only the living but also the dead. A display case (FM, Hall # 10) announced that "tea is thought to have been employed first by Buddhist monks as an aid to meditation." Vessels for making tea and other consumables have been found in tombs dating from the Han dynasty (206 BC–220 AD).

monetary or other extrinsic reward. Play usually takes the form of a game engaging two or more people, but often it is displayed by a solitary person playing a video or card game. Play has further been characterized as being egalitarian, informal, recreational, leisurely, and more or less pleasurable (see Sutton-Smith and Kelly-Byrne, 1984, pp. 305–10).

Defining play. As with most definitions, one must establish the criteria, both inclusionary and exclusionary, that will satisfy one's definition of play. For example, does a baseball game played for money become work rather than play? If the game becomes painful, even bloody, does it cease to be play? On the other hand, does hard work that is no longer remunerated, but simply recreational and voluntary, become a form of play for the worker? The character of play varies in part from culture to culture, from one social class to another, and from generation to generation. Should we fashion a definition of play that is so general that it fits any time, place, and culture and thus may be stripped of most of its specificity? We are free, of course, to fashion several narrower definitions of play that correspond with the cultural norms that govern it in various societies, but a definition needs to be sufficiently general to transcend these numerous socio-historical domains and yet specific enough to give substance and meaning to the term.

Multicultural aspects of play. Play, of course, is socially constructed, and when we view children, say, at play in our own neighborhood we have little doubt what is occurring. There is a subjective certainty among observers when watching children engaged in a game of football and apparently enjoying it that they are engaged in play (Martin, 1984). On the other hand, we easily recognize when it ceases to be play, turns into a "rumble," and becomes rather sinister.

Play is normally fun-oriented, voluntary, often spontaneous, intrinsically motivated, often staged, imaginative, even fantastical. But not all properties of play as normally understood in Western society are present at any one moment. Play may appear to be spontaneous and fun-oriented in children, but it can also have been shaped by the need of their parents to get them off their hands, or by the ulterior motives of parents to improve their children's social and psychomotor skills. The culture of play-training, like the culture of schooling and work apprenticeships, can also be influenced by economic or religious values that fluctuate over time. Although play such as puppies or other vertebrates engage in has genetic determinants, among humans it is a largely socio-psychologically constructed activity. As such it goes without saying that the character of such activities varies across cultures and fluctuates over time. Any one or more of the properties of play that exist at one time and in one place may be absent in others. According to *The Dictionary of Psychology* (Corsini, 1999) play consists of "activities freely sought and pursued for the sake of enjoyment" (p. 732). This definition is too sparse to be useful in this text. But any definition of play that would satisfy criteria that are universal and ageless appears to be a goal beyond reach. We are not dealing with an invariant and universal phenomenon. Delving deeper, let us take for example one criterion for play: *pleasurable*. Sutton-Smith and Kelly-Byrne (1984) wrote:

> Historical anthropology is rife with bloody and brutal but playful customs. There were the Eskimos who competed at twisting each other's ears off; or who attached a piece of leather to each other's testicles to see who could outpull the other in a tug-of-war, all the while smiling and if possible laughing to indicate their playful intent ... There were the Mayans who played ball games in which the captain of the losing team had his head chopped off; there were Indian tribes in which masses playing at lacrosse injured, brained and brutalized each other ... (p. 310)

These playful activities do not meet mainstream Western standards for leisured pleasure. There are other manifestations of play that are as extreme in terms of positive emotions as the above are in negative emotions. Csikszentmihalyi's influential writings on his construct, *flow* (for example, 1991), has contributed to this useful perspective, although this notion may have so diluted the notion of play that the ancient distinction between play and work no longer exists. An example is the surgeon who goes on vacation to Acapulco with his wife and, after several days, reports to a local hospital for the pleasure of working while his wife goes to the beach (Sutton-Smith and Kelly-Byrne, 1984, p. 308). Should we add that many individuals collect and sell coins and rare currencies for a living, pleasurable and remunerative activities deriving in some cases from childhood hobbies? One can even make a career of play to earn a living, such that it may evolve into a compulsory rather than a simply leisure activity. Clearly it is possible to find expressions of "play" that invalidate any conceivable definition of it.

The remainder of this section will skirt around extreme expressions of play.

Developmental aspects of play

Developmental psychologists refer to play as *ludic* activity that is pursued throughout life, most intensely in retirement or childhood when leisure was more abundant than it is today. The treatment of this subject was until recently largely confined to childhood. The reason posited by some is that play among children is of critical importance for the acquisition of skills – spatial, psychomotor, social, and simply cognitive – that will ensure their well being in adulthood. Anomalously, although human development is regarded as a life-span process, the study of play as it affects adult development seems to have been neglected. Yet developmental considerations associated with aging and retarding physical decline are also at play in one's choice of leisure activity. The complementary view is that play is any activity engaged in across the life span that gives pleasure and is perceived to enhance one's sense of well being. Playing cribbage with one's spouse, rock climbing, practicing the violin, or collecting coins and stamps would be, in this perspective, a play activity, independent of its developmental advantages.

We have an archetypal impulse to ritualize, to engage in mock battle, and to formalize rule-governed competitive activities. This leads us to pit ourselves against one another in games that develop cognitive and psychomotor skills. Such games are regulated and spatially organized as in football, baseball, rugby, and other team sports. These have all the properties of tribal warfare as one can witness in international football, rugby, or cricket matches. (Indeed, the rules of war in the West formerly had many of the ritual properties of games.) Normally, however, team contests are without consequence except for the psychological elation (and bruising) that come with scoring and being scored against. Vicarious participation in games played by teams with which one identifies can appear as intense as actual play. The seriousness of this activity is revealed in the passionate rooting for a home team that manifests as hooliganism. Even reasonable fan loyalty bespeaks an identification of the non-players with "their" team, such that they personally resonate to its successes and defeats as if they were their own.

Personality traits and play. From the point of view of the personologist, it is of interest to understand play in terms of the subjective pleasure it brings to individuals, the gratification they get in exercising a combination of cognitive and, in many cases, psychomotor skills, the tension release, the expectation, possibly the exhilaration of winning, the pleasure of struggle they experience in agonistic displays and play-fighting, often accompanied by screaming and laughter (Aldis, 1975), and the simple joy of socializing with their peers. The tendency, for example, to be shy or socially bold, anxious or self-composed, risk-taking or cautious, competitive and self-reliant or group-oriented will not only affect the amount of time one allocates to play as opposed to other leisure activities but the kind of play in which one chooses to engage. Individuals who are highly competitive will engage in play where they can assert their dominance. Those who are shy will seek to play, if at all, with only a few intimate friends, or even by themselves. Studious children will more often prefer to read a mystery novel, which can be construed as vicarious, fantasy play.

Life-span perspective. Although play is an activity that is pursued throughout life, most intensely in those periods, such as in retirement (a Western socio-economic notion) or childhood, when leisure is normally, but not always, more abundant, the treatment of this subject is largely confined to the early stages of life and is found almost exclusively in child psychology literature. As noted above the reason for this is that play among children is of critical importance for the acquisition of life skills. Although Vygotsky (cited in Bruner, 1973) averred that "a child's play must always be interpreted as the imaginary illusory realization of unrealizable desires" (p. 302), desires nevertheless that can be fulfilled in adulthood, it is obviously more than this; it is the immature mimicry of the adult roles that the nature of the species prescribes. Such play serves critical developmental purposes.

Edward Bordin (1994) wrote, "when we assume two powerful imperatives, those of work and play, one can gain a better understanding of human behavior" (p. 94). This dichotomous assertion appears to be a truncation of the reality. These two "assumptions" are not "more essential" to an understanding of human behavior than sexual pursuits, or social and group activities (say, in Thanksgiving dinners). Play is evident in all aspects of life, whether it involves vigorous psychomotor activity, as in playing football, or engaging in a sedentary card game or a game of cribbage. Its importance is reflected in the popular Latin aphorism that the masses can be satisfied by bread and circuses. It can be symbolic, as when children or adults engage in improvisational and "pretend" drama, and passive and vicarious as when hundreds of millions watch the Superbowl or the Mondiale. It can be solitary, as when one plays the slot machines, or social as in a board game. The play instinct is manifest in such diverse activities as telling jokes, teasing, windsurfing, or playing solitaire, working on cross-word puzzles, engaging in foot races, or watching a hammer-throwing contest.

Personological implications. Moreno referred to the data that are generated in psychodrama as "constituting surplus reality." This is what imagination in play brings into existence: a highly variable feature of personality in any population. Some have it to such a degree that it leads them to embrace a comedic, theatrical career. Others pursue game-playing with such seriousness that it propels them into professional sports to bask and sweat under the Klieg lights – and legions of others would do so if they had the psychophysical, world-class (or regional) talent necessary to achieve such eminence. When individuals fashion a career in sports or theater, however, the line between play and work becomes, of course, fuzzy.

Playfulness that is enmeshed with party activities is characterized by humorous, witty, bantering exchanges that trigger laughter and occasional physical displays of camaraderie. It is related to the teasing that is a common feature of children's play (Keltner *et al.*, 2001). There are aspects of playfulness that would evidently load heavily on the superfactor, extraversion, but there are important aspects of playfulness, alluded to above, that a description of extraversion leaves in the shadows.

Playfulness and ethology. Playfulness is evident not just in humans but in many other vertebrates (Burghardt, 2005; Pellegrini and Smith, 2005a, 2005b), although the complexity and the prevalence of play across species appear to be correlated with the size of the brain, measured in Encephalization Quotients (EQ)[6] (Iwaniuk, Nelson, and Pellis, 2001). Fagen (1981) observed that a highly developed cerebral cortex provides the neurological capability for play, which does not *in se* demonstrate that playfulness is genetically programmed. Play, as one would expect, is usually social. It can involve an adult with a child, one or more children with other children, or an adult with other adults. It can be as simple as two lion cubs "rough-housing" under the watch of their mother or as complex and elaborate as the rule-governed games in which adults and children engage.

The buildup of action-specific energy results in spontaneous behavior that drains away such energy, even when the behavior does not immediately maintain the conditions for life (see Lorenz, 1981, for a classic ethological expression of this principle). The need to discharge this excess energy and, additionally, to escape from boredom drives highly evolved organisms of all life stages to engage in apparently voluntary, uncoerced activities. We can characterize that activity as play. That play serves to discharge surplus energy or to developmentally promote skills that will be of value in other, later contexts does not diminish the notion that it fundamentally serves to stimulate the CNS in ways one can control and

[6] EQ may be defined as "brain volume of the animal in question divided by brain volume of a standard comparison animal belonging to the same class, correcting for body size. If a cat is arbitrarily assigned a standard mammalian encephalization quotient of 1.0, then a rat scores 0.4, a chimpanzee 2.5, a dolphin 6.0, and a human 6.3." Available at: www.encyclopedia.com/doc/1O87-encephalizationquotient.html, accessed July 17, 2008.

preferably enjoy. The reader will note that the advantages to play can be conceptualized as immediate and, at another level, as more distal. The more distal purposes are developmental and to some extent species-specific, although all are both biology-based and environmentally conditioned (see Burghardt, 1984, pp. 5–41, for a fuller discussion of these issues).

Creativity

Creativity and intelligence[7]

A child of both playfulness and necessity, creativity is increasingly being studied as a personality trait in its own right, although it has most often been regarded as an aspect of intelligence (Dellas and Gaier, 1970). This may account for the relative lack of attention that it has received in the personological literature. In Guilford's (1956) multifactorial model of intelligence, however, creativity is considered to be a cognitive aspect of intelligence with which each of us is more or less richly endowed. Zeidner (1995, p. 307), on the other hand, has argued that creativity reflects a behavioral inclination, in short a *trait*, as well as an aptitude reflecting skill and knowledge. Educational psychologists, many of whom are also personologists, equate it with divergent thinking, that is, thinking outside the box (indeed, with restructuring the "box," if not breaking it) and with the generation of new ideas, technical inventions, scientific schemas, and artistic forms. Beyond a consideration of intelligence, Simonton (2000) wrote that the personality profile of creative persons generally comprises several other attributes, such as a disposition "to be independent, nonconformist, unconventional, even bohemian ... [further] they are likely to have wide interests, greater openness to new experiences, a more conspicuous behavioral and cognitive flexibility, and more risk-taking boldness" (p. 153).

Numerous studies and, most notably, many conducted at the Institute for Personality Assessment and Research (Berkeley, CA) have established the personological aspect of creativity (Barron, 1969). Martindale (2001) implicitly endorsed this distinction when he observed that for individuals endowed with a "disposition toward originality" there is a correlation between intelligence and creativity. Although it is generally recognized that there is a weak correlation between creativity and intelligence among those who have an IQ above 120 (Eysenck, 1995), it is clear that some

[7] Although *creativity* and *giftedness* are thriving domains of scholarly investigation, there are many textbooks of personality psychology that neglect them.

creative accomplishments cannot be authored, nor even understood, by those whose IQ does not greatly exceed that level.[8]

Martindale (2001) reviewed the accomplishments of one of the pre-eminent geniuses of modern science, Thomas Young.[9] Compared with other highly creative scientists, Young did not publish much, which is not surprising given that the scientific journal was not the ubiquitous instrument for disseminating knowledge that it is today. Martindale observed that the custom of counting publications to measure academic stature would not fairly assess Young's worth. "Young serves as a valuable reminder that we had better read the publications rather than counting them" (p. 343).

In view of Young's prodigious accomplishments (see footnote 9, *A Prometheus for Our Times*) one must ask why his renown has dimmed over the years (limited as it is to the scientists who specialize in the fields to which Young made signal contributions), unlike the renown of Galileo, Newton, Planck, Einstein, and Bohr. One can allege he engaged in a poor time-investment strategy (*cf.* Sternberg and Lubart, 1991) as he pursued

[8] Creative aptitudes are unevenly distributed. Einstein, a theoretical physicist of the highest rank, admitted he could not understand the mathematics of Nobel laureate Schrödinger. I'm not aware that Schrödinger fully comprehended Einstein's theories of relativity, though he certainly understood the Riemannian mathematics that is at the base of his General Theory.

[9] *A Prometheus for Our Times.* Martindale (2001) observed the following about a creative person whom history has largely forgotten:

> Most people have never heard of Thomas Young. The name is familiar, though, to scholars in a variety of disciplines. Psychologists know of his trichromatic theory of color vision and his explanation of color blindness as well as that he discovered astigmatism and was the first to demonstrate that the crystalline lens of the eye is responsible for accommodation. Physicists know him as the founder of the modern wave theory of light, which he supported with his double-slit experiment and with which he was able to explain Newton's rings (colors seen on thin films such as soap bubbles), as well as a variety of other phenomena; as the father of interferometry and spectography; as the inventor of the diffraction grating; as the first to use and define energy in its modern form; for his coefficient of elasticity, which is still used today; and for a variety of other discoveries. Chemists know that he was the first to attempt to measure the diameter of a molecule and the first to offer a satisfactory explanation of capillary action. Oceanographers are aware of his theory of tides, which vastly improved on Newton's oversimplified explanation. Workers in disciplines as diverse as physiology and seismology know that he invented the smoke kymograph. Physicians are aware of his theory of circulation of the blood and that he offered the first systematic medical nosology or classification of diseases. Acousticians are aware of his work on resonant frequencies. Actuarial scientists know of his foundational work on the value of annuities. Linguists know that he, rather than Champollion, was the first to decode the meaning of the Egyptian hieroglyphs on the Rosetta Stone and are aware of his important work on the history of Proto-Indo-European languages as well as his anticipation of the international phonetic alphabet. I have listed only the most important of Thomas Young's discoveries. Though Thomas Young was not a great poet, Edmund Burke praised the renditions of English verse into ancient Greek that he composed. (p. 342)

his multiple interests, spreading himself too thin. This moots the question as to whether creative individuals are generally driven less by achieving renown than by a passion for satisfying their curiosity about specific aspects of the real world. Nabokov's lifelong passion for lepidoptery was inspired not by a drive for fame but by the inordinate pleasure that the study of butterflies gave him. (Indeed, he is reputed to have observed that the two most intense pleasures that were afforded him in life were writing and lepidoptery.) Perhaps Young, like Nabokov, simply enjoyed pursuing solutions to scientific problems in the many fields that fascinated him, independent of the esteem of his peers or of later generations. In any event, Sternberg and Lubart recommend that in choosing a problem in which to invest time and energy one "buy low and sell high" (p. 683). Those who are most successful in advancing their sciences avoid the ceiling effect of investing their energies in the investigation of mature paradigms. Thomas Young was not a victim of that effect.

Wallach and Kogan (1965) appear to have demonstrated the virtual independence of creativity and intelligence. However, current research indicates that a modicum of high–normal intelligence correlates positively with and is indeed necessary for creative thinking (Zeidner, 1995). There is, nonetheless, only a weak correlation between IQ above 120 and creativity. People across the IQ spectrum tend to resort to their ingrained cognitive structures and habitual patterns when faced with the normal challenges of daily life, and still more so when faced with a serious, sudden crisis. They call up easily retrievable schemas, the spontaneous, conventional responses that surface when they do not have the time to ruminate or to playfully generate and manipulate old and novel variables. Creative people do this as well, but when they have the leisure, they toy with new schemas and paradigms and engage in thought (*gedanken*) experiments. Though respecting the work of their predecessors, they dare to challenge time-honored paradigms. The iconic example of this is the work of Copernicus who overturned the geocentric theory of Ptolemy.[10]

The eager exploration of one's environment doubtless has an evolutionary basis. Survival (still more, prosperity) demands as thorough an understanding of one's environment, physical and social, as one can muster. One even sees this in lowly vertebrates, such as rats in a Skinner box, who poke their noses in every corner, under every leaf and straw, to

[10] The subtlety and brilliance of this ancient astronomer's reasoning is apparent in a cursory overview of his *Almagest*. One does not have to do the math to appreciate it. On the other hand, Ptolemy had illustrious Greek predecessors who were heliocentrist opponents of the geocentric position. He gave them short shrift (Taliaferro, 1952, p. 1).

tease out the threats and bounties of their niche. The exploration of novel situations meets a number of needs, one of which is an outlet for an activity drive as well as a certain minimal level of arousal.

Creative people have "a disposition toward originality" argued Barron (1955) and do not perseverate in antiquated theory. This platitude does not explain creativity. Trait psychologists state the obvious: that creative people are open to new experiences, to new ways of configuring old data, as well as using old paradigms for new data. This is resonant with the schemas of assimilation and accommodation by which Piaget described cognitive development in children. They are curious; they love to find patterns in nature, analogies in disparate disciplines, and simplicity in complex arrays of data (McCrae, 1987). Most individuals find comfort working within the paradigms in which they have invested large parts of their professional lives. For that reason revolutions in science are usually started by the young. Science, as Paul Samuelson is reputed to have said, only advances funeral by funeral. On the other hand, "If it ain't broke don't fix it" is an aphorism that is anathema to innovators, who are interested in improving on state-of-the-art technologies, whether broken or not. They are willing to run the risk of degrading something in the interests of trying to improve it, of engaging, in economic terms, in creative destruction.

Antecedents of creative acts

Highly creative people need, of course, to have acquired state-of-the-art competencies in their area of specialization in order to advance it. Even paradigm "busters" need to have fully understood the properties and limitations of the paradigm they are discarding and to have discovered the fatal anomalies that compromise it. With each advance of science, making creative improvements requires a longer, more intense apprenticeship. Designing a more efficient guidance system or engine propellant for a spaceship requires a greater depth of education than making the proverbial better mousetrap. Not surprising is the fact that specialization in any profession now requires that individuals get a doctorate in their field of expertise, do residencies and post-doctoral internships ("postdocs"), and extend their training, especially in medical settings and bench-based laboratories, until they are well into their thirties.

Creative individuals rarely generate a scientific advance *ab ovo*. They build on the achievements of others. Mlodinow (2003), in his fascinating personal account of the last years of the Nobel (physics) laureate, Richard Feynman, had the opportunity to observe the great man when Mlodinow was a junior academic at Caltech. He observed that:

the creative mind has a vast attic. That homework problem you did in college, that intriguing but seemingly pointless paper you spent a week deciphering as a post-doc, that offhand remark of a colleague, all are stored in hope chests somewhere up in a creative person's brain, often to be picked through and applied by the subconscious at the most unexpected moments. (p. 82)

Referring to the many apparently random coincidences that favor scientific discoveries, Louis Pasteur observed (and Seneca, it is reputed, as well) "Chance favors the prepared mind" (Weisberg, 1993, p. 79). One can only discover the resemblances in diverse phenomena or generate a novel solution to a wisely chosen problem if one has the cognitive tools and the bank of ideas that allow one to configure in a different way the ideas one already possesses or the novel ideas one chances upon.

Personality correlates of creativity

"The creation of something new is not accomplished by the intellect but by the play instinct acting from inner necessity. The creative mind plays with the objects it loves" (attributed to Jung, 2003). As the Nobel physics laureate, Richard Feynman, observed, "An important part of the creative process is play. At least for some scientists" (Mlodinow (2003). But he cautioned that this was hard to maintain, as one got older. The emphasis on the playful context for creative thinking suggests that the trait of playfulness is entwined with the personality trait, creativity, and with another intelligence factor, curiosity. Whether we wish to categorize curiosity as a personality trait, an intelligence factor, or both depends on theoretical bias and taste, rather than the intrinsic properties of this variable.

Curiosity, motivation, drivenness, openness to new ideas, independence, optimism, and other personality traits – not to mention a modicum of intelligence – are important contributors to the emergence of creative ideas. A balanced view of creativity, still more of genius, requires in Eysenck's model a consideration of molecular genetics and neurological and hormonal variables (Vernon, 1997), principally dopamine and serotonin, which can enhance and inhibit creative thought. Too much or too little of the latter, of course, has profound consequences not only for creativity, and the enhancement or reduction of latent inhibition, but for mood states and expression of personological variables. The intriguing aspect of this model is that the condition for high creativity and genius is the presence of certain neurotransmitters and, indeed, of testosterone in certain proportions. Too much or too little of any neurotransmitter results, of course, in anomalies. Nyborg (1997), for example, states, "Too much dopamine leads to a high P [psychoticism] score, lack of inhibition … a flat association gradient, a too wide horizon for orderly combination of remote elements, low creativity, and acute psychosis" (p. 427). Genetic and biological variables are as basic to creativity as they are to intelligence, although these two human properties are, beyond some minima, unrelated.

A favorable social environment

Whether creative thinkers gain wide acceptance or not depends on the gatekeepers of their disciplinary domain. If journal editors and editorial boards do not find the work of a young researcher congenial to their own paradigmatic biases, they tend to raise the bar for publication. Research funding depends on the prevailing *Zeitgeist* and the perceived social needs of the community funding the research. Further, researchers who pursue a line of investigation that is at odds with influential scientists in their own institution may find their job at risk. If they find support among their colleagues, they can go on to achieve their goals. As Mlodinow observed of Schwarz who was pursuing research in "string theory" at Caltech, Schwarz was there at the sufferance of Murray Gell-Mann, a Nobel laureate whose views clashed dramatically with those of Richard Feynman.

Though the quality of the social environment is critical to creative thinking, all the ideas, discussions, ideational infrastructure, and emotional support that a positive milieu affords will not generate a single creative idea. Archimedes was alone in his tub when he made his personal discovery for measuring the displacement of water by a body. Eight physicists in a swimming pool would have been unlikely to achieve such an insight. And if the insight did occur there, it would have arisen as a spark in one person's mind, not simultaneously in eight or even two. Neither a piano concerto by Mozart, the masterwork of Cervantes, nor Maxwell's laws of physics could have been conceived or created by a group, as much as bits and pieces, scraps of melodies, and vague analogies that others supplied allowed the marvelous alchemy to occur. Groups of scientists are productive as they go about the business of doing what Thomas Kuhn referred to as "normal science." Research teams do not engage in the discovery phase of science so much as in the confirmatory and testing phase (Knorr-Cetina, 1981). Although several individuals are capable of independently discovering or inventing a new law, principle, or theory, as in the invention of the calculus by Newton and Leibniz, they can rarely do so as a single mind. They can, however, working together, make alternating contributions to a final product, as in the discovery of the double helix of DNA by Francis Crick, James Watson, Maurice Wilkins, and Rosalind Franklin (consult Wertheimer, 2007).

Creativity, aesthetics, and human nature

Sigmund Koch (1956) argued early on in his career that the mechanistic models of the Human that behaviorists endorse failed to accurately portray the Human "as he creates or loves or plays or responds to the aesthetic surfaces of the human and natural environment" (p. 65). Before qualitative research methodologies found their place in the sun, Koch averred that an understanding of the arts could not be achieved by the (otherwise valuable) experimental approaches that had come to dominate American psychology. It was necessary, in his view, to explore the sources and context of creativity *in conversation* with established creative individuals. Laboratory-based, experimental methods – pinning the butterfly to the wall – were inappropriate in his view. These convictions crystallized in the 1960s as he worked in the Ford Foundation's Program in the Humanities and the Arts (Franklin, 2001). To understand the psychic processes in which artists engage as they create their works of art, he inaugurated the Boston University Aesthetics Research Project in the 1980s. In this setting, he conducted a series of interviews with distinguished creative artists to investigate the underpinnings of the theories of motivation, aesthetics, and creativity that he had been elaborating since the 1950s. Koch was the exemplar of a movement to understand creativity and other human experience through idiographic, case-study methods, collectively referred to as qualitative rather than experimental.[11]

Although the creative idea always erupts like a light beam in one individual's intellect, it is actually the product of the distributed knowledge fermenting in a socio-cultural crucible (Knorr-Cetina, 1981; Wallace and Gruber, 1989). Were clones of James Clerk Maxwell, Albert Einstein, or Neils Bohr to be born and raised in a Stone Age society of Papua New Guinea, one would expect that they would make significant contributions to the knowledge base of their society, but nothing on the scale of their historic achievements. It is easy to romanticize aesthetic and creative works, for they are among the most distinctive of human activities, and those who have achieved eminence in the arts, be it music or sculpture or architecture, have assumed roles of cultural icons. There is an understandable reluctance to subject them to scientific scrutiny. The appeal of the mystical and the assimilation of the arts to the domain of the Muses can cloud the need that the lepidopterist has, not only to study the butterfly in its various bucolic habitats and travels, but also to examine

[11] A colleague, Bruce M. Shore, reminds me that there are journals reporting rigorously conducted research that abound in the field of creativity and giftedness. Among the former are: *Creativity Research Journal* and *Journal of Creative Behavior*. Among the latter are: *Journal for the Education of the Gifted*, *Gifted Child Quarterly*, and *High Ability Studies*.

it under the microscope. Those who believe in the unity of knowledge and who are convinced that a basically biophysical theory will eventually explain creativity in the arts – that most distinctive of human activities – would argue that both a qualitative and a scientistic approach to this subject is necessary.

Creativity and mental disorder

That genius is often correlated with madness is folklore psychology. The notion is also supported in part by scientific psychology (Nyborg, 1997). No one disputes the fact that the "mad" think in unusual, aberrant, novel, and unconventional but not always constructively creative ways. Their madness by definition allows them, if it does not oblige them, to shatter paradigms. Well known is the fact that there have been individuals of high intelligence, like Friedrich Nietzsche or Ezra Pound, who lapsed into serious mental disorder whether in their youth or later in life and yet were able to generate remarkable products in the fields, say, of psychology, mathematics, physics, and poetry. The vividness[12] of these examples leads one to conclude that this is more frequent than it truly is.

The mad-genius hypothesis. The folkloric belief that there is a disproportionate number of creative geniuses among the "mad" may spring from the vivid historical examples of great artists and scientists who have suffered from psychotic or other serious cognitive disorders. Much is made in print and film media of the cognitive disorders of Vincent Van Gogh, Jackson Pollack, John Nash, or Willem de Kooning. Their various disorders are not the norm among geniuses. That the psychotic think in unusual and novel patterns is self-evident, for their "madness" by definition often leads them to think in unconventional ways (this conclusion can be reached by a number of logical pathways). That this may occasionally lead gifted but deranged individuals to make socially valuable contributions to society is indisputable. But to conclude that these colorful, often isolated, individuals are representative of geniuses is to engage in popular heuristics that lead to invalid conclusions. That there is a higher correlation of high creativity to madness than to sanity seems a mischievous notion. Because the divergent thinking of the disordered genius resembles the creativity of great scientists, poets, mathematicians, and playwrights is not grounds for postulating that geniuses are

[12] The *vividness heuristic* is a well-established principle of inferential reasoning. Individuals tend to exaggerate the importance of vivid cues when solving inferential problems. Bland and pallid cues tend to be overlooked (for example, Nisbett and Ross, 1980).

disproportionately afflicted with psychopathology.[13] On the other hand, an opposing view, somewhat attenuated, has some merit.

Some folkloric beliefs can turn out to be veridical. Eysenck (1995) in an exceptionally persuasive book argued that high scores on his *Psychoticism* scale correlated with creativity, crediting Eugen Bleuler in modern times with the inspiration for this principle (Rushton, 1997, pp. 406–9). Rushton (1997) replicated other studies (for example, Woody and Claridge, 1977) that demonstrated significant levels of correlation between psychoticism and creativity, and, of course, of IQ and creativity. Nevertheless, creative persons can be justly characterized as "independent, nonconformist, even bohemian, and they are likely to have wide interests, greater openness to new experiences, a more conspicuous behavioral and cognitive flexibility, and more risk-taking boldness" (Simonton, 2000, p. 153; also see Andreasen, 2005, for a discursive treatment of the neuropsychological bases of creativity). Even though there is some evidence that creativity shows up in various domains among the moderately personality disordered (Eysenck, 1995), the web of causal factors linking them, and their directionality, is far from clear.

Curiosity

Creativity springs more from an interest in what *might be* than in what already exists. Curiosity, a personality trait that is allied to creativity, encompasses an interest in both. Various theories explaining the root causes of curiosity differentially shape the definition of this trait. For example, a drive theory of curiosity postulating an instinctive impulsion to acquire knowledge defines it differently than one that explains it by extrinsic, situation-induced reactions. The former can lead to a definition of curiosity as an appetite or passion for new knowledge – a property of many organisms. The extrinsic motivation inherent to the latter approach leads to a definition of curiosity as a response to a changing or novel environment affording rewards and punishments, positive and negative reinforcements.[14] Necessity is, indeed, often the mother of invention. And needless to emphasize, nature and nurture are both implicated in this trait.

[13] This is evocative of the *representativeness heuristic* (see Nisbett and Ross, 1980), a device pervasive in human affairs that disposes us to attribute a property or a trait to individuals on the basis of a groundless stereotype – and without consideration of the *baseline rates* for the incidence of such traits among different groups.

[14] A negative reinforcement in behaviorist theories is any reduction in painful or unpleasant experiences that is a consequence of a specific behavior.

Historical perspectives. A pantheon of historically significant scholars have studied and written about curiosity. Both Aristotle and Cicero noted that people are intrinsically motivated to acquire knowledge independent of any profit they might reap from it. Cicero interpreted the myth of Ulysses and the Sirens as a parable about curiosity, as Loewenstein (1994) noted in his lucid review of the literature bearing on curiosity. In Cicero's view, Ulysses and his men heard the tantalizing songs of the Sirens and lured by curiosity explored the rocky shores whence they came. The classical position of Hellenic Greeks that curiosity is a virtue was greatly constricted by Saint Augustine ([400 BCE] 1991a, 1991b), who argued in his fifth-century *Confessions* and *On Christian Doctrine* that the pious faithful must focus fixedly on the afterlife, shun the vain pursuit of knowledge for its own sake, and not "venture heedlessly upon the pursuit of the branches of learning that are in vogue beyond the pale of the Church ..." (p. 737). This view in part shaped Western thought for a thousand years and nearly brought to shipwreck the careers of many scientists of the Renaissance, most notably Galileo, whom the Vatican confined to his home for the last years of his life (Sobel, 2000).[15]

Modern views

The preferred modern view of curiosity has been that there is an innate human penchant for exploring the opportunities and the threats afforded by the world in which we live, as well as simply seeking novel experiences. Classical scholars like Immanuel Kant, Jeremy Bentham, and William James espoused this view, affirmed that it was an essential feature of human nature, and observed that its expression and intensity varied from person to person. Modern personologists take it as axiomatic that love of learning is lifelong, but that there are individual differences in the domains of inquiry to which individuals are drawn, as well as in the level of commitment they have to such inquiry.

D. E. Berlyne (for example, 1950, 1954, 1960, 1966) is arguably the most important twentieth-century student of curiosity. He pioneered experimentation on this trait during the 1950s, and by his analyses catalyzed widespread research and interest in it for over a generation. Among the distinctions that he made in defining curiosity was a foundational one

[15] Dava Sobel (2000) has written a charming account of Galileo's life and that of his daughter, a nun in the order of Poor Clares, inspired in great measure by the correspondence that this daughter carried on with her father through much of his life. His conflicts with the institutional Church and their obstruction of his work are cleverly illumined in this excellent history.

between (a) *perceptual* curiosity that orients one to novel ambient stimuli, and which is evident in all vertebrates and other organisms, and (b) *epistemic* curiosity – the wish to understand – which is most characteristic of the complex learning patterns of humans.

Berlyne also distinguished between curiosity which is wide-ranging and general (he termed it *diversive*) and the curiosity that was *specific*, bounded by the parameters of the discipline that are immediately important to the enquirer. Among the many instruments that were developed to assess individual curiosity, the *Ontario Test of Intrinsic Motivation* (*cf.* Day, 1971, 1982), designed to capture both specific and general interests, is of particular interest. The specificity–diversity dimensions postulated by Berlyne appeared to be supported by this true–false, 110-item instrument when it was factor analyzed. The nature of the diversity factor, however, proved to be problematic. Whether it simply describes an arousal-oriented drive or is a reflection of general curiosity about one's ambient world is still controversial. How the nature of the curiosity exhibited by the individual who studies several languages, who reads voraciously in various genres, and is torn between studying law, architecture, and cytogenetics differs from that exhibited by the individual whose only interest is, say, astrophysics, has yet to be resolved.

Theories of curiosity. Motivational explanations for human behavior are pervasive in all societies. Humans have sought information about the motives that were driving the behavior of others, especially as this can affect their own well being. One does not have to be paranoid to engage in this survivalist strategy. Organisms that take no heed of the intentions of others or who succumb to a naive gullibility are, as history shows, truly in peril. This may be the basis for the pervasive *Fundamental Attribution Error*, classically formulated by Lee Ross (1977; see Ross and Nisbett, 1991, on *dispositionism*), that observers of the behavior of others are more prone to seek a psychic than a situational explanation for it.

Drives (Corsini, 1999) are physiologically based motivational systems that spur activities that can enhance one's well being. The drive to access information – call it curiosity – that endues persons with the power and resources to manage their environment is different from the drives of hunger or sex in that it derives less from tissue needs than cognitional needs. Those who view curiosity as a drive (see Loewenstein, 1994) are divided as to whether curiosity is a homeostatic drive, like hunger, that is internally generated to stabilize the *internal milieu*, or a stimulus-induced response, such as spotting clues to the presence of danger. The more plausible theory is that curiosity is both, that it is an instinct (McDougall, 1918) motivating organisms to explore their environment but modulated and activated by cues in that environment. A more recent theory that considers the systemic nature of organisms living in a complex world is the *Confluence Theory of Curiosity*, which affirms that there are multiple causal pathways that can explain curious behavior. It is similar to the theory of creativity proposed by Sternberg and Lubart (1991). This theory appears to have greater explanatory power than those theories that are uni-principled.

Ethology and curiosity. Whether or not curiosity is a primary disposition deriving from a fundamental need (say, for security or for food) is still debated. Studies of behaviors in monkeys and rats that resemble curiosity activities in humans support the position that a disposition analogous to human curiosity is biologically programmed in all vertebrates. Harlow (1950) and his associates (Harlow, Harlow, and Meyer, 1950) demonstrated that Rhesus monkeys engaged in puzzle solving for which there was no cued reward. For example, they worked to open a latch that they were given no reason to expect led to anything – which of course does not mean that they did not imagine it did. They appeared to engage in such searches and puzzle solving for the "fun of it" rather than for any extrinsic reward. Butler (1957) and Butler and Alexander (1955) confirmed that monkeys rapidly learned to identify the particular window of their enclosure that could be opened to give them a view on another, larger room. They appeared to like "a room with a view." Like other primates, humans will engage in near meaningless activity for no other reason than to ward off boredom and thus achieve a moderate level of arousal (Gleitman, 1996, p. 93).

Drive-theoretic approaches. Theories of curiosity correspond to the broader scientific paradigm within which its various authors work. Behaviorally oriented psychologists have favored a psychomechanical approach to curiosity-driven behavior that is ultimately under the control of environmental agents. Self-efficacy theorists (for example, Bandura, 1982; White, 1959) aver that people pursue learning because it enhances feelings of mastery and accomplishment, as well as economic prospects and public recognition. Freudians, on the other hand, have pursued a uni-principled, libidinal explanation for curiosity, affirming that it is a sublimation of the normal interest in sexual matters they had as infants, but which was suppressed during the socialization process.

Behaviorists, focused on reflex arcs and mechanistic explanations of behavior, broadened the notion of curiosity to include the *orienting response*. This response is triggered by any novelty in the environment – the crackling of brush in the distance, a strange smell up wind, possibly of a predator, the unexpected movement of an object in one's visual field. The evolutionarily adaptive aspect of this attentional response – especially its importance for survival – argues for its cognitive–perceptual as well as for its emotional character.[16] That it is evident in infra-human species as well as in humans strongly suggests that it is an innate, biologically based capability – albeit triggered by environmental stimuli.

Curiosity: the drive and the triggers. A distinction needs to be made between the personality trait, curiousness, and the conditions that serve as catalysts for its operation. Just as the human propensity to engage in sexual behavior can be triggered by an astonishingly varied array of social and other environmental cues, so curiosity can be instigated by myriad conditions. For example, when individuals are confronted by a baffling situation, events that violate their expectations, generate ambiguity, or ignite concerns about their future, they are generally moved to investigate. This investigative mode is tinged with either pleasure or fear, occasionally by both, depending on how one construes the potential of the situation for good or ill. Festinger's (1957) theory of *cognitive dissonance*, which evokes the need to resolve ambiguity and inconsistencies in our public and private worlds, tangentially relates to the construct curiosity. The state of cognitive dissonance often spurs efforts to resolve this dissonance by a further exploration of the problem.

These troubling conditions are not the cause of curiosity but the occasion for this disposition to be activated. Analogously, the presence of food is not the cause of hunger but the condition for it to be operationalized, and the presence of "sex objects" is not the cause of the sex drive but (normally) the occasion for it to be activated. Similarly, a variety of conditions, which should not be confused with curiosity, can stimulate this disposition. The need to mentate (say, to daydream, think, problem solve, or plan), like the accumulation of so much action-specific energy directed toward an objective, is an organismic *trait*. The trait exists even when it is dormant; in the case of curiosity it only becomes the "state of curiosity" when it is activated. Loewenstein (1994) summed up this view when he wrote, "Need for cognition has become widely viewed as a personality trait rather than a psychological state subject to situational influences" (p. 85).

[16] As cognitional and emotional dynamics are intricately enmeshed, both are implicated in any discussion of curiosity. Care must be taken not to take the mental construct as a real "thing." One can variously define this construct so as to capture different aspects of our knowledge orientation, in both its affective and cognitive modes.

Seeking information for its own sake. That evolutionary psychologists regard curiosity in vertebrates as a pervasively selected (Darwinian) adaptation improving the chances of survival does not account *directly* for the fact that individual primates, as noted above among Rhesus monkeys, seem to seek out new information for the sheer fun of it (Asch, Patton, and Hershey, 1990). People seek puzzles to solve – semantic, spatial, mathematical, and logical – which bring no monetary reward or social recognition. Leonard Mlodinow (2003), a theoretical physicist, stated as a student, "I like thinking hard (I always loved doing math problems)." He continued, as if explaining Berlyne's diversity dimension of curiosity, "but this doesn't explain why I like physics ... it applies to a lot of pursuits" (p. 154). Addressing as well as solving difficult problems is a pleasurable experience for millions whether it is a board game, a card game, a "treasure" hunt, or a crossword puzzle. What appeals is the quest – with the hope for success. That some abandon the search before they fully succeed does not invalidate the fact that they derived pleasure from the activity, before they decided that the game was no longer worth the candle.

There are great individual differences in this voluntary exposure to problematic situations (such as occur in a game of bridge or in the challenge of learning a second or third language sufficient to allow one to converse with native speakers of those languages). Personal tastes will dictate the specific domain of investigation; native ability will limit the levels one strives to reach. Organisms continue to engage in a behavior long after it has ceased to serve any useful purpose. This is what Allport referred to as "functional autonomy." Long after the itching powder that had been placed on a rat's ear had disappeared, the ear scratching continued. Has the human penchant for pursuing knowledge generalized to domains that are not relevant to its evolutionarily selected purpose of ensuring survival and promoting its well being? Curiosity has driven us to understand the history of our species, to find the distance to remote galaxies, to determine the origins of the universe, and to achieve a *Theory of Everything*. This drive appears to exist even where its scientific character is masked in the search for juicy gossip, baseball statistics, or trivial information useful in board games. It is manifest, of course, in humans' obsession with the hereafter and the means to please a Cosmic Deity.

Sexual behavior and personality

Sexual behaviors spring from deep biopsychological roots that have assured the survival and procreation of many species. Among humans, these behaviors are overlaid with myriad cultural prescriptions and constraints and are subject, in form and intensity, to wide individual personality differences. As important as sexual behavior is in understanding personality and the social systems that prevail in any culture, a focus on human sexuality is absent from many personality textbooks (consult the "subject index" of a random sample of personality textbooks). When it is addressed in some detail it is often treated in the context of a developmental psychology, usually Freudian, which has more historical than contemporary significance.

Personality psychology's interest in a scientific treatment of human sexual behavior received its major contemporary impetus from the comprehensive and intensive studies of Alfred Kinsey (1894–1956) and his associates (Kinsey, Pomeroy, and Martin, [1948] 1998; Kinsey *et al.*, 1953), although Havelock Ellis, W. H. Masters and V. E. Johnson, and John Money are also rightly regarded as pioneers in the study of human sexuality. Kinsey proved to be a controversial figure in sexology. An obituary for Kinsey published in *The New York Times*[17] reported that in his influential study "fourteen researchers interviewed 5,300 white men and boys and 5,940 women and girls, each interview encompassing up to 300 extremely personal questions. The interviewers even examined the sex life of babies as young as 2 years, asking the questions of their mothers."[18] Not least contentious were the team's findings that male virility begins to decline at the age of 17 and that homosexual activities, considered at that time to be psychosocially aberrant, were more prevalent than commonly supposed. Kinsey was interested in establishing the population norms for various sexual attitudes and activities, a complex of variables he labeled *socio-sexual*. He and his team of researchers left an enduring legacy. The *Kinsey Institute for Research in Sex, Gender and Reproduction* continues the work he began as a professor at Indiana University, Bloomington.

From this seam of data arose the personological finding that there is great individual variation in all aspects of sexual behavior: frequency of various sexual acts, both social and private, overt and fantasized, licit and illicit (depending on the culture). Other inferences bore on incidence of extramarital affairs, prevalence of homosexual encounters and relationships, number of concurrent partners, degree of emotional involvement necessary to motivate a couple to engage in sexual intercourse, or sexual encounters that are unpremeditated, unemotional, inconsequential, depersonalized, and lacking in personal investment. Although promiscuous sexual behavior is subject to a multitude of cultural constraints, much

[17] August 26, 1956, available at: www.english.upenn.edu/~afilreis/50s/kinsey.html.

[18] Some may question how representative of the White American population of the day were respondents to a sex survey that probed the most intimate bedroom details of lives and activities about which people typically do not talk to total strangers.

of the variance in such behavior can be accounted for by personality traits that have genetic determinants. The sensationalized accounts of the sexually illicit conduct of many political figures, from Thomas Jefferson to Bill Clinton, have been scientifically explained by innate dispositions, on the one hand (*cf.* Watson and Clark, 1993), and socio-cultural factor as well as a pathologized childhood, on the other hand (*cf.* Pinker, 2002, p. 179). These explanations appear cogent; but they are not, nor could they be, mutually exclusive.

Sexual penchants, like other personality traits, are neurobiologically grounded. The Big-Three Model of personality traits developed and popularized by Eysenck (see Chapter 5 on traits, above) is predicated on the principle that these mega-traits – introversion–extraversion, neuroticism, and psychoticism – represent heritable dispositions. Refinements of this model by Clark and Watson (1999) have resulted in a re-labeling of the psychoticism factor as disinhibition versus constraint factor (DvC); and the extraversion factor as extraversion/positive emotionality (E/PE). That patterns of sexual behavior are correlated with DvC is intuitively plausible, and empirical evidence supports this notion. Watson and Clark (1993) have demonstrated, for example, that there is a correlation between their measure of DvC and the number of sexual partners in the previous year. Attitudes toward casual sex and promiscuity, willingness to engage in *risky* sexual behaviors, and penchant for novel expressions of sex are clearly related to a variety of personality traits, not least of which are religious compunctions (Zuckerman, Tushup, and Finner, 1976; *cf.* Zuckerman 1997), conservatism (Eysenck, 1976), and self-esteem. These moderating forces can leave individuals feeling conflicted, all the more so as they sense the potential for social harm – not only to their partners and themselves, but to third parties.

Sex drive. Large individual differences exist for each sex relative to the amount of sexual activity in which individuals are inclined to engage – or about which they fantasize and dream. Clearly, the production of testosterone, estrogen, and other gonadal hormones correlates in some measure with the level of interest and engagement in sexual activities. This is most evident during adolescence when surges of gonadal hormones course through the body and precipitate sexual arousal that is profoundly distracting, if not imperative. A dominant view has been that the within-sex variability in socio-sexual and solitary sexual activity at all ages of life is a function in part of the levels of various sex hormones the body is capable of producing (Aron *et al.*, 2005; Eysenck, 1976; Kelley, 1978; Reiss, 1982;

Simpson and Gangestad, 1991). Simpson and Gangestad, using their *Sociosexual Orientation Inventory* (SOI) explored the sex-drive hypothesis to determine, among other matters, which indicators of interest in sex (like willingness to engage in premarital sex, have multiple partners, or engage in sex within or outside of an emotional relationship) either co-vary or are independent.

There are, of course, a number of moderating variables that alter patterns of socio-sexuality, such as religious scruples, social values, and personality variables such as DvC and E/PE. The willingness of adolescents and younger adults to engage in sexual activities without emotional ties is most evident in their masturbatory acts. They do not normally develop a romantic interest in themselves, and their solitary sex will diminish as social outlets, licit or illicit, with or without emotional commitments, appear on their social horizon. On the other hand, the fact that willingness to engage in uncommitted sex does not co-vary with frequency of sex within a committed relationship is not a compelling argument against the sex-drive hypothesis. Sexual activities and ruminations are driven by the production of oxytocin, serotonin, dopamine, vasopressin, brain opioids, adrenocorticosteroids, and maturing physiology, but do not account for all of the variance in sexual interest. These variables merely complement rather than invalidate findings that personality traits differentially influence socio-sexual patterns of behavior.

Socio-sexuality. The construct, socio-sexuality, captures the commonalities of sexual and sex-related attitudes exhibited by individuals in any culture. In simplest terms it can be defined as the degree to which one is either inhibited or promiscuous. This construct, a dimension within DvC, can have a sub-set of behavioral expressions. At one end of this continuum are "unrestricted individuals [who] are more willing to engage in uncommitted sex ... and require less time, less sexual exclusivity, less investment, less commitment, and weaker emotional ties with their partners prior to having sex with them" (Simpson and Gangestad, 1991, p. 872) than those who are more conservative and inhibited. Assuming this construct represents a personality trait, it does not differ from other traits in that it is expressed differently in situations and social contexts that have widely varying demand characteristics. Uninhibited Westerners who have immigrated, for example, to a *Sharia*-observant Muslim nation will feel constrained to alter, or at a minimum conceal, expression of their sexual mores in the host society. Or very religious but characterologically disinhibited Roman Catholics living in traditional Catholic communities will feel obliged to remain in troubled marriages rather than divorce and remarry. Age and habituation are other moderator variables that alter the socio-sexual patterns governed by this trait-like disposition.

Sex and evolutionary psychology. In an evolutionary perspective (discussed in Chapter 4 on biology and personality), those behavior patterns that favor reproductive success would have been selected. These selection pressures would have been different for women than for men, which accounts for the inter-sex differences evident in the mate-selection criteria that prevail in each sex. Given the constraints that motherhood places on the female, especially during pregnancy and breastfeeding, the criteria she uses in mate selection to ensure the viability of the family and household bear primarily on fidelity, emotional investment, and socio-legal commitment, not to mention financial and physical resources. These criteria evoke considerations of emotional bonding that transcend the libidinal drive of either partner. Indeed, sexual desire can exist, as we have seen, without a smidgeon of affection. On the other hand, individuals can become passionately enamored of another without the desire to have sex with that person (Diamond, 1998; Nardi, 1999; Williams, 1992). This occurs between both same-gender and other-gender pairs.

The analysis of these phenomena in an evolutionary perspective is puzzling as it is evident that emotional bonding (as well as sexual desire) are mediated by biological processes. The analysis is further complicated by the fact that individuals can variously be attracted sexually to members of one gender but become infatuated with members of a different gender. This 2 × 2 matrix is further complicated by the finding that social conditioning can overlay and modify the biological determinants of sexual orientation (see Diamond, 2003). As we have seen, a mere exposure effect can promote a liking for an object, even a preference that formerly did not exist. Indeed, even bizarre and hazardous situations, such as occur on the battlefield, can promote an affectional bond between soldiers that rivals that of any other relationship. The current state of research makes any endorsement of a socio-sexual model hazardous.

Romance and concupiscence. Sexual desire, the biologically mediated appetite that humans have for sexual intercourse with a partner of their choice, has been assumed to be the foundation for romantic love. This assumption has been called into question by Diamond (2003). She avers that the emotional affection associated with physical desire is not simply the result of socio-cultural factors and their evolution but is, like the sex drive itself, a neurochemically mediated process. Affectional bonds in mammalian species and the "distinct behaviors and intense feelings associated with [them]" (p. 174) are also biologically grounded, share in part a common neurochemical base, and *can* mediate sexual desire. Although the influence of sex-drive and affectional bonds appears to be bidirectional, these motivators can exist independently in any relationship. In other words, individuals can be emotionally attached to other humans – soldiers, say, in a theater of war – with no interest whatsoever in a physically sexual expression. Likewise, others can yearn for a sexual liaison – for example, like footloose adventurers in a singles bar – without a shred of emotional commitment.

The interaction of these two personality dimensions, on the other hand, can generate the foundation for a lasting marital relationship. Both, of course, are supplemented by cognitive variables, such as a community of interests – political and religious, financial security, dependent children and elders, common socio-recreational outlets, and general emotional comfort. All the variables that contribute to durability wax and wane in intensity, not least the ardor of sexual desire.

Altruism

The sources of *altruism* have always been of interest to evolutionary psychologists. The socio-biologists of the 1970s (for example, Dawkins, [1976] 2006; Wilson, 1975) noted that environmentalists would consider it a central problem in their paradigm. That organisms have through evolution become motivated primarily to propagate their personal and "selfish genes" might not appear to be easily reconcilable with the innumerable cases of individuals who have spontaneously risked their lives for those who are not blood relatives, or even to rescue, say, a pet dog. That individuals sacrifice their personal welfare and even their lives for others would, of course, also be a puzzle to those who considered individuals as governed only by a pleasure–pain principle. Still more puzzling is the fact that millions of humans have chosen to remain childless to more effectively work among the sick, the poverty-stricken, and the educationally disadvantaged of other ethno-racial groups. This apparent problem has been clearly explicated by Steven Pinker (2002b) in his book, *The Blank Slate*.

Behavioral geneticists argue that although *Homo sapiens* like other species underwent dramatic changes in the remote past that favored our survival, individual differences appear in the *expression* of our genetic endowment. Moreover, the conditions that favored the propagation of certain genes in past epochs may no longer exist. They are variably expressed today in environments that no longer call for them. Millions of newborns appear yearly on the face of the planet, but not like mint-fresh pennies, indistinguishable from each other in appearance or behavioral potential. Each is unique. Contingencies will arise in the here-and-now of their development that will influence them to act in ways that are only partially consistent with genetically determined dispositions.

Altruism is a concept allied to that of nurturance. Expressions of both can be found among nonhuman primates and other vertebrates (Warneken *et al.*, 2007). How these constructs are related depends, of course, on how they are defined. If altruism simply refers to "unselfish behavior that favorably affects the survival, comfort, and state of mind of others" (Corsini, 1999, p. 38), then altruism characterizes the nurturant behavior of many animals toward their young. Bears, felines, and humans, among many other animals, engage in fierce behavior to protect their young, often at great risk to themselves. They also engage in demonstrations of affection, hugging and enfolding their young, soothing them when they have been hurt, nursing them when they are hungry and frightened. Defending one's abode where one's offspring and other kin reside is as evident among, say, bears with their cubs, as among humans with their children. The evolutionary roots of altruism and nurturance go deep into the common ancestry of all vertebrates.

Reber (1995) defined altruism as referring only to the "elevation of the welfare, happiness, interests or even the survival of others above one's own" (p. 26). Even in that light, the construct cannot be construed as a uniquely human trait. One unambiguously discovers it in chimpanzees, for example, as Warneken *et al.* (2007) have demonstrated. In humans it shows up early in life independent of the culture in which the individual is born and raised, which further suggests that such nurturance is innate. Pinker (2002b) has observed that "children as young as a year-and-a-half spontaneously give toys, proffer help, and try to comfort adults or other children who are visibly distressed" (p. 188; see also, Zahn-Wexler *et al.*, 1992). But the distinction between altruism and nurturance hinges on several other criteria.

Nurturance vis-à-vis altruism. Arnold Buss (1988) analyzed the components of nurturance and reduced them to instrumental, cognitive, and affective elements. These traditional categories, given that they are thoroughly intermeshed, are useful only within limits. Buss' view is that the help that nonhuman primates give to their young as well as to others involves no cost to them except in a situation where they place themselves in danger. As altruism, in his view, requires that there be a *cost to the benefactor*, it is found only among nonhumans in the nurturant and protective behavior directed to the safety and well being of their infants. There is no evidence that animals engage in other kinds of altruistic behavior, such as depriving themselves of food when they observe serious need in others. Humans, on the other hand, will deprive themselves of food, money, leisure, health, freedom, even a kidney or a portion of their liver to help others – in some instances for those who are not their own children or other close kin.

Nurturance that has costs. When one considers that all behavior requires an expenditure of time and energy (minimal as it may be), the assumption that some nurturance has no costs cannot be sustained. Any help or service one provides to others requires some mentation, empathy, energy, and time, however slight these may be (*cf.* Loewenstein and Small, 2007). Thus, one can argue that there is a cost to all activity dedicated to the service of others, even such pleasurable activities as nursing one's young. But Buss (1988) excludes nursing the primate infant from the category *nurturance* as it involves "physiological processes rather than psychological nurturance" (p. 143). This is a risky assertion as it is difficult to separate affective and psychological processes from the physiological dimensions of any deliberate act, such as occurs when an infant chimpanzee, piglets, or bear cubs are suckled by their parent.

Beyond cavil is the statement that caregiving to needy persons who cannot possibly recompense the caregiver is altruistic. Those who care for dementing relatives, legions of whom suffer from Alzheimer's or Lewy Body Disorder, are altruistic in a classic sense. The costs involve intense psychological distress, fatigue, burnout, financial loss, and decrements to a sense of self (Nelham, 2005). Similarly individuals who donate, say, a kidney that another may live or escape the constraints of frequent dialysis, serve in centers of contagion and risk their own health, or enter an inferno to rescue children and invalids (assuming material compensation is not the primary motivation for their behavior) are likewise considered altruistic. Individual differences exist in this trait, as in all others. In some it may appear to be totally absent.

Individual differences in altruism. Altruism, like other personality traits, is a continuous variable, and individual differences are as variable here as they are elsewhere. There is historical evidence that individuals will risk and even sacrifice their own lives to save the lives of others. Such behavior is clearly at one pole of the continuum; at the other is placing one's own welfare *above* that of all others in every respect, no matter the disparity in the benefits to either. The former has been reported, for example, in the exploits of those in the armed forces who have been posthumously awarded the Victoria Cross or Congressional Medal of Honor for having, say, thrown themselves on a hand grenade to protect their comrades. The literature on psychopathic personality disorder is replete with instances near the opposite pole. In the extreme, psychopaths may even take pleasure in observing the suffering of animals and humans and, indeed, causing it.

Empathy–joy hypothesis. That selfish motivation is at the root of altruistic behavior is a philosophical notion proposed by a number of researchers (for example, Smith, Keating, and Stotland, 1989). Citing the seventeenth-century moralist, La Bruyère, who had written that "the most delicious pleasure is to cause that of other people" (p. 90) they promoted the *empathy–joy hypothesis.* It stipulates that those who see others in distress may become empathically concerned and seek to reduce that distress in view of the happiness it will bring both the beneficiary and themselves. Like La Bruyère, proponents of this hypothesis affirm that individuals derive pleasure in varying degrees by reducing the suffering of others (see Wallach and Wallach, 1983). Adam Smith expressed this same idea in his *Theory of Moral Sentiments* (1759) when he wrote, "How selfish soever man may be supposed there are evidently some principles in his nature which interest him in the fortunes of others, and render their happiness necessary to him, although he derives nothing from it except the pleasure of seeing it" (p. 641).

Empathy–altruism hypothesis. A comparable position has been taken by Batson *et al.* (1991), whose *empathy–altruism hypothesis* emphasizes the prosocial feelings of witnesses as they view the sufferings of others, even of dumb animals. The thought, for example, of an animal caught in a leg-hold trap excites empathy in many individuals and motivates them to work for the abolition of this method of harvesting fur. This human impulse is much stronger, and is urged moreover by moral codes, when a child or a spouse is a victim of physical and psychological abuse. That the empathy–altruism studies of Batson *et al.* (1991) explain benevolent acts does not invalidate the empathic–joy hypothesis. Both positions appear to be true, empirically as well as intuitively, for altruism and egoism are inextricably enmeshed.[19]

[19] *Is Caregiving to Animals an Altruistic Behaviour?* Leatrice Spevack (2003), in an article on care of pets, wrote that Candice Ptolemy, a university administrator, not only attends to the needs of her own disabled pets, she also works for an animal rescue agency. Describing herself, Ptolemy said:

> If you're a real, real rescuer, it's in your blood – you cannot walk away from it. I wish I was addicted to heroin – I could get on methadone. But it's a passion. And compassion. Every single animal has value. They all have the right to live and be loved.

The costs to her are staggering. Her day begins at 5.30 am and ends near midnight. She spends between $500 and $800 a month on veterinarian bills alone. She says:

As footnote 19 suggests, altruism would seem to extend to disinterested human concern for animal welfare as well as fellow humans. Perhaps it extends to concern for the health of the entire planet and all suffering and endangered species.

Cognition and emotion in altruism. Nurturant behavior involves both cognitive and affective mechanisms. Without rehearsing the debate sparked by the James–Lange theory of emotions (discussed in Chapter 9), both emotion and understanding are involved in the altruistic act. Understanding the needs of others normally, if not necessarily, precedes the steps one takes to help them. As no cognition can be entirely devoid of affect, the empathy that people experience when witnessing the suffering of others, whether human or animal, and which appears to be a precursor of a genuinely altruistic act (Hoffman, 1981), is necessarily freighted with emotions. The anticipation of empathic joy in the relief of suffering or in the pleasure that one brings to others is a powerful motivation. That one can also envisage a reward for the altruistic act, and engage in it for that purpose, would seem to diminish its moral luster, without, however, erasing it completely.

Individuals who give help to others in the hopes of receiving help themselves (possibly the evolutionary root of this virtue) are not popularly considered altruistic; those who anonymously help others are. Whether one considers acts of mixed motivation virtuous or not will depend, on the one hand, on how much self-interest is perceived to be involved in helping, and, on the other hand, on the degree to which one considers such acts either genetically driven and spontaneous or fully deliberate acts of free will. Meshed with this issue is the theory of *reciprocal altruism*, considered evolutionarily selected and adaptive by some (Dawkins, [1976] 2006; Pinker, 2002b; Wilson, 1975, 1999), spiritually founded by others, or indeed both. Further complicating the issue are the findings that psychologically needy people are less inclined to give help where no expectation of recompense or a future payback exists (Romer, Gruder,

I can't take vacations. I can never buy myself anything new, and if I do it gets pee-ed on, poo-ed on or ripped up. I've had plastic surgery on my face from dog bites. My first cat bit me so badly that it severed the nerve between my thumb and my forefinger on my right hand.

Spevack reports similar devotion toward a cat with cerebellar hypoplasia, a disorder that prevents it from eating or drinking on its own, or running or jumping – even its walking is more like stumbling. Photographer, Roy Norgrove, has only been able to take three vacations over the past 11 years, two of them weekends, as he has to give care to this animal several times each day. Nonetheless, he says, "Even on my worst days – lonely, debts, no work – to see that defenceless, smoochable little face looking up at me melts away all the troubles … I would truly do it all again."

and Lizzadro, 1986). Obviously one's views on human nature and the wellsprings of moral behavior will influence one's judgment of what constitutes an altruistic act.

Complicating this issue are the varying levels of self-interest, interacting with the perceived costliness to the altruist of a helping act. Although we must admit that the absolutely selfless human may, like Popper's hypothetical black swan, exist (they do, in fact, exist, as I've seen them), it is reasonable to assume that all individuals will get varying levels of satisfaction, both internal and external, psychological and material, calculated and implicit, joy-conferring and sadness-relieving, when they help another person, whether friend, acquaintance, kin, or stranger.

Socio-biology and altruism. A socio-biological perspective on altruism is predicated on the notion that altruism has been selected in evolution. The virtue is at the service of the species, and of individuals driven, for better or for worse, to transmit their genes – indirectly their selves – into a continuous future existence (Dawkins, [1976] 2006; Wilson, 1975). Helping others who share one's genes may *not* be considered altruistic if absolute disinterest and selflessness is a criterion for this virtue. If one is not a purist in this respect the question arises as to how much selfishness is permitted to silt into the helping act before it is no longer altruistic. The thesis of a selfish gene precludes at some level of mentation the thoroughly selfless act. A polar stance on this issue is that there are no truly unselfish acts and that altruism is a moralistic construct that springs from codes of conduct founded on religio-philosophic rather than scientific principles.

Wilson (1978) argued that altruism is only apparently such for it ultimately boils down to mutual "back-scratching." Reciprocal altruism is the notion that members of a society help each other because the costs are less than the potential benefits. Every generous act is an insurance premium that creates the expectation that "I may need their help some day, so I bank on the hope they'll reciprocate." The extent to which this calculation enters our conscious mind is difficult to assess in any instance. Even should it become measurable, it would vary from individual to individual, and would be influenced by the situation, personal resources, prevailing mood at the moment, perceived chances of success, and estimate of the other's real need.

Altruism and charity. Independent of its biological and social roots, altruistic behavior is evident in the multitude of charitable organizations that depend on the selfless gifts of legions of sympathizers. Volunteerism is a way of life for many in the West and elsewhere, who give of their time, skills, and energy to support secular as well as faith-based organizations. The large numbers of volunteers who flooded to southern Manhattan from various parts of the United States to help survivors in the wake of the 9/11 tragedy hardly needs to be further documented. It is an example of the spontaneous compassion and nurturance that have always been in evidence in society, as self-interested as some of these helpers may have been. That this has become a state-endorsed imperative is evident in the Good Samaritan laws that have been enacted in many jurisdictions.[20] And, of course, many nations have structured their tax laws to encourage such altruism by allowing a unit of benefit to be purchased and given away at a tax-levied discount. Indeed, many individuals will only make contributions to institutions that can assure them of a tax deduction. This, of course, gets them more bang for their buck – humans seek bargains even for gifts that create a net loss to them.

Spirituality

Religious beliefs typically implicate notions of preternatural agents that exercise causal roles in the conduct of human affairs. Among deists God is simply conceived as the "ground of our being," a noninterventionist agent who created whatever exists *qua* universe and set it in motion to unwind as it may. In the earliest recorded histories of most cultures, evidence exists of religious belief in a supernatural world in which one or more divinities dabble in the affairs of "men." Currently, the practice of such religions is globally manifest in an assortment of varieties. Even within secularized strata of contemporary Western societies, religiousness[21] remains a powerful force. The domestic and foreign policy of many nations, as well as transnational political movements, are influenced by the underpinning belief systems of theocratic and even secular regimes. The prevalence of such behavior varies greatly, of course, from culture to culture and individual to individual. Religious differences frequently instigate inter-ethnic rivalries (Eidelson and Eidelson, 2003) that boil up into

[20] These laws received publicity at the time of the tragic death of Princess Diana in a Paris automobile accident. Some *paparazzi* took pictures of the victims rather than coming to their aid. They were charged with violating the law obliging them to take reasonable measures to succor citizens in dire need.

[21] As the term *religiosity* is often used to connote pretentious and false piety, the term *religiousness* will be used to connote a sincere, unaffected piety, in order to avoid ambiguity.

violence.[22] On the other hand, religion has inspired sublime works of art, magnificent architectural structures, music, poetry, sculpture, paintings, and, most importantly, charitable organizations inspired by religious sentiment and doctrine. So though millions of humans have been brutalized and killed in religious strife, large numbers have been saved from poverty and disease through the generosity of faith-based organizations. Independent of the real or fictive basis for these powerful religious movements, it is important for psychology to study them in an empirical fashion in order to enhance their potential for good and to reduce their potential for evil.[23]

Religiousness has so pervasively dominated socio-political movements and personal mentation through the ages that some theorists aver that it is a sixth *major* personality trait (Piedmont, 1999; *cf.* MacDonald, 2000) and have been moved to label humans *Homo religiosus*[24] (Albright and Ashbrook, 2001). Human beings are the only species, to our knowledge, that generate theological ideas to explain their origins and the unfolding of events in their daily lives. Reasons enough exist to generate a religion that gives meaning to human life. The hope and the conviction that humans are not alone in the universe help people to cope with the reality of death and the press of an unforgiving, often brutal, environment. Religion

[22] Richard Dawkins (echoing Nobel laureate, Steven Weinberg) averred on Channel 4 (United Kingdom), January 9, 2006 that in a world without religion good people would do good deeds; evil people would do evil deeds. In a world with religion we witness, additionally, good people doing evil deeds.

[23] *The Many Faces of Religion.* In a review of Pascal Boyer's (2001) book on religion that appeared in the *Guardian Unlimited Books* (February 13, 2002), the (unnamed) reviewer wrote:

> Any argument about religion, whether conducted in the seminar room or the saloon bar, is likely to hit the buffers not just because people hold different religious beliefs but because they disagree about what should or should not be counted as an instance of religion in the first place. Nobody will query the inclusion of what goes on at High Mass in Notre Dame or on the prayer mats of the Islamic faithful or in a Hindu temple or at a Merina death ritual in Madagascar. But what about initiation ceremonies, hero cults (including Elvis worship), charter myths, civil weddings, national anthems, silences in memory of the dead, charms, talismans and amulets, taboos on bodily fluids, spiritualism, oneiromancy, rain dances, Christmas presents, oaths and curses, apotropaic rituals in the face of physical danger, Wordsworthian nature worship (or present-day environmentalism), Confucian respect for authority, Neoplatonist metaphysics, Pythagorean reverence for number and harmony, Wittgensteinian mysticism, Freudian psychoanalysis, autonomist political theory, Kipling's as well as Socrates's references to a personal "demon," and the mild fascination with the occult shared by Pliny the Elder, John Buchan, and generations of ghost story enthusiasts and horror movie buffs?

[24] The penchant to re-label *Homo sapiens sapiens* (the very smart hominid) according to the aptitude-under-study – considered to be specific to humans – is evident in the use of the terms *Homo faber* (man the builder), *Animal rationalis* (the rational animal), *Homo politicus* (the political species), *Homo ludens* (the playful animal), and *Homo oeconomicus* (man as economic agent).

enlarges the minuscule stage on which they make a few ephemeral appearances and dignifies lives that flare vividly for a few moments and then wink out in the immensity of a limitless universe. That there may be a naturalistic and biological explanation for this does not diminish the cogency of religious feelings experienced by myriad communities of humans. The debate is still in its infancy as we have not yet well established the proportions in which genetic endowment and environmental variables influence the development of various elements of lifestyle. The extent to which preoccupation with religion and the sacred is a product of reasonable assessments of evidence, the engine of intuition and emotion propelling our search for solace and peace, or more simply the mechanical unfolding of genetic processes needs further study.

Fred Hoyle, the renowned astrophysicist, said "I have always thought it curious that, while most scientists claim to eschew religion, it actually dominates their thoughts more than it does the clergy" (quoted in Davies, 1992, p. 223). Scientists in the modern sense, people whose lives are largely absorbed in finding rational and verifiable explanations of the universe, are nevertheless confronted like the rest of humanity with a universe whose origins are, at least for the present, difficult to comprehend. This tension between a philosophical commitment to rationalist empiricism, on the one hand, and the incommensurables of the universe, on the other hand, troubles many of them.

Julian Jaynes (1990), in a widely acclaimed doctoral thesis submitted to Yale University (and initially rejected by his academic department), wrote:

> We have our houses of gods, which record our births, define us, marry us, and bury us, receive our confessions and intercede with the gods to forgive us our trespasses. Our laws are based upon values which without their divine pendancy would be empty and unenforceable. Our national mottoes and hymns of state are usually divine invocations. Our kings, presidents, judges and officers begin their tenures with oaths to the now silent deities taken upon the writings of those who have last heard them. (p. 317)

Although the tide of divine voices that filled our minds in past eras has been overwhelmed by the scientific and secular voices of the past few centuries, the religious discourse of the past is still in evidence in the press and the other news media of the Americas. The adversarial posture that science and religion frequently take vis-à-vis each other is most evident in the controversies between *creationists* and *evolutionists*.[25] Whether the

[25] One can argue that the concepts of creationism and human evolution are not mutually exclusive.

positions taken by the preponderance of these spokesmen spring largely from personality factors disposing them to respond favorably to religious vis-à-vis secular variables or from an objective assessment of the evidence for either position is a question under study.

Personality and religion. Just as extraversion or emotional lability are personality traits that pervade every aspect of individuals' lives, so religiousness is a personological impulsion that shapes individuals' lifestyle. Hill and Pargament (2003) support that view when they quote the *semioticist* (see *symbologist*), Mircea Eliade: "Religious man can live only in a sacred world, because it is only in such a world that he participates in being, that he has a real existence" (p. 68). Scientists, such as Edward O. Wilson, who have adopted an empiricist position relative to religiousness, believe that it is a trait that has been evolutionarily selected. Their view is that it is so tightly woven into the fabric of the human psyche that many people feel, and act, as if caught in a numinous dragnet. A clear expression of this conviction is voiced in Francis Thompson's poem:

> I fled Him, down the nights and down the days;
> I fled Him, down the arches of the years;
> I fled Him, down the labyrinthine ways
> Of my own mind; and in the mist of tears
> I hid from Him, and under running laughter.
> Up vistaed hopes I sped;
> And shot, precipitated,
> Adown Titanic glooms of chasmed fears,
> From those strong Feet that followed, followed after.[26]

These mystical sentiments are in sharp distinction with the view of others who prefer to take a rationalist approach to understanding the world (for example, Wilson, 1999). Of course, religiousness and materialism, like all psychosocial traits, are variables that are distributed along a continuum. They may be conceptualized as two poles of a single dimension within which individuals typically find a fluctuating balance that is comfortable for them during various stages of their life (see Cloninger, 2004b, for a monistic view of the universe that attempts to reconcile these positions).

Individual differences. Personality psychology has taken notice of the individual differences that are manifest in the religious activities in which people engage (Emmons and McCullough, 1999). Although

[26] These are the first lines of Francis Thompson's poem, *The Hound of Heaven*.

religion has been variously conceptualized across recent millennia, in our era William James ([1902] 1961) defined it as the "feelings, acts, experiences of individual men in their solitude ... in relation to whatever they may consider divine" (p. 42). He sub-titled his classic work on religious experience, *A Study in Human Nature*. There was no doubt in his mind that men and women's preoccupation throughout history with the transpersonal, and their pursuit of the numinous, is intrinsic to human nature and reveals itself in their personality. He was aware that this pursuit often morphed into an obsession with immortality, "eternal salvation," feelings of "chosenness" and divine predilection (*cf.* Eidelson and Eidelson, 2003), communing with the Ultimate Rescuer, and with abandonment-of-self to Divine Providence (for example, de Caussade, [1751] 2001).

Miller and Thoresen (2003) wrote that "William James and others throughout the 20th century related the spiritual to a person's character, personality, or disposition, often with an emphasis on the person's social and emotional style and manner of living ..." (p. 27). Restricting the study of spiritual practices to what is observable or to what can be inferred from well-controlled studies, two conclusions become evident. First, there are wide individual differences in preferences for verbalizing religious beliefs, pious sentiments, transcendent and deity-based moral values, and ethical objectives in private rather than in the context of group religious services. For some, religion and spirituality are very private matters. For others, they involve active participation in social, religious, and militant movements. Second, individuals differ in the levels of their commitment and zeal to disseminating their beliefs among those who are nonbelievers or simply have different beliefs. Third, Paloutzian, Richard, and Rambo (1999) have found that religious conversion has little measurable impact on measures of the *Big Five* superfactors: to wit, extraversion, emotional lability, agreeableness, openness, and conscientiousness. On the other hand, they found that embracing a religion had profound effects on life goals, attitudes, prosocial (and anti-social) behaviors, and social values – expressions of stable personality traits that could profoundly shape the presentation of self to the everyday world.

Freud ([1928] 1964, p. 71) postulated that religion is a delusional system that engenders "obsessive neurosis," disabling guilt, strangled affect, repressed sexuality, intolerance, and a preoccupation with the hereafter leading to alienation from the work-a-day world. Hyper-religiosity can easily be detected in some mental disorders and may even be a sign of neurological disease, but the selective evolutionary advantage that a common religious belief system may confer on tribal societies suggests that a belief in the transcendent is instinctual and thus natural. A moderate position is taken by Edward O. Wilson (1999, pp. 260–90), who contrasted the position of those who have a belief in a divinity who intervenes on a daily basis in the affairs of humanity, and those who do not. He characterized the former as *transcendentalists*, the latter as *empiricists*. Admittedly there is a middle ground, but, as an empiricist himself, he examined the pros and cons of these polar positions.

If, indeed, religiousness is instinctual and, like all traits of human nature, expressed differentially in each human, it is not surprising to find that some people are highly religious, given to organized, prayerful, and meditative worship, concerned about pleasing invisible deities, and prone to the ritualistic and even sacrificial rites that flourished in primitive societies.

The spiritual and religious impulse operates as a powerful motivator, often out of our awareness, especially among fundamentalist believers – and in every domain of life: political, socio-economic, military, and familial. Others, by contrast, are prone to a more secularist, rationalist, pragmatic, make-the-best-of-this-world lifestyle. Both groups can live principled lives, and, indeed, both groups have evolved moral codes that have well served the common good. Religions have evolved moral codes that have assured the stability and welfare of the societies in which they have prospered. Secular deists like the founders of the American republic developed their own code of natural law, the self-evident principles that continue to this day to move their citizenry, if not every politician, to ever greater respect for the dignity and rights of each. If culture is personality writ large, a way of being in the world, there are wide personality differences in the way various societies choose to express their religious beliefs and to interface with other societies that wish to live differently.

For some people, worshiping in common is a moral imperative. Others do not feel obliged to engage in mosque-, temple-, or church-based social activities. Although one's involvement in public rather than private expressions of spirituality would seem to be related to the constructs of introversion and extraversion, well-designed studies of this question and others have been absent. Miller and Thoresen (2003) have stated, however, that anecdotal evidence suggests that religion has been associated with disabling guilt, morbid anxiety about displeasing god(s), compulsive rituals such as endless repetitions of formulaic prayer, intolerance of dissident opinions, and inflexibility. On the other hand, religiousness has been correlated with greater enjoyment of life (Brady *et al.*, 1999), particularly when physical ailments were already diminishing quality of life.

Ethics and religion. One's penchant to find an explanation for phenomena by appealing to transcendent laws governing the universe as well as to a divinity (or divinities) that has not only created the universe but decreed its physical laws and codes of conduct has been evident since pre-history. The enmeshment of religion with ethics is self-evident. Even those who shun organized religion and formal codes of morality tend to comply with the spiritual templates that underpin much of our civil and criminal codes. That such a penchant is a personality factor, implicated in the supertrait, *conscientiousness*, is apparent.

There is a broad chasm between an individualistic spirituality and a collectivist religiousness, the first prized by those who prefer their relationship with the *numinous* to be a private matter, the second by those who find their spiritual fulfillment principally in a community of faith. The conviction that at the end of time everything will have worked out for the best, fore-ordained or at least tolerated by divine providence, contributes to a peace of mind that is evident in both groups – and by their presentation of self to others.

Spirituality is a trait found in all cultures and, as far as we can tell, in all times (Brown, 1991; Elkins, 2001). Elkins avers that it is infused with a "mysterious energy," which puts him at odds with Edward O. Wilson (1999, pp. 266–90) and other scientists who, in the interests of a unified theory of all natural disciplines, insist that there be an *empirical* basis for the principles of a psychology of religion. Some psychologists would argue that spirituality, or the religious experience, cannot be treated exclusively in a scientific manner, as it deals fundamentally with incommensurables. The past decade, however, has seen numerous studies that have donned a scientific mantle. Spirituality and religious beliefs have been woven into the weft of all societies since pre-history; this constitutes, sociologically, politically, and psychologically, an empirical reality that can and should be subject to empirical analysis.

Clearly there are other conceptualizations of spirituality, among which are natural and humanistic spirituality. Though these are largely secularized versions of religion, they nevertheless leave the door open to paranormal, supernatural, and numinous possibilities. The existential angst generated by thoughts of physical and cognitive decline terminating in death, entwined with the dreadful specters of famine, pestilence, and war that assail our senses and minds daily, drives people to seek a more secure psychological space in which to live out their days. That religion derives in part from the response to such existential anxieties may account for the emotional dynamics that propel people to embrace their preferred version of it, to identify with it, and to propagate it. This understanding of our common destiny underpins the compassion and love of neighbor that are ultimate objectives of most if not all religions (Emmons and Paloutzian, 2003).

A rich assortment of religious experiences has been explained (for example, Brown, 1991, pp. 113–15) by sleep deprivation, music-entrancement, prolonged fasting, ingestion of hallucinogenic drugs, brain pathology, and engagement in prolonged rhythmic activities, such as dancing in a pulsating sound-and-light-drenched environment surrounded by others similarly enraptured. That the mystical experience of "acidheads" mimics the experiences of traditional mystics, dervishes, and contemplatives is not, of course, *prima facie* evidence that all religious experiences are naturally induced. This does suggest that there are common psychobiological processes, whether catalyzed by natural or supernatural agents, which are activated in all of these religious experiences. On the other hand, it has yet to be demonstrated that the rapture experienced by individuals in the tent of a charismatic evangelist is fundamentally different from that which dancers experience at a rave. It may never be.

Measurement. The principle that measurement is essential to a scientific understanding of the world is part of a contemporary philosophy of science. The pedigree of the science on which the modern world is built can be traced back to Pythagoras, not to religious prophets or other spokesmen of the numinous. Especially since the era of Galileo (1564–1642) and Descartes (1596–1650), this science has been built on mathematics and measurement. The psychology of religion, like any other psychology, can define itself as a science only if the instruments devised for its study have sound psychometric properties. Not surprisingly, a plethora of measures of religiousness has been developed over the past generation (Hill and Hood, 1999), most of which are self-report, paper-and-pencil instruments. The limitations of research delving into personalized databases of an introspective character using units of analysis that cannot be crisply defined are obvious. To expect that the deep wells of belief and passionate conviction can be tapped in a reliable way by asking subjects to think about and report their feelings – to be translated into checks on a piece of paper – does not meet conventional standards of scientific rigor. Even the most conscientious pursuit of accurate data in psychology is subject to unconscious distortion of which the participants are variously unaware (for example, Bargh and Chartrand, 1999). Widely accepted dual-process models of motivation, that is, schemas involving conscious and unconscious sources of catalysis, suggest that plumbing the religious, mystical, and spiritual ideation of human beings is fraught, to say the least, with scientific difficulties.

Conclusion. Recent history continues to demonstrate that religion is "very important" to humans generally and to Americans, specifically (Gallup, 2007). Proselytizing fervor has not abated over the centuries, nor has the rhetoric employed in its service changed. Indeed, large segments of the American public have shown an increased interest in spiritual issues in recent decades (see Powell, Shahabi, and Thoresen, 2003). What is now different is the enhanced political, communicational, and psychological leverage available to those who wish others to think and ultimately worship as they do.

The relevance of belief systems and religious behavior to health matters has been the subject of a recent issue of the *American Psychologist* (2003). Needless to emphasize, billions of human beings who are not Americans have similar, some even stronger, religious affiliations. Psychological studies of the religious impulse and its differential interaction with health dynamics, on the one hand, and its enmeshment with political movements, on the other hand, will certainly take on increasing importance in the twenty-first century. This consideration is a segue to the more somber issue of grief and mourning.

Mourning becomes us all

The end of life, like the start of life, is fraught with heavy emotion and expectations for the future. These are externalized in myriad funeral rituals that have evolved over the centuries. In their diversity they reveal different values, worldviews, religious schemas, and cosmologies, but in all cases profound and stressful expressions of loss. There are individual differences in the impact that the death of a loved one has on survivors, and in their willingness to expose their interior response to public gaze. This loss is so poignant and immobilizing for some that they need to engage friends or professional mourners to be the public face of their anguish and grief. Tugged in many directions by the expectations of family, friends, and the larger community while enmeshed in memories and searing realizations of what life without a child, a spouse, or a parent will be, the bereaved ask others to manage the pomp and obsequies for them.

An Asian perspective. Although Chinese societies have hired professional mourners from time immemorial for this purpose, the practice is apparently on the decline. Nevertheless the *Taipei Times* (October 16, 2005) has reported on one group called the *Filial Daughters Band*, that engages commercially in vicarious mourning. "Grieving relatives are often too weary or too numb to shed the requisite amount of tears, so rather than leave a void, they hire groups like the *Filial Daughters Band* to perform their mournful stuff." In a municipal funeral home in the outskirts of Taipei, the leader of this thriving group, a 22-year-old woman, climbed to the stage and began an exhausting routine intended to reflect the deepest emotions of the bereaved family. She launched "into her signature high-pitched, heart-wrenching wailing while pounding the floor and crawling on her knees to express grief for a dead stranger."[27]

A mainland version of this culture of mourning is reported in *Harper's Magazine* (Yiwu, November, 2005, pp. 15–20). Liao Yiwu interviewed a contemporary 70-year-old professional mourner, Li Changgeng, who regrets the Westernization of ancient Chinese rituals. He traces them to the founder of his profession, Confucius, who played the *suona* to support his family. (The suona is a traditional wind instrument accoutred with a brass bell, used at funerals and weddings.) A segment from the interview follows:

LIAO YIWU: How do you manage to wail and howl over a stranger?

LI CHANGGENG: I entered the mourning profession at the age of twelve. My teacher forced me to practice the basic suona tunes, as well as to learn how to wail and chant. Having a solid foundation in the basics enables a performer to improvise and to produce an earth-shattering effect.

Our wailing sounds more authentic than that of the relatives of the deceased. Most people who have lost family burst into tears and begin wailing upon seeing the body, but their wailing doesn't last. Soon they're overcome with grief, and they either suffer from shock or pass out. But we control our emotions and improvise with great ease. We can wail as long as is requested.

Li Changgeng described his move to Sichuan Province, and the growth of his career, which entailed learning many funereal tunes. He continued:

> These songs have been performed for hundreds of years, passed down from generation to generation. There are specific instructions on when to hit the high notes or drop to the low notes, when to use a cracked voice or to be high-pitched, when to wail with the effect of a dry throat or to cry with tears, when to make your body tremble with great sadness or to sound like you're losing your voice. It has to be very precise.

The complexities of the rituals, which are only alluded to here, depend on the individual preferences of the bereaved.

A Western perspective. In the West, we have the recent example of an intelligent, sensitive, and creative woman, Joan Didion (2005), who left a public record of her passage through a period of mourning for her equally gifted husband, John Gregory Dunne. He died at home immediately after their hospital visit to their comatose daughter, Quintana Roo.

[27] The Korean film, *Secret Sunshine* (2007), portrays various cultural expressions of mourning even within the same community and family. Johnson and Dunn (2008) acclaim it as a "stunning picture of the complexities of grief, meaning, and religious community."

It is evident that the cultural norms and religious heritage that pervaded Joan Didion's family life conditioned the grieving process for her. Layered into those influences were her unique personality and personal history, which shaped the way she lived through the ordeals of ambulances, physicians, autopsy, waking, funeral, and subsequent re-ordering of her socio-domestic life. She was characterized by a social worker as "a cool customer," no doubt low on an emotional lability scale. The funeral at the Cathedral of St. John the Divine on the upper West Side of Manhattan transpired in a subdued, decorous, and ritual manner. All the actors in the poignant, painful drama were self-composed, controlled, undemonstrative.

But there is a common, nomothetic side to these two disparate accounts of bereavement and mourning. In the recesses of their psyche those who have lost, say, a young child, experience "depth charges" detonating in their sleep and in solitary moments, when memories of lost opportunities to engage the deceased are triggered by a scrap of music, a view of their undisturbed bedroom, a scent, a mislaid note. The repercussions are profound and somatic. Didion (p. 28) quotes Eric Lindemann (1944), who was chief of psychiatry at Massachusetts General Hospital, on the occasion of the 1942 Cocoanut Grove fire, which killed hundreds of youths: he reported the family survivors experienced "sensations of somatic distress occurring in waves lasting from twenty minutes to an hour at a time, a feeling of tightness in the throat, choking with shortness of breath, need for sighing, and an empty feeling in the abdomen, lack of muscular power, and an intense subjective distress described as tension or mental pain." Though these, in the aggregate, appear to be pathological signs, they are, irrespective of the ethnic arena in which they are played out, universal signs of normal grieving. The neuro-endocrinological template that supports grieving is the same for all humans – and, we can speculate, resembles the grieving and depression that other vertebrates appear to experience at the loss of close ones. Of course, temperament and personal history, as well as the "symbiotic" character of the dissolved relationship, moderate these expressions of our human nature.

We thus circle back to the theme with which this chapter began – dinner with friends and family, the sacred meals, and the commemorative festivals. Whether the bereaved hire nuns and monks to chant Buddhist scriptures to hasten the deceased along the path of an improved afterlife or celebrate Eucharist in an Episcopalian cathedral praying that flights of angels loft their loved ones to their final rest, the psychic wellsprings are the same. The angst-driven pursuit of meaning and immortality shapes the socio-religious rituals with which we surround the death of our loved ones. And as we each in due time take our place in the endless recycling of our carbon-based biosphere we know that we will support the life that gave rise to our own moments of joy and those relationships that gave beauty to our personal existence. As for those left behind, they will cast our shadow through the branching friendships and family ties we have made possible.

References

Abe, J. A. and Izard, C. E. 1999. "A longitudinal study of emotion expression and personality relations in early development," *Journal of Personality and Social Psychology*, 77: 566–77.

Abelson. R. P. 1985. "A variance explanation paradox: when a little is a lot," *Psychological Bulletin*, 97: 129–33.

Abramson, L. Y. and Alloy, L. B. 1981. "Depression, non-depression, and cognitive illusions: a reply to Schwartz," *Journal of Experimental Psychology*, 110: 436–47.

Ackerman, P. L. and Beier, M. E. 2003. "Intelligence, personality, and interests in the career choice process," *Journal of Career Assessment*, 11(2): 205–18.

Adams, H. [1900] 2000. *The Education of Henry Adams.* New York: Houghton Mifflin Harcourt.

Adler, A. [1931] 1958. *What Life Should Mean to You.* New York: Capricorn Books.

Adler, A. [1927] 1959. *Understanding Human Nature.* New York: Premier Books.

Adler, L., Wedekind, D., Pitz, J., Weniger, G., and Huether, G. 1997. "Endocrine correlates of personality traits: a comparison between emotionally stable and emotionally labile healthy young men," *Neuropsychobiology*, 35(4): 205–10.

Adolphs, R. and Damasio, A. R. 1998. "The human amygdala in social judgment," *Nature*, 393: 470–74.

Adolphs, R., Tranel, D., and Damasio, A. R. 1994. "Impaired recognition of emotion in facial expressions following bilateral damage to the human amygdala," *Nature*, 372: 669–72.

Agonito, R. 1977. *History of Ideas on Woman: A Source Book.* New York: Perigee Books.

Aiken, L. R. 2000. *Psychological Testing and Assessment,* 10th edn. Needham Heights, MA: Allyn and Bacon.

Ainsworth, M. D. S. 1962. "The effects of maternal deprivation: a review of findings and controversy in the context of research strategy," in *Deprivation of Maternal Care: A Reassessment of its Effects, Health Papers, 14.* Geneva: World Health Organization.

Ainsworth, M. D. S., Blehar, M. C., Waters, E., and Wall, S. 1978. *Patterns of Attachment: A Psychological Study of the Strange Situation.* Hillsdale, NJ: Erlbaum.

Albright, C. R. and Ashbrook, J. B. 2001. *Where God Lives in the Human Brain.* Napierville, IL: Sourcebooks.

Aldis, O. 1975. *Play Fighting.* New York: Academic Press.

Allen, J. R. and Allen, B. A. 1995. "Narrative theory, redecision therapy, and postmodernism," *Transactional Analysis Journal*, 25: 327–34.

Allen, T. W. 1971. "The individual psychology of Alfred Adler: an item of history and a promise of a revolution," *The Counseling Psychologist*, 3(1): 3–24.

Allport, G. W. [1937] 1961. *Personality: A Psychological Interpretation*, rev. edn. New York: Henry Holt

Allport, G. W. 1955. *Becoming: Basic Considerations for a Psychology of Personality.* New Haven, CT: Yale University Press.

Allport, G. W. 1960. *Personality and Social Encounter: Selected Essays.* Boston, MA: Beacon Press.

Allport, G. W. 1967. "Autobiography," in E. G. Boring and G. Lindzey (eds.), *A History of Psychology in Autobiography.* New York: Appleton-Century-Crofts, vol. 5, pp. 1–25.

Allport, G. W. 1968. *The Person in Psychology: Selected Essays.* Boston, MA: Beacon Press.

Allport, G. W. and Odbert, H. S. 1936. "Trait-names: a psycho-lexical study," *Psychological Monographs*, 47:1 (whole No. 211).

Allport, G. W. and Vernon, P. E. 1933. *Studies in Expressive Movement.* New York: Macmillan.

American Educational Research Association, American Psychological Association, and National Council on Measurement in Education 1999. *Standards for Educational and Psychological Testing.* Washington, DC: American Psychological Association.

American Psychiatric Association 1980. *Diagnostic and Statistical Manual of Mental Disorders*, 3rd edn. Washington, DC.

American Psychological Association 1975. "Report of the task force on sex bias and sex-role stereotyping in psychotherapeutic practice," *American Psychologist*, 30(12): 1169–75.

American Psychological Association 2002. "Ethical principles of psychologists and code of conduct," *American Psychologist*, 57: 1060–73.

American Psychological Association 2007. "Paul Rozin: award for distinguished scientific contributions," *American Psychologist*, 62(8): 751–54.

Anastasi, A. 1988. *Psychological Testing*, 6th edn. New York: Macmillan.

Anastasi, A. and Urbina, S. 1997. *Psychological Testing*, 7th edn. Upper Saddle River, NJ: Prentice-Hall.

Andreasen, N. C. 2005. *The Creating Brain: The Neuroscience of Genius.* New York: Danna Press.

Angyal, A. 1951. "A theoretical model for personality studies," *Journal of Personality*, 20: 131–42.

Ansbacher, H. L. and Ansbacher, R. R. (eds.) 1956. *The Individual Psychology of Alfred Adler: A Systematic Presentation in Selections from his Writings.* New York: Basic Books.

Antelman, S. M. and Caggiula, A. R. 1977. "Tails of stress-related behavior: a neuropharmacological model," in I. Hanin and E. Usdin (eds.), *Animal Models in Psychiatry and Neurology.* New York: Pergamon, pp. 227–45.

Archer, J. 1996. "Sex differences in social behavior: are the social role and evolutionary explanations compatible?," *American Psychologist*, 51: 909–17.

Arendt, H. 1958. *The Human Condition.* Chicago: University of Chicago Press.

Ariès, P. 1962. *Centuries of Childhood: A Social History of Family Life,* trans. R. Baldick. New York: Knopf.

Aristotle [*c.* 320 BC] 1991. *Nichomachean Ethics,* trans. W. D. Ross *et al.*, in Encyclopædia Britannica edn., *Great Books of the Western World.* University of Chicago Press, vol. 8.

Arnold, M. B. 1960. *Emotion and Personality.* New York: Columbia University Press.

Aron, A., Fisher, H., Mashek, D. J., Strong, G., Li, H. F., and Brown, L. L. 2005. "Reward, motivation, and emotion systems associated with early-stage intense romantic love," *Journal of Neurophysiology,* 94(1): 327–37.

Asch, D. A., Patton, J. P., and Hershey, J. C. 1990. "Knowing for the sake of knowing," *Medical Decision Making,* 10: 47–57.

Augustine [400 CE] 1912. *On the Trinity,* trans. A. West Haddan, in Benjamin Rand (ed.), *The Classical Psychologists.* Boston, MA: Houghton Mifflin.

Augustine [400 CE] 1991a. *The Confessions,* trans. R. S. Pine-Coffin, Chicago, IL: Encyclopædia Britannica, Inc., vol. 16, pp. 1–159.

Augustine [400 CE] 1991b. *On Christian Doctrine,* trans. J. F. Shaw, 1961. Chicago, IL: Encyclopædia Britannica, Inc., vol. 16, pp. 701–84.

Ausubel, D. P., Sullivan, E. V., and Ives, S. W. 1980. *Theory and Problems of Child Development.* New York: Grune & Stratton.

Baldwin, J. M. [1897] 1906. *Mental Development in the Child and the Race,* 3rd edn. New York: Macmillan.

Balog, A. 1997. "Inner-directed actions," *Sociological Perspectives,* 40: 33–59.

Baltes, P. B. 1983. "Life-span developmental psychology: observations on history and theory revisited," in R. M. Lerner (ed.), *Developmental Psychology: Historical and Philosophical Perspectives.* Hillsdale, NJ: Erlbaum, pp. 70–111.

Baltes, P. B. 1987. "Theoretical propositions of life-span developmental psychology: on the dynamics between growth and decline," *Developmental Psychology,* 23(5): 611–26.

Baltes, P. B. 1997. "On the incomplete architecture of human ontogeny: selection, optimization, and compensation as foundation of developmental theory," *American Psychologist,* 52(4): 366–80.

Baltes, P. B. and Graf, P. 1996. "Psychological aspects of aging: facts and frontiers," in D. Magnusson (ed.), *The Lifespan Development of Individuals: Behavioural, Neurobiological and Psychosocial Perspectives.* Cambridge University Press, pp. 427–60.

Baltes, P. B., Reuter-Lorenz, P. A., and Rösler, F. 2006a. "Prologue: biocultural co-constructivism as a theoretical metascript," in P. B. Baltes, P. A. Reuter-Lorenz, and F. Rösler (eds.), pp. 3–39.

Baltes, P. B., Reuter-Lorenz, P. A., and Rösler, F. (eds.) 2006b. *Lifespan Development and the Brain: The Perspective of Biocultural Co-constructivism.* Cambridge University Press.

Baltes, P. B. and Staudinger, U. M. (2000). "Wisdom: a metaheuristic (pragmatic) to orchestrate mind and virtue toward excellence," *American Psychologist,* 55: 122–36.

Bandura, A. 1969. *Principles of Behavior Modification.* New York: Holt, Rinehart & Winston.

Bandura, A. 1982. "Self-efficacy mechanism in human agency," *American Psychologist*, 37: 122–47.

Bandura, A. 1986. *Social Foundations of Thought and Action: A Social Cognitive Theory*. Englewood Cliffs, NJ: Prentice-Hall.

Bandura, A. 1998. "Exploration of fortuitous determinants of life paths," *Psychological Inquiry*, 9(2): 95–99.

Bandura, A. 1999. "Social cognitive theory of personality," in L. A. Pervin and O. P. John (eds.), pp. 154–96.

Bargh, J. A., and Chartrand, T. L. 1999. "The unbearable automaticity of being," *American Psychologist*, 54: 462–79.

Barlow, G. 1977. "Modal action patterns," in T. A. Sebeok (ed.), *How Animals Communicate*. Bloomington, IN: Indiana University Press, pp. 98–134.

Bar-On, R. 1997. *The Bar-On Emotional Quotient Inventory: A Measure of Emotional Intelligence*. Toronto: Multi-Health Systems.

Baron-Cohen, S. 2005. "The neuropsychology of autism and pervasive developmental disorders – The extreme male brain theory: an expert interview with Simon Baron-Cohen." Medscape Psychiatry and Mental Health, December 14. Available at: www.medscape.com/viewarticle/518449.

Baron-Cohen, S., Knickmeyer, R. C., and Belmonte, M. K. 2005. "Sex differences in the brain: implications for explaining autism," *Science*, 310(5749): 819–23.

Barrett, L. F. and Russell, J. 1998. "Independence and bipolarity in the structure of current affect," *Journal of Personality and Social Psychology*, 74(4): 967–84.

Barron, F. X. 1955. "The disposition toward originality," *Journal of Abnormal and Social Psychology*, 51: 478–85.

Barron, F. X. 1969. *Creative Person and Creative Process*. New York: Holt, Rinehart & Winston.

Batson, C. D., Batson, J. G., Slingsby, J. K., Harrell, K. L., Peekna, H. M., and Todd, R. M. 1991. "Empathic joy and the empathy–altruism hypothesis," *Journal of Personality and Social Psychology*, 61: 413–26.

Baumeister, R. F. 1988. "Should we stop studying sex differences altogether?," *American Psychologist*, 43: 1092–95.

Baumeister, R. F. 1994. "The crystallization of discontent in the process of major life change," in T. F. Heatherton and J. L. Weinberger (eds.), pp. 281–97.

Baumeister, R. F. 1999. "On the interface between personality and social psychology," in L. A. Pervin and O. P. John (eds.), pp 347–66.

Baumeister, R. F. and Twenge, J. 2003. "The social self," in T. Millon and M. J. Lerner (eds.), *Personality and Social Psychology, Vol. 5: The Handbook of Psychology*. New York: Wiley, pp. 327–52.

Bayer, R. and Spitzer, R. L. 1985. "Neurosis, psychodynamics, and DSM-III," *Archives of General Psychiatry*, 42: 187–96.

Becker, H. 1952. "Science, culture, and society," *Philosophy of Science*, 19: 273–87.

Beer, J. M., Arnold, R. D., and Loehlin, J. C. 1998. "Genetic and environmental influences on MMPI factor scales: joint model fitting to twin and adoption data," *Journal of Personality and Social Psychology*, 74: 818–27.

Belsky, J. K. 1990. *The Psychology of Aging: Theory, Research, and Interventions*. Pacific Grove, CA: Brooks/Cole.

Belsky, J. and Nezworski, T. (eds.) 1988. *Clinical Implications of Attachment.* Hillsdale, NJ: Erlbaum.

Bem, S. L. 1974. "The measurement of psychological androgyny," *Journal of Consulting and Clinical Psychology*, 42: 155–62.

Bem, S. L. 1977. "On the utility of alternative procedures for assessing psychological androgyny," *Journal of Consulting and Clinical Psychology*, 45: 196–205.

Bem, S. L. 1993. *The Lenses of Gender: Transforming the Debate on Sexual Inequality.* New Haven, CT: Yale University Press.

Bem, D. J. and Allen, A. 1974. "On predicting some of the people some of the time: the search for cross-situational consistencies in behavior," *Psychological Review*, 81: 506–20.

Benedict, R. 1946. *The Chrysanthemum and the Sword: Patterns of Japanese Culture.* Cambridge, MA: Riverside Press.

Benedict, R. [1932] 1949. *Patterns of Culture.* London: Routledge & Kegan.

Benenson, J. F. 1993. "Greater preference among females than males for dyadic interaction in early childhood," *Child Development*, 64: 544–55.

Ben-Porath, Y. S. and Butcher, J. N. 1991. "The historical development of personality assessment," in C. E. Walker (ed.), *Clinical Psychology: Historical and Research Foundations.* New York: Plenum, pp. 121–56.

Berlyne, D. E. 1950. "Novelty and curiosity as determinants of exploratory behavior," *British Journal of Psychology*, 41: 68–80.

Berlyne, D. E. 1954. "A theory of human curiosity," *British Journal of Psychology*, 45: 180–91.

Berlyne, D. E. 1960. *Conflict, Arousal, and Curiosity.* New York: McGraw Hill.

Berlyne, D. E. 1966. "Curiosity and exploration," *Science* 153(3731): 25–33.

Bernal, M. 1987. *Black Athena: The Afro-Asiatic Roots of Classical Civilization.* London: Free Association Books, vol. 1.

Bernard, C. 1869. "Des effets physiologiques de la morphine et leur combinaison avec ceux du chloroforme," *Bulletin Thérapeutique*, 77: 241–56. Paris: Baillière.

Bernard, J. 1950. "Can science transcend culture?," *Scientific Monthly*, 7: 268–73.

Bernstein, J. 1993. *Cranks, Quarks, and the Cosmos: Writings on Science.* New York: Basic Books.

Berscheid, E. 1985. "Interpersonal attraction," in G. Lindzey and E. Aronson (eds.), *Handbook of Social Psychology*, 3rd edn. New York: Random House, vol. II, pp. 413–84.

Beutler, L. E., Williams, R. E., Wakefield, P. J., and Entwhistle, S. R. 1995. "Bridging scientist and practitioner perspectives in clinical psychology," *American Psychologist*, 50(12): 984–94.

Bhattacharya, M. and Chatterjee, R. 2000. "Collaborative innovation as a process for cognitive development," *Journal of Interactive Learning Research*, 11(3–4): 295–312.

Birren, J. E. and Schaie, K. W. 1996. *Handbook of the Psychology of Aging*, 3rd edn. San Diego, CA: Academic Press.

Blackburn, R., Renwick, S. J. D., Donnelly, J. P., and Logan, C. 2004. "Big Five or big two: superordinate factors in the NEO Five-Factor Inventory and the Antisocial Personality Questionnaire," *Personality and Individual Differences*, 5(37): 957–70.

Blatner, A. 2001. "Psychodrama," in R. J. Corsini (ed.), *Handbook of Innovative Therapies,* 2nd edn. New York: Wiley, pp. 530–53.

Block, J. 1995a. "A contrarian view of the five-factor approach to personality description," *Psychological Bulletin,* 117: 187–215.

Block, J. 1995b. "Going beyond the five factors given: rejoinder to Costa and McCrae (1995) and Goldberg and Saucier (1995)," *Psychological Bulletin,* 117(2): 226–29.

Bloom, H. 1998. *Shakespeare: The Invention of the Human.* New York: Riverhead Books/Penguin Putnam.

Bohan, J. S. (ed.) 1992. "Prologue: reviewing psychology, replacing women – an end searching for a means," in J. S. Bohan (ed.), *Seldom Seen, Rarely Heard: Women's Place in Psychology.* Boulder, CO: Westview Press, pp. 9–54.

Bordin, E. S. 1994. "Intrinsic motivation and the active self: convergence from a psychodynamic perspective," in M. L. Savickas and R. W. Lent (eds.), *Convergence in Career Development Theories.* Palo Alto, CA: Consulting Psychologists Press, pp. 53–61.

Boring, E. G. 1950. *A History of Experimental Psychology,* 2nd edn. Englewood Cliffs, NJ: Prentice-Hall.

Bornstein, M. H. 1989. "Sensitive periods in development: structural characteristics and causal interpretations," *Psychological Bulletin,* 105: 179–97.

Bornstein, R. F. 1992. "Subliminal mere exposure effects," in R. F. Bornstein and T. S. Pittman (eds.), *Perception without Awareness: Cognitive, Clinical, and Social Perspectives.* New York: Guilford, pp. 191–210.

Bornstein, R. F. and Masling, J. 1985. "Orality and latency of volunteering to serve as experimental subjects: a replication," *Journal of Personality Assessment,* 49: 306–10.

Borsboom, D., Mellenbergh, G. J., and van Heerden, J. 2004. "The concept of validity," *Psychological Review,* 111(4): 1061–71.

Bottome, P. 1939. *Alfred Adler, Apostle of Freedom.* London: Faber & Faber.

Bourdieu, P. 1984. *Distinction: A Social Critique of the Judgement of Taste,* trans. R. Nice, London: Routledge.

Bower, G. 1981. "Mood and memory," *American Psychologist,* 36: 129–48.

Bowers, K. S. and Farvolden, P. 1996. "Revisiting a century-old Freudian slip: from suggestion disavowed to the truth repressed," *Psychological Bulletin,* 119: 355–80.

Bowlby, J. 1946. *Forty-four Juvenile Thieves: Their Characters and Home Life.* Paris: Baillière.

Bowlby, J. 1969. *Attachment and Loss, Vol. 1: Attachment.* New York: Basic Books.

Bowlby, J. 1973. *Attachment and Loss, Vol. 2: Separation: Anxiety and Anger.* New York: Basic Books.

Boyer, P. 2001. *Religion Explained: The Human Instincts that Fashion Gods.* London: Heinemann.

Brabeck, M. and Brown, L. 1997. "Feminist theory and psychological practice," in J. Worell and N. Johnson (eds.), *Shaping the Future of Feminist Psychology: Education, Research and Practice.* Washington, DC: American Psychological Association, pp. 15–35.

Brady, M. J., Peterman, A. H., Fitchett, G., Mo, M., and Cella, D. 1999. "A case for including spirituality in quality of life measurement in oncology," *Psycho-oncology*, 8: 417–28.

Breuer, J. and Freud, S. [1895] 1960. *Studies in Hysteria.* Oxford: Beacon.

Brieger, P. and Marneros, A. 1997. "Dysthymia and cyclothymia: historical origins and contemporary development," *Journal of Affective Disorders*, 45(3): 117–26.

Brim, O. G. 1958. "Family structure and sex role learning by children: a further analysis of Helen Koch's data," *Sociometry*, 21: 1–16.

Brizendine, L. 2006. *The Female Brain.* New York: Random House.

Brody, H. 1987. *Stories of Sickness.* New Haven, CT: Yale University Press.

Bromley, D. G. 1988. *Falling from the Faith: Causes and Consequences of Religious Apostasy.* Newbury Park, CA: Sage.

Broughton, J. M. and Freeman-Moir, D. J. 1982. *The Cognitive Developmental Psychology of James Mark Baldwin: Current Theory and Research in Genetic Epistemology.* Norwood, NJ: Ablex.

Brown, D. E. 1991. *Human Universals.* New York: McGraw-Hill.

Bruner, J. 1973. *Going Beyond the Information Given.* New York: Norton.

Bruner, J. 1986. *Actual Minds, Possible Worlds.* Cambridge, MA: Harvard University Press.

Bruner, J. 1987. "Life as narrative," *Social Research* 54(1): 11–32.

Bugental, J. F. T. 1963. "Humanistic psychology: a new breakthrough," *American Psychologist*, 18: 563–67.

Bullock, T. H. 1958. "Evolution of neurophysiological mechanisms," in A. Roe and G. G. Simpson (eds.), pp. 165–77.

Burghardt, G. M. 1984. "On the origins of play," in P. K. Smith (ed.), *Play in Animals and Humans.* Oxford: Basil Blackwell, pp. 5–41.

Burghardt, G. M. 2005. *The Genesis of Animal Play: Testing the Limits.* Cambridge, MA: MIT Press.

Burnet, J. 1920. *Early Greek Philosophy*, 3rd edn. London: A. & C. Black.

Buss, A. 1988. *Personality: Evolutionary Heritage and Human Distinctiveness.* Hillsdale, NJ: Erlbaum.

Buss, A. and Plomin, R. 1975. *A Temperament Theory of Personality Development.* New York: Wiley-Interscience.

Buss, D. M. 1985. "Human mate selection," *American Scientist*, 73: 47–51.

Buss, D. M. 1995. "Psychological sex differences: origins through sexual selection," *American Psychologist*, 50(3): 164–68.

Buss, D. M. 1996. "Paternity uncertainty and the complex repertoire of human mating strategies," *American Psychologist*, 51(2): 161–62.

Buss, D. M. 1999. "Human nature and individual differences," in L. A. Pervin and O. P. John (eds.), pp. 31–56.

Buss, D. M. and Schmitt, D. P. 1993. "Sexual strategies theory: an evolutionary perspective on human mating," *Psychological Review*, 100:, 204–32.

Buss, D. M., Shackelford, T. K., Choe, J., Buunk, B. P., and Dijkstra, P. 2000. "Distress about mating rivals," *Personal Relationships*, 7(3): 235–43.

Butcher, J. N. (ed.) 1996. *International Adaptations of the MMPI-2: Research and Clinical Applications.* Minneapolis, MN: University of Minnesota Press.

Butler, R. A. 1957. "The effect of deprivation of visual incentives on visual exploration in monkeys," *Journal of Comparative and Physiological Psychology*, 50: 177–79.

Butler, R. A. and Alexander, H. M. 1955. "Daily patterns of visual exploratory behavior in the monkey," *Journal of Comparative and Physiological Psychology*, 48: 247–49.

Cacioppo, J. T., Hawkley, L. C., Rickett, E. M., and Masi, C. M. 2005. "Sociality, spirituality, and meaning making: Chicago health, aging, and social relations studies," *Review of General Psychology*, 9(2): 143–55.

Cairns, R. B. 1994. "The making of a developmental science: the contributions and intellectual heritage of James Mark Baldwin," in R. D. Parke, P. A. Ornstein, J. J. Rieser, and C. Zahn-Waxler (eds.), pp. 127–43.

Campbell, D. T. and Fiske, D. W. 1959. "Convergent and discriminant validation by the multitrait–multimethod matrix," *Psychological Bulletin*, 56: 81–105.

Campbell, J. [1968] 1991. *The Masks of God: Primitive Mythology, Vol. 1; Oriental Mythology, Vol. 2; Occidental Mythology, Vol. 3; Creative Mythology, Vol. 4.* New York: Viking.

Cannon, W. B. 1927. "The James–Lange theory of emotions: a critical examination and an alternative theory," *American Journal of Psychology*, 39: 106–24.

Cantor, N. F. 1993. *The Civilization of the Middle Ages*, rev. edn. New York: HarperCollins.

Caprara, G. V. and Cervone, D. 2000. *Personality: Determinants, Dynamics, and Potentials.* New York: Cambridge University Press.

Carlson, N. R. 1998. *Physiology of Behavior.* Boston: Allyn and Bacon.

Carmichael, C. M. and McGue, M. 1994. "A longitudinal family study of personality change and stability," *Journal of Personality*, 62: 1–20.

Carson, A. 2008. Personal communication.

Caspi, A. and Roberts, B. W. 1999. "Personality continuity and change across the life course," in L. A. Pervin and O. P. John (eds.), pp. 300–26

Caspi, A. and Silva, P. A. 1995. "Temperamental qualities at age 3 predict personality traits in young adulthood: longitudinal evidence from a birth cohort," *Child Development*, 66: 486–98.

Cassidy, S. B. and Driscoll, D. J. 2009. "Prader–Willi syndrome," *European Journal of Human Genetics*, 17: 3–13.

Cattell, R. B. 1943. "The measurement of adult intelligence," *Psychological Bulletin*, 40: 153–93.

Cattell, R. B. 1950. *Personality: A Systematic, Theoretical, and Factual Study.* New York: McGraw-Hill.

Cattell, R. B. 1952. *Factor Analysis: An Introduction and Manual for the Psychologist and Social Scientist.* New York: Harper.

Cattell, R. B. 1965. *The Scientific Analysis of Personality.* Hammondsworth: Penguin Books.

Cattell, R. B. 1971. *Abilities: Their Structure, Growth and Action.* New York: Houghton Mifflin.

Caussade, J-P. de [1751] 2001. *Abandonment to Divine Providence.* Grand Rapids, MI: Christian Classics Ethereal Library.

Champagne, F. A. and Curley, J. P. 2005. "How social experiences influence the brain," *Current Opinion in Neurobiology*, 15(6): 704–9.

Chandrasekhar, S. 1985. *Ellipsoidal Figures of Equilibrium*. New Haven, CT: Yale University Press.

Charmaz, K. 1995. "Grounded theory," in J. A. Smith, R. Harré, and L. Van Langenhove (eds.), *Rethinking Methods in Psychology*. London: Sage, pp. 27–49.

Chaucer, G. [1386] 1990. *The Canterbury Tales,* in Encyclopædia Britannica edn., *Great Books of the Western World*. University of Chicago Press, vol. 19, pp. 275–476.

Cheng, A. T. A. 2001. "Case definition and culture: are people all the same?," *British Journal of Psychiatry*, 179: 1–3. Available at: http://bjp.rcpsych.org/cgi/reprint/179/1/1.pdf, accessed July 13, 2008.

Chodoff, P., Friedman, P. B., and Hamburg, D. A. 1964. "Stress, defenses and coping behavior: observations in parents of children with malignant disease," *American Journal of Psychiatry*, 120: 743–49.

Chodorow, N. 1978. *The Reproduction of Mothering: Psychoanalysis and the Sociology of Gender*. Berkeley, CA: University of California Press.

Chomsky, N. 1957. *Syntactic Structures*. The Hague: Mouton.

Christopherson, E. R. 1989. "Injury control," *American Psychologist*, 44: 237–41.

Churchland, P. M., and Churchland. P. S. 1998. "Recent work on consciousness: philosophical, theoretical, and empirical," in *On the Contrary: Critical Essays, 1987–1997*. Cambridge, MA: MIT Press.

Churchland, P. S. 1996. "The hornswoggle problem," *Journal of Consciousness Studies*, 3: 402–8.

Cicchetti, D. and Barnett, D. 1991. "Attachment organization in maltreated preschoolers," *Development and Psychopathology*, 3: 397–411.

Clark, L. A. and Watson, D. 1999. "Temperament: a new paradigm for trait psychology," in L. A. Pervin and O. P. John (eds.), pp. 399–423.

Clark, M. S. and Isen, A. M. 1982. "Toward understanding the relationship between feeling states and social behavior," in A. H. Hastorf and A. M. Isen (eds.), *Cognitive Social Psychology*. New York: Elsevier, pp. 73–108.

Cloninger, C. R. 2004a. *Feeling Good: The Science of Well-being*. Oxford University Press.

Cloninger, C. R. 2004b. *The Biological Sciences of the West and the Psychology of the East: An Integrated Approach to Psychotherapy*. New York: Oxford University Press.

Cloninger, C. R, Adolfsson, R., and Svrakic, N. M. 1996. "Mapping genes for human personality," *Nature Genetics*, 12: 3–4.

Cloninger, C. R., Svrakic, D. M., and Przybeck, T. R. 1993. "A psychobiological model of temperament and character," *Archives of General Psychiatry*, 50: 975–90.

Cloninger, S. C. 1996. *Personality: Description, Dynamics, and Development*. New York: Freeman.

Cohen, R. J., Montague, P., Nathanson, L. S., and Swerdlik, M. E. 1988. *Psychological Testing: An Introduction to Tests and Measurement*. Mountain View, CA: Mayfield.

Cole, M. 1991. "Conclusion," in L. B. Resnick, J. M. Levine, and S. D. Teasley (eds.), pp. 398–417.

Coleman, J. C., Butcher, J. N., and Carson, C. 1984. *Abnormal Psychology and Modern Life,* 7th edn. Glenview, IL: Scott, Foresman.

Colvin, C. R. and Brown, J. D. 1994. "Do positive illusions foster mental health? An examination of the Taylor and Brown formulation," *Psychological Bulletin,* 116(1): 3–20.

Colvin, C. R., Brown, J. D., and Funder, D. C. 1995. "Overly positive self-evaluations and personality: negative implications for mental health," *Journal of Personality and Social Psychology,* 68(6): 1152–62.

Cooley, C. H. [1902] 1922. *Human Nature and the Social Order,* rev. edn. New York: Charles Scribner and Sons.

Corsini, R. J. 1997. Personal communication.

Corsini, R. J. 1999. *The Dictionary of Psychology.* Philadelphia, PA: Brunner/Mazel.

Corsini, R. J. 2000. "Introduction," in R. J. Corsini and D. Wedding (eds.), *Current Psychotherapies,* 6th edn. Itasca, IL: F. E. Peacock, pp. 1–14.

Corsini, R. J. 2002. Personal communication.

Cosmides, L. and Tooby, J. 1999. "Toward an evolutionary taxonomy of treatable conditions," *Journal of Abnormal Psychology,* 108: 453–64.

Costa, Jr., P. T. and McCrae, R. R. 1985. *The NEO Personality Inventory Manual.* Odessa, FL: Psychological Assessment Resources, p. 2.

Costa, Jr., P. T. and McCrae, R. R. 1988. "Personality in adulthood: a six-year longitudinal study of self reports and spousal ratings on the NEO Personality Inventory," *Journal of Personality and Social Psychology,* 54: 853–63.

Costa, Jr., P. T. and McCrae, R. R. 1989. "Personality continuity and the changes of adult life," in M. Storandt and G. R. VandenBos (eds.), *The Adult Years: Continuity and Change.* Washington DC: American Psychological Association, pp. 5–77.

Costa, Jr., P. T. and McCrae, R. R. 1992. *Revised NEO Personality Inventory (NEO-PI-R) and NEO Five-Factor Inventory (NEO-FFI) Professional Manual.* Odessa, FL: Psychological Assessment Resources.

Crago, M. B. 1988. "Cultural context in communicative interaction of Inuit children," Ph.D. thesis, School of Human Communication Disorders, McGill University.

Crandall, B. F. and Spence, M. A. 1974. "Linkage relations of the phenylcarbamide locus," *Human Heredity,* 24: 247–52.

Cronbach, L. J. 1980. "Validity on parole: how can we go straight?," *New Directions for Testing and Measurement,* 5: 99–108.

Cronbach, L. J. and Meehl, P. E. 1955. "Construct validity in psychological tests," *Psychological Bulletin,* 52: 281–302.

Cross, S. E. and Markus, H. R. 1999. "The cultural constitution of personality," in L. A. Pervin and O. P. John (eds.), pp. 378–96.

Crowne, D. P. and Marlowe, D. 1964. *The Approval Motive.* New York: Wiley.

Csikszentmihalyi. M. 1991. *Flow: The Psychology of Optimal Experience.* New York: HarperCollins.

Cummings, N. A. 2005. "Expanding a shrinking economic base: the right way, the wrong way, and the mental health way," in R. H. Wright and N. A. Cummings (eds.), *Destructive Trends in Mental Health: The Well-intentioned Path to Harm.* New York: Routledge, pp. 87–111.

Cunningham, K. A. and Napier, T. C. 2008. "Anne Elizabeth Kelley (1954–2007)," *American Psychologist*, 63(7): 615–16.

Cushman, P. 1992. "Psychotherapy to 1992: a historically situated interpretation," in D. K. Freedheim (ed.), *History of Psychotherapy: A Century of Change*. Washington, DC: American Psychological Association, pp. 21–64.

Cushman, P. 1995. *Constructing the Self, Constructing America*. New York: Addison-Wesley.

Damasio, A. R. 1994. *Descartes' Error: Emotion, Reason, and the Human Brain*. New York: Avon Books.

Damasio, A. R. 1999. *The Feeling of What Happens: Body and Emotion in the Making of Consciousness*. New York: Harcourt.

Damasio, A. R. 2000. "A neurobiology for consciousness," in T. Metzinger (ed.), *Neural Correlates of Consciousness: Empirical and Conceptual Questions*. Cambridge, MA: MIT Press, pp. 111–20.

Damasio, A. 2003. *Looking for Spinoza: Joy, Sorrow, and the Feeling Brain*. Toronto: Harvest/Harcourt.

Damasio, H., Grabowski, T., Frank, R., Galaburda, A. M., and Damasio, A. R. 1994. "The return of Phineas Gage: the skull of a famous patient yields clues about the brain," *Science*, 264:, 1102–5.

Danziger, K. 1997. *Naming the Mind: How Psychology found its Language*. Thousand Oaks, CA: Sage.

Darwin, C. [1872] 1965. *The Expression of the Emotions in Man and Animals*. University of Chicago Press.

Darwin, C. [1872] 1972. *The Descent of Man, Selection in Relation to Sex*. New York: D. Appleton.

Davies, M., Stankov, L., and Roberts, R. D. 1998. "Emotional intelligence: in search of an elusive construct," *Journal of Personality and Social Psychology*, 75(4): 989–1015.

Davies, P. 1992. *The Mind of God: The Scientific Basis for a Rational World*. New York: Simon & Schuster.

Dawes, R. M. 1993. "Prediction of the future versus an understanding of the past: a basic asymmetry," *American Journal of Psychology*, 106: 1–24.

Dawes, R. M. 1994. *House of Cards: Psychology and Psychotherapy Built on Myth*. New York: Free Press.

Dawkins, R. [1976] 2006. *The Selfish Gene*. Oxford University Press.

Day, H. 1971. "The measurement of specific curiosity," in H. Day, D. Berlyne, and D. Hunt (eds.), *Intrinsic Motivation: A New Direction in Education*. Toronto: Holt, Rinehart & Winston, pp. 99–112.

Day, H. 1982. "Curiosity and the interested explorer," *Performance and Instruction*, 21: 19–22.

Deary, I. J., Thorpe, G., Wilson, V., Starr, J. M., and Whalley, L. J. 2003. "Population sex differences in IQ at age 11: the Scottish mental survey 1932," *Intelligence*, 31(6): 533–42.

Deaux, K. and LaFrance, M. 1998. "Gender," in D. T. Gilbert and S. T. Fiske (eds.), *The Handbook of Social Psychology*, 4th edn. New York: McGraw-Hill, vol. 2, pp. 788–827.

Delgado, J. M. R. 1963. "Cerebral heterostimulation in a monkey colony," *Science*, 141(whole No. 3576): 161–63.
Delgado, J. M. R. 2000. "Neuronal imprinting of human values," *International Journal of Psychophysiology*, 35(2–3): 237–46.
Dellas, M. and Gaier, E. L. 1970. "Identification of creativity: the individual," *Psychological Bulletin*, 73: 55–73.
Dennett, D. 1991. *Consciousness Explained.* Boston, MA: Little, Brown.
Dewsbury, D. A. 1978. "What is (was?) the 'Fixed Action Pattern?'," *Animal Behaviour*, 26: 310–11.
Diamond, J. 1993. *The Third Chimpanzee: The Evolution and Future of the Human Animal.* New York: HarperPerennial.
Diamond, L. M. 1998. "Development of sexual orientation among adolescent and young adult women," *Developmental Psychology*, 34: 1085–95.
Diamond, L. M. 2003. "What does sexual orientation orient? A behavioral model distinguishing romantic love and sexual desire," *Psychological Review*, 110: 173–92.
Diamond, S. 1957. *Personality and Temperament.* New York: Harper & Row.
Didion, J. 2005. *The Year of Magical Thinking.* London: Fourth Estate.
Digman, J. M. 1997. "Higher-order factors of the Big Five," *Journal of Personality & Social Psychology*, 73: 1246–56.
Digman, J. M. and Takemoto-Chock, N. K. 1981. "Factors in the natural language of personality: reanalysis and comparison of six major studies," *Multivariate Behavioral Research*, 16: 149–70.
Dinesen, I. (transcribed by David Schimpf, available at: cw.mariancollege.edu/dschimpf/keylinesfrombabettesfeast.htm, accessed November 19, 2005) www.press.uchicago.edu/Misc/Chicago/243230.html#copyright.
Dobzhansky, T. 1964. *Heredity and the Nature of Man.* New York: New American Library.
Dodge, R. 1931. *Conditions and Consequences of Human Variability.* New Haven, CT: Yale Univesity Press.
Dostoevsky, F. [1864] 1993. *Notes from Underground,* trans. R. Pevear and L. Volokhonsky. New York: Vintage Classics.
Draper, T. W. 1995. "Canine analogs of human personality factors," *Journal of General Psychology*, 122: 241–52.
Dreikurs, R. 1950. *Fundamentals of Adlerian Psychology.* Chicago, IL: Alfred Adler Institute.
Drewery, W., Winslade, J., and Monk, G. 2000. "Resisting the dominating story: toward a deeper understanding of narrative therapy," in R. A. Neimeyer (ed.), pp. 243–63.
Dumont, F. 1991. "Expertise in psychotherapy: inherent liabilities of becoming experienced," *Psychotherapy: Theory/Research/Practice and Training*, 28(3): 422–28.
Dumont, F. 1993. "Inferential heuristics in clinical problem formulation: selective review of their strengths and weaknesses," *Professional Psychology: Research and Practice*, 24: 196–205.
Dumont, F. 2000. "Personality theory," in E. F. Borgatta, and R. J. V. Montgomery (eds.), *Encyclopedia of Sociology*, 2nd edn. New York: Macmillan Reference, vol. 3, pp. 2082–90.

Dumont, F. 2005. "The biological sciences of the West and the monistic psychology of the East: A modern synthesis. Critique of '*Feeling Good: The Science of Well-Being*,' by C. Robert Cloninger," *PsycCRITIQUES: Contemporary Psychology: APA Review of Books*, pp. 1–8 in e-version, January 27.

Dumont, F. and Carson, A. 1995. "Precursors of vocational psychology in ancient civilizations," *Journal of Counseling and Development*, 73(4): 371–78.

Dumont, F. and Fitzpatrick, M. 2001. "The real relationship: schemas, stereotypes, and personal history," *Psychotherapy: Theory/Research/Practice and Training*, 38: 12–20.

Dumont, F. and Smith, D. 1996. "Projectives and their infirm research base," *Professional Psychology: Research and Practice*, 27: 419–21.

Dunn, J. and Plomin, R. 1990. *Separate Lives: Why Siblings are so Different.* New York: Basic Books.

Dupré, J. 2000. "The fight for science and reason," *The Sciences*, 40(2): 40–45.

Durham, W. H. C. 1991. *Co-evolution: Genes, Culture and Human Diversity.* Stanford University Press.

Dwairy, M. and Van Sickle, T. D. 1996. "Western psychotherapy in traditional Arabic societies," *Clinical Psychology Review*, 16(3): 231–49.

Eagly, A. H. 1987. *Sex Differences in Social Behavior: A Social Role Interpretation.* Hillsdale, NJ: Erlbaum.

Eagly, A. H. 1995. "The science and politics of comparing women and men," *American Psychologist*, 50: 145–58.

Eagly, A. H. 1996. "Differences between women and men: their magnitude, practical importance, and political meaning," *American Psychologist*, 51(2): 158–59.

Eagly, A. H. 1997. "Sex differences in social behavior: comparing social role theory and evolutionary psychology," *American Psychologist*, 52(12): 1380–83.

Eagly, A. H. and Wood, W. 1999. "The origins of sex differences in human behavior: evolved dispositions versus social roles," *American Psychologist*, 54(6): 408–23.

Economist, The 2008. "Your call is important to us," p. 14, March 8.

Edelman, G. M. 1992. *Bright Air, Brilliant Fire.* New York: Basic Books.

Edelman, G. M. and Tononi, G. 2000. *A Universe of Consciousness: How Matter Becomes Imagination.* New York: Basic Books.

Edwards, A. L. 1990. *Edwards Personality Inventory: R*, rev. edn. Seattle, WA: Author.

Egeland, B. R., Carlson, E., and Sroufe, L. A. 1993. "Resilience as process," *Development & Psychopathology*, 5(4): 517–28.

Ehrlich, E. 2000. *You've got Ketchup on your Muumuu.* New York: Henry Holt.

Eidelson, R. J. and Eidelson, J. I. 2003. "Dangerous ideas: five beliefs that propel groups toward conflict," *American Psychologist*, 58: 182–92.

Ekman, P. 1973. *Darwin and Facial Expression. A Century of Research in Review.* New York: Academic Press.

Ekman, P. 1984. "Expression and nature of emotion," in K. Scherer and P. Ekman (eds.), *Approaches to Emotion.* Hillsdale, NJ: Erlbaum, pp. 319–43.

Ekman, P. and Davidson, R. J. 1994. *The Nature of Emotion: Fundamental Questions.* New York: Oxford University Press

Ekman, P., Friesen, W. V., and Ellsworth, P. 1972. *Emotion in the Human Face.* New York: Pergamon.

Eldredge, N. and Gould, S. J. 1972. "Punctuated equilibria: an alternative to phyletic gradualism," in T. J. Schopf (ed.), *Models in Paleobiology.* San Francisco, CA: Freeman, Cooper, pp. 82–115.

Eldredge, N. and Tattersall, I. 1982. *The Myths of Human Evolution.* New York: Columbia University Press.

Elkins, D. N. 2001. "Beyond religion: toward a humanistic spirituality," in K. G. Schneider, J. T. Bugental, and J. F. Pierson (eds.), *The Handbook of Humanistic Psychology: Leading Edges in Theory, Research, & Practice.* Thousand Oaks, CA: Sage, pp. 201–12.

Ellenberger, H. F. 1958. "A clinical introduction to psychiatric phenomenology and existential analysis," in R. May, E. Angel, and H. F. Ellenberger (eds.), *Existence.* New York: Basic Books, pp. 92–194.

Ellenberger, H. F. 1970. *The Discovery of the Unconscious: The History and Evolution of Dynamic Psychiatry.* New York: Basic Books.

Elwell, F. 1996. *Verstehen: Max Weber's HomePage,* July 16, 2001, accessed September 8, 2008 http://www.faculty.rsu.edu/~felwell/Theorists/Weber/Whome.htm.

Emmerich, W. 1959. "Parental identification in young children," *Genetic Psychology Monographs, 1959,* 60: 257–308.

Emmons, R. A. and McCullough, M. E. (eds.) 1999. "Religion in the psychology of personality," *Journal of Personality,* 67(6): whole issue.

Emmons, R. A., and Paloutzian, R. F. 2003. "The psychology of religion," in S. T. Fiske, D. L. Schacter, and C. Zahn-Waxler (eds.), *Annual Review of Psychology.* Palo Alto, CA: Annual Reviews, pp. 377–402.

Erikson, E. H. 1963. *Childhood and Society.* New York: Norton.

Erikson, E. H. 1969. "Growth and crises of the healthy personality," in H-M. Chiang and A. H. Maslow (eds.), *The Healthy Personality: Readings.* New York: Van Nostrand Reinhold, pp. 30–34.

Exner, J. E. 2002. *The Rorschach: Basic Foundations and Principles of Interpretation: Vol. 1.* Hoboken, NJ: Wiley.

Eysenck, H. J. 1967. *The Biological Basis of Personality.* Springfield, IL: Charles C. Thomas.

Eysenck, H. J. 1970. *The Structure of Personality.* London: Metheun.

Eysenck, H. J. 1976. *Sex and Personality.* London: Open Books.

Eysenck, H. J. 1978. "An exercise in mega-silliness," *American Psychologist,* 33: 517.

Eysenck, H. J. 1979. *The Nature and Measurement of Intelligence.* New York: Springer-Verlag.

Eysenck, H. J. 1990a. "Biological dimensions of personality," in L. A. Pervin (ed.), *Handbook of Personality: Theory and Research.* New York: Guilford, pp. 244–76.

Eysenck, H. J. 1990b. "Genetic and environmental contributions to individual differences: the three major dimensions of personality," *Journal of Personality,* 58: 245–61.

Eysenck, H. J. 1993. "From DNA to social behaviour: conditions for a paradigm of personality research," in J. Hettema and I. J. Deary (eds.), *Foundations of Personality.* Dordrecht: Kluwer Academic, pp. 55–66.

Eysenck, H. J. 1995. *Genius: The Natural History of Creativity.* Cambridge University Press.

Eysenck, H. J. and Eysenck, M. W. 1985. *Personality and Individual Differences.* New York: Plenum Press.

Eysenck, H. J. and Eysenck, S. B. G. 1968. *Eysenck Personality Inventory.* San Diego, CA: Educational and Industrial Testing Service.

Eysenck, H. J. and Eysenck, S. B. G. 1975. *Manual of the Eysenck Personality Questionnaire.* San Diego, CA: Educational and Industrial Testing Service.

Fagen, R. M. 1981. *Animal Play Behavior.* New York: Oxford University Press.

Fagot, B. I. 1991. "Peer relations in boys and girls from two to seven," Paper presented at the biennial meeting of the Society for Research in Child Development, Seattle, WA.

Farley, F. 2000. "Hans Eysenck (1916–1997)," *American Psychologist,* 55(6): 674–75.

Fausto-Sterling, A. 2000. "The five sexes, revisited," *The Sciences.* The New York Academy of Sciences, 40(4): July/August, 18–23.

Featherman, D. L. 1983. "Life-span perspectives in social science research," in P. B. Baltes and O. G. Brim, Jr. (eds.), *Life-span Development and Behavior.* New York: Academic Press, vol. 5, pp. 1–57.

Feingold, A. 1992. "Sex differences in variability in intellectual abilities: a new look at an old controversy," *Review of Educational Research,* 62: 61–84.

Feingold, A. 1994. "The additive effects of group differences in central tendency and variability are important in distributional comparisons between groups," *American Psychologist,* 50: 5–13.

Feltham, C. 2007. *What's Wrong with Us? The Anthropathology Thesis.* Chichester: Wiley.

Ferreira, A. J. 1965. "Emotional factors in prenatal environment: a review," *Journal of Nervous and Mental Disorders,* 141: 108–18.

Festinger, L. 1957. *A Theory of Cognitive Dissonance.* Evanston, IL: Row, Peterson.

Feynman, R. P. 1974. *Cargo Cult Science: Caltech Commencement Address,* available at: http://www.softpanorama.org/Skeptics/IT_skeptic/cargo_cult_programming.shtml.

Fischer, G. H. and Molenaar, I. W. 1995. *Rasch Models: Foundations, Recent Developments, and Applications.* New York: Springer.

Fiske, D. W. 1949. "Consistency of the factorial structures of personality ratings from different sources," *Journal of Abnormal and Social Psychology,* 44: 329–44.

Fiske, S. T. 1993. "Controlling other people: the impact of stereotyping on power," *American Psychologist,* 48: 621–28.

Fitch, W. T. 2005. "The evolution of music in comparative perspective," *Annals of the New York Academy of Sciences,* 1060(1): 29–49.

Forgas, J. P. 1990. "Affective influences on individual and group judgments," *European Journal of Social Psychology,* 20: 441–53.

Foucault, M. 1980. *Power/Knowledge: Selected Interviews and other Writings.* New York: Pantheon Books.

Foucault, M. [1965] 1988a. *Madness and Civilization: A History of Insanity in the Age of Reason.* New York: Vintage Books.

Foucault, M. 1988b. *The Final Foucault.* Cambridge, MA: MIT Press.

Frankl, V. E. 1959. *Man's Search for Meaning: An Introduction to Logotherapy.* New York: Washington Square Press.

Frankl, V. E. 1965. *The Doctor and the Soul.* New York: Alfred A. Knopf.

Frankl, V. E. 1970. "Forerunner of existential psychiatry," *Journal of Individual Psychology*, 26: 10–16.

Franklin, M. B. 2001. "The artist speaks: Sigmund Koch on aesthetics and creative works," *American Psychologist*, 56: 445–52.

Frederiksen, N. 1965. "Response set scores as predictors of performance," *Personnel Psychology*, 18: 225–44.

Freud, S. [1932] 1952. *New Introductory Lectures on Psycho-Analysis,* trans. W. J. H. Sprott, in Encyclopædia Britannica edn., *Great Books of the Western World.* University of Chicago Press, vol. 54.

Freud, S. [1905] 1953a. *Three Essays on Sexuality. The Standard Edition of the Complete Psychological Works of Sigmund Freud: Vol. VII,* trans. and ed. J. Strachey. London: Hogarth.

Freud, S. [1905] 1953b. *The Case of Little Hans. The Standard Edition of the Complete Psychological Works of Sigmund Freud: Vol. VII,* trans. and ed. J. Strachey. London: Hogarth.

Freud, S. 1955. *The Standard Edition of the Complete Psychological Works of Sigmund Freud: Vol. II,* trans. and ed. J. Strachey. London: Hogarth.

Freud, S. [1925] 1959. *An Autobiographical Study. The Standard Edition of the Complete Psychological Works of Sigmund Freud: Vol. XX,* trans. and ed. J. Strachey. London: Hogarth Press, pp. 7–74.

Freud, S. [1930] 1961. *Civilization and its Discontents. The Standard Edition of the Complete Psychological Works of Sigmund Freud: Vol. XXI.* London: Hogarth, pp. 59–145.

Freud, S. [1931] 1962. *Female Sexuality. The Standard Edition of the Complete Psychological Works of Sigmund Freud: Vol. XXI,* trans. and ed. J. Strachey. London: Hogarth.

Freud, S. [1928] 1964. *The Future of an Illusion.* Garden City, NY: Doubleday.

Frodi, A., Macaulay, J., and Thorne, P. R. 1977. "Are women always less aggressive than men? A review of the experimental literature," *Psychological Bulletin*, 84(4): 634–60.

Fromm, E. 1947. *Man for Himself.* New York: Rinehart.

Frye, N. 1992. *Power of Words: Being a Second Study of the Bible and Literature.* Toronto: Penguin Books.

Gallup 2007. Questions and answers about Americans' religion, available at: http://www.gallup.com/poll/103459/Questions-Answers-About-Americans-Religion.aspx.

Gallup, G. G. 1970. "Chimpanzees: self-recognition," *Science*, 167: 86–87.

Gallup, G. G. 1977. "Self-recognition in primates: a comparative approach to the bidirectional properties of consciousness," *American Psychologist*, 32: 329–38.

Gantt, W. H. 1966. "Reflexology, schizokinesis, and autokinesis," *Conditional Reflex*, 1: 57–68.

Garb, H. N. 1998. "Recommendations for training in the use of the thematic apperception test (TAT)," *Professional Psychology: Research and Practice*, 29(6): 621–22.

Gardner, H. 1993. *Multiple Intelligences: The Theory in Practice.* New York: Basic Books.

Gardner, M. 2000. "Introduction," in *The Annotated Alice,* L. Carroll, *Alice's Adventures in Wonderland and Through the Looking Glass.* New York: Norton.

Garmezy, N. 1983. "Stressors of childhood," in N. Garmezy and M. Rutter (eds.), *Stress, Coping and Development in Children.* New York: McGraw-Hill.

Garrison, A. 1997. "Adaptationism, mental health, and therapeutic outcome," *Psychotherapy,* 34(2): 107–14.

Gazzaniga, M. S. 1997. "What are brains for?," in R. L. Solso (ed.), *Mind and Brain Sciences in the 21st Century.* Cambridge, MA: MIT Press, pp. 157–71.

Geary, D. C. 1998. *Male, Female: The Evolution of Human Sex Differences.* Washington, DC: American Psychological Association.

Gentry, T. A., Polzine, K. M., and Wakefield, J. A. 1985. "Human genetic markers associated with variation in intellectual abilities and personality," *Personality & Individual Differences,* 16(1): 111–13.

Gesell, A. 1954. "The ontogenesis of infant behavior," in L. Carmichael (ed.), *Manual of Child Psychology,* 2nd edn. New York: Wiley, pp. 335–73.

Gieser, L. and Stein, M. I. (eds.) 1999. *Evocative Images: The Thematic Apperception Test and the Art of Projection.* Washington, DC: American Psychological Association.

Gigerenzer, G. 1991. "From tools to theories: a heuristic of discovery in cognitive psychology," *Psychological Review,* 98(2): 254–67.

Gilligan, C. 1982. *In a Different Voice.* Cambridge, MA: Harvard University Press.

Gilligan, S. G. and Bower, G. H. 1984. "Cognitive consequences of emotional arousal," in C. Izard, J. Kagan, and R. Zajonc (eds.), *Emotions, Cognition, and Behavior.* New York: Cambridge University Press, pp. 547–88.

Glass, G. V, McGraw, B., and Smith, M. L. 1981. *Meta-analysis in Social Research.* Beverly Hills, CA: Sage.

Gleitman, H. 1996. *Basic Psychology,* 4th edn. New York: Norton.

Glenn, C. G., Driscoll, D. J., Thomas, P. Y., and Nicholls, R. D. 1997. "Genomic imprinting: potential function and mechanisms revealed by the Prader–Willi and Angelman syndromes," *Molecular Human Reproduction,* 3: 321–32.

Glover, D. and Gatos, A. 2001. "Chimp researcher is the Martha Stewart of the jungle," *The Saturday Post.* Toronto, November 3.

Goethe, J. W. von [1832] 1952. *Faust: Parts One and Two.* Chicago, IL: Encyclopædia Britannica, Inc.

Golby, A. J., Gabrieli, J. D. E., Chiao, J. Y., and Eberhardt, J. L. 2001. "Differential responses in the fusiform region to same-race and other-race faces," *Nature Neuroscience,* 4: 845–50.

Goldberg, L. R. 1990. "An alternative 'description of personality': the structure of phenotypic personality traits. The Big-Five factor structure," *Journal of Personality and Social Psychology,* 59: 1216–29.

Goldberg, L. R. 1993. "The structure of phenotypic personality traits," *American Psychologist,* 48: 26–34.

Goldschmidt, R. 1933. "Some aspects of evolution," *Science,* 78: 539–47.

Goldsmith, H. H., Buss, K. A., and Lemery, K. S. 1997. "Toddler and childhood temperament: expanded content, stronger genetic evidence, new evidence for the importance of environment," *Developmental Psychology,* 33: 891–905.

Goldstein, K. 1939. *The Organism: A Holistic Approach to Biology Derived from Pathological Data in Man.* New York: American Book Company.

Goldstone A. P. 2004. "Prader–Willi syndrome: advances in genetics, pathophysiology, and treatment," *Trends in Endocrinology and Metabolism*, 15(1): 12–20.

Goleman, D. 1995. *Emotional Intelligence: Why it can Matter more than IQ.* New York: Bantam.

Goodall, J. 1986. *The Chimpanzees of Gombe.* Cambridge, MA: Harvard University Press.

Gould, S. J. 1977a. *Ever since Darwin.* New York: Norton.

Gould, S. J. 1977b. *Ontogeny and Phylogeny.* Cambridge, MA: Harvard University Press.

Gould, S. J. 1981. *The Mismeasure of Man.* New York: Norton.

Gould, S. J. and Eldredge, N. 1977. "Punctuated equilibria: the tempo and mode of evolution reconsidered," *Paleobiology*, 3: 115–51.

Grawe, K. 2007. *Neuropsychotherapy: How the Neurosciences Inform Effective Psychotherapy.* Mahwah, NJ: Erlbaum.

Gray, J. A. 1982. *The Neuropsychology of Anxiety: An Enquiry into the Functions of the Septal-hippocampal System.* Oxford University Press.

Greenberg, L. 1993. "Emotion and change processes in psychotherapy," in M. Lewis and J. Haviland-Jones (eds.), *Handbook of Emotions.* New York: Guilford, pp. 499–508.

Greenberg, L. S. and Safran, J. D. 1987. *Emotion in Psychotherapy: Affect, Cognition, and the Process of Change.* New York: Guilford Press.

Greenberg, L. and Safran, J. 1989. "Emotion in psychotherapy," *American Psychologist*, 44: 19–29.

Greenhalgh, T. and Hurwitz, B. 1998. *Narrative-based Medicine.* London: BMJ Books.

Gross, J. J. 1999. "Emotion and emotion regulation," in L. A. Pervin and O. P. John (eds.), pp. 525–52.

Guilford, J. P. 1956. "The structure of intellect," *Psychological Bulletin*, 53: 267–93.

Guilford, J. P. 1959. "Three faces of intellect," *American Psychologist*, 14: 469–79.

Güntürkün, O. 2006. "Letters on nature and nurture," in P. B. Baltes, P. A. Reuter-Lorenz, and F. Rösler (eds.), pp. 379–97.

Gupta, U. and Singh, P. 1982. "Exploratory studies in love and liking and types of marriage," *Indian Journal of Applied Psychology*, 19(2): 92–97.

Gustafsson, J. E. and Undheim, J. O. 1996. "Individual differences in cognitive functions," in D. C. Berliner and R. C. Calfee (eds.), *Handbook of Educational Psychology.* New York: Macmillan Library Reference, pp. 186–242.

Gutmann, D. L. 1987. *Reclaimed Powers: Toward a New Psychology of Men and Women in Later Life.* New York: Basic Books.

Hall, C. S., Lindzey, G., and Campbell, J. B. 1998. *Theories of Personality.* New York: Wiley.

Hall, E. T. 1966. *The Hidden Dimension.* Garden City, NY: Doubleday.

Hall, E. T. 1969. *The Silent Language.* Greenwich, CT: Fawcett.

Hall, G. S. 1922. *Senescence: The Last Half of Life.* New York: Appleton.

Halpern, D. F. 1994. "Stereotypes, science, censorship, and the study of sex differences," (Special issue). *Feminism and Psychology*, 4: 523–30.
Halpern, D. F. 1997. "Sex differences in intelligence: implications for education," *American Psychologist*, 52(10): 1091–102.
Hamilton, E. 1957. *The Echo of Greece*. New York: Norton.
Hamilton, E. [1958] 1993. *The Greek Way*. New York: Norton.
Hare-Mustin, R. T. and Marecek, J. 1988. "The meaning of difference: gender theory, postmodernism, and psychology," *American Psychologist*, 43(6): 455–64.
Hare-Mustin, R. T. and Marecek, J. 1989. "Thinking about postmodernism and gender theory," *American Psychologist*, 44(10): 1333–34.
Harlow, H. F. 1950. "Learning and satiation of response in intrinsically motivated complex puzzle performance in monkeys," *Journal of Comparative and Physiological Psychology*, 43: 289–94.
Harlow, H. F. 1958. "The nature of love," *American Psychologist*, 13: 673–85.
Harlow, H. F. 1959. "Love in infant monkeys," *Scientific American*, 200(6): 68–74.
Harlow, H. F. and Harlow, M. K. 1962. "Social deprivation in monkeys," *Scientific American*, 207: 136–46.
Harlow, H. F. and Harlow, M. K. 1966. "Learning to love," *American Scientist*, 54: 244–72.
Harlow, H. F. and Harlow, M. K. 1969. "Effects of various mother–infant relationships on Rhesus monkey behaviors," in B. M. Foss (ed.), *Determinants of Infant Behaviour*. London: Methuen, vol. 4, pp. 15–36.
Harlow, H. F., Harlow, M. K., and Meyer, D. I. 1950. "Learning motivated by a manipulation drive," *Journal of Experimental Psychology*, 40: 228–34.
Harré, R. 1984. *Personal Being: A Theory for Individual Psychology*. Cambridge, MA: Harvard University Press.
Hart, D. and Karmel, M. P. 1996. "Self-awareness and self-knowledge in humans, apes, and monkeys," in A. E. Russon, K. A. Bard, and S. T. Parker (eds.), *Reaching into Thought: The Minds of the Great Apes*. Cambridge University Press, pp. 325–47.
Hartley, R. E. 1959. "Sex-role pressures and socialization of the male child," *Psychological Reports*, 5: 457–68.
Hartley, R. E. 1960. "Children's concepts of male and female roles," *Merrill-Palmer Quarterly*, 6: 83–91.
Hartshorne, H. and May, M. A. 1928. *Studies in the Nature of Character: Vol. 1. Studies in Deceit*. New York: Macmillan.
Hartshorne, H. and May, M. A. 1929. *Studies in the Nature of Character: Vol. 2. Studies in Service and Self-control*. New York: Macmillan.
Hatano, G. and Inagaki, K. 1991. "Sharing cognition through collective comprehension activity," in L. B. Resnick, J. M. Levine, and S. D. Teasley (eds.), pp. 331–48.
Hatfield, E., Caccioppo, J. T., and Rapson, R. L. 1994. *Emotional Contagion*. New York: Cambridge University Press.
Havens, L. L. 1993. "Twentieth century psychiatry: a view from the sea," *American Journal of Psychiatry*, 138(10): 1279–87.
Havighurst, R. A. 1973. "History of developmental psychology: socialization and personality development through the life span," in P. B. Baltes and

K. W. Schaie (eds.), *Life-span Developmental Psychology: Personality and Socialization*. New York: Academic Press, pp. 3–24.

Heatherton, T. F. and Nichols, P. A. 1994. "Conceptual issues in assessing whether personality can change," in T. F. Heatherton and J. L. Weinberger (eds.), pp. 3–18.

Heatherton, T. F. and Weinberger, J. L (eds.) 1994. *Can Personality Change?* Washington, DC: American Psychological Association.

Hebb, D. O. 1949. *The Organization of Behavior: A Neuropsychological Theory*. New York: Wiley.

Heidbreder, E. 1933. *Seven Psychologies*. New York: Appleton-Century-Crofts.

Helgesen, S. 1990. *The Female Advantage: Women's Ways of Leadership*. New York: Doubleday.

Helson, R. 1990. "Creativity in women: outer and inner views over time," in M. A. Runco and R. S. Albert (eds.), *Theories of Creativity*. Newbury Park, CA: Sage, pp. 46–58.

Helson, R., Mitchell, V., and Moane, G. 1984. "Personality and patterns of adherence and non-adherence to the social clock," *Journal of Personality and Social Psychology*, 46: 1079–96.

Helson, R. and Moane, G. 1987. "Personality change in women from college to mid-life," *Journal of Personality and Social Psychology*, 53: 176–86.

Helson, R. and Stewart, A. 1994. "Personality change in adulthood," in T. F. Heatherton and J. L. Weinberger (eds.), pp. 201–25.

Henson, R. K. 2006. "Effect–size measures and meta-analytic thinking in counseling psychology research," *The Counseling Psychologist*, 34(5): 601–29.

Hess, E. H. 1959. "Imprinting," *Science*, 130: 133–41.

Hess, E. H. 1965. "Attitude and pupil size," *Scientific American*, 212(4): 46–65.

Hilgard, E. R. 1987. *Psychology in America: A Historical Survey*. San Diego, CA: Harcourt Brace Jovanovitch.

Hill, P. C. and Hood, Jr., R. W., 1999. *Measures of Religiosity*. Birmingham, AL: Religious Education Press.

Hill, P. C. and Pargament, K. I. 2003. "Advances in the conceptualization and measurement of religion and spirituality: implications for physical and mental health research," *American Psychologist*, 28: 64–74.

Hobbes, T. [1651] 1990. *Leviathan*, in Encyclopædia Britannica edn., *Great Books of the Western World*. University of Chicago Press, vol. 21, pp. 41–283.

Hoff, P. 1998. "Emil Kraepelin and forensic psychiatry," *International Journal of Law and Psychiatry*, 21: 343–53.

Hoffman, L. W. 1991. "The influence of the family environment on personality: accounting for sibling differences," *Psychological Bulletin*, 11(2): 187–203.

Hoffman, L. W., Paris, S., Hall, E., and Schell, R. 1988. *Developmental Psychology Today*. New York: Random House.

Hoffman, M. L. 1981. "Is altruism part of human nature?," *Journal of Personality and Social Psychology*, 40, 121–37.

Hollands, J. 2001. *Same Game, Different Rules: How to get Ahead without Being a Bully Broad, Ice Queen or Ms. Understood*. New York: McGraw-Hill.

Horenczyk, G. 1996. "The actualization balance of ethnic identity," *Journal of Personality and Social Psychology*, 70(4): 836–43.

Horney, K. 1937. *The Neurotic Personality of Our Time*. New York: Norton.
Horney, K. 1939. *New Ways in Psychoanalysis*. New York: Norton.
Horney, K. 1950. *Neurosis and Human Growth*. New York: Norton.
Horney, K. 1967. *Feminine Psychology*. New York: Norton
Huizinga, J. [1938] 1971. *Homo Ludens: A Study in the Play-elements in Culture*, trans. R. F. C. Hull. Boston, MA: Beacon Press.
Hyde, J. S. and Plant, E. A. 1995. "Magnitude of psychological gender differences: another side to the story," *American Psychologist*, 50: 159–61.
Hyman, R. B. 1988. "Four stages of adulthood: an exploratory study of growth patterns of inner-direction and time-competence in women," *Journal of Research in Personality*, 22: 117–27.
Illich, I. 1976. *Limits to Medicine*. New York: Penguin.
Isen, A. M. 1984. "Toward understanding the role of affect in cognition," in R. S. Wyer and T. K. Srull (eds.), *Handbook of Social Cognition*. Hillsdale, NJ: Erlbaum, vol. 3, pp. 179–236.
Iwaniuk, A. N., Nelson, J. E., and Pellis, S. M. 2001. "Do big-brained animals play more? Comparative analyses of play and relative brain size in mammals," *Journal of Comparative Psychology*, 115: 29–41.
Izard, C. E. 1992. "Basic emotions, relations among emotions, and emotion–cognition relations," *Psychological Review*, 99: 561–65.
Izard, C. E. 1994. "Innate and universal facial expressions: evidence from developmental and cross-cultural research," *Psychological Bulletin*, 115: 288–99.
Izard, C. E. 2001. "Emotional intelligence or adaptive emotions," *Emotion*, 1: 249–57.
Jacklin, C. N. and Maccoby, E. E. 1978. "Social behavior at 33 months in same-sex and mixed-sex dyads," *Child Development*, 49: 557–69.
James, W. [1892] 1963a. *Psychology*. Greenwich, CT: Fawcett.
James, W. [1902] 1963b. *The Varieties of Religious Experience: A Study in Human Nature*. New York: University Books.
James, W. [1890] 1990. *The Principles of Psychology*, in Encyclopædia Britannica edn., *Great Books of the Western World*. Chicago University Press.
Jaynes, J. 1990. *The Origin of Consciousness in the Breakdown of the Bicameral Mind*. Boston: Houghton Mifflin.
Jensen, A. R. [1987] 2003. "Individual differences in the Hick paradigm," in P. Vernon (ed.), *Speed of Information Processing and Intelligence*. Norwood, NJ: Ablex, pp. 101–75.
Jilek, W. G. 1995. "Emil Kraepelin and comparative sociocultural psychiatry," *European Archives of Psychiatry & Clinical Neuroscience*, 245(4–5): 231–38.
John, O. P., Robins, R. W., and Pervin, L. A. 2008. *Handbook of Personality: Theory and Research*, 3rd edn. New York: Guilford.
John, O. P. and Srivasta, S. 1999. "The Big Five trait taxonomy: history, measurement, and theoretical perspectives," in L. A. Pervin and O. P. John (eds.), pp. 102–38.
Johnson, A. J. and Dunn, T. 2008. "Walking in the valley of the shadow of death," *PsycCRITIQUES*, 53(36): article 9.
Jones, E. 1955. *The Life and Work of Sigmund Freud*. New York. Basic Books, vol. 2.
Jones, M. 1969. *Social Psychiatry in Practice: The Idea of the Psychiatric Community*. Harmondsworth: Penguin Books.

Jones, M. 1984. "From therapeutic communities to social systems to futurism," *American Journal of Social Psychiatry*, 4(2): 5–9.

Jones, S. 1999. *Almost like a Whale.* London: Black Swan.

Jones, S. 2001. *Almost like a Whale: The Origin of Species Updated.* London: Black Swan.

Jourard, S. 1971. *The Transparent Self*, rev. edn. New York: Van Nostrand Reinhold.

Jung, C. G. 1933. *Modern Man in Search of a Soul.* New York: Harcourt.

Jung, C. G. 1957. *The Undiscovered Self.* New York: Mentor Books.

Jung, C. G. [1921] 1971. *Psychological Types. Collected Works.* Princeton University Press, vol. 6.

Jung, C. G. 2003. Available at: www.quotationspage.com/subjects/creativity.

Kagan, J. 1994. *Galen's Dream.* New York: Basic Books.

Kagan, J. and Moss, H. A. 1962. *Birth to Maturity: A Study in Psychological Development.* New York: Wiley.

Kagan, J. and Reznick, J. S. 1986. "Shyness and temperament," in W. H. Jones, J. M. Cheek, and S. R. Briggs (eds.), *Shyness: Perspectives on Research and Treatment.* New York: Plenum, pp. 81–90.

Kahill, S. 1984. "Human figure drawing in adults: an update of the empirical evidence. 1967–1982," *Canadian Psychology*, 25: 269–92.

Kallmann, F. J. 1946. "The genetic theory of schizophrenia: an analysis of 691 schizophrenic twin index families," *American Journal of Psychiatry*, 103: 309–22.

Kandel, E. R. 1996. "Zellülare Grundlagen von Lernen und Gedächtnis (Cellular mechanisms of learning and memory)," in E. R. Kandel, J. H. Schwarz, and T. M. Jessell (eds.), *Neurowissenschaften*, Heidelberg: Spectrum Akademischer Verlag.

Kandel, E. 2007. *In Search of Memory: The Emergence of a New Science of Mind.* New York: Norton.

Kaplan, R. M. and Saccuzzo, D. P. 1997. *Psychological Testing: Principles, Applications, and Issues*, 5th edn. Belmont, CA: Wadsworth/Thomson.

Kaplan, R. M. and Saccuzzo, D. P. 2005. *Psychological Testing: Principles, Applications, and Issues*, 6th edn. Belmont, CA: Wadsworth/Thomson.

Katz, H. E., Russ, S. W., and Overholser, J. C. 1993. "Sex differences, sex roles, and projection on the TAT: matching stimulus to examinee gender," *Journal of Personality Assessment*, 60(1): 186–91.

Kazdin, A. E. 2008. "Evidence-based treatment and practice: new opportunities to bridge clinical research and practice, enhance the knowledge base, and improve patient care," *American Psychologist*, 63(3): 146–59.

Kelley, J. 1978. "Sexual permissiveness: evidence for a theory," *Journal of Marriage and the Family*, 45: 455–68.

Kelloniemi, H., Ek, E., and Laitinen, J. 2005. "Optimism, dietary habits, body mass index and smoking among young Finnish adults," *Appetite*, 45: 169–76.

Keltner, D., Capps, L., Kring, A., Young, R. C., and Heery, E. A. 2001. "Just teasing: a conceptual analysis and empirical review," *Psychological Bulletin*, 127: 229–48.

Kemperman, R. F. J. (affiliated with Kema, I. P., Knegtering, H., Bischoff, R., and Muskiet, F. A. J.) 2007. "Nutrition and biomarkers in psychiatry," Doctoral thesis, University of Groningen. Available at: www.google.com/search?q=Kemperman%2C+Kema%2C+Knegtering%2C+Bischoff%2C+Muskiet+&btnG=Search&meta, accessed September 7, 2009.

Kihlstrom, J. F. 1994. "Hypnosis, delayed recall, and the principles of memory," *The International Journal of Clinical and Experimental Hypnosis*, 42: 337–45.

Kimmel, H. L. and Lester, D. 1987. "Personalities of those who can taste phenylthiocarbamide," *Psychological Reports*, 61: 586.

Kimura, D. 1999. *Sex and Cognition*. Cambridge, MA: MIT Press.

King, D. B. and Wertheimer, Michael 2005. *Max Wertheimer & Gestalt Theory*. New Brunswick, NJ: Transaction Publishers.

Kinsey, A., Pomeroy, W., Martin, C., and Gebhard, P. 1953. *Sexual Behavior in the Human Female*. Philadelphia, PA: Saunders.

Kinsey, A. C., Pomeroy, W. B., and Martin, C. E. [1948] 1998. *Sexual Behavior in the Human Male*. Bloomfield, IN: Indiana University Press.

Klaus, M., Kennell, J., and Klaus, P. 1996. *Bonding: Building the Foundation of Secure Attachment and Independence*. Reading, MA: Addison-Wesley.

Klein, D. B. 1970. *A History of Scientific Psychology: Its Origins and Philosophical Backgrounds*. New York: Basic Books.

Kleinfeld, J. S. and Yerian, S. (eds.) 1995. *Gender Tales: Tensions in the Schools*. New York: St. Martin's Press.

Kleinman, A. 1988. *The Illness Narratives: Suffering, Healing and the Human Condition*. New York: Basic Books.

Kline, P. 1986. "Psychotherapy and Freudian psychology: the negative view," in S. Mogdil and C. Mogdil (eds.), *Hans Eysenck: Consensus and Controversy*. London: Falmer.

Kline, P. 1995. "A critical review of the measurement of personality and intelligence," in D. H. Soklofske and M. Zeidner (eds.), *International Handbook of Personality and Intelligence*. New York: Plenum Press, pp. 505–24.

Knorr-Cetina, K. D. 1981. *The Manufacture of Knowledge*. New York: Pergamon.

Koch, H. L. 1955. "Some personality correlates of sex, sibling position and sex of sibling among five- and six-year-old children," *Genetic Psychology Monographs*, 52: 3–50.

Koch H. L. 1957. "The relation in young children between characteristics of their playmates and certain attributes of their siblings," *Child Development*, 28: 175–202.

Koch, S. 1956. "Behavior as 'intrinsically' regulated: work notes toward a pre-theory of phenomena called 'motivational,'" in M. R. Jones (ed.), *Nebraska Symposium on Motivation*, 4. Lincoln, NE: University of Nebraska Press, pp. 42–87.

Koch, S. 1971. "Reflections on the state of psychology," *Social Research*, 38: 669–709.

Koch, S. 1981. "The nature and limits of psychological knowledge: lessons of a century qua 'science,'" *American Psychologist*, 36(3): 257–69.

Kohlberg, L. and Kramer, R. 1969. "Continuities and discontinuities in childhood and adult moral development," *Human Development*, 12: 93–120.

Kohlberg, L., Levine, C., and Hewer, A. 1983. *Moral Stages: A Current Formulation and a Response to Critics*. Basel: Karger.

Kohut, H. and Seitz, P. F. D. 1963. "Psychoanalytic theory of personality," in J. M. Wepman and R. W. Heine (eds.), *Concepts of Personality*. Chicago, IL: Aldine, pp. 113–41.

Kraepelin, E. 1899. *Psychiatrie: Ein Lehrbuch für Studirende und Aertze,* 6th edn. Leipzig: Barth.

Krantz, D. L. 1998. "Chances' continuing challenges," *Psychological Inquiry*, 9(2): 116–17.

Kreppner, K. 1998. "William L. Stern: a neglected founder of developmental psychology," in R. D. Parke, P. A Ornstein, J. J. Rieser, and C. Zahn-Waxler (eds.), pp. 311–31.

Kretschmer, E. 1925. *Physique and Character: An Investigation of the Nature of Constitution and of the Theory of Temperament,* trans. W. J. H. Sprott. New York: Harcourt Brace.

Krosnick, J. A. and Sedikides, C. 1990. "Self-monitoring and self-protective biases in use of consensus information to predict one's own behavior," *Journal of Personality and Social Psychology*, 58(4): 718–28.

Krueger, J. I., Vohs, K. D., and Baumeister, R. F. 2008. "Is the allure of self-esteem a mirage after all?," *American Psychologist*, 63(1): 64–65.

Kuhn, T. 1970. *The Structure of Scientific Revolutions.* University of Chicago Press.

Kuhn, T. S. 1989. "Logic of discovery or psychology of research?," in I. Lakatos and A. Musgrave (eds.), pp. 1–23.

LaFrenière, P., Strayer, F. F., and Gauthier, R. 1984. "The emergence of same-sex affiliative preferences among pre-school peers: a developmental/ethological perspective," *Child Development*, 55: 1958–65.

Lakatos, I. 1970. "Falsification and the methodology of scientific research programmes," in I. Lakatos and A. Musgrave (eds.), *Criticism and the Growth of Knowledge*. Cambridge University Press, pp. 91–196.

Lakoff, G. and Johnson, M. 1980. *Metaphors We Live By*. University of Chicago Press.

Lankton, S. R. and Lankton, C. H. 1983. *The Answer Within: A Clinical Framework of Ericksonian Hypnotherapy*. New York: Brunner/Mazel.

Lasko, J. K. 1954. "Parent behavior toward first and second children," *Genetic Psychology Monographs*, 49: 97–137.

Laurence, J.-R., Day, D., and Gaston, L. 1998. "From memories of abuse to the abuse of memories," in S. J. Lynn and K. M. McConkey (eds.), *Truth in Memory*. New York: Guilford, pp. 323–46.

Lazarus, R. S. 1991. *Emotion and Adaptation*. Oxford University Press.

Le Scolan, N., Hausberger, M., and Wolff, A. 1997. "Stability over situations in temperamental traits of horses as revealed by experimental and scoring approaches," *Behavioural Processes*, 41: 257–66.

Leahey, T. H. 1987. *A History of Psychology: Main Currents in Psychological Thought*. Englewood Cliffs, NJ: Prentice-Hall.

Leahey, T. H. 2000. *A History of Psychology: Main Currents in Psychological Thought,* 5th edn. Upper Saddle River, NJ: Prentice Hall.

Leaper, C. and Ayres, M. M. 2007. "A meta-analytic review of gender variations in adults' language use: talkativeness, affiliative speech, and assertive speech," *Personality and Social Psychology Review*, 11(4): 328–63.

Leary, D. E. 2001. "One big idea, one ultimate concern: Sigmund Koch's critique of psychology and hope for the future," *American Psychologist*, 56: 425–32.

LeDoux, J. 1996. *The Emotional Brain: The Emotional Underpinnings of Emotional Life.* New York: Simon & Schuster.

LeDoux, J. 2003. *Synaptic Self: How our Brains Become Who We Are.* New York: Penguin Books.

Lefkowitz, M. R. and Fant, M. B. 1982. *Women's Life in Greece and Rome.* Baltimore, MD: Johns Hopkins University Press.

Lenneberg, E. H. 1967. *Biological Foundations of Language.* New York: Wiley.

Levine, R. A. and Campbell, D. T. 1972. *Ethnocentrism: Theories of Conflict, Ethnic Attitudes, and Group Behavior.* New York: Wiley.

Levinson, D. J. 1978. *The Seasons of a Man's Life.* New York: Knopf.

Levinson, D. J. 1986. "A conception of adult development," *American Psychologist*, 41(1): 3–13.

Levitt, N. 1999. *Prometheus Bedeviled: Science and the Contradictions of Contemporary Culture.* Piscataway, NJ: Rutgers University Press.

Lewin, R. 1988. "A lopsided look at evolution," *Science*, 241: 291–93.

Lewis, M. and Brooks-Gunn, J. 1979. *Social Cognition and the Acquisition of Self.* New York: Plenum Press.

Lieberman, M. A., Yalom, I. D., and Miles, M. B. 1973. "Hazards of encounter groups," *Encounter Groups: First Facts.* New York: Basic Books, pp. 167–210.

Lilienfeld, S. O. and Marino, L. 1995. "Essentialism revisited: evolutionary theory and the concept of mental disorder," *Journal of Abnormal Psychology*, 104(33): 421–29.

Lilienfeld, S. O., Wood, J. M., and Garb, H. N. 2000. "The scientific status of projective techniques," *Psychological Science in the Public Interest*, 1(2): 27–66.

Lilla, M. 2008. *The Stillborn God: Religion, Politics, and the Modern West.* New York: Alfred A. Knopf.

Linnenbrink, E. A. and Pintrich, P. R. 2002. "Achievement goal theory and affect: an asymmetrical, bidirectional model," *Educational Psychologist*, 37(2): 69–78.

Lippert, E. 2008. *Cargo Cult Science*, available at: www.softpanorama.org/Skeptics/IT_skeptic/cargo_cult_programming.shtml, accessed May 9, 2008.

Lipsey, M. W. and Wilson, D. B. 1993. "The efficacy of psychological, educational, and behavioral treatment: confirmation from meta-analysis," *American Psychologist*, 48: 1181–209.

Lipsitt, L. P. 2002a. "Behavior develops – minds don't snap and nerves don't break," *Brown University News Service*, April, 2002.

Lipsitt, L. P. 2002b. Personal communication. email, May 4, 2002.

Lipton, P. 2004. "Genetic and generic determinism: a new threat to free will," in D. Rees and S. Rose (eds.), *The New Brain Sciences: Perils and Prospects.* Cambridge University Press, pp. 88–100.

Liu, J. H. and Liu, S. H. 1999. "Interconnectedness and Asian social psychology," in T. Sugiman, M. Karasawa, J. H. Liu, and C. Ward (eds.), *Progress in Asian Social Psychology. Theoretical and Empirical Contributions.* Seoul: Kyoyook-Kwqahak-Sa Publishing, vol. 2, pp. 9–31.

Lloyd, B. and Smith C. 1986. "The effects of age and gender on social behaviour in very young children," *British Journal of Social Psychology*, 25: 33–41.

Locke, J. [1690] 1990. "Concerning human understanding," in Encyclopædia Britannica edn., *Great Books of the Western World.* University of Chicago Press, vol. 33, pp. 85–395.

Loevinger, J. 1976. *Ego Development.* San Francisco, CA: Jossey-Bass.

Loewenstein, G. 1994. "The psychology of curiosity: a review and reinterpretation," *Psychological Bulletin,* 116: 75–98.

Loewenstein, G. and Small, D. A. 2007. "The scarecrow and the tin man: the vicissitudes of human sympathy and caring," *Review of General Psychology,* 11: 112–26.

Loftus, E. F. 1979. "The malleability of human memory," *American Scientist,* 67: 312–20.

Loftus, E. F. 1993. "The reality of repressed memories," *American Psychologist,* 48: 518–37.

Loftus, E. F. 2003. "Make-believe memories," *American Psychologist,* 58(1): 867–73.

Loftus, E. F., Donders, K., Hoffman, H. G. and Schooler, J. W. 1989. "Creating new memories that are quickly assessed and confidently held," *Memory and Cognition,* 17: 607–16.

Loftus, E. F. and Ketcham 1994. *The Myth of Repressed Memory: False Memories and Accusations of Sexual Abuse.* New York: St. Martin's Press.

Logue, A. W. 1986. *The Psychology of Eating and Drinking.* New York: Freeman.

Lorenz, K. Z. 1952. *King Solomon's Ring.* London: Methuen

Lorenz, K. Z. 1960. "Imprinting," in R. C. Birney and R. C. Teevan (eds.), *Instinct: An Enduring Problem in Psychology: Selected Readings.* Princeton, NJ: Van Nostrand, pp. 52–64.

Lorenz, K. Z. 1981. *The Foundations of Ethology.* New York: Springer-Verlag.

Lowry, R. 1971. *The Evolution of Psychological Theory: 1650 to the Present.* Chicago, IL: Aldine, Atherton.

Lyman, S. M., Scott, M. B., Smeltzer, L., Waltman, J. L., Jones, S. E., Guerrero, L. K., Andersen, P. A. Crusco, A. H., Wetzel, C. G., Sigelman, C. K., and Adams, R. M. 1999. "Contact codes: proxemics and haptics," in L. K. Guerrero and J. A. DeVito (eds.), *The Nonverbal Communication Reader: Classic and Contemporary Readings,* 2nd edn. Prospect Heights, IL: Waveland Press, pp. 175–223.

Lytton, H. 1977. "Do parents create, or respond to, differences in twins?," *Developmental Psychology,* 13: 456–59.

Maccoby, E. E. 1998. *The Two Sexes: Growing up Apart, Coming Together.* Cambridge, MA: Belknap Press of Harvard University Press.

Maccoby, E. E. and Jacklin, C. N. 1974. *The Psychology of Sex Differences.* Stanford University Press.

MacDonald, D. A. 2000. "Spirituality: description, measurement, and relation to the five-factor model of personality," *Journal of Personality,* 68(1): 153–97.

Machover, K. 1949. *Personality Projection in the Drawing of the Human Figure.* Springfield, IL: Charles C. Thomas.

Maher, B. A. and Maher, W. B. 1994. "Personality and psychopathology: a historical perspective," *Journal of Abnormal Psychology,* 103: 72–77.

Mahoney, M. 1976. *Scientist as Subject: The Psychological Imperative.* Cambridge, MA: Ballinger.

Mahoney, M. 1991. *Human Change Processes: The Scientific Foundations of Psychotherapy.* New York: Basic Books.
Mahoney, M. J. (ed.) 1995. *Cognitive and Constructive Psychotherapies: Theory, Research, and Practice.* New York: Springer.
Mahrer, A. A. 2000. "Philosophy of science and the foundations of psychotherapy," *American Psychologist,* 55: 1117–25.
Mahrer, A. R. 1989. *Experiencing: A Humanistic Theory of Psychology and Psychiatry.* University of Ottawa Press.
Mahrer, A. R. 1996. *The Complete Guide to Experiential Psychotherapy.* New York: Wiley.
Mannheim, K. 1936. *Ideology and Utopia.* New York: Harvest Books.
Manrubia, S. C., Derrida, B., and Zanette, D. H. 2003. "Genealogy in the era of genomics," *American Scientist,* 91: 158–65.
Marazziti, D., Akiskal, H. S., Rossi, A., and Cassano, G. B. 1999. "Alteration of the platelet serotonin transporter in romantic love," *Psychological Medicine,* 29(3): 741–45.
Markus, H. R. and Herzog, A. R. 1992. "The role of the self-concept in aging," in K. W. Schaie and M. P. Lawton (eds.), pp. 110–43.
Markus, H. and Kunda, Z. 1986. "Stability and malleability in the self-concept," *Journal of Personality and Social Psychology,* 51: 858–66.
Marlo, H. and Wagner, M. K. 1999. "Expression of negative and positive events through writing: implications for psychotherapy and health," *Psychology & Health,* 14(2): 193–215.
Marsella, A. J. 1998. "Toward a global-community psychology: meeting the needs of a changing world," *American Psychologist,* 53, 1282–91.
Marsella, A. J., Dubanoski, J., Hamada, W. C., and Morse, H. 2000. "The measurement of personality across cultures: historical, conceptual, and methodological issues and considerations," *American Behavioral Scientist,* 44(1): 41–62.
Marsella, A. J. and Yamada, A. M. 2000. "Culture and mental health: an overview of issues, research, and directions," in I. Cuella and F. Paniagua (eds.), *Handbook of Multicultural Assessment and Therapy.* New York: Academic Press, pp. 3–27.
Martel, Y. 2002. "Giving life to Pi," *Arts and Life: National Post,* Toronto, November 9.
Martell, R. F., Lane, D. M., and Emrich, C. 1996. "Male–female differences: a computer simulation," *American Psychologist,* 51(2): 157–58.
Martin, C. L. 1987. "A ratio measure of sex stereotyping," *Journal of Personality and Social Psychology,* 52: 489–99.
Martin, J. and Sugarman, J. 2000. "Between the modern and the postmodern: the possibility of self and progressive understanding in psychology," *American Psychologist,* 55: 397–406.
Martin, P. 1984. "The whys and wherefores of play in cats," in P. K. Smith (ed.), pp. 71–94.
Martin, R. P., Wisenbaker, J., and Huttunen, N. 1994. "Review of factor analytic studies of temperament measures based on the Thomas–Chess structural model: implications for the Big Five," in C. F. Halverson, G. A. Kohnstamm,

and R. P. Martin (eds.), *The Developing Structure of Temperament and Personality from Infancy to Adulthood.* Hillsdale, NJ: Erlbaum, pp. 157–72.

Martindale, C. 2001. "Oscillations and analogies: Thomas Young, MD, FRS, genius," *American Psychologist*, 56: 342–45.

Marx, M. H. (ed.) 1963. *Theories in Contemporary Psychology.* Toronto: Collier-Macmillan.

Mascie-Taylor, C. G., McManus, I. C., MacLarnon, A. M., and Lanigan, P. M. 1983. "The association between phenylthiocarbamide (PTC) tasting ability and psychometric variables," *Behavior Genetics*, 13(2): 191–96.

Masling, J. 1986. "Orality, pathology, and interpersonal behavior," in J. Masling (ed.), *Empirical Studies of Psychoanalytic Studies.* New York: Analytic Press, vol. 2, pp. 73–106.

Maslow, A. H. 1954. *Motivation and Personality.* New York: Harper.

Maslow, A. H. 1962. *Toward a Psychology of Being.* Princeton, NJ: Van Nostrand.

Maslow, A. H. 1968. *Toward a Psychology of Being.* Princeton, NJ: Van Nostrand.

Maslow, A. H. 1970. *Motivation and Personality.* New York: Harper.

Matlin, M. W. 1993. *The Psychology of Women*, 2nd edn. Fort Worth, TX: Harcourt, Brace.

Mayer, J. D., Salovey, P., and Caruso, D. 2000. "Competing models of emotional intelligence," in R. J. Sternberg (ed.), *Handbook of Human Intelligence,* 2nd edn. New York: Cambridge University Press, pp. 396–420.

Mayne, T. J. and Ramsey, J. 2001. "The structure of emotion: a nonlinear dynamic systems approach," in T. J. Mayne and G. A. Bonanno (eds.), *Emotion: Current Issues and Future Directions.* New York: Guilford, pp. 1–38.

McAdams, D. P. 1994. *The Person: An Introduction to Personality Psychology*, 2nd edn. New York: Harcourt Brace.

McAdams, D. P. 1996. "Personality, modernity, and the storied self: a contemporary framework for studying persons," *Psychological Inquiry*, 7: 295–321.

McAdams, D. P. 1999. "Personal narratives and the life story," in L. A. Pervin and O. P. John (eds.), pp. 478–500.

McClearn, G. E. 1962. "The inheritance of behavior," in L. Postman (ed.), *Psychology in the Making: Histories of Selected Research Problems.* New York: Knopf, pp. 144–252.

McClelland, D. C. 1958. "Methods of measuring human motivation," in J. W. Atkinson (ed.), *Motives in Fantasy, Action, and Society.* New York: Van Nostrand, pp. 12–13.

McClelland, D. C. 1961. *The Achieving Society.* New York: Van Nostrand.

McClelland, D. C. 1965. "*N* achievement and entrepreneurship: a longitudinal study," *Journal of Personality and Social Psychology*, 1: 389–92.

McClelland, D. C. 1985. *Human Motivation.* Glenview, IL: Scott, Foresman.

McClelland, D. C., Atkinson, J. W., Clark, R. A., and Lowell, E. L. 1953. *The Achievement Motive.* New York: Appleton-Century-Crofts.

McCrae, R. R. 1987. "Creativity, divergent thinking, and openness to experience," *Journal of Personality and Social Psychology*, 52: 1258–65.

McCrae, R. R. and Costa, Jr., P. T. 1990. *Personality in Adulthood.* New York: Guilford.

McCrae, R. R. and Costa, P. T. 1996. "Toward a new generation of personality theories: theoretical contexts for the five-factor model," in J. S. Wiggins (ed.), *The Five-Factor Model of Personality: Theoretical Perspectives*. New York: Guilford, pp. 51–87.

McCrae, R. R. and Costa, P. T. 1999. "A five-factor theory of personality," in L. A. Pervin and O. P. John (eds.), pp. 139–53.

McCrea, R. R., Costa, P. T., del Pilar G. H., Rolland, J.-P., and Parker, W. D. 1998. "Cross-cultural assessment of the five-factor model – the revised NEO personality inventory," *Journal of Cross-Cultural Psychology*, 29: 171–88.

McCulloch, W. S. 1965. *Embodiments of Mind*. Cambridge, MA: MIT Press.

McDougall, W. 1918. *An Introduction to Social Psychology*. Boston, MA: Luce.

McDowell, N. 1991. *From the Field Notes of Margaret Mead and Reo Fortune*. Washington, DC: Smithsonian Institution Press.

McGrew, K. and Flanagan, D. 1998. *The Intelligence Test Desk Reference: Gf–Gc Cross-battery Assessment*. Boston, MA: Allyn & Bacon.

McGue, M., Bacon, S., and Lykken, D. T. 1993. "Personality stability and change in early adulthood: a behavioral genetic analysis," *Developmental Psychology*, 29: 96–109.

Mead, G. H. 1934. *Mind, Self and Society*, C. W. Morris (ed.). University of Chicago Press.

Mead, M. [1935] 2001. *Sex and Temperament in Three Primitive Societies*. New York: HarperCollins.

Meehl, P. E. 1945. "The dynamics of 'structured' personality tests," *Journal of Clinical Psychology*, 1: 296–303.

Meehl, P. E. 1954. *Clinical versus Statistical Prediction*. Minneapolis, MN: University of Minnesota Press.

Meehl, P. E. 1960. "The cognitive activity of the clinician," *American Psychologist*, 15: 19–27.

Meehl, P. E. 1973. *Psychodiagnosis Selected Papers*. New York: Norton.

Meehl, P. E. 1978. "Theoretical risks and tabular asterisks: Sir Karl, Sir Ronald, and the slow progress of soft psychology," *Journal of Consulting and Clinical Psychology*, 46: 806–34.

Meehl, P. E. 1995. "Bootstraps taxometrics: solving the classification problem in psychopathology," *American Psychologist*, 50(4): 266–75.

Mehrabian, A. 1971. *Silent Messages*. Belmont, CA: Wadsworth.

Messick, S. 1989. "Validity," in R. L. Linn (ed.), *Educational Measurement*. Washington, DC: American Council on Education and National Council on Measurement in Education, pp. 13–103.

Miller, J., Kranzler, J., Liu, Y., Schmalfuss, I., Theriaque, D. W., Shuster, J. J., Hatfield, A., Mueller, O. T., Goldstone, A. P., Sahoo, T., Beaudet, A. L., and Driscoll, D. J. 2006. "Neurocognitive findings in Prader–Willi syndrome and early-onset morbid obesity," *Journal of Pediatrics*, 149: 192–98.

Miller, W. R. and C'deBaca, J. 1994. "Quantum change: toward a psychology of transformation," in T. F. Heatherton and J. L. Weinberger (eds.), pp. 253–80.

Miller, W. R. and Thoresen, C. E. 2003. "Spirituality, religion, and health: an emerging research field," *American Psychologist*, 58: 24–35.

Millon, T. 1994. *Millon Index of Personality Styles (MIPS) Manual.* San Antonio, TX: Psychological Corporation.
Millon, T. 1996. *Personality and Psychopathology: Building a Clinical Science.* New York: Wiley.
Millon, T. M. and Davis R. D. 1996. *Disorders of Personality: DSM-IV^{TM} and Beyond,* 2nd edn. New York: Wiley.
Minsky, M. 1986. *The Society of Mind.* New York: Simon & Schuster.
Mischel, W. 1968. *Personality and Assessment.* New York: Wiley
Mischel, W. 1973. "Toward a cognitive social learning reconceptualization of personality," *Psychological Review,* 80: 252–83.
Mischel, W. 1984. "Convergences and challenges in the search for consistency," *American Psychologist,* 39: 351–64.
Mischel, W. and Shoda, Y. 1995. "A cognitive–affective system theory of personality: reconceptualizing situations, dispositions, dynamics, and invariance in personality structure," *Psychological Review,* 102: 246–68.
Mischel, W. and Shoda, Y. 1999. "Integrating dispositions and processing dynamics within a unified theory of personality: the cognitive–affective personality system," in L. A. Pervin and O. P. John, (eds.), pp. 197–218.
Mlodinow, L. 2003. *Feynman's Rainbow: A Search for Beauty in Physics and in Life.* New York: Warner Books.
Moghaddam, F. M. 1998. *Social Psychology: Exploring Universals Across Cultures.* New York: Freeman.
Monk, G., Winslade, J., Crocket, K., and Epston, D. (eds.) 1997. *Narrative Therapy in Practice: The Archeology of Hope.* San Francisco, CA: Jossey-Bass.
Monod, J. 1971. *Chance and Necessity: An Essay on the Natural Philosophy of Modern Biology,* trans. A. Wainhouse. New York: Vintage.
Montagu, A. 1981. *Growing Young.* New York: McGraw-Hill.
Montagu, A. 1986. *Touching: The Human Significance of the Skin.* New York: Harper & Row.
Montaigne, M. de [1580] 1990. *Essays, book one,* in Encyclopædia Britannica edn., *Great Books of the Western World.* University of Chicago Press, vol. 23, pp. 47–198.
Moore, J. L. and Rocklin, T. R. 1998. "The distribution of distributed cognition: multiple interpretations and uses," *Educational Psychology Review,* 10(1): 97–113
More, T. [1516] 1989. *Utopia,* eds. and trans. George M. Logan and Robert M. Adams. Cambridge University Press.
Morgan, C. D. and Murray, H. A. 1935. "A method for investigating fantasies," *Archives of Neurology and Psychiatry,* 34: 289–304.
Morizot, J. and Le Blanc, M. 2003. "Continuity and change in personality traits from adolescence to midlife: a 25-year longitudinal study comparing representative and adjudicated men," *Journal of Personality,* 71(5): 705–55.
Mosak, H. 1995. "Adlerian psychotherapy," in R. J. Corsini and D. Wedding (eds.), *Current Psychotherapies,* 5th edn. Itasca, IL: F. E. Peacock, pp. 54–97.
Most, R. B. and Zeidner, M. 1995. "Constructing personality and intelligence instruments," in D. H. Saklofske and M. Zeidner, pp. 475–503.
Mowrer, O. H. 1976. "Therapeutic groups and communities in retrospect and prospect," in *Proceedings of the First World Conference on Therapeutic*

Communities, Norrköping, Sweden, September. Montreal: Portage Press, pp. 2–62.
Murphy, G. 1949. *Historical Introduction to Modern Psychology*. New York: Harcourt Brace.
Murphy, G. 1968. *Psychological Thought from Pythagoras to Freud: An Informal Iintroduction*. New York: Harcourt Brace and World.
Murphy, G. and Kovach, J. K. 1972. *Historical Introduction to Modern Psychology*, 2nd edn. New York: Harcourt Brace Jovanovitch.
Murphy, R. F. 1987. *The Body Silent*. New York: Henry Holt.
Murray, H. [1938] 2007. *Explorations in Personality*. Oxford University Press.
Murray, H. A. 1940. "What should psychologists do about psychoanalysis?," *Journal of Abnormal and Social Psychology*, 49: 150–75.
Murray, H. A. 1958. "Drive, time strategy, measurement, and our way of life," in G. Lindzey (ed.), *Assessment of Human Motives*. New York: Holt, Rinehart & Winston, pp. 183–96.
Myers, B. J. 1984. "Mother–infant bonding: rejoinder to Kennell and Klaus," *Developmental Review*, 4(3): 283–88.
Myers, I. B., McCaulley, M. H., Quenk, N. L., and Hammer, A. L. 1998. *MBTI Manual (A Guide to the Development and Use of the Myers Briggs Type Indicator)*, 3rd edn. Mountain View, CA: Consulting Psychologists Press
Myers, I. B. and Myers, P. B. 1995. *Gifts Differing: Understanding Personality Type*, Palo Alto, CA: Davies-Black.
Nardi, P. 1999. *Gay Men's Friendships*. University of Chicago Press.
Nash, J. 1978. *Developmental Psychology: A Psychobiological Approach*, 2nd edn. Englewood Cliffs, NJ: Prentice-Hall.
Neimeyer, R. A. and Raskin, J. D. (eds.) 2000. *Constructions of Disorder: Meaning-making Frameworks for Psychotherapy*. Washington, DC: American Psychological Association
Nelham, C. 2005. "A narrative analysis exploring the effects of long-term care-giving on the female caregiver's sense of self," Ph.D. thesis (Supervisor, F. Dumont) McGill University, Montreal.
Nelson, C. A. 2006. "Neurobehavioral development in the context of biocultural co-constructivism," in P. B. Baltes, P. A. Reuter-Lorenz, and F. Rösler (eds.), pp. 61–81.
Neugarten, B. L. 1977. "Personality and aging," in J. E. Birren and K. W. Schaie (eds.), *Handbook of the Psychology of Aging*. New York: Van Nostrand Reinhold, pp. 626–49.
Neugarten, B. L. 1979. "Time, age, and the life cycle," *American Journal of Psychiatry*, 136: 887–94.
Newcomb, T. M. 1929. *Consistency of Certain Extrovert[sic]–Introvert Behavior Patterns in 51 Problem Boys*. New York: Teachers College, Columbia University.
Newell, A., Shaw, J. C., and Simon, H. A. 1958. "Elements of a theory of human problem solving," *Psychological Review*, 65: 151–66.
Nicholas, B. and Gillett, G. 1997. "Doctors' stories, patients' stories: a narrative approach to teaching medical ethics," *Journal of Medical Ethics*, 23: 295–99.
Nisbett, R. E. and Ross, L. 1980. *Human Inference: Strategies and Shortcomings of Social Judgment*. Englewood Cliffs, NJ: Prentice-Hall.

Nisbett, R. E. and Wilson, T. D. 1978. "Telling more than we can know: verbal reports on mental processes," *Psychological Review*, 84: 231–59.

Nugent, J. 2002. "Why is English so hard?," available at: www.accentia.org/english_hard.htm, accessed on April 23, 2002.

Nyborg, H. 1994. *Sex, Body, Mind, and Society: The Physicological Approach.* New York: Praeger.

Nyborg, H. 1997. "Molecular creativity, genius and madness," in H. Nyborg (ed.), *The Scientific Study of Human Nature.* New York: Elsevier Science, pp. 422–61.

O'Hanlon, W. H. and Weiner-Davis, M. 1989. *In Search of Solutions: A New Direction in Psychotherapy.* New York: Norton.

Ochsner, K. N. and Lieberman, M. D. 2001. "The emergence of social cognitive neuroscience," *American Psychologist*, 56: 717–34.

Olsack, L., Perreault, S., and Moghaddam, F. M. 1997. "Similarity and intergroup relations," *International Journal of Intercultural Relations*, 21: 113–23.

Ondaatje, M. 1993. *The English Patient.* London: Picador.

Onstad, K. 2003. "Movies in review," *The National Post,* Toronto, pp. B1–B2, January 3.

Ortony, A. and Turner, T. J. 1990. "What's basic about basic emotions?," *Psychological Review*, 97: 315–31.

Ozer, D. J. 1999. "Four principles for personality assessment," in L. A. Pervin and O. P. John (eds.), pp. 671–88.

Paloutzian, R. F., Richardson, J. T., and Rambo, L. R. 1999. "Religious conversion and personality change," *Journal of Personality and Social Psychology*, 67: 1047–80.

Panksepp, J. 1998. *Affective Neuroscience: The Foundations of Human and Animal Emotions.* Oxford University Press.

Parke, R. D., Ornstein, P. A., Rieser, J. J., and Zahn-Waxler, C. 1994. "The past as prologue: an overview of a century of developmental psychology," in R. D. Parke, P. A. Ornstein, J. J Rieser, and C. Zahn-Waxler (eds.), *A Century of Developmental Psychology.* Washington, DC: American Psychological Association, pp. 1–75.

Parry, A. 1997. "Why we tell stories: the narrative construction of reality," *Transactional Analysis Journal*, 27: 118–27.

Parsons, T. 1955. "The American family: its relations to personality and the social structure," in T. Parsons and R. F. Bales (eds.), *Family Socialization and Interaction Process,* Glencoe, IL: Free Press, pp. 3–21.

Patterson, C. H. and Watkins, Jr., C. E. 1996. *Theories of Psychotherapy.* New York: HarperCollins.

Pavlidis, I., Eberhardt, N. L., and Levine, J. A. 2002. "Seeing through the face of deception," *Nature,* 415: 35.

Pedersen, P. B. 2000. "Multiculturalism as a generic and permanent perspective of global psychology," presented at the American Psychological Association convention, Washington, DC, August 4–8.

Pedersen, P. and Marsella, A. J. 1982. "The ethical crisis in cross-cultural counseling and psychotherapy," *Professional Psychology*, 13: 492–500.

Pellegrini, A. D. and Smith, P. K. 2005a. "Play in great apes and humans," in A. D. Pellegrini and P. K. Smith (eds.), *The Nature of Play: Great Apes and Humans.* New York: Guilford, pp. 3–12.

Pellegrini, A. D. and Smith, P. K. 2005b. "Conclusion," in A. D. Pellegrini and P. K. Smith (eds.), pp. 285–97.

Pennebaker, J. W., Kiecolt-Glaser, J. K., and Glaser, R. 1988. "Disclosure of traumas and immune function: health implications for psychotherapy," *Journal of Consulting and Clinical Psychology*, 56: 239–45.

Penrose, R. 1994. *Shadows of the Mind.* New York: Oxford University Press.

Perls, F. S. 1969. *Gestalt Therapy Verbatim.* Moab, UT: Real People Press.

Perls, F. S., Hefferline, R. F., and Goodman, P. 1951. *Gestalt Therapy.* New York: Julian Press.

Pervin, L. A. (ed.) 1990. "Biological dimensions of personality," in *Handbook of Personality: Theory and Research.* New York: Guilford, pp. 244–76.

Pervin, L. A. 1996. *The Science of Personality.* New York: Wiley.

Pervin, L. A. 1999. "Epilogue: constancy and change in personality theory and research," in L. A. Pervin and O. P. John (eds.), *Handbook of Personality: Theory and Research,* 2nd edn. New York: Guilford, pp. 689–704.

Peterson, B. S. 2008. Personal communication.

Peterson, D. R. 1995. "The reflective educator," *American Psychologist,* 50(12): 975–83.

Pickering, A. D. and Gray, J. A. 1999. "The neuroscience of personality," in L. A. Pervin and O. P. John (eds.), pp. 277–99.

Piedmont, R. L. 1999. "Does spirituality represent the sixth factor of personality? Spiritual transcendence and the five-factor model," *Journal of Personality,* 67(6): 985–1013.

Piedmont, R. L., McCrae, R. R., Riemann, R., and Alois Angleitner, A. 2000. "On the invalidity of validity scales: evidence from self-reports and observer ratings in volunteer samples," *Journal of Personality and Social Psychology,* 78(3): 582–93.

Pike, Kenneth L. 1967. *Language in Relation to a Unified Theory of the Structure of Human Behavior,* 2nd edn. The Hague: Mouton (1st edn. in three volumes, 1954, 1955, 1960).

Pinker, S. 1997. *How the Mind Works.* New York: Norton.

Pinker, S. 2002a. "A biological understanding of human nature: a talk with Steven Pinker," September 9, 2002, available at: www.edge.org/3rd_culture/pinker_blank/pinker_blank_print.html, accessed September 28, 2002.

Pinker, S. 2002b. *The Blank Slate: The Modern Denial of Human Nature.* New York: Penguin.

Pitcher, E. G. and Schultz, L. H. 1983. *Boys and Girls at Play: The Development of Sex Roles.* South Hadley, MA: Bergin & Garvey.

Plato [*c.* 370 BCE] 1991. *The Republic,* trans. B. Jowett, in Encyclopædia Britannica edn., *Great Books of the Western World.* University of Chicago Press, vol. 6.

Plessen, K. J., Bansal, R., Zhu, H., Whiteman, R., Amat, J., Quackenbush, G. A., Martin, L., Durkin, K., Blair, C., Royal, J., Hugdahl, K., and Peterson, B. S. 2006. "Hippocampus and amygdala morphology, in attention-deficit/hyperactivity disorder," *Archives of General Psychiatry,* 7(63): 795–807.

Plomin, R. 1991. "Genetic risk and psychosocial disorders: links between the normal and the abnormal," in M. Rutter and P. Casaer (eds.), *Biological*

Risk Factors for Psychosocial Disorders. Cambridge University Press, pp. 101–138.

Plomin, R. 2003. "50 years of DNA: what it has meant to psychological science," *American Psychological Society Observer*, 16(4): 7–8.

Plomin, R. and Caspi, A. 1999. "Behavioral genetics and personality," in L. A. Pervin and O. P. John (eds.), pp. 251–76.

Plomin, R. and Daniels, D. 1987. "Why are children in the same family so different from one another?," *Behavior and Brain Sciences*, 10: 1–60.

Plomin, R. and Rutter, M. 1998. "Child development, molecular genetics, and what to do with genes once they are found," *Child Development*, 69: 1223–42.

Plomin, R., Willerman, L., and Loehlin, J. C. 1976. "Resemblance and appearance and the equal environments assumption in twin studies of personality traits," *Behavior Genetics*, 6: 43–52.

Plutchik, R. 1976. "Emotions and attitudes related to being overweight," *Journal of Clinical Psychology*, 32: 21–24.

Plutchik, R. 1994. *The Psychology and Biology of Emotion.* New York: HarperCollins.

Polanyi, M. 1958. *Personal Knowledge: Towards a Post-critical Philosophy.* University of Chicago Press.

Polanyi, M. 1965. "The structure of consciousness," *Brain*, 88: 799–810.

Pope, B. 1953. "Socio-economic contrasts in children's peer culture prestige values," *Genetic Psychology Monographs*, 48: 157–220.

Pope, K. S. and Vasquez, M. J. T. 2007. *Ethics in Psychotherapy and Counseling: A Practical Guide,* 3rd edn. San Francisco, CA: Jossey-Bass.

Popper, K. 1963. *Conjectures and Refutations: The Growth of Scientific Knowledge.* New York: Basic Books.

Popper, K. R. 1982. "The place of mind in nature," in R. Q. Elvee (ed.), *Mind in Nature.* New York: Harper & Row, pp. 31–59.

Potter, H. 2008. "Bernini: the man of many heads," available at: www.nytimes.com/2008/08/08/arts/design/08bern.html?pagewanted=2, accessed August 7, 2008.

Powell, L. H., Shahabi, L., and Thoresen, C. 2003. "Religion and spirituality. Links to physical health," *American Psychologist*, 58: 36–52.

Pratt, C. C. 1939. *The Logic of Modern Psychology.* New York: Macmillan.

President's Commission on Excellence in Special Education 2002. "A new era: revitalizing special education for children and their families," available at: www.ed.gov/inits/commissionsboards/whspecialeducation/index.html, accessed June 8, 2008.

Prior, H., Schwarz, A., and Güntürkün. O. 2008, "Mirror-induced behavior in the magpie (*Pica pica*): evidence of self-recognition," published August 19, 2008, available at: doi:10.1371/journal.pbio.0060202; http://biology.plosjournals.org/archive/1545–7885/6/8/pdf, accessed February 6, 2009.

Proust, M. 1913. "Remembrance of things past," *Swann's Way: Within a Budding Grove,* Vol. 1, trans. C. K. S. Monorieff and T. Kilmartin. New York: Vintage.

Rabban, M. 1950. "Sex role identity in young children in two diverse social groups," *Genetic Psychology Monographs*, 42: 81–158.

Ramachandra, V. S. and Blakeslee, S. 1998. *Phantoms in the Brain: Human Nature and the Architecture of the Mind.* London: Fourth Estate.

Raskin, N. J. and Rogers, C. R. 2000. "Person-centered therapy," in R. J. Corsini and D. Wedding (eds.), pp. 133–67.

Reber, R. S. 1995. *The Penguin Dictionary of Psychology,* 2nd edn. London: Penguin.

Redding, R. E. 2001. "Sociopolitical diversity in psychology: the case for pluralism," *American Psychologist*, 56: 205–15.

Reiss, I. L. 1982. "Trouble in paradise: the current status of sexual science," *Journal of Sex Research*, 18: 97–113.

Resnick, L. B., Levine, J. M., and Teasley, S. D. (eds.) 1991. *Perspectives on Socially Shared Cognition.* Washington, DC: American Psychological Association.

Reuning, K. 1941. *Joy and Freude; A Comparative Study of the Linguistic Field of Pleasurable Emotions in English and German.* Swarthmore, PA: distributed by Swarthmore College Bookstore.

Richert, A. J. 2002. "The self in narrative therapy: thoughts from a humanistic/existential perspective," *Journal of Psychotherapy Integration*, 12(1): 77–104.

Ridley, C. 1995. *Overcoming Unintentional Racism in Counseling: A Practitioner's Guide to Intentional Intervention.* Thousand Oaks, CA: Sage.

Rieber, R. W. 1998. "The assimilation of psychoanalysis in America: from popularization to vulgarization," in R. W. Rieber and K. D. Salzinger (eds.), *Psychology: Theoretical–Historical Perspectives.* Washington, DC: American Psychological Association, pp. 355–98.

Riemann, R., Angleitner, A., and Strelau, J. 1977. "Genetic and environmental influences on personality: a study of twins reared together using the self- and peer report NEO–FFI scales," *Journal of Personality*, 65: 449–75.

Roberts, B. W. 1997. "Plaster or plasticity: are adult work experiences associated with personality change in women?," *Journal of Personality*, 65: 205–32.

Roberts, B. W. and DelVecchio, W. F. 2000. "The rank-order consistency of personality traits from childhood to old age: a quantitative review of longitudinal studies," *Psychological Bulletin*, 126: 3–25.

Robins, R. W., Norem, J. K., and Cheek, J. M. 1999. "Naturalizing the self," in L. A. Pervin and O. P. John (eds.), pp. 443–77.

Roe, A. 1961. "The psychology of the scientist," *Science*, 134: 456–59.

Rogers, C. R. 1942. *Counseling and Psychotherapy: Newer Concepts in Practice.* Boston, MA: Houghton Mifflin.

Rogers, C. R. 1947. "Some observations on the organization of personality," *American Psychologist*, 2: 358–68.

Rogers, C. R. 1951. *Client-centered Therapy: Its Current Practice, Implications, and Theory.* Boston, MA: Houghton Mifflin.

Rogers, C. R. 1955. "Persons or science? A philosophical question," *American Psychologist*, 10: 267–78.

Rogers, C. R. 1959. "A theory of therapy, personality, and interpersonal relationships, as developed in a client-centered framework," in S. Koch, (ed.), *Psychology: A Study of a Science.* New York: McGraw-Hill, vol. 3, pp. 184–256.

Rogers, C. R. 1961. *On Becoming a Person.* Boston, MA: Houghton Mifflin.

Rogers, C. R. 1980. *A Way of Being.* Boston, MA: Houghton Mifflin.

Rogers, C. R. and Skinner, B. F. 1956. "Some issues concerning the control of human behavior: a symposium," *Science*, 124: 1057–66.

Rohde, D. L. T., Olson, S., and Chang, J. T. 2004. "Modelling the recent common ancestry of all living humans," *Nature*, 431: 562–66.

Romer, D., Gruder, C. L., and Lizzadro, T. 1986. "A person–situation approach to altruistic behavior," *Journal of Personality and Social Psychology*, 51: 1001–12.

Rosanoff, A. J. 1935. "The etiology of manic-depressive syndromes with special reference to their occurrence in twins," *American Journal of Psychiatry*, 91: 725–62.

Rosanoff, A. J., Handy, L. M., Plesset, I. R., and Brush, S. 1934. "The etiology of so-called schizophrenic psychoses with special reference to their occurrence in twins," *American Journal of Psychiatry*, 91: 247–86.

Rosch, E. H. 1973. "Natural categories," *Cognitive Psychology*, 4: 328–50.

Rosen, S. (ed.) 1982. *My Voice Will Go with You: The Teaching Tales of Milton H. Erickson*. New York: Norton.

Rosenfeld, Z. 2008. "Feeling and form: from Plato to Penrose, the music is the message," *Harper's*, June, pp. 89–94.

Rosenthal, R. 1990. "How are we doing in soft psychology?," *American Psychologist*, 45: 775–77.

Rosenthal, R. 1991. *Meta-analytic Procedures for Social Research*, rev. edn. Newbury Park, CA: Sage.

Ross, L. 1977. "The intuitive psychologist and his shortcomings: distortions in the attribution process," in L. Berkowitz (ed.), *Advances in Experimental Social Psychology*. New York: Academic, vol. 10, pp. 174–221.

Ross, L. and Nisbett, R. E. 1991. *The Person and the Situation: Perspectives of Social Psychology*. New York: McGraw-Hill.

Rossini, E. D. and Moretti, R. J. 1997. "Thematic Apperception Test (TAT) interpretation: practice recommendations from a survey of clinical psychology doctoral programs accredited by the American Psychological Association," *Professional Psychology: Research and Practice*, 28: 393–98.

Rosten, L. 1968. *The Joys of Yiddish*. New York: Washington Square Press.

Rosten, L. 1999. *The New Joys of Yiddish*. New York: Galahad Press.

Rotter, J. B. 1966. "Generalized expectancies for internal versus external control of reinforcement," *Psychological Monographs: General and Applied*, 80(1): 1–28.

Rotter, J. B. 1975. "Some problems and misconceptions related to the construct of internal versus external control of reinforcement," *Journal of Consulting and Clinical Psychology*, 43(1): 56–67.

Rotter, J. B. 1990. "Internal versus external control of reinforcement: a case history of a variable," *American Psychologist*, 45(4): 489–93.

Rubenstein, R. E. 2003. *Aristotle's Children: How Christians, Muslims, and Jews Rediscovered Ancient Wisdom and Illuminated the Dark Ages*. New York: Harcourt.

Rubin, Z. and McNeil, E. B. 1987. *Psychology: Being Human*, 4th edn. New York: Harper & Row.

Ruddick, S. 1989. *Maternal Thinking: Toward a Politics of Peace*. New York: Ballantine.

Rushdie, S. 2005. "Karma chameleon," *Saturday Guardian, Review*, pp. 15–16, December 17.

Rushton, J. P. 1997. "(Im)pure genius – psychoticism, intelligence, and creativity," in H. Nyborg (ed.), pp. 404–21.

Russell, J. A. 1980. "A circumplex model of affect," *Journal of Personality and Social Psychology*, 39: 1161–78.

Russell, J. A. and Barrett, L. F. 1999. "Core affect, prototypical emotional episodes, and other things called emotion: dissecting the elephant," *Journal of Personality and Social Psychology*, 76(5): 805–19.

Russell, M. T. and Karol, D. 1994. *The 16PF Fifth Edition Administrator's Manual.* Champaign, IL: Institute for Personality and Ability Testing.

Rutter, M. and Silberg, J. 2002. "Gene–environment interplay in relation to emotional and behavioral disturbance," *Annual Review of Psychology.* Palo Alto, CA: Annual Reviews, vol. 53, pp. 463–90.

Ryan, R. M. 1985. "Thematic Apperception Test," in D. J. Keyser and R. C. Sweetland (eds.), *Test Critiques.* Kansas City, MO: Test Corporation of America, vol. 2, pp. 799–814.

Ryan, R. M. and Deci, E. L. 2000. "Self-determination theory and the facilitation of intrinsic motivation, social development, and well-being," *American Psychologist*, 55(1): 68–78.

Rychlak, J. 1968. *A Philosophy of Science for Personality Theory.* Boston, MA: Houghton Mifflin.

Sacks, O. 1996. *An Anthropologist on Mars: Seven Paradoxical Tales.* Toronto: Vintage Canada.

Sacks, O. 2007. *Musicophilia: Tales of Music and the Brain.* London: Picador.

Safran, J. D. and Greenberg L. S. (eds.) 1991. *Emotions, Psychotherapy, and Change.* New York: Guilford.

Saklofske, D. H. and Zeidner, M. (eds.) 1995. "Preface," *International Handbook of Personality and Intelligence.* New York: Plenum Press, pp. xvii–xix

Salomon, G. (ed.) 1993. *Distributed Cognitions: Psychological and Educational Considerations.* Cambridge University Press.

Salomon, G. (ed.) 1997. *Distributed Cognitions: Psychological and Educational Considerations.* New York: Cambridge University Press.

Salovey, P., Detweiler-Bedell, B. T., Detweiler-Bedell, J. B., and Mayer, J. D. 2008. "Emotional intelligence," in M. Lewis, J. M. Haviland-Jones, and L. F. Barrett (eds.), *Handbook of Emotions,* 3rd edn. New York: Guilford, pp. 533–47.

Salovey, P. and Mayer, J. D. 1990. "Emotional intelligence," *Imagination, Cognition, and Personality*, 9: 185–211.

Salovey, P. and Mayer, J. D. 1994. "Some final thoughts about personality and intelligence," in R. J. Sternberg and P. Ruzgis (eds.), *Personality and Intelligence.* New York: Cambridge University Press, pp. 303–18.

Scarr, S. 1968. "Environmental bias in twin studies," *Eugenics Quarterly*, 15(1): 34–40.

Scarr, S. 1992. "Developmental theories for the 1990s: development and individual differences," *Child Development*, 63: 1–19.

Scarr, S. 1993. "Developmental theories for the 1990s: development and individual differences," in M. Gauvin and M. Cole (eds.), *Readings on the Development of Children.* New York: Freeman, pp. 46–62.

Schachter, S. and Singer, J. 1962. "Cognitive, social and physiological determinants of emotional state," *Psychological Review*, 69: 379–99.

Scheier, M. F. and Carver, C. S. 1993. "On the power of positive thinking: the benefits of being optimistic," *Current Directions in Psychological Science*, 2: 26–30.

Schofield, V. W. 1981. "Complementary and conflicting identities: images of interaction in an interracial school," in S. A. Asher and J. M. Gottman (eds.), *The Development of Children's Friendships*. New York: Cambridge University Press, pp. 53–90.

Scott, J. P. 1963. "The process of primary socialization in canine and human infants," *Monograph: Social Research in Child Development*, 28(1): No. 85.

Scruton, R. 2006. "A carnivore's credo," *Harper's*, vol. 312, No. 1872, pp. 21–24, 26, May.

Searle, J. R. 1992. *The Rediscovery of the Mind*. Cambridge, MA: MIT Press.

Sedikides, C. and Skowronski, J. J. 1997. "The symbolic self in evolutionary context," *Personality and Social Psychology Review*, 1: 80–102.

Seligman, M. E. P. 1975. *Helplessness*. San Francisco, CA: Freeman.

Seligman, M. E. P. and Csikszentmihalyi, M. 2000. "Positive psychology: an introduction," *American Psychologist*, 55(1): 5–14.

Shaffer, D. R. 2000. *Social and Personality Development*, 4th edn. Belmont, CA: Wadsworth.

Shazer, S. de 1988. *Clues: Investigating Solutions in Brief Therapy*. New York: Norton.

Sherif, M. [1961] 1988. *The Robbers' Cave Experiment: Intergroup Conflict and Cooperation*. Toronto: Wesleyan University Press/University Press of New England.

Sherif, M., White, B. J., and Harvey, O. J. 1955. "Status in experimentally produced groups," *American Journal of Sociology*, 60: 370–79.

Sherwin, B. B. 1994. "Sex hormones and psychological functioning in postmenopausal women," *Experimental Gerontology*, 29: 423–30.

Shorter, E. 1997. *A History of Psychiatry: From the Era of the Asylum to the Age of Prozac*. Toronto: Wiley.

Shorter, E. 2008. Personal communication.

Shweder, R. A. 1991. *Thinking Through Cultures: Expeditions in Cultural Psychology*. Cambridge, MA: Harvard University Press.

Shweder, R. A. and Sullivan, M. A. 1993. "Cultural psychology: who needs it?," *Annual Review of Psychology*, 44: 497–523.

Siever, L. J. and Davis, K. L. 1991. "A psychobiological perspective on the personality disorders," *American Journal of Psychiatry*, 148: 1647–58.

Simonov, P. V. 1997. "Brain mechanisms of emotions," *Neuroscience and Behavioural Physiology*, 27(4): 405–13.

Simonton, D. K. 2000. "Creativity: cognitive, personal, developmental, and social aspects," *American Psychologist*, 55: 151–58.

Simpson, J. A. and Gangestad, S. W. 1991. "Individual differences in sociosexuality: evidence for convergent and discriminant validity," *Journal of Personality and Social Psychology*, 60: 870–83.

Sines, J. O. 1970. "Actuarial versus clinical prediction in psychopathology," *British Journal of Psychiatry*, 116: 129–44.

Singer, J. A. and Salovey, P. 1988. "Mood and memory: evaluating the network theory of affect," *Clinical Psychology Review*, 8: 211–51.
Skinner, B. F. 1938. *The Behavior of Organisms*. New York: Appleton-Century-Crofts.
Skinner, B. F. 1983. "Intellectual self-management in old age," *American Psychologist*, 38: 239–44.
Skolnick, N. J. and Zuckerman, M. 1979. "Personality change in drug abusers: a comparison of therapeutic community and prison groups," *Journal of Consulting & Clinical Psychology*, 47: 768–70.
Slochower, J. and Kaplan, S. P. 1980. "Anxiety, perceived control, and eating in obese and normal weight persons," *Appetite*, 1: 75–83.
Smith, D. and Dumont, F. 1995. "A cautionary study: unwarranted interpretations of the Draw-A-Person Test," *Professional Psychology: Research and Practice*, 26: 298–303.
Smith, J. L. 2002. *The Psychology of Food and Eating*. New York: Palgrave.
Smith, K. D., Keating, J. P., and Stotland, E. 1989. "Altruism revisited: the effect of denying feedback on a victim's status to empathic witnesses," *Journal of Personality and Social Psychology*, 57: 641–50.
Smuts, B. 1995. "The evolutionary origins of patriarchy," *Human Nature*, 6: 1–32.
Smuts. J. C. 1926. *Holism and Evolution*. New York: Macmillan.
Snow, C. 1987. "Relevance of the notion of a critical period to language acquisition," in M. H. Bornstein (ed.), *Sensitive Periods in Development: Interdisciplinary Perspectives*. Hillsdale, NJ: Erlbaum, pp. 183–209.
Sobel, D. 2000. *Galileo's Daughter: A Historical Memoir of Science, Faith, and Love*. New York: Penguin.
Society for Psychotherapy Research 2007. *Book of Abstracts*, 38th International Meeting, Ulm, Germany, June 20–23: Author.
Sorabji, R. 2006. *Self: Ancient and Modern Insights about Individuality, Life, and Death*. University of Chicago Press.
Sorum, P. C. 1994. "Patient as author, physician as critic: insights from contemporary literary theory," *Archives of Family Medicine*, 3: 549–56.
Spence, K. W. 1963. "Types of constructs in psychology," in M. H. Marx (ed.), pp. 162–78.
Spengler, P. M. and Strohmer, D. C. 1994. "Clinical judgmental biases: the moderating role of counselor cognitive complexity and counselor client preferences," *Journal of Counseling Psychology*, 41: 8–17.
Sperry, R. W. 1958. "Developmental basis of behavior," in A. Roe and G. G. Simpson (eds.), pp. 128–39.
Sperry, R. W. 1993. "The impact and promise of the cognitive revolution," *American Psychologist*, 48: 878–85.
Sperry, R. W. 1995. "The future of psychology," *American Psychologist*, 50: 505–6.
Spevack, L. 2003. "Lifetime commitment," *National Post*, p. B4, July 17.
Spitzer, R. L. and Williams, J. B. W. 1987. "Introduction," *Diagnostic and Statistical Manual of Mental Disorders*, rev. 3rd edn. Washington, DC: American Psychiatric Association, pp. xvii–xxvii.
Stannard, D. E. 1992. *American Holocaust: Columbus and the Conquest of the New World*. New York: Oxford University Press.

Sternberg, R. J. 1985. *Beyond IQ: A Triarchic Theory of Human Intelligence.* New York: Cambridge University Press.

Sternberg, R. J. 1988. *The Triarchic Mind.* New York: Viking.

Sternberg, R. J. 1990. *Wisdom: Its Nature, Origins, and Development.* New York: Cambridge University Press.

Sternberg, R. J. and Lubart, T. I. 1991. "Investing in creativity," *American Psychologist*, 51: 677–88.

Stone, A. A. 1997. "Where will psychoanalysis survive?," *Harvard Magazine*, 35–39.

Strauss, A. and Corbin, J. 1990. *Basics of Qualitative Research: Grounded Theory, Procedures and Techniques.* London: Sage.

Strogatz, S. 2004. *Sync: The Emerging Science of Spontaneous Order.* New York: Penguin.

Strohmer, D. C. and Shivy, V. A. 1994. "Bias in counselor hypothesis testing: testing the robustness of counselor confirmatory bias," *Journal of Counseling and Development*, 73: 191–97.

Strohmer, D. C., Shivy, V. A., and Chiodo, A. L. 1990. "Information processing strategies in counselor hypothesis testing: the role of selective memory and expectancy," *Journal of Counseling Psychology*, 37: 465–72.

Sullivan, H. S. 1953. *The Interpersonal Theory of Psychiatry.* New York: Norton.

Sulloway, F. J. 1979. *Freud, Biologist of the Mind: Beyond the Psychoanalytic Legend.* New York: Basic Books.

Sulloway, F. J. 1996. *Born to Rebel: Birth Order, Family Dynamics, and Creative Lives.* New York: Pantheon.

Suomi, S. J., Novak, M. A., and Well, A. 1996. "Aging in Rhesus monkeys: different windows on behavioral continuity and change," *Developmental Psychology*, 32: 1116–28.

Sutton-Smith, B. and Kelly-Byrne, D. 1984. "The idealization of play," in P. K. Smith (ed.), *Play in Animals and Humans.* New York: Basil Blackwell, pp. 305–21.

Swim, J. K. 1994. "Perceived versus meta-analytic effect sizes: an assessment of the accuracy of gender stereotypes," *Journal of Personality and Social Psychology*, 66: 21–36.

Taliaferro, R. C. 1952. "Introduction," C. Ptolemaeus, *The Almagest, Part 1*, in Encyclopædia Britannica edn., *Great Books of the Western World.* Chicago University Press.

Tamres, L. K., Janicki, D., and Helgeson, V. S. 2002. "Sex differences in coping behavior: a meta-analytic review and an examination of relative coping," *Personality and Social Psychology Review*, 6(1): 2–30.

Taylor, C. 1989. *Sources of the Self: The Making of the Modern Identity.* Cambridge, MA: Harvard University Press.

Taylor, E. 1998. "William James on the demise of positivism in American psychology," in R. W. Rieber and K. D. Salzinger (eds.), pp. 101–34.

Taylor, J. A. 1953. "A personality scale of manifest anxiety," *Journal of Abnormal Psychology*, 48: 285–90.

Taylor, S. E. 1983. "Adjustment to threatening events: a theory of cognitive adaptation," *American Psychologist*, 38: 1161–73.

Taylor, S. E. 1989. *Positive Illusions: Creative Self-deception and the Healthy Mind.* New York: Basic Books.

Taylor, S. E. and Brown J. D. 1988. "Illusion and well being: a social psychological perspective on mental health," in R. F. Baumeister (ed.), *The Self in Social Psychology.* Philadelphia: Psychology Press/Taylor & Francis, pp. 43–68.

Taylor, S. E. and Brown, J. D. 1994. "Positive illusions and well-being revisited: separating fact from fiction," *Psychological Bulletin,* 116: 21–27.

Teilhard de Chardin, P. 1959. *The Phenomenon of Man.* London: Collins.

Tellegen, A., Lykken, D. T., Bouchard, Jr., T. J., Wilcox, K. J., and Rich, S. 1988. "Personality similarity in twins reared apart and together," *Journal of Personality and Social Psychology,* 54: 1031–39.

Terman, L. M. 1916. *The Measurement of Intelligence: An Explanation of and a Complete Guide for the Use of the Stanford Revision and Extension of the Binet–Simon Intelligence Scale.* Boston, MA: Houghton Mifflin.

Thabes, V. 1997. "Survey analysis of women's long-term, postdivorce adjustment," *Journal of Divorce & Remarriage,* 27: 163–75.

Theophrastus [*c.* fourth century BCE] 1929. *The Characters,* trans. J. M. Edmonds. London: Heinemann.

Thomas, A. and Chess, S. 1977. *Temperament and Development.* New York: Brunner/Mazel.

Thomas, A. and Chess, S. 1980. *The Dynamics of Psychological Development.* New York: Brunner/Mazel.

Thomas, A., Chess, S., and Birch, H. 1968. *Temperament and Behavior: Disorders in Children.* New York University Press.

Thomas, L. 1979. *The Medusa and the Snail: More Notes of a Biology Watcher.* New York: Viking Penguin

Thomas, L. 1990. *A Long Line of Cells: Collected Essays.* New York: Penguin.

Thorndike, E. L. 1920. "Intelligence and its uses," *Harper's,* vol. 140, pp. 227–35.

Thornhill, R. and Thornhill, N. W. 1983. "Human rape: an evolutionary analysis," *Ethology & Sociobiology,* 4(3): 137–73.

Thurstone, L. L. 1938. *Primary Mental Abilities.* University of Chicago Press.

Tinbergen, N. 1939. "On the analysis of social organization among vertebrates, with special reference to birds," *American Midland Naturalist,* 21: 210–34.

Tinbergen, N. 1951. *The Study of Instinct.* Oxford University Press.

Tinbergen, N. 1953. *Social Behaviour in Animals.* London: Methuen.

Tobin, J. J. 1989. *Preschool in Three Cultures: Japan, China, and the United States.* New Haven, CT: Yale University Press.

Tolstoy, L. [1886] 1987. *The Death of Ivan Ilyich.* New York: Bantam Books.

Tolstoy, L. [1869] 1990. *War and Peace,* in C. Fadiman and P. W. Goetz (eds.), Encyclopaedia Britannica edn., *Great Books of the Western World.* Chicago University Press, vol. 51.

Tomkins, S. S. 1962. *Affect, Imagery, Consciousness.* New York: Springer.

Tononi, G. and Edelman, M. 1998. "Neuroscience: consciousness and complexity," *Science,* 282: 1846–51.

Tooby, J. and Cosmides, L. 1990. "On the universality of human nature and the uniqueness of the individual: the role of genetics and adaptation," *Journal of Personality,* 58: 17–68.

Tooby, J. and Cosmides, L. 1992. "Psychological foundations of culture," in J. Barkow, L. Cosmides, and J. Tooby (eds.), *The Adapted Mind.* New York: Oxford University Press, pp. 19–136.

Torrey, E. F. 1992. *Freudian Fraud.* New York: HarperCollins.

Triandis, H. C. 1978. "Some universals of social behavior," *Personality and Social Psychology Bulletin*, 4: 1–16.

Triandis, H. C., Bontempo, R., Villareal, M. J., Asai, M., and Lucca, N. 1988. "Individualism and collectivism: cross-cultural perspectives on self–ingroup relationships," *Journal of Personality and Social Psychology*, 54(2): 323–38.

Triandis, H. C., McCusker, C., and Hui, C. H. 1990. "Multimethod probes of individualism and collectivism," *Journal of Personality and Social Psychology*, 59(5): 1006–20.

Trivers, R. L. 1972. "Parental investment and sexual selection," in B. Campbell (ed.), *Sexual Selection and the Descent of Man, 1871–1971.* Chicago, IL: Aldine, pp. 136–79.

Tupes, E. C. and Christal, R. C. 1992. "Recurrent personality factors based on trait ratings," *Journal of Personality*, 60: 225–51.

Turner, J. E., Husman, J., and Schallert, D. L. 2002. "The importance of students' goals in their emotional experience of academic failure: investigating the precursors and consequences of shame," *Educational Psychologist*, 37: 79–89.

Tversky, A. 1977. "Features of similarity," *Psychological Review*, 84: 327–52.

Tweedie, D. F. 1961. *Logo-therapy and the Christian Faith.* Grand Rapids, MI: Baker Book House.

United States Department of Justice 1995. *Crime in the United States.* Washington, DC: US Government Printing Office.

Vaihinger, H. 1925. *The Philosophy of As If.* New York: Harcourt, Brace & World.

Vaillant, G. E. 1977. *Adaptation to Life.* Boston, MA: Little, Brown.

Vernon, P. A. 1997. "Behavioral genetic and biological approaches to intelligence," in H. Nyborg (ed.), pp. 240–58.

Virbila, S. I. 2005. "Las Vegas, lighted by stars," *Los Angeles Times, The Review*, available at: http://www.calendarlive.com/dining/cl-fo-review28dec28,0,209082.story?coll=cl-home-more-channels, December 25.

Von Krafft-Ebing, R. [1886] 1937. *Pathologia Sexualis.* New York: Physicians and Surgeons Books.

Vosler, N. R. and Page-Adams, D. 1996. "Predictors of depression among workers at the time of a plant closing," *Journal of Sociology & Social Welfare*, 23(4): 25–42

Voyer, D., Voyer, S., and Bryden, M. P. 1995. "Magnitude of sex differences in spatial abilities: a meta-analysis and consideration of critical variables," *Psychological Bulletin*, 117(2): 250–70.

Vygotsky, L. S. [1925] 1997. *The Collected Works of L. S. Vygotsky*, Vol. 3: *Problems of the Theory and History of Psychology.* New York: Plenum Press.

Wade, T. C. and Baker, T. B. 1977. "Opinions and use of psychological tests," *American Psychologist*, 32: 874–82.

Waehler, C. A. 1997. "Drawing bridges between science and practice," *Journal of Personality Assessment*, 69: 482–87.

Wakefield, J. C. 1992. "The concept of mental disorder: on the boundary between biological facts and social values," *American Psychologist*, 4: 373–88.
Wakefield, J. C. 1999. "Evolutionary versus prototype analyses of the concept of disorder," *Journal of Abnormal Psychology*, 108: 374–99.
Wallace, D. B. and Gruber, H. E. (eds.) 1989. *Creative People at Work: Twelve Cognitive Case Studies*. New York: Oxford University Press.
Wallach, M. A. and Kogan, N. 1965. "A new look at the creativity–intelligence distinction," *Journal of Personality*, 33: 348–69.
Wallach, M. A. and Wallach, L. W. 1983. *Psychology's Sanction for Selfishness: The Error of Egoism in Theory and Therapy*. New York: Freeman.
Walsh, R. 1999. "Asian contemplative disciplines: common practices, clinical applications, and research findings," *The Journal of Transpersonal Psychology*, 31(2): 83–107.
Walsh, R. 2000. "Asian psychotherapies," in R. J. Corsini and D. Wedding (eds.), pp. 407–44.
Ward, J. 2006. *The Student's Guide to Cognitive Neuroscience*. Hove: Psychology Press (Taylor & Francis).
Warde, A. 1997. *Consumption, Food and Taste: Culinary Antinomies and Commodity Culture*. London: Sage.
Warneken F., Hare B., Melis A. P., Hanus D., and Tomasello, M. 2007. "Spontaneous altruism by chimpanzees and young children," *PLoS Biol* 5(7): e184, available at: www.doi:10.1371/journal.pbio.0050184doi:10.1371/journal.pbio.0050184.
Watson, D. and Clark, L. A. 1993. "Behavioral disinhibition versus constraint: a dispositional perspective," in D. M. Wegner and J. W. Pennebaker (eds.), *Handbook of Mental Control*. Upper Saddle River, NJ: Prentice Hall, pp. 506–27.
Watson, J. B. (1925). *Behaviorism*. New York: Norton.
Watson, J. B. [1924] 1930. *Behaviorism*, rev. edn. Chicago University Press.
Watson, R. I. 1963. *The Great Psychologists: From Aristotle to Freud*. Philadelphia: Lippincott.
Watzlawick, P. 1974. *Change: Principles of Problem Formulation and Problem Resolution*. New York: Norton.
Weber, M. [1904] 1930. *The Protestant Ethic and the Spirit of Capitalism* (*Die Protestantische Ethik und der Geist das Kapitalismus*), trans. Talcott Parsons. New York: Charles Scribner.
Wedding, D. and Boyd, M. A. 1998. *Movies and Mental Illness: Using Films to Understand Psychopathology*. Boston, MA: McGraw-Hill.
Weiner, I. 1989. "On competence and ethicality in psychodiagnostic assessment," *Journal of Personality Assessment*, 53: 827–31.
Weisberg, R. 1993. *Creativity: Beyond the Myth of Genius*. New York: Freeman.
Wertheimer, M. [1925] 1938. "Gestalt theory," in W. D. Ellis (ed.), *A Sourcebook of Gestalt Psychology*. London: Routledge & Kegan Paul.
Wertheimer, M. 2007. "The case of the purloined picture: Rosalind Franklin and the keystone of the double helix," in E. A. Gavin, A. Clamar, and M. A. Siderits (eds.) *Women of Vision: Their Psychology, Circumstances and Success*. New York: Springer, pp. 259–71.

Wertheimer, Michael 2008. Personal communication.
Westen, D. and Gabbard, G. O. 1999. "Psychoanalytic approaches to personality," in L. A. Pervin and O. P. John, pp. 57–101.
White, M. and Epston, D. 1990. *Narrative Means to Therapeutic Ends*. New York: Norton.
White, R. W. 1959. "Motivation reconsidered: the concept of competence," *Psychological Review*, 66: 297–333.
White, S. H. 1992. "G. Stanley Hall: from philosophy to developmental psychology," *Developmental Psychology*, 28: 25–34.
Whyte, L. L. 1960. *The Unconscious Before Freud*. New York: Basic Books.
Widiger, T. A., Trull, T. S., Clarkin, J. F., Sanderson, C., and Costa, P. T. 1994. "A description of the DSM-III-R and DSM-IV with the five-factor model of personality," in P. T. Costa and T. A. Widiger (eds.), *Personality Disorders and the Five-factor Model of Personality*. Washington, DC: American Psychological Association, pp. 41–56.
Widiger, T. A., Verheul, R., and van den Brink, W. 1999. "Personality and psychopathology," in L. A. Pervin and O. P. John (eds.), pp. 347–66.
Wiener, N. 1954. *The Human Use of Human Beings*. Boston, MA: Houghton Mifflin.
Wiggins, J. S. 1962. "Strategic, method, and stylistic variance in the MMPI," *Psychological Bulletin*, 59: 224–42.
Wiggins, J. S. 1968. "Personality structure," *Annual Review of Psychology*. Palo Alto, CA: Annual Reviews, vol. 19, pp. 293–50.
Wilber, K. 1986. "The spectrum of development," in K. Wilber, J. Engler, and D. P. Brown (eds.), *Transformations of Consciousness: Conventional and Contemplative Perspectives on Development*. Boston, MA: Shambhala, pp. 65–106.
Williams, J. E. and Best, D. L. 1982. *Measuring Sex Stereotypes: A Multi-nation Study*. Beverly Hills, CA: Sage.
Williams, W. L. 1992. "The relationship between male–male friendship and male–female marriage," in P. Nardi (ed.), *Men's Friendships*. Newbury Park, CA: Sage, pp. 187–200.
Wilson, E. O. 1975. *Sociobiology: The New Synthesis*. Cambridge, MA: Harvard University Press.
Wilson, E. O. 1978. *On Human Nature*. Cambridge, MA: Harvard University Press.
Wilson, E. O. 1999. *Consilience: The Unity of Knowledge*. New York: Vintage, p. 234.
Wilson, T. D. and Nisbett, R. E. 1978. "The accuracy of verbal reports about the effects of stimuli on evaluations and behavior," *Social Psychology*, 41(2): 118–31.
Winter, D. G. and Barenbaum, N. B. 1999. "History of modern personality theory and research," in L. A. Pervin and O. P. John, pp. 3–27.
Wolf, A. P. and Huang, C.-S. 1980. *Marriage and Adoption in China, 1845–1945*. Stanford University Press.
Wolman, B. B. 1968. "The historical role of Johann Friedrich Herbart," in B. B. Wolman (ed.), *Historical Roots of Contemporary Psychology*. New York: Harper & Row, pp. 29–46.

Woodworth, R. S. 1920. *Personal Data Sheet.* Chicago: Stoelting.
Woody, E. and Claridge, G. 1977. "Psychoticism and thinking," *British Journal of Social and Clinical Psychology*, 16: 241–48.
Yalom, I. 1980. *Existential Psychotherapy.* New York: Basic Books.
Yalom, I. D. 1995. *The Theory and Practice of Group Psychotherapy.* New York: Basic Books.
Yang, K. S. 1999. "Towards an indigenous Chinese psychology: a selective review of methodological, theoretical, and empirical accomplishments," *Chinese Journal of Psychology*, 41(2): 181–211.
Zahn-Wexler, C., Radke-Yarrow, M., Wagner, E., and Chapman, M. 1992. "Development of concern for others," *Developmental Psychology*, 28: 126–36.
Zajonc, R. 1980. "Feeling and thinking: preferences need no inferences," *American Psychologist*, 35: 151–75.
Zajonc, R. B. and Mullaly, P. R. 1997. "Birth order: reconciling conflicts effects," *American Psychologist*, 52: 685–99.
Zeidner, M. 1995. "Personality trait correlates of intelligence," in D. H. Saklofske and M. Zeidner (eds.), pp. 299–319.
Zeig, J. K. (ed.) 1980. *Teaching Seminar with Milton H. Erickson, M.D.* New York: Brunner/Mazel.
Zuckerman, M. 1987. "All parents are environmentalists until they have their second child," *Behavioral and Brain Sciences*, 10: 42–44.
Zuckerman, M. 1997. "The psychobiological basis of personality," in H. Nyborg (ed.), pp. 3–16.
Zuckerman, M., Tushup, R., and Finner, S. 1976. "Sexual attitudes and experience: attitude and personality correlates and changes produced by a course in sexuality," *Journal of Consulting and Clinical Psychology*, 44: 7–19.
Zuriff, G. E. 1999. "Ideology over reason," *American Psychologist*, 54(1): 71.

Author index

Subject index

Lightning Source UK Ltd.
Milton Keynes UK
UKOW01f0101240917
309766UK00016B/288/P

9 780521 133265